Third Edition

THEORY OF VIBRATION WITH APPLICATIONS

William T. Thomson, *Professor Emeritus*

Department of Mechanical and Environmental Engineering
University of California
Santa Barbara, California

CBS

CBS Publishers & Distributors Pvt. Ltd.

New Delhi • Bengaluru • Chennai • Kochi • Kolkata • Mumbai
Hyderabad • Uttarakhand • Nagpur • Patna • Pune • Jharkhand

ISBN: 81-239-0883-0

First Indian Edition: 1990
Reprint: 1999, 2002

Reprinted in India by CBS Publishers & Distributors in arrangement with Unwin Hyman Ltd. U.K.

Published by **Satish Kumar Jain** and produced by **Varun Jain** for
CBS Publishers & Distributors Pvt. Ltd.,
4819/XI Prahlad Street, 24 Ansari Road, Daryaganj, New Delhi - 110002
delhi@cbspd.com, cbspubs@airtelmail.in • www.cbspd.com
Ph.: 23289259, 23266861, 23266867 • Fax: 011-23243014

Corporate Office: 204 FIE, Industrial Area, Patparganj, Delhi - 110 092
Ph: 49344934 • Fax: 011-49344935
E-mail: publishing@cbspd.com • publicity@cbspd.com

Branches:
- ***Bengaluru:*** 2975, 17th Cross, K.R. Road, Bansankari 2nd Stage, Bengaluru - 70 • Ph: +91-80-26771678/79 • Fax: +91-80-26771680 E-mail: cbsbng@gmail.com, bangalore@cbspd.com
- ***Chennai:*** No. 7, Subbaraya Street, Shenoy Nagar, Chennai - 600030 Ph: +91-44-26681266, 26680620 • Fax: +91-44-42032115 E-mail: chennai@cbspd.com
- ***Kochi:*** Ashana House, 39/1904, A.M. Thomas Road, Valanjambalam, Ernakulum, Kochi • Ph: +91-484-4059061-65 Fax: +91-484-4059065 • E-mail: cochin@cbspd.com
- ***Kolkata:*** 6-B, Ground Floor, Rameshwar Shaw Road, Kolkata - 700014 Ph: +91-33-22891126/7/8 • E-mail: kolkata@cbspd.com
- ***Mumbai:*** 83-C, Dr. E. Moses Road, Worli, Mumbai - 400018 Ph: +91-9833017933, 022-24902340/41 • E-mail: mumbai@cbspd.com

Representatives:
- Hyderabad: 0-9885175004
- Patna: 0-9334159340
- Jharkhand: 0-9811541605
- Nagpur: 0-9021734563
- Pune: 0-9623451994
- Uttarakhand: 0-9716462459

Printed at:
Neekunj Print Process, Delhi

CONTENTS

PREFACE

This book is another updating revision of the Theory of Vibration with Applications, 2nd Edition, 1981, which has been revised over the years since 1948. In keeping with the continuing advances in modern technology and pedagogy, additional changes have been made in the subject matter, arrangement of topics, mode of presentation, and emphasis. To keep the book to a reasonable size, material of lesser interest has been deleted to make space for more timely topics. Attempts have been made to improve the clarity of presentation and retain the simplicity of the earlier editions.

In the first four chapters, which deal with single degree of freedom systems, the simple physical approach of the previous edition has been adhered to and, hopefully, improved upon. In Chapter 2, the Rayleigh Energy Method for conservative systems is supplemented by the Virtual Work Method, which extends to non-conservative systems. Frequency Response concepts using complex algebra have been added to the third chapter. The section on computer calculation still maintains its place of importance.

Chapter 5 has been condensed somewhat, with the intent of providing an introduction to the multi-degree of freedom systems and the more general discussion of properties of vibrating systems in Chapter 6. The latter chapter has been rearranged to separate the flexibility and stiffness procedures. It also leads to the free-body basis as an introduction to the Finite Element approach for generating the stiffness matrix and a brief insight into the modal summation technique.

Chapter 7 deals with generalized coordinates, virtual work and Lagrange's formulation of the vibration equation. The chapter has been moved forward in the text to take earlier advantage of the basic nature of this important subject. Generalized coordinates and Lagrange's theory will greatly broaden and strengthen the student's understanding of vibration theory and the entire field of dynamics. Early aquisition of this knowledge should be beneficial.

Chapters 8 and 9 deal with normal modes of continuous systems and the method of modal summation, which is one of the most useful procedures in vibration. Modal summation need not be limited to lower modes only and can be applied in the vicinity of excitation frequencies which may be in the higher frequency range. Prior knowledge of Lagrangian generalized coordinates fortifies this area.

Chapter 10, introducing the Finite Element Method, is a major addition to the text. It is illustrated by examples of simple systems, to facilitate the easy understanding of the assembly procedure of the element matrices. Often the small number of elements used will lead to poor accuracy which can be remedied by using many elements of smaller dimensions. The strength of the finite element method lies in solving the problem with computer programs which will handle large numbers of degrees of freedom. The method is used extensively in industry and some understanding of its basic background should be a part of those interested in the field of vibration or structural analysis.

Numerical methods for estimating the fundamental and few higher modes of vibration are presented in Chapters 11 and 12. Of these, the Rayleigh-Ritz, Dunkerly, Holzer, Myklestad, and Transfer Matrices represent classic examples. These methods are easily programmed for digital computation and are included due to their simplicity.

Chapter 13 treats dynamic systems subject to random excitation. Such excitation are not deterministic and must be examined from probability and statistical points of view. In many cases, the probability density of the random excitation is normally distributed, which facilitates the computation. The point of view taken here is that, given a random record, an autocorrelation can be easily determined from which the spectral density and mean square response can be calculated. Fourier methods play a large part in this study.

The last chapter of the book is a brief discussion of simple non-linear systems. Since superposition is not applicable to non-linear systems, analytical procedures of the previous sections are not readily available. The method of the phase plane offers some understanding of such systems. However, graphical methods for the phase plane solution have been superceded by the digital computer which encounters no difficulty with nonlinear problems. When the nonlinearities are small, certain classical equations have yielded to approximate solutions by the perturbation or iteration techniques.

William T. Thomson

THE SI SYSTEM OF UNITS

THE SI SYSTEM OF UNITS

The English system of units that has dominated the United States from historical times is now being replaced by the SI system of units. Major industries throughout the United States either have already made, or are in the process of making the transition, and engineering students and teachers must deal with the new SI units as well as the present English system. We present here a short discussion of the SI units as they apply to the vibration field and outline a simple procedure to convert from one set of units to the other.

The basic units of the SI system are:

Units	*Name*	*Symbol*
Length	Meter	m
Mass	Kilogram	kg
Time	Second	s

The following quantities pertinent to the vibration field are derived from these basic units:

Force	Newton	N ($= kg \cdot m/s^2$)
Stress	Pascal	Pa ($= N/m^2$)
Work	Joule	J ($= N \cdot m$)
Power	Watt	W ($= J/s$)
Frequency	Hertz	Hz ($= 1/s$)

Moment of a force	$N \cdot m\ (= kg \cdot m^2/s^2)$
Acceleration	m/s^2
Velocity	m/s
Angular velocity	$1/s$
Moment of inertia (area)	$m^4\ (mm^4 \times 10^{-12})$
Moment of inertia (mass)	$kg \cdot m^2\ (kg \cdot cm^2 \times 10^{-4})$

Because the meter is a large unit of length, it will be more convenient to express it as the number of millimeters multiplied by 10^{-3}. Vibration instruments, such as accelerometers, are in general calibrated in terms of $g = 9.81$ m/s^2, and hence expressed in nondimensional units. It is advisable to use nondimensional presentation whenever possible.

In the English system the weight of an object is generally specified. In the SI system, it is more common to specify the mass, a quantity of matter which remains unchanged with location.

In working with the SI system, it is advisable to think directly in SI units. This will require some time but the following round numbers will help to develop a feeling of confidence in the use of SI units.

The newton is a smaller unit of force than the pound. One pound of force is equal to 4.4482 newtons or approximately four and a half times the value for the pound. (An apple weighs approximately $\frac{1}{4}$ lb or approximately 1 newton).

One inch is 2.54 cm, or 0.0254 meter. Thus the acceleration of gravity, which is 386 in/s^2 in the English system, becomes $386 \times 0.0254 = 9.81$ m/s^2 or approximately 10 m/s^2.

TABLE OF APPROXIMATE EQUIVALENTS

1 lb	$\cong$	4.5 N
Accel. of gravity g	$\cong$	10 m/s^2
Mass of 1 slug	$\cong$	15 kg
1 ft	$\cong$	1/3 m

SI conversion. A simple procedure to convert from one set of units to another follows: Write the desired SI units equal to the English units, and put in canceling unit factors. For example, suppose we wish to convert torque in the English units into the SI units, we proceed as follows:

Example 1

$$[\text{Torque SI}] = [\text{Torque English}] \times [\text{multiplying factors}]$$

$$[\text{N} \cdot \text{m}] = [\not{\text{lb}}\ \not{\text{in.}}]\left(\frac{\text{N}}{\not{\text{lb}}}\right)\left(\frac{\text{m}}{\not{\text{in.}}}\right)$$

$$= [\text{lb in.}](4.448)(0.0254)$$

$$= [\text{lb in.}](0.1129)$$

THE SI SYSTEM OF UNITS

THE SI SYSTEM OF UNITS

The English system of units that has dominated the United States from historical times is now being replaced by the SI system of units. Major industries throughout the United States either have already made, or are in the process of making the transition, and engineering students and teachers must deal with the new SI units as well as the present English system. We present here a short discussion of the SI units as they apply to the vibration field and outline a simple procedure to convert from one set of units to the other.

The basic units of the SI system are:

Units	*Name*	*Symbol*
Length	Meter	m
Mass	Kilogram	kg
Time	Second	s

The following quantities pertinent to the vibration field are derived from these basic units:

Force	Newton	N $(= kg \cdot m/s^2)$
Stress	Pascal	Pa $(= N/m^2)$
Work	Joule	J $(= N \cdot m)$
Power	Watt	W $(= J/s)$
Frequency	Hertz	Hz $(= 1/s)$

Moment of a force	$N \cdot m\ (= kg \cdot m^2/s^2)$
Acceleration	m/s^2
Velocity	m/s
Angular velocity	$1/s$
Moment of inertia (area)	$m^4\ (mm^4 \times 10^{-12})$
Moment of inertia (mass)	$kg \cdot m^2\ (kg \cdot cm^2 \times 10^{-4})$

Because the meter is a large unit of length, it will be more convenient to express it as the number of millimeters multiplied by 10^{-3}. Vibration instruments, such as accelerometers, are in general calibrated in terms of $g = 9.81$ m/s^2, and hence expressed in nondimensional units. It is advisable to use nondimensional presentation whenever possible.

In the English system the weight of an object is generally specified. In the SI system, it is more common to specify the mass, a quantity of matter which remains unchanged with location.

In working with the SI system, it is advisable to think directly in SI units. This will require some time but the following round numbers will help to develop a feeling of confidence in the use of SI units.

The newton is a smaller unit of force than the pound. One pound of force is equal to 4.4482 newtons or approximately four and a half times the value for the pound. (An apple weighs approximately $\frac{1}{4}$ lb or approximately 1 newton).

One inch is 2.54 cm, or 0.0254 meter. Thus the acceleration of gravity, which is 386 in/s^2 in the English system, becomes $386 \times 0.0254 = 9.81\ m/s^2$ or approximately 10 m/s^2.

TABLE OF APPROXIMATE EQUIVALENTS

1 lb	$\cong$	4.5 N
Accel. of gravity g	$\cong$	$10\ m/s^2$
Mass of 1 slug	$\cong$	15 kg
1 ft	$\cong$	1/3 m

SI conversion. A simple procedure to convert from one set of units to another follows: Write the desired SI units equal to the English units, and put in canceling unit factors. For example, suppose we wish to convert torque in the English units into the SI units, we proceed as follows:

Example 1

$$[\text{Torque SI}] = [\text{Torque English}] \times [\text{multiplying factors}]$$

$$\begin{aligned}[N \cdot m] &= [\not{lb}\ \not{in}.]\left(\frac{N}{\not{lb}}\right)\left(\frac{m}{\not{in}.}\right) \\ &= [\text{lb in.}](4.448)(0.0254) \\ &= [\text{lb in.}](0.1129)\end{aligned}$$

CONVERSION FACTORS*
U.S.-BRITISH UNITS TO SI UNITS

To Convert From	*To*	*Multiply By*
(Acceleration)		
foot/second^2 (ft/s^2)	metre/second^2 (m/s^2)	3.048×10^{-1}*
inch/second^2 (in./s^2)	metre/second^2 (m/s^2)	2.54×10^{-2}*
(Area)		
foot^2 (ft^2)	metre^2 (m^2)	9.2903×10^{-2}
inch^2 (in.^{2})	metre^2 (m^2)	6.4516×10^{-4}*
yard^2 (yd^2)	metre^2 (m^2)	8.3613×10^{-1}
(Density)		
pound mass/inch^3 (lbm/in.^{3})	kilogram/metre^3 (kg/m^3)	2.7680×10^4
pound mass/foot^3 (lbm/ft^3)	kilogram/metre^3 (kg/m^3)	1.6018×10
(Energy, Work)		
British thermal unit (BTU)	joule (J)	1.0551×10^3
foot-pound force (ft · lbf)	joule (J)	1.3558
kilowatt-hour (kw · h)	joule (J)	3.60×10^6*
(Force)		
kip (1000 lbf)	newton (N)	4.4482×10^3
pound force (lbf)	newton (N)	4.4482
ounce force	newton (N)	2.7801×10^{-1}
(Length)		
foot (ft)	metre(m)	3.048×10^{-1}*
inch (in.)	metre (m)	2.54×10^{-2}*
mile (mi), (U.S. statute)	metre (m)	1.6093×10^3
mile (mi), (international nautical)	metre (m)	1.852×10^3*
yard (yd)	metre (m)	9.144×10^{-1}*
(Mass)		
pound · mass (lbm)	kilogram (kg)	4.5359×10^{-1}
slug (lbf · s^2/ft)	kilogram (kg)	1.4594×10
ton (2000 lbm)	kilogram (kg)	9.0718×10^2
(Power)		
foot-pound/minute (ft · lbf/min)	watt (W)	2.2597×10^{-2}
horsepower (550 ft · lbf/s)	watt (W)	7.4570×10^2
(Pressure, stress)		
atmosphere (std) (14.7 lbf/in.^{2})	newton/metre^2 (N/m^2 or Pa)	1.0133×10^5
pound/foot^2 (lbf/ft^2)	newton/metre^2 (N/m^2 or Pa)	4.7880×10
pound/inch^2 (lbf/in.^{2} or psi)	newton/metre^2 (N/m^2 or Pa)	6.8948×10^3
(Velocity)		
foot/minute (ft/min)	metre/second (m/s)	5.08×10^{-3}*
foot/second (ft/s)	metre/second (m/s)	3.048×10^{-1}*
knot (nautical mi/h)	metre/second (m/s)	5.1444×10^{-1}
mile/hour (mi/h)	metre/second (m/s)	4.4704×10^{-1}*
mile/hour (mi/h)	kilometre/hour (km/h)	1.6093
mile/second (mi/s)	kilometre/second (km/s)	1.6093
(Volume)		
foot^3 (ft^3)	metre^3 (m^3)	2.8317×10^{-2}
inch^3 (in.^{3})	metre^3 (m^3)	1.6387×10^{-5}

*Exact value

Source: Meriam, J. L., *Dynamics*, 2*nd Ed*. (*SI Version*). New York: John Wiley and Sons, Inc., 1975.

The International System of Units (SI), July 1974, National Bureau of Standards, Special Publication 330.

Example 2

$$[\text{Moment of inertia SI}] = [\text{Moment of inertia, Eng}] \times [\text{mult. factors}]$$

$$\left[\begin{array}{c}\text{kg}\cdot\text{m}^2\\ = \text{N}\cdot\text{m}\cdot\text{s}^2\end{array}\right] = [\text{lb in. s}^2]\left(\frac{\text{N}}{\text{lb}}\cdot\frac{\text{m}}{\text{in.}}\right)$$

$$= [\text{lb in. s}^2](4.448 \times 0.0254)$$

$$= [\text{lb in. s}^2](0.1129)$$

Example 3

Modulus of Elasticity: E

$$[\text{N/m}^2] = \left[\frac{\text{lb}}{\text{in}^2}\right]\left(\frac{\text{N}}{\text{lb}}\right)\left(\frac{\text{in}}{\text{m}}\right)^2$$

$$= \left[\frac{\text{lb}}{\text{in}^2}\right](4.448)\left(\frac{1}{0.0254}\right)^2$$

$$= \left[\frac{\text{lb}}{\text{in}^2}\right](6894.7)$$

$$\text{E of steel N/m}^2 = (29 \times 10^6 \text{ lb/in.}^2)(6894.7) = \underline{\underline{200 \times 10^9 \text{ N/m}^2}}$$

Example 4

Spring Stiffness: K

$$[\text{N/m}] = [\text{lb/in.}] \times (175.13)$$

Mass: M

$$[\text{kg}] = [\text{lb}\cdot\text{sec}^2/\text{in.}] \times (175.13)$$

1

OSCILLATORY MOTION

The study of vibration is concerned with the oscillatory motions of bodies and the forces associated with them. All bodies possessing mass and elasticity are capable of vibration. Thus most engineering machines and structures experience vibration to some degree, and their design generally requires consideration of their oscillatory behavior.

Oscillatory systems can be broadly characterized as *linear* or *nonlinear*. For linear systems the principle of superposition holds, and the mathematical techniques available for their treatment are well developed. In contrast, techniques for the analysis of nonlinear systems are less well known, and difficult to apply. However, some knowledge of nonlinear systems is desirable, since all systems tend to become nonlinear with increasing amplitude of oscillation.

There are two general classes of vibrations—free and forced. *Free vibration* takes place when a system oscillates under the action of forces inherent in the system itself, and when external impressed forces are absent. The system under free vibration will vibrate at one or more of its *natural frequencies*, which are properties of the dynamical system established by its mass and stiffness distribution.

Vibration that takes place under the excitation of external forces is called *forced vibration*. When the excitation is oscillatory, the system is forced to vibrate at the excitation frequency. If the frequency of excitation coincides with one of the natural frequencies of the system, a condition of *resonance* is encountered, and dangerously large oscillations may result. The failure of major structures such as bridges, buildings, or airplane wings is an awesome possibility under resonance. Thus, the calculation of the natural frequencies is of major importance in the study of vibrations.

Vibrating systems are all subject to *damping* to some degree because energy is dissipated by friction and other resistances. If the damping is small, it has very little influence on the natural frequencies of the system, and hence the calculations for the natural frequencies are generally made on tne basis of no damping. On the other hand, damping is of great importance in limiting the amplitude of oscillation at resonance.

The number of independent coordinates required to describe the motion of a system is called *degrees of freedom* of the system. Thus a free particle undergoing general motion in space will have three degrees of freedom, while a rigid body will have six degrees of freedom, i.e., three components of position and three angles defining its orientation. Furthermore, a continuous elastic body will require an infinite number of coordinates (three for each point on the body) to describe its motion; hence its degrees of freedom must be infinite. However, in many cases, parts of such bodies may be assumed to be rigid, and the system may be considered to be dynamically equivalent to one having finite degrees of freedom. In fact, a surprisingly large number of vibration problems can be treated with sufficient accuracy by reducing the system to one having a few degrees of freedom.

1.1 HARMONIC MOTICN

Oscillatory motion may repeat itself regularly, as in the balance wheel of a watch, or display considerable irregularity, as in earthquakes. When the motion is repeated in equal intervals of time τ, it is called *periodic mction*. The repetition time τ is called the *period* of the oscillation, and its reciprocal, $f = 1/\tau$, is called the *frequency*. If the motion is designated by the time function $x(t)$, then any periodic motion must satisfy the relationship $x(t) = x(t + \tau)$.

The simplest form of periodic motion is *harmonic motion*. It can be demonstrated by a mass suspended from a light spring, as shown in Fig. 1.1-1. If the mass is displaced from its rest position and released, it will oscillate up and down. By placing a light source on the oscillating mass, its motion can be

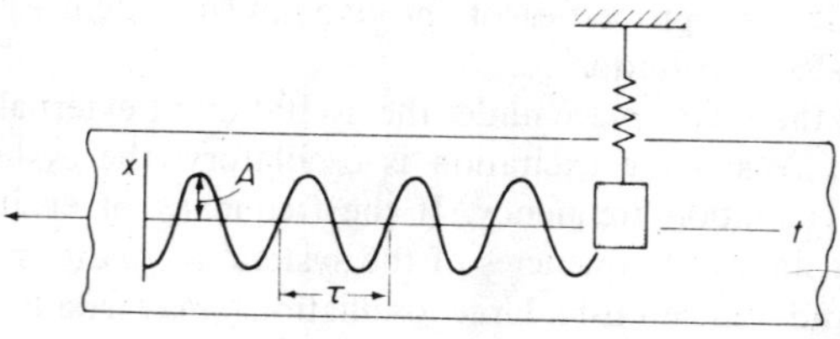

Figure 1.1-1. Recording of harmonic motion.

1

OSCILLATORY MOTION

The study of vibration is concerned with the oscillatory motions of bodies and the forces associated with them. All bodies possessing mass and elasticity are capable of vibration. Thus most engineering machines and structures experience vibration to some degree, and their design generally requires consideration of their oscillatory behavior.

Oscillatory systems can be broadly characterized as *linear* or *nonlinear*. For linear systems the principle of superposition holds, and the mathematical techniques available for their treatment are well developed. In contrast, techniques for the analysis of nonlinear systems are less well known, and difficult to apply. However, some knowledge of nonlinear systems is desirable, since all systems tend to become nonlinear with increasing amplitude of oscillation.

There are two general classes of vibrations—free and forced. *Free vibration* takes place when a system oscillates under the action of forces inherent in the system itself, and when external impressed forces are absent. The system under free vibration will vibrate at one or more of its *natural frequencies*, which are properties of the dynamical system established by its mass and stiffness distribution.

Vibration that takes place under the excitation of external forces is called *forced vibration*. When the excitation is oscillatory, the system is forced to vibrate at the excitation frequency. If the frequency of excitation coincides with one of the natural frequencies of the system, a condition of *resonance* is encountered, and dangerously large oscillations may result. The failure of major structures such as bridges, buildings, or airplane wings is an awesome possibility under resonance. Thus, the calculation of the natural frequencies is of major importance in the study of vibrations.

Vibrating systems are all subject to *damping* to some degree because energy is dissipated by friction and other resistances. If the damping is small, it has very little influence on the natural frequencies of the system, and hence the calculations for the natural frequencies are generally made on tne basis of no damping. On the other hand, damping is of great importance in limiting the amplitude of oscillation at resonance.

The number of independent coordinates required to describe the motion of a system is called *degrees of freedom* of the system. Thus a free particle undergoing general motion in space will have three degrees of freedom, while a rigid body will have six degrees of freedom, i.e., three components of position and three angles defining its orientation. Furthermore, a continuous elastic body will require an infinite number of coordinates (three for each point on the body) to describe its motion; hence its degrees of freedom must be infinite. However, in many cases, parts of such bodies may be assumed to be rigid, and the system may be considered to be dynamically equivalent to one having finite degrees of freedom. In fact, a surprisingly large number of vibration problems can be treated with sufficient accuracy by reducing the system to one having a few degrees of freedom.

1.1 HARMONIC MOTICN

Oscillatory motion may repeat itself regularly, as in the balance wheel of a watch, or display considerable irregularity, as in earthquakes. When the motion is repeated in equal intervals of time τ, it is called *periodic mction*. The repetition time τ is called the *period* of the oscillation, and its reciprocal, $f = 1/\tau$, is called the *frequency*. If the motion is designated by the time function $x(t)$, then any periodic motion must satisfy the relationship $x(t) = x(t + \tau)$.

The simplest form of periodic motion is *harmonic motion*. It can be demonstrated by a mass suspended from a light spring, as shown in Fig. 1.1-1. If the mass is displaced from its rest position and released, it will oscillate up and down. By placing a light source on the oscillating mass, its motion can be

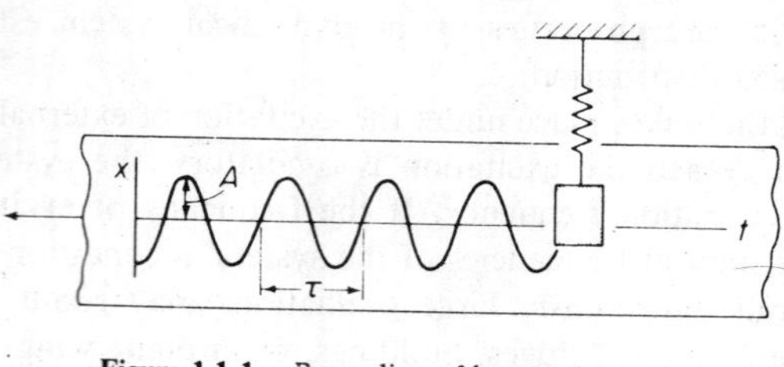

Figure 1.1-1. Recording of harmonic motion.

in Eq. (1.2-1), we obtain

$$\begin{aligned} x(t) &= \frac{a_0}{2} + \sum_{n=1}^{\infty} \left[\tfrac{1}{2}(a_n - ib_n)e^{i\omega_n t} + \tfrac{1}{2}(a_n + ib_n)e^{-i\omega_n t} \right] \\ &= \frac{a_0}{2} + \sum_{n=1}^{\infty} \left[c_n e^{i\omega_n t} + c_n^* e^{-i\omega_n t} \right] \\ &= \sum_{n=-\infty}^{\infty} c_n e^{i\omega_n t} \end{aligned} \tag{1.2-4}$$

where

$$\begin{aligned} c_0 &= \tfrac{1}{2} a_0 \\ c_n &= \tfrac{1}{2}(a_n - ib_n) \end{aligned} \tag{1.2-5}$$

Substituting for a_n and b_n from Eq. (1.2-3), we find c_n to be

$$\begin{aligned} c_n &= \frac{1}{\tau} \int_{-\tau/2}^{\tau/2} x(t)(\cos \omega_n t - i \sin \omega_n t)\, dt \\ &= \frac{1}{\tau} \int_{-\tau/2}^{\tau/2} x(t) e^{-i\omega_n t}\, dt \end{aligned} \tag{1.2-6}$$

Some computational effort can be minimized when the function $x(t)$ is recognizable in terms of the even and odd functions

$$x(t) = E(t) + O(t) \tag{1.2-7}$$

An even function $E(t)$ is symmetric about the origin so that $E(t) = E(-t)$, i.e., $\cos \omega t = \cos(-\omega t)$. An odd function satisfies the relationship $O(t) = -O(-t)$, i.e., $\sin \omega t = -\sin(-\omega t)$. The following integrals are then helpful:

$$\begin{aligned} \int_{-\tau/2}^{\tau/2} E(t) \sin \omega_n t\, dt &= 0 \\ \int_{-\tau/2}^{\tau/2} O(t) \cos \omega_n t\, dt &= 0 \end{aligned} \tag{1.2-8}$$

When the coefficients of the Fourier series are plotted against frequency ω_n, the result is a series of discrete lines called the *Fourier spectrum*. Generally plotted are the absolute values $|2c_n| = \sqrt{a_n^2 + b_n^2}$ and the phase $\phi_n = \tan^{-1} b_n/a_n$, an example of which is shown in Fig. 1.2-2.

With the aid of the digital computer, harmonic analysis today is efficiently carried out. A computer algorithm known as the *fast Fourier transform*† (FFT) is commonly used to minimize the computation time.

†See J. S. Bendat & A. G. Piersol, *Random Data* (New York: John Wiley & Sons, 1971), pp. 305–306.

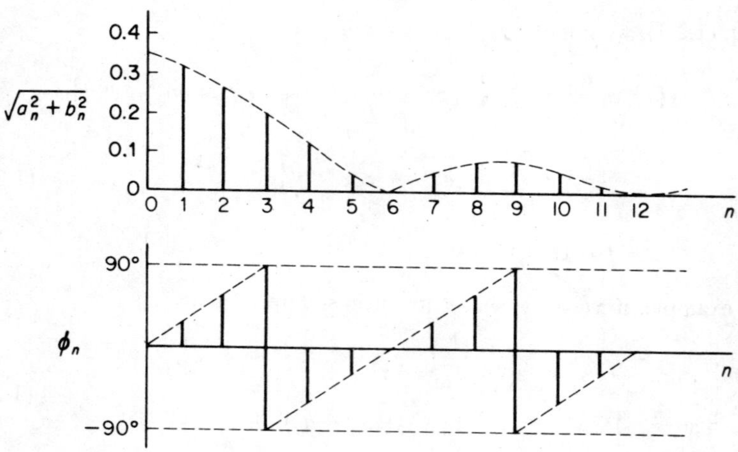

Figure 1.2-2. Fourier spectrum for pulses shown in Prob. 1-16, $k = \frac{1}{3}$.

1.3 VIBRATION TERMINOLOGY

Certain terminologies used in the vibration need to be represented here. The simplest of these are the *peak value* and the *average value*.

The peak value generally indicates the maximum stress that the vibrating part is undergoing. It also places a limitation on the "rattle space" requirement.

The average value indicates a steady or static value somewhat like the DC level of an electrical current. It can be found by the time integral

$$\bar{x} = \lim_{T\to\infty} \frac{1}{T}\int_0^T x(t)\, dt \tag{1.3-1}$$

For example, the average value for a complete cycle of a sine wave, $A \sin t$, is zero; whereas its average value for a half-cycle is

$$\bar{x} = \frac{A}{\pi}\int_0^{\pi} \sin t\, dt = \frac{2A}{\pi} = 0.637\text{ A}$$

It is evident that this is also the average value of the rectified sine wave shown in Fig. 1.3-1.

The square of the displacement generally is associated with the energy of the vibration for which the mean square value is a measure. The *mean square value* of a time function $x(t)$ is found from the average of the squared values, integrated over some time interval T:

$$\overline{x^2} = \lim_{T\to\infty} \frac{1}{T}\int_0^T x^2(t)\, dt \tag{1.3-2}$$

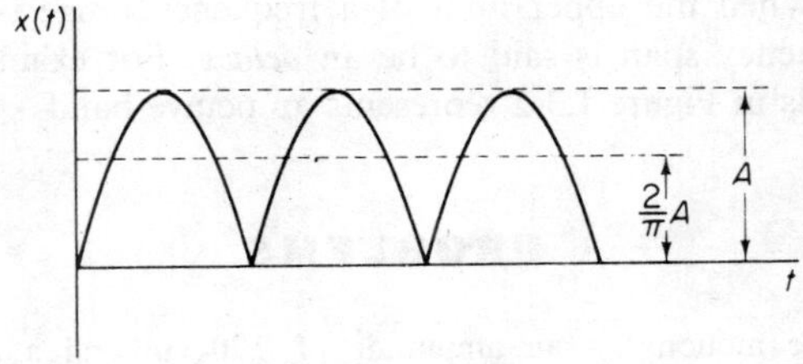

Figure 1.3-1. Average value of a rectified sine wave.

For example, if $x(t) = A \sin \omega t$, its mean square value is

$$\overline{x^2} = \lim_{T\to\infty} \frac{A^2}{T}\int_0^T \frac{1}{2}(1 - \cos 2\omega t)\, dt = \frac{1}{2}A^2$$

The *root mean square* (rms) value is the square root of the mean square value. From the previous example, the rms of the sine wave of amplitude A is $A/\sqrt{2} = 0.707$ A. Vibrations are commonly measured by rms meters.

Decibel: The decibel is a unit of measurement that is frequently used in vibration measurements. It is defined in terms of a power ratio.

$$\begin{aligned} \text{Db} &= 10 \log_{10}\left(\frac{p_1}{p_2}\right) \\ &= 10 \log_{10}\left(\frac{x_1}{x_2}\right)^2 \end{aligned} \tag{1.3-3}$$

The second equation results from the fact that power is proportional to the square of the amplitude or voltage. The decibel is often expressed in terms of the first power of amplitude or voltage as

$$\text{Db} = 20 \log_{10}\left(\frac{x_1}{x_2}\right) \tag{1.3-4}$$

Thus an amplifier with a voltage gain of 5 has a decibel gain of

$$20 \log_{10}(5) = +14$$

Because the decibel is a logarithmic unit, it compresses or expands the scale.

Band	*Frequency range (Hz)*	*Frequency Bandwidth*
1	10–20	10
2	20–40	20
3	40–80	40
4	200–400	200

Figure 1.3-2.

Octave: When the upper limit of a frequency range is twice its lower limit, the frequency span is said to be an *octave*. For example, each of the frequency bands in Figure 1.3-2 represents an octave band.

PROBLEMS

1-1 A harmonic motion has an amplitude of 0.20 cm and a period of 0.15 s. Determine the maximum velocity and acceleration.

1-2 An accelerometer indicates that a structure is vibrating harmonically at 82 cps with a maximum acceleration of 50 g. Determine the amplitude of vibration.

1-3 A harmonic motion has a frequency of 10 cps and its maximum velocity is 4.57 m/s. Determine its amplitude, its period, and its maximum acceleration.

1-4 Find the sum of two harmonic motions of equal amplitude but of slightly different frequencies. Discuss the beating phenomena that result from this sum.

1-5 Express the complex vector $4 + 3i$ in the exponential form $Ae^{i\theta}$.

1-6 Add two complex vectors $(2 + 3i)$ and $(4 - i)$ expressing the result as $A\angle\theta$.

1-7 Show that the multiplication of a vector $z = Ae^{i\omega t}$ by i rotates it by 90°.

1-8 Determine the sum of two vectors $5e^{i\pi/6}$ and $4e^{i\pi/3}$ and find the angle between the resultant and the first vector.

1-9 Determine the Fourier series for the rectangular wave shown in Fig. P1-9.

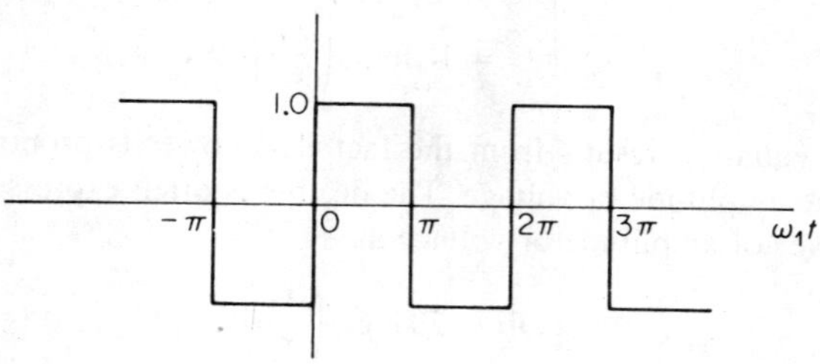

Figure P1-9.

1-10 If the origin of the square wave of Prob. 1-9 is shifted to the right by $\pi/2$, determine the Fourier series.

1-11 Determine the Fourier series for the triangular wave shown in Fig. P1-11.

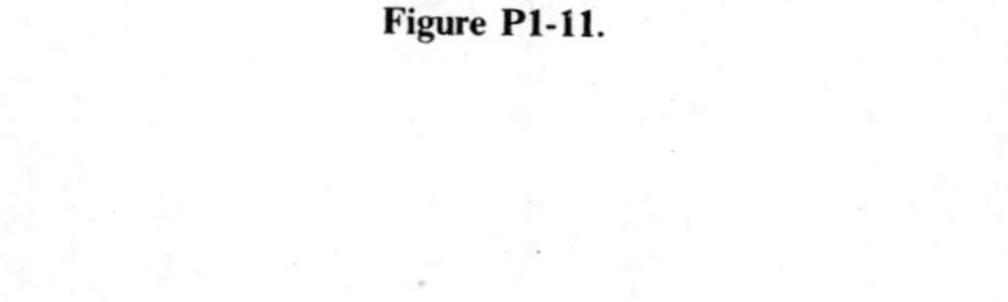

Figure P1-11.

1-12 Determine the Fourier series for the sawtooth curve shown in Fig. P1-12. Express the result of Prob. 1-12 in the exponential form of Eq. (1.2-4).

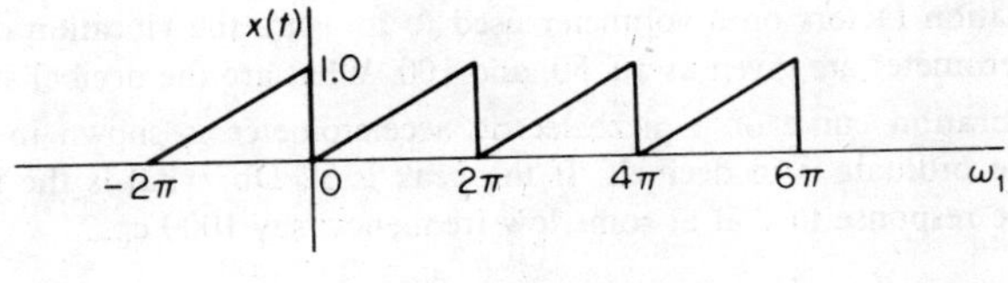

Figure P1-12.

1-13 Determine the rms value of a wave consisting of the positive portions of a sine wave.

1-14 Determine the mean square value of the sawtooth wave of Prob. 1-12. Do this two ways, from the squared curve and from the Fourier series.

1-15 Plot the frequency spectrum for the triangular wave of Prob. 1-11.

1-16 Determine the Fourier series of a series of rectangular pulses shown in Fig. P1-16. Plot c_n and ϕ_n versus n when $k = \frac{2}{3}$.

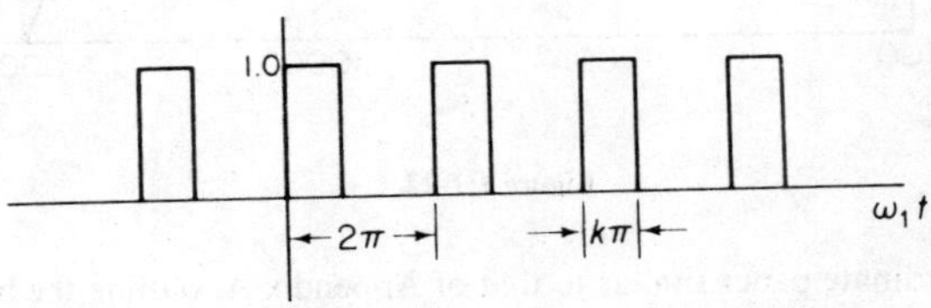

Figure P1-16.

1-17 Write the equation for the displacement s of the piston in the crank-piston mechanism shown in Fig. P1-17, and determine the harmonic components and their relative magnitudes. If $r/l = \frac{1}{3}$, what is the ratio of the second harmonic compared to the first?

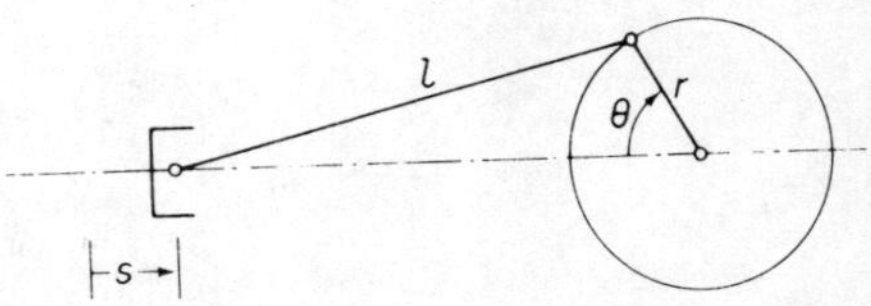

Figure P1-17.

1-18 Determine the mean square of the rectangular pulse shown in Fig. P1-18 for $k = 0.10$. If the amplitude is A, what would an rms voltmeter read?

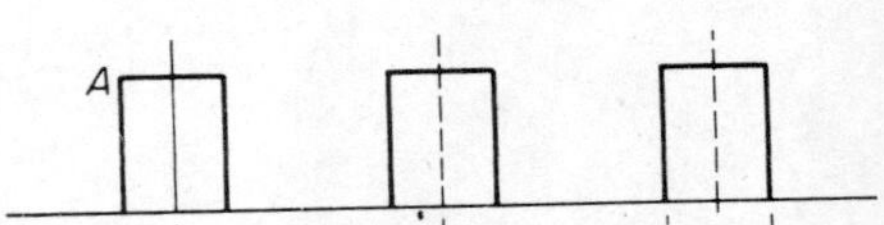

Figure P1-18.

1-19 Determine the mean square value of the triangular wave of Fig. P1-11.

1-20 An rms voltmeter specifies an accuracy of ± 0.5 Db. If a vibration of 2.5 mm rms is measured, determine the millimeter accuracy as read by the voltmeter.

1-21 Amplification factors on a voltmeter used to measure the vibration output from an accelerometer are given as 10, 50, and 100. What are the decibel steps?

1-22 The calibration curve of a piezoelectric accelerometer is shown in Fig. P1-22 where the ordinate is in decibels. If the peak is 32 Db, what is the ratio of the resonance response to that at some low frequency, say 1000 cps?

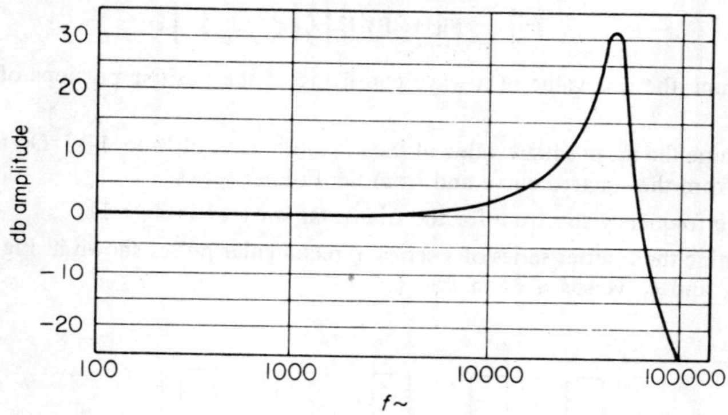

Figure P1-22.

1-23 Using coordinate paper similar to that of Appendix A, outline the bounds for the following vibration specifications. Max. acceleration = 2 g, max. displacement = 0.08 in, min. and max. frequencies: 1 Hz and 200 Hz.

2 FREE VIBRATION

All systems possessing mass and elasticity are capable of free vibration, or vibration which takes place in the absence of external excitation. Of primary interest for such a system is its natural frequency of vibration. Our object here is to learn to write its equation of motion and evaluate its natural frequency, which is mainly a function of the mass and stiffness of the system.

Damping in moderate amounts has little influence on the natural frequency and may be neglected in its calculation. The system can then be considered to be conservative and the principle of conservation of energy offers another approach to the calculation of the natural frequency. The effect of damping is mainly evident in the diminishing of the vibration amplitude with time. Although there are many models of damping, only those which lead to simple analytic procedures are considered in this chapter.

2.1 VIBRATION MODEL

The basic vibration model of a simple oscillatory system consists of a mass, a massless spring, and a damper. The mass is considered to be lumped and measured in the SI system as kilograms. In the English system the mass is $m = w/g$ lb $\cdot$ s^2/in.

The spring supporting the mass is assumed to be of negligible mass. Its force-deflection relationship is considered to be linear, following Hooke's law, $F = kx$, where the stiffness k is measured in Newtons/meter or pounds/inch.

The viscous damping, generally represented by a dashpot, is described by a force proportional to the velocity, or $F = c\dot{x}$. The damping coefficient c is measured in Newtons/meter/second or pounds/inch/second.

2.2 EQUATIONS OF MOTION — NATURAL FREQUENCY

Figure 2.2-1 shows a simple undamped spring-mass system, which is assumed to move only along the vertical direction. It has one degree of freedom (DOF), since its motion is described by a single coordinate x.

When placed into motion, oscillation will take place at the natural frequency f_n, which is a property of the system. We now examine some of the basic concepts associated with the free vibration of systems with one degree of freedom.

Newton's second law is the first basis for examining the motion of the system. As shown in Fig. 2.2-1 the deformation of the spring in the static equilibrium position is Δ, and the spring force $k\Delta$ is equal to the gravitational force w acting on the mass m:

$$k\Delta = w = mg \tag{2.2-1}$$

Measuring the displacement x from the static equilibrium position, the forces acting on m are $k(\Delta + x)$ and w. With x chosen to be positive in the downward direction, all quantities—force, velocity, and acceleration—are also positive in the downward direction.

We now apply Newton's second law of motion to the mass m

$$m\ddot{x} = \Sigma F = w - k(\Delta + x)$$

and since $k\Delta = w$, we obtain

$$m\ddot{x} = -kx \tag{2.2-2}$$

It is evident that the choice of the static equilibrium position as reference for x has eliminated w, the force due to gravity, and the static spring force $k\Delta$ from

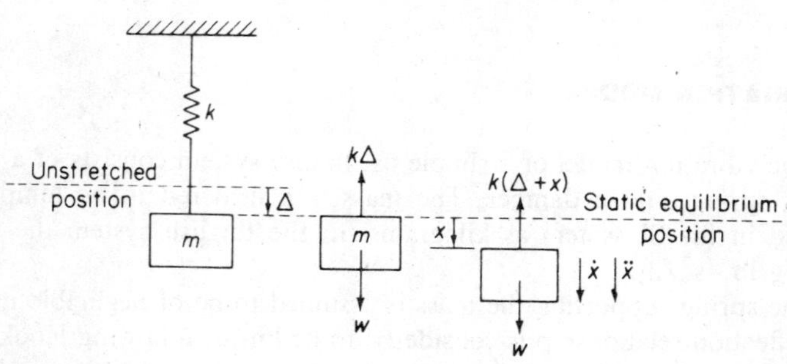

Figure 2.2-1. Spring-mass system and free-body diagram.

the equation of motion, and the resultant force on m is simply the spring force due to the displacement x.

Defining the circular frequency ω_n by the equation

$$\omega_n^2 = \frac{k}{m} \tag{2.2-3}$$

Eq. (2.2-2) may be written as

$$\ddot{x} + \omega_n^2 x = 0 \tag{2.2-4}$$

and we conclude by comparison with Eq. (1.1-6) that the motion is harmonic. Equation (2.2-4), a homogeneous second-order linear differential equation, has the following general solution:

$$x = A \sin \omega_n t + B \cos \omega_n t \tag{2.2-5}$$

where A and B are the two necessary constants. These constants are evaluated from initial conditions $x(0)$ and $\dot{x}(0)$, and Eq. (2.2-5) can be shown to reduce to

$$x = \frac{\dot{x}(0)}{\omega_n} \sin \omega_n t + x(0) \cos \omega_n t \tag{2.2-6}$$

The natural period of the oscillation is established from $\omega_n \tau = 2\pi$, or

$$\tau = 2\pi \sqrt{\frac{m}{k}} \tag{2.2-7}$$

and the natural frequency is

$$f_n = \frac{1}{\tau} = \frac{1}{2\pi} \sqrt{\frac{k}{m}} \tag{2.2-8}$$

These quantities may be expressed in terms of the statical deflection Δ by observing Eq. (2.2-1), $k\Delta = mg$. Thus Eq. (2.2-8) may be expressed in terms of the statical deflection Δ as

$$f_n = \frac{1}{2\pi} \sqrt{\frac{g}{\Delta}} \tag{2.2-9}$$

Note that τ, f_n and ω_n depend only on the mass and stiffness of the system, which are properties of the system.

Although our discussion was in terms of the spring-mass system of Fig. 2.2-1, the results are applicable to all single-DOF systems including rotation. The spring may be a beam or torsional member and the mass may be replaced by a mass moment of inertia. A table of values for the stiffness k for various types of springs is presented at the end of the chapter.

Example 2.2-1

A $\frac{1}{4}$-kg mass is suspended by a spring having a stiffness of 0.1533 N/mm. Determine its natural frequency in cycles per second. Determine its statical deflection.

Solution: The stiffness is

$$k = 153.3 \text{ N/m}$$

Substituting into Eq. (2.2-8), the natural frequency is

$$f = \frac{1}{2\pi}\sqrt{\frac{k}{m}} = \frac{1}{2\pi}\sqrt{\frac{153.3}{0.25}} = 3.941 \text{ Hz}$$

The statical deflection of the spring suspending the $\frac{1}{4}$-kg mass is obtained from the relationship $mg = k\Delta$

$$\Delta = \frac{mg}{k_{\text{N/mm}}} = \frac{0.25 \times 9.81}{0.1533} = 16.0 \text{ mm}$$

Example 2.2-2

Determine the natural frequency of the mass M on the end of a cantilever beam of negligible mass shown in Fig. 2.2-2.

Solution: The deflection of the cantilever beam under a concentrated end force P is

$$x = \frac{Pl^3}{3EI} = \frac{P}{k}$$

where EI is the flexural rigidity. Thus the stiffness of the beam is $k = 3EI/l^3$ and the natural frequency of the system becomes

$$f_n = \frac{1}{2\pi}\sqrt{\frac{3EI}{Ml^3}}$$

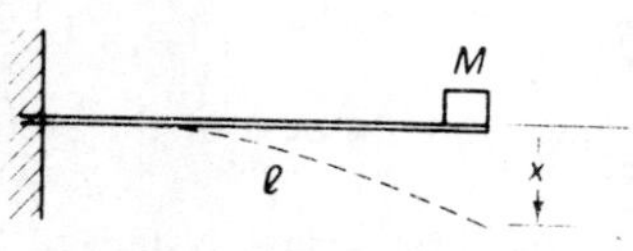

Figure 2.2-2.

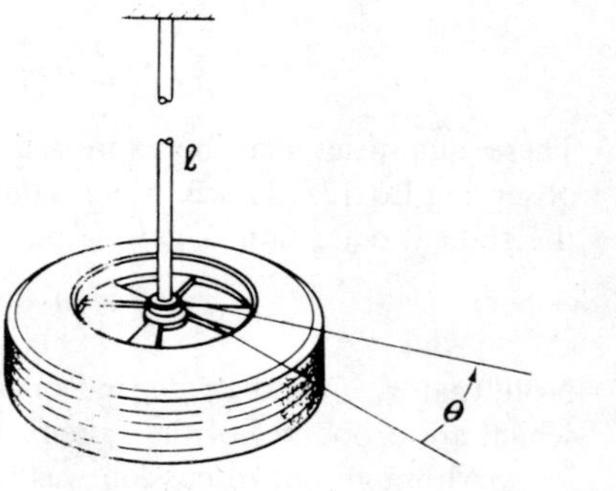

Figure 2.2-3.

Example 2.2-3

An automobile wheel and tire are suspended by a steel rod 0.50 cm in diameter and 2 m long as shown in Fig. 2.2-3. When the wheel is given an angular displacement and released, it makes 10 oscillations in 30.2 s. Determine the polar moment of inertia of the wheel and tire.

Solution: The rotational equation of motion corresponding to Newton's equation is

$$J\ddot{\theta} = -K\theta$$

where J is the rotational mass moment of inertia, K is the rotational stiffness, and θ is the angle of rotation in radians. Thus the natural frequency of oscillation is equal to

$$\omega_n = 2\pi \frac{10}{30.2} = 2.081 \text{ rad/s}$$

The torsional stiffness of the rod is given by the equation $K = GI_p/l$, where $I_p = \pi d^4/32$ = polar moment of inertia of the circular cross-sectional area of the rod, l = length, and $G = 80 \times 10^9 \text{ N/m}^2$ = shear modulus of steel.

$$I_p = \frac{\pi}{32}(0.5 \times 10^{-2})^4 = 0.006136 \times 10^{-8} \text{ m}^4$$

$$K = \frac{80 \times 10^9 \times 0.006136 \times 10^{-8}}{2} = 2.455 \text{ N} \cdot \text{m/rad}$$

Substituting into the natural frequency equation, the polar moment of inertia of the wheel and tire is

$$J = \frac{K}{\omega_n^2} = \frac{2.455}{2.081^2} = 0.567 \text{ kg} \cdot \text{m}^2$$

Example 2.2-4

Figure 2.2-4 shows a uniform bar pivoted about point o with springs of equal stiffness k at each end. The bar is horizontal in the equilibrium position with spring forces P_1 and P_2. Determine the equation of motion and its natural frequency.

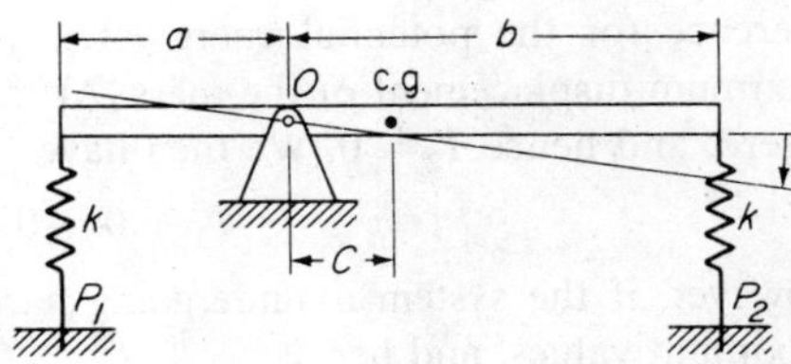

Figure 2.2-4.

Solution: Under rotation θ, the spring force on the left is decreased while that on the right is increased. With J_o as the moment of inertia of the bar about o, the moment equation about o is

$$\sum M_o = (P_1 - ka\theta)a + mgc - (P_2 + kb\theta)b = J_o\ddot{\theta}$$

However,

$$P_1 a + mgc - P_2 b = 0$$

in the equilibrium position, and hence we need to consider only the moment of the forces due to displacement θ, which is

$$\sum M_o = (-ka^2 - kb^2)\theta = J_o\ddot{\theta}$$

Thus the equation of motion can be written as

$$\ddot{\theta} + \frac{k(a^2 + b^2)}{J_o}\theta = 0$$

and by inspection the natural frequency of oscillation is

$$\omega_n = \sqrt{\frac{k(a^2 + b^2)}{J_o}}$$

2.3 ENERGY METHOD

In a conservative system the total energy is constant, and the differential equation of motion can also be established by the principle of conservation of energy. For the free vibration of an undamped system, the energy is partly kinetic and partly potential. The kinetic energy T is stored in the mass by virtue of its velocity, whereas the potential energy U is stored in the form of strain energy in elastic deformation or work done in a force field such as gravity. The total energy being constant, its rate of change is zero as illustrated by the following equations:

$$T + U = \text{constant} \tag{2.3-1}$$

$$\frac{d}{dt}(T + U) = 0 \tag{2.3-2}$$

If our interest is only in the natural frequency of the system, it can be determined by the following considerations. From the principle of conservation of energy we can write

$$T_1 + U_1 = T_2 + U_2 \tag{2.3-3}$$

where $_1$ and $_2$ represent two instances of time. Let $_1$ be the time when the mass is passing through its static equilibrium position and choose $U_1 = 0$ as reference for the potential energy. Let $_2$ be the time corresponding to the maximum displacement of the mass. At this position, the velocity of the mass is zero, and hence $T_2 = 0$. We then have

$$T_1 + 0 = 0 + U_2 \tag{2.3-4}$$

However, if the system is undergoing harmonic motion, then T_1 and U_2 are maximum values, and hence

$$T_{max} = U_{max} \tag{2.3-5}$$

The preceding equation leads directly to the natural frequency.

Example 2.3-1

Determine the natural frequency of the system shown in Fig. 2.3-1.

Solution: Assume that the system is vibrating harmonically with amplitude θ from its static equilibrium position. The maximum kinetic energy is

$$T_{max} = \frac{1}{2}J\dot{\theta}^2 + \frac{1}{2}m\left(r_1\dot{\theta}\right)^2$$

The maximum potential energy is the energy stored in the spring which is

$$U_{max} = \frac{1}{2}k\left(r_2\theta\right)^2$$

Equating the two, the natural frequency is

$$\omega_n = \sqrt{\frac{kr_2^2}{J + mr_1^2}}$$

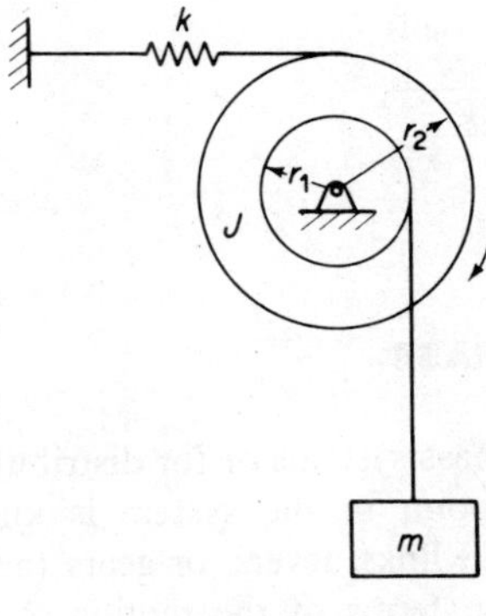

Figure 2.3-1.

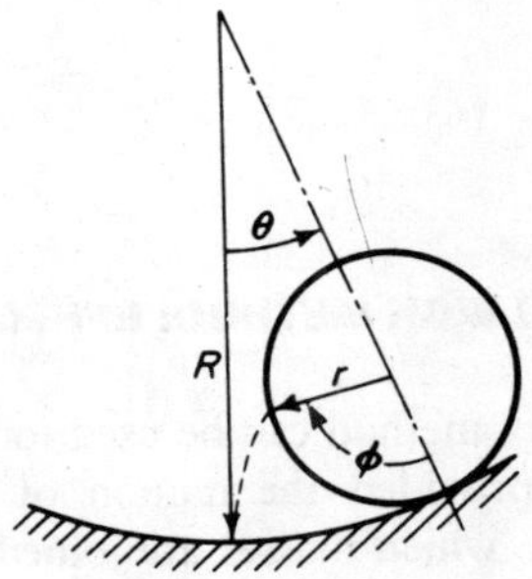

Figure 2.3-2.

The student should verify that the loss of potential energy of m due to position $r_1\theta$ is canceled by the work done by the equilibrium force of the spring in the position $\theta = 0$.

Example 2.3-2

A cylinder of weight w and radius r rolls without slipping on a cylindrical surface of radius R, as shown in Fig. 2.3-2. Determine its differential equation of motion for small oscillations about the lowest point. For no slipping, we have $r\phi = R\theta$.

Solution: In determining the kinetic energy of the cylinder, it must be noted that both translation and rotation take place. The translational velocity of the center of the cylinder is $(R - r)\dot{\theta}$, whereas the rotational velocity is $(\dot{\phi} - \dot{\theta}) = (R/r - 1)\dot{\theta}$, since $\dot{\phi} = (R/r)\dot{\theta}$ for no slipping. The kinetic energy may now be written as

$$T = \frac{1}{2}\frac{w}{g}\left[(R - r)\dot{\theta}\right]^2 + \frac{1}{2}\frac{w}{g}\frac{r^2}{2}\left[\left(\frac{R}{r} - 1\right)\dot{\theta}\right]^2$$

$$= \frac{3}{4}\frac{w}{g}(R - r)^2\dot{\theta}^2$$

where $(w/g)(r^2/2)$ is the moment of inertia of the cylinder about its mass center.

The potential energy referred to its lowest position is

$$U = w(R - r)(1 - \cos\theta)$$

which is equal to the negative of the work done by the gravity force in lifting the cylinder through the vertical height $(R - r)(1 - \cos\theta)$.

Substituting into Eq. (2.3-2)

$$\left[\frac{3}{2}\frac{w}{g}(R - r)^2\ddot{\theta} + w(R - r)\sin\theta\right]\dot{\theta} = 0$$

and letting $\sin\theta = \theta$ for small angles, we obtain the familiar equation for harmonic motion

$$\ddot{\theta} + \frac{2g}{3(R - r)}\theta = 0$$

By inspection, the circular frequency of oscillation is

$$\omega_n = \sqrt{\frac{2g}{3(R - r)}}$$

2.4 RAYLEIGH METHOD: EFFECTIVE MASS

The energy method can be used for multimass systems or for distributed mass systems, provided the motion of every point in the system is known. In systems in which masses are joined by rigid links, levers, or gears the motion of the various masses can be expressed in terms of the motion $\dot{x}$ of some specific point and the system is simply one of a single DOF, since only one coordinate is necessary. The kinetic energy can then be written as

$$T = \frac{1}{2} m_{\text{eff}} \dot{x}^2 \tag{2.4-1}$$

where m_{eff} is the *effective mass* or an equivalent lumped mass at the specified point. If the stiffness at that point is also known, the natural frequency can be calculated from the simple equation

$$\omega_n = \sqrt{\frac{k}{m_{\text{eff}}}} \tag{2.4-2}$$

In distributed mass systems such as springs and beams, a knowledge of the distribution of the vibration amplitude becomes necessary before the kinetic energy can be calculated. Rayleigh† showed that with a reasonable assumption for the shape of the vibration amplitude, it is possible to take into account previously ignored masses and arrive at a better estimate for the fundamental frequency. The following examples illustrate the use of both of these methods.

Example 2.4-1

Determine the effect of the mass of the spring on the natural frequency of the system shown in Fig. 2.4-1.

Solution: With $\dot{x}$ equal to the velocity of the lumped mass m, we will assume the velocity of a spring element located a distance y from the fixed end to vary linearly with y as follows

$$\dot{x}\frac{y}{l}$$

†John [illegible]tt, Lord Rayleigh, *The Theory of Sound*, Vol. 1, 2n[illegible]v. ed. (New York: Dover Pub[illegible]37), pp. 109–110.

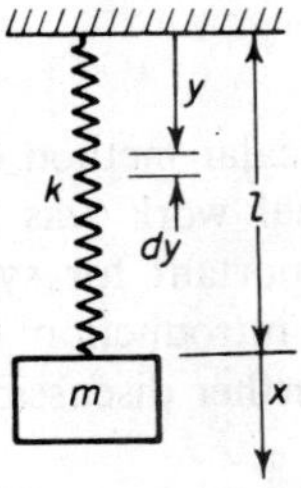

Figure 2.4-1. Effective mass of spring.

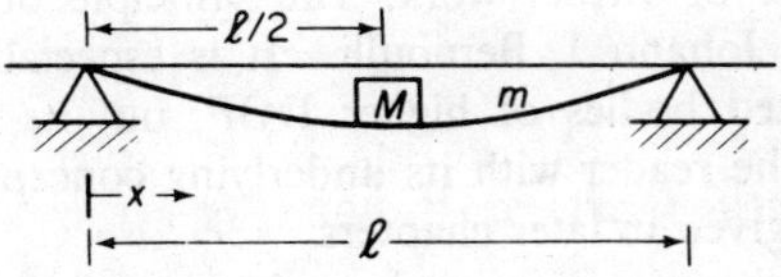

Figure 2.4-2.

The kinetic energy of the spring may then be integrated to

$$T_{\text{add}} = \frac{1}{2}\int_0^l \left(\dot{x}\frac{y}{l}\right)^2 \frac{m_s}{l}\,dy = \frac{1}{2}\frac{m_s}{3}\dot{x}^2$$

and the effective mass is found to be one-third the mass of the spring. Adding this to the lumped mass, the revised natural frequency is

$$\omega_n = \sqrt{\frac{k}{m + \frac{1}{3}m_s}}$$

Example 2.4-2

A simply supported beam of total mass m has a concentrated mass M at midspan. Determine the effective mass of the system at midspan and find its fundamental frequency. The deflection under the load due to a concentrated force P applied at midspan is $Pl^3/48\,EI$. (See Fig. 2.4-2 and table of stiffness at end of chapter.)

Solution: We will assume the deflection of the beam to be that due to a concentrated load at midspan or

$$y = y_{\max}\left[\frac{3x}{l} - 4\left(\frac{x}{l}\right)^3\right] \qquad \left(\frac{x}{l} \leqslant \frac{1}{2}\right)$$

The maximum kinetic energy of the beam itself is then

$$T_{\max} = \frac{1}{2}\int_0^{l/2} \frac{2m}{l}\left\{\dot{y}_{\max}\left[\frac{3x}{l} - 4\left(\frac{x}{l}\right)^3\right]\right\}^2 dx = \frac{1}{2}(0.4857\ \text{m})\,\dot{y}_{\max}^2$$

The effective mass at midspan is then equal to

$$m_{\text{eff}} = M + 0.4857\ \text{m}$$

and its natural frequency becomes

$$\omega_n = \sqrt{\frac{48\,EI}{l^3(M + 0.4857\ \text{m})}}$$

2.5 PRINCIPLE OF VIRTUAL WORK

We now complement the energy method by another scalar method based on the principle of virtual work. The principle of virtual work was first formulated by Johann J. Bernoulli.† It is especially important for systems of interconnected bodies of higher DOF, but its brief introduction here will familiarize the reader with its underlying concepts. Further discussion of the principle is given in later chapters.

The principle of virtual work is associated with the equilibrium of bodies, and may be stated as follows: *If a system in equilibrium under the action of a set of forces is given a virtual displacement, the virtual work done by the forces will be zero.*

The terms used in this statement are defined as follows: (1) A virtual displacement δr is an imaginary infinitesimal variation of the coordinate given instantaneously. The virtual displacement must be compatible with the constraints of the system. (2) Virtual work δW is the work done by all the active forces in a virtual displacement. Since there is no significant change of geometry associated with the virtual displacement, the forces acting on the system are assumed to remain unchanged for the calculation of δW.

The principle of virtual work as formulated by Bernoulli is a static procedure. Its extension to dynamics was made possible by D'Alembert (1718–1783) who introduced the concept of the inertia force. Thus inertia forces are included as active forces when dynamic problems are considered.

Example 2.5-1

Using the virtual work method, determine the equation of motion for the rigid beam of mass M loaded as shown in Fig. 2.5-1.

Solution: Draw the beam in the displaced position θ and place the forces acting on it, including the inertia and damping forces. Give the beam a virtual displacement $\delta\theta$ and determine the work done by each force.

$$\text{Inertia force } \delta W = -\left(\frac{Ml^2}{3}\ddot{\theta}\right)\delta\theta$$

$$\text{Spring force } \delta W = -\left(k\frac{l}{2}\theta\right)\frac{l}{2}\,\delta\theta$$

$$\text{Damper force } \delta W = -(cl\dot{\theta})\,l\,\delta\theta$$

$$\text{Uniform load } \delta W = \int_0^l (p_0 f(t)\,dx)\,x\,\delta\theta = p_0 f(t)\frac{l^2}{2}\,\delta\theta$$

Summing the virtual work and equating to zero gives the differential equation of motion:

$$\left(\frac{Ml^2}{3}\right)\ddot{\theta} + (cl^2)\dot{\theta} + k\frac{l^2}{4}\theta = p_0\frac{l^2}{2}f(t)$$

†Johann J. Bernoulli (1667–1748), Basel, Switzerland.

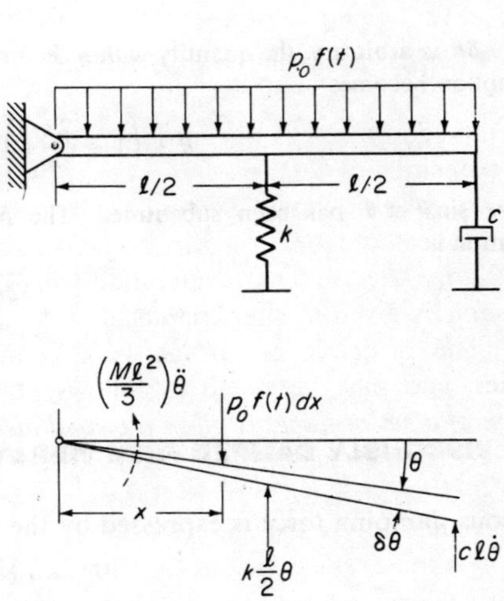

Figure 2.5-1.

Example 2.5-2

Two simple pendulums are connected together with the bottom mass restricted to vertical motion in a frictionless guide, as shown in Fig 2.5-2. Since only one coordinate θ is necessary, it represents an interconnected single-DOF system. Using the virtual work method, determine the equation of motion and its natural frequency.

Solution: Sketch the system displaced by a small angle θ and place on it all forces including inertia forces. Next give the coordinate θ a virtual displacement $\delta\theta$. Due to this displacement, m_1 and m_2 will undergo vertical displacements of $l\delta\theta \sin\theta$ and $2l\delta\theta \sin\theta$, respectively. (The acceleration of m_2 can easily be shown to be $2l(\ddot{\theta}\sin\theta + \dot{\theta}^2\cos\theta)$, and its virtual work will be an order of infinitesimal smaller than that for the gravity force and can be neglected). Equating the virtual work to zero, we have

$$\delta W = -(m_1 l\ddot{\theta})\,l\delta\theta - (m_1 g)\,l\delta\theta\sin\theta - (m_2 g)\,2l\,\delta\theta\sin\theta = 0$$

$$= -\left[m_1 l\ddot{\theta} + (m_1 + 2m_2)\,g\sin\theta\right] l\,\delta\theta = 0$$

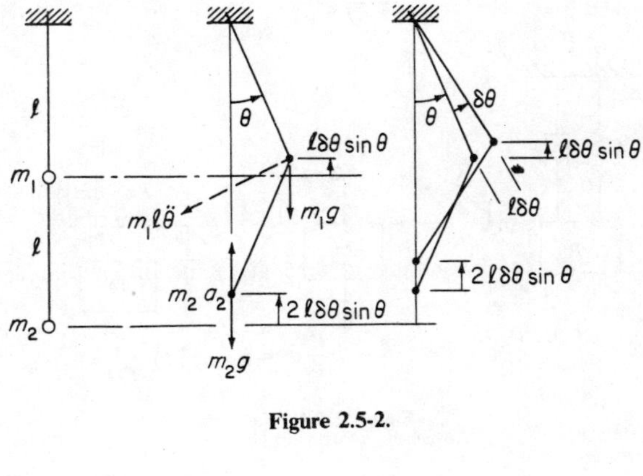

Figure 2.5-2.

Since $\delta\theta$ is arbitrary, the quantity within the brackets must be zero. Thus the equation of motion becomes

$$\ddot{\theta} + \left(1 + \frac{2m_2}{m_1}\right)\frac{g}{l}\theta = 0$$

where $\sin\theta \cong \theta$ has been substituted. The natural frequency from the preceding equation is

$$\omega_n = \sqrt{\left(1 + \frac{2m_2}{m_1}\right)\frac{g}{l}}$$

2.6 VISCOUSLY DAMPED FREE VIBRATION

Viscous damping force is expressed by the equation

$$F_d = c\dot{x} \tag{2.6-1}$$

where c is a constant of proportionality. Symbolically it is designated by a dashpot, as shown in Fig. 2.6-1. From the free-body diagram the equation of motion is seen to be

$$m\ddot{x} + c\dot{x} + kx = F(t) \tag{2.6-2}$$

The solution of the above equation has two parts. If $F(t) = 0$, we have the homogeneous differential equation whose solution corresponds physically to that of *free-damped vibration*. With $F(t) \neq 0$, we obtain the particular solution that is due to the excitation irrespective of the homogeneous solution. We will first examine the homogeneous equation that will give us some understanding of the role of damping.

With the homogeneous equation

$$m\ddot{x} + c\dot{x} + kx = 0 \tag{2.6-3}$$

the traditional approach is to assume a solution of the form

$$x = e^{st} \tag{2.6-4}$$

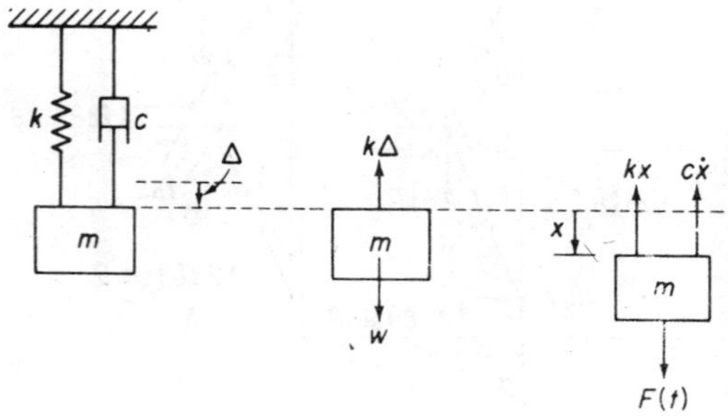

Figure 2.6-1.

where s is a constant. Upon substitution into the differential equation, we obtain

$$(ms^2 + cs + k)e^{st} = 0$$

which is satisfied for all values of t when

$$s^2 + \frac{c}{m}s + \frac{k}{m} = 0 \tag{2.6-5}$$

Equation (2.6-5), which is known as the *characteristic equation*, has two roots

$$s_{1,2} = -\frac{c}{2m} \pm \sqrt{\left(\frac{c}{2m}\right)^2 - \frac{k}{m}} \tag{2.6-6}$$

Hence, the general solution is given by the equation

$$x = Ae^{s_1 t} + Be^{s_2 t} \tag{2.6-7}$$

where A and B are constants to be evaluated from the initial conditions $x(0)$ and $\dot{x}(0)$.

Equation (2.6-6) substituted into (2.6-7) gives

$$x = e^{-(c/2m)t}\left(Ae^{\sqrt{(c/2m)^2 - k/m}\,t} + Be^{-\sqrt{(c/2m)^2 - k/m}\,t}\right) \tag{2.6-8}$$

The first term, $e^{-(c/2m)t}$, is simply an exponentially decaying function of time. The behavior of the terms in the parentheses, however, depends on whether the numerical value within the radical is positive, zero, or negative.

When the damping term $(c/2m)^2$ is larger than k/m, the exponents in the above equation are real numbers and no oscillations are possible. We refer to this case as *overdamped*.

When the damping term $(c/2m)^2$ is less than k/m, the exponent becomes an imaginary number, $\pm i\sqrt{k/m - (c/2m)^2}\,t$. Since

$$e^{\pm i\sqrt{k/m - (c/2m)^2}\,t} = \cos\sqrt{\frac{k}{m} - \left(\frac{c}{2m}\right)^2}\,t \pm i \sin\sqrt{\frac{k}{m} - \left(\frac{c}{2m}\right)^2}\,t$$

the terms of Eq. (2.6-8) within the parentheses are oscillatory. We refer to this case as *underdamped*.

In the limiting case between the oscillatory and nonoscillatory motion, $(c/2m)^2 = k/m$, and the radical is zero. The damping corresponding to this case is called *critical damping*, c_c.

$$c_c = 2m\sqrt{\frac{k}{m}} = 2m\omega_n = 2\sqrt{km} \tag{2.6-9}$$

Any damping can then be expressed in terms of the critical damping by a nondimensional number ζ, called the *damping ratio*

$$\zeta = \frac{c}{c_c} \tag{2.6-10}$$

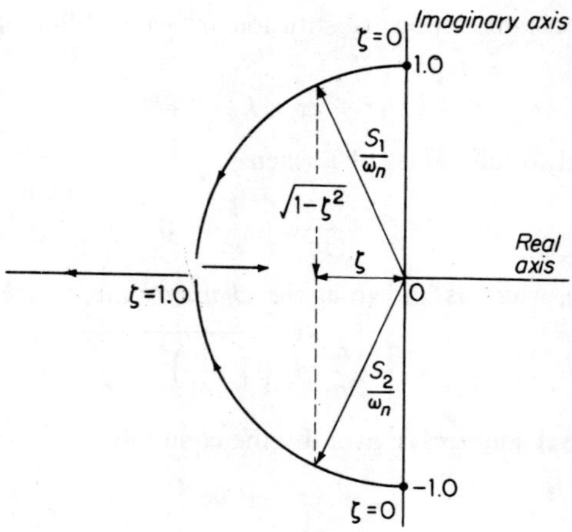

Figure 2.6-2.

and we can also express $s_{1,2}$ in terms of ζ as follows.

$$\frac{c}{2m} = \zeta\left(\frac{c_c}{2m}\right) = \zeta\omega_n$$

Equation (2.6-6) then becomes

$$s_{1,2} = \left(-\zeta \pm \sqrt{\zeta^2 - 1}\right)\omega_n \tag{2.6-11}$$

The three cases of damping discussed here now depend on whether ζ is greater than, less than, or equal to unity. Furthermore, the differential equation of motion can now be expressed in terms of ζ and ω_n as

$$\ddot{x} + 2\zeta\omega_n\dot{x} + \omega_n^2 x = \frac{1}{m}F(t) \tag{2.6-12}$$

This form of the equation for single-DOF systems will be found to be helpful in identifying the natural frequency and the damping of the system. We will frequently encounter this equation in the modal summation for multi-DOF systems.

Figure 2.6-2 shows Eq. (2.6-11) plotted in a complex plane with ζ along the horizontal axis. If $\zeta = 0$, Eq. (2.6-11) reduces to $s_{1,2}/\omega_n = \pm i$ so that the roots on the imaginary axis correspond to the undamped case. For $0 \le \zeta \le 1$, Eq. (2.6-11) can be rewritten as

$$\frac{s_{1,2}}{\omega_n} = -\zeta \pm i\sqrt{1 - \zeta^2}$$

The roots s_1 and s_2 are then conjugate complex points on a circular arc converging at the point $s_{1,2}/\omega_n = -1.0$. As ζ increases beyond unity, the

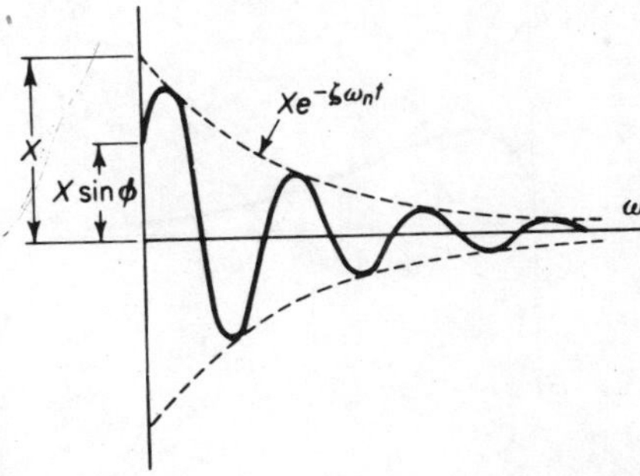

Figure 2.6-3. Damped oscillation $\zeta < 1.0$.

roots separate along the horizontal axis and remain real numbers. With this diagram in mind, we are now ready to examine the solution given by Eq. (2.6-8).

Oscillatory motion. [$\zeta < 1.0$ (Underdamped Case).] Substituting Eq. (2.6-11) into (2.6-7), the general solution becomes

$$x = e^{-\zeta\omega_n t}\left(Ae^{i\sqrt{1-\zeta^2}\,\omega_n t} + Be^{-i\sqrt{1-\zeta^2}\,\omega_n t}\right) \tag{2.6-13}$$

The above equation can also be written in either of the following two forms

$$x = Xe^{-\zeta\omega_n t}\sin\left(\sqrt{1-\zeta^2}\,\omega_n t + \phi\right) \tag{2.6-14}$$

$$= e^{-\zeta\omega_n t}\left(C_1 \sin\sqrt{1-\zeta^2}\,\omega_n t + C_2 \cos\sqrt{1-\zeta^2}\,\omega_n t\right) \tag{2.6-15}$$

where the arbitrary constants X, ϕ, or C_1, C_2 are determined from initial conditions. With initial conditions $x(0)$ and $\dot{x}(0)$, Eq. (2.6-15) can be shown to reduce to

$$x = e^{-\zeta\omega_n t}\left(\frac{\dot{x}(0) + \zeta\omega_n x(0)}{\omega_n\sqrt{1-\zeta^2}}\sin\sqrt{1-\zeta^2}\,\omega_n t + x(0)\cos\sqrt{1-\zeta^2}\,\omega_n t\right) \tag{2.6-16}$$

The equation indicates that the *frequency of damped oscillation* is equal to

$$\omega_d = \frac{2\pi}{\tau_d} = \omega_n\sqrt{1-\zeta^2} \tag{2.6-17}$$

Figure 2.6-3 shows the general nature of the oscillatory motion.

Nonoscillatory motion. [$\zeta > 1.0$ (Overdamped Case).] As ζ exceeds unity, the two roots remain on the real axis of Fig. 2.6-2 and separate, one increasing and the other decreasing. The general solution then becomes

$$x = Ae^{\left(-\zeta+\sqrt{\zeta^2-1}\right)\omega_n t} + Be^{\left(-\zeta-\sqrt{\zeta^2-1}\right)\omega_n t} \tag{2.6-18}$$

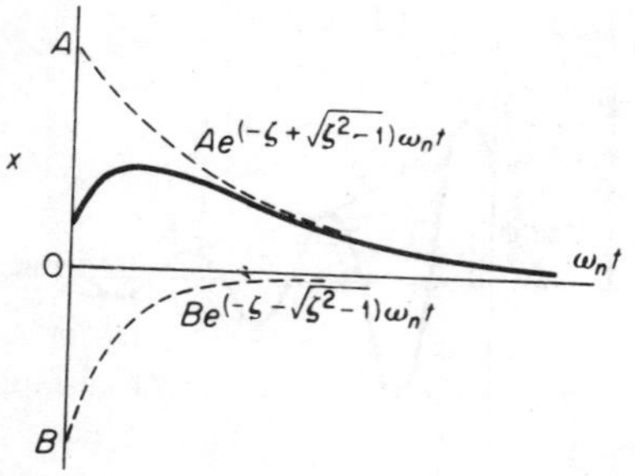

Figure 2.6-4. Aperiodic motion $\zeta > 1.0$.

where

$$A = \frac{\dot{x}(0) + \left(\zeta + \sqrt{\zeta^2 - 1}\right)\omega_n x(0)}{2\omega_n\sqrt{\zeta^2 - 1}}$$

and

$$B = \frac{-\dot{x}(0) - \left(\zeta - \sqrt{\zeta^2 - 1}\right)\omega_n x(0)}{2\omega_n\sqrt{\zeta^2 - 1}}$$

The motion is an exponentially decreasing function of time as shown in Fig. 2.6-4, and is referred to as *aperiodic*.

Critically damped motion. [$\zeta = 1.0$.] For $\zeta = 1$, we obtain a double root $s_1 = s_2 = -\omega_n$, and the two terms of Eq. (2.6-7) combine to form a single term, which is lacking in the number of constants required to satisfy the two initial conditions.

The correct general solution is

$$x = (A + Bt)e^{-\omega_n t} \tag{2.6-19}$$

which for the initial conditions $x(0)$ and $\dot{x}(0)$ becomes

$$x = \left\{x(0) + \left[\dot{x}(0) + \omega_n x(0)\right]t\right\}e^{-\omega_n t}$$

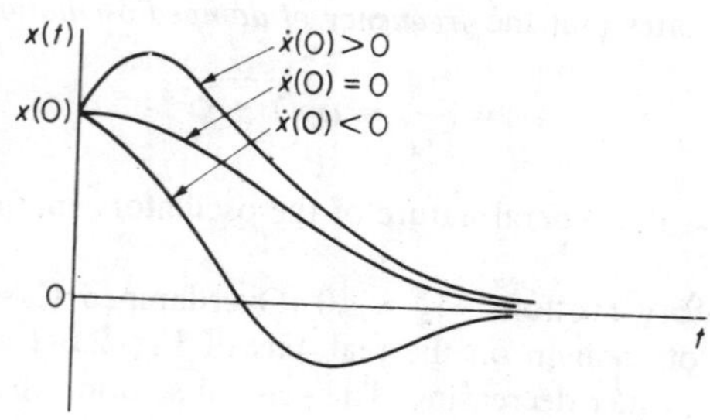

Figure 2.6-5. Critically damped motion $\zeta = 1.0$.

This can also be found from Eq. (2.6-16) by letting $\zeta \to 1$. Figure 2.6-5 shows three types of response with initial displacement $x(0)$.

2.7 LOGARITHMIC DECREMENT

A convenient way to determine the amount of damping present in a system is to measure the rate of decay of free oscillations. The larger the damping, the greater will be the rate of decay.

Consider a damped vibration expressed by the general equation (2.6-14)

$$x = Xe^{-\zeta\omega_n t}\sin\left(\sqrt{1-\zeta^2}\,\omega_n t + \phi\right)$$

which is shown graphically in Fig. 2.7-1. We introduce here a term called *logarithmic decrement*, which is defined as the natural logarithm of the ratio of any two successive amplitudes. The expression for the logarithmic decrement then becomes

$$\delta = \ln\frac{x_1}{x_2} = \ln\frac{e^{-\zeta\omega_n t_1}\sin\left(\sqrt{1-\zeta^2}\,\omega_n t_1 + \phi\right)}{e^{-\zeta\omega_n(t_1+\tau_d)}\sin\left(\sqrt{1-\zeta^2}\,\omega_n(t_1+\tau_d) + \phi\right)} \tag{2.7-1}$$

and since the values of the sines are equal when the time is increased by the damped period τ_d, the preceding relation reduces to

$$\delta = \ln\frac{e^{-\zeta\omega_n t_1}}{e^{-\zeta\omega_n(t_1+\tau_d)}} = \ln e^{\zeta\omega_n\tau_d} = \zeta\omega_n\tau_d \tag{2.7-2}$$

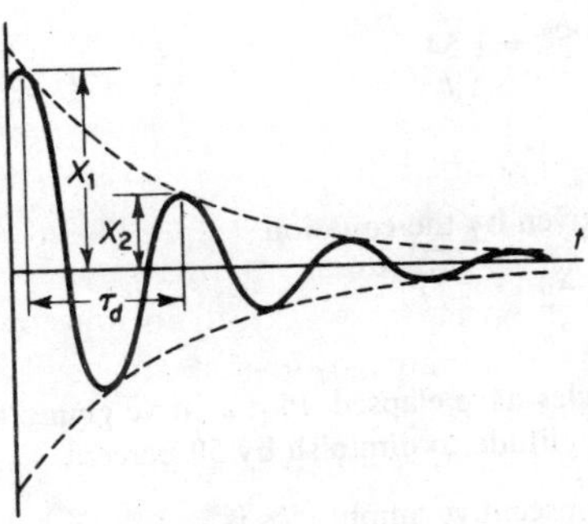

Figure 2.7-1. Rate of decay of oscillation measured by the logarithmic decrement.

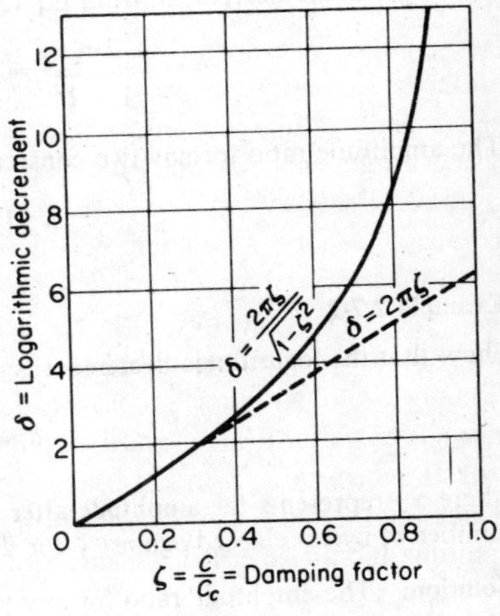

Figure 2.7-2. Logarithmic decrement as function of ζ.

Substituting for the damped period, $\tau_d = 2\pi/\omega_n\sqrt{1-\zeta^2}$, the expression for the logarithmic decrement becomes

$$\delta = \frac{2\pi\zeta}{\sqrt{1-\zeta^2}} \tag{2.7-3}$$

which is an exact equation.

When ζ is small, $\sqrt{1-\zeta^2} \cong 1$, and an approximate equation

$$\delta \cong 2\pi\zeta \tag{2.7-4}$$

is obtained. Figure 2.7-2 shows a plot of the exact and approximate values of δ as a function of ζ.

Example 2.7-1

The following data are given for a vibrating system with viscous damping: $w = 10$ lb, $k = 30$ lb/in., and $c = 0.12$ lb/in./s. Determine the logarithmic decrement and the ratio of any two successive amplitudes.

Solution: The undamped natural frequency of the system in radians per second is

$$\omega_n = \sqrt{\frac{k}{m}} = \sqrt{\frac{30 \times 386}{10}} = 34.0 \text{ rad/s}$$

The critical damping coefficient c_c and damping factor ζ are

$$c_c = 2m\omega_n = 2 \times \tfrac{10}{386} \times 34.0 = 1.76 \text{ lb/in./s}$$

$$\zeta = \frac{c}{c_c} = \frac{0.12}{1.76} = 0.0681$$

The logarithmic decrement, from Eq. (2.7-3), is

$$\delta = \frac{2\pi\zeta}{\sqrt{1-\zeta^2}} = \frac{2\pi \times 0.0681}{\sqrt{1-0.0681^2}} = 0.429$$

The amplitude ratio for any two consecutive cycles is

$$\frac{x_1}{x_2} = e^{\delta} = e^{0.429} = 1.54$$

Example 2.7-2

Show that the logarithmic decrement is also given by the equation

$$\delta = \frac{1}{n}\ln\frac{x_0}{x_n}$$

where x_n represents the amplitude after n cycles have elapsed. Plot a curve giving the number of cycles elapsed against ζ for the amplitude to diminish by 50 percent.

Solution: The amplitude ratio for any two consecutive amplitudes is

$$\frac{x_0}{x_1} = \frac{x_1}{x_2} = \frac{x_2}{x_3} = \cdots = \frac{x_{n-1}}{x_n} = e^{\delta}$$

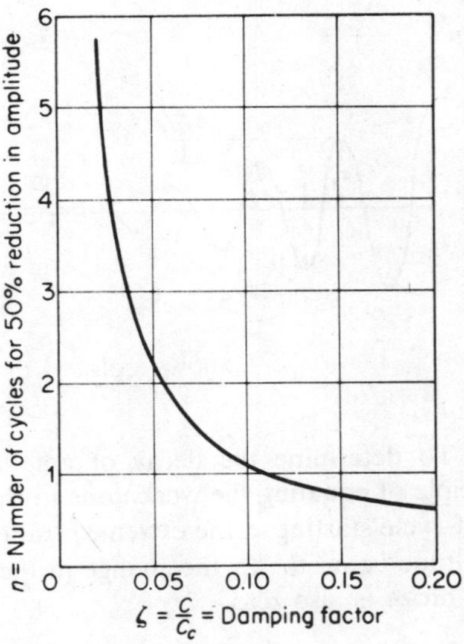

Figure 2.7-3.

The ratio x_0/x_n can be written as

$$\frac{x_0}{x_n} = \left(\frac{x_0}{x_1}\right)\left(\frac{x_1}{x_2}\right)\left(\frac{x_2}{x_3}\right)\cdots\left(\frac{x_{n-1}}{x_n}\right) = (e^{\delta})^n = e^{n\delta}$$

from which the required equation is obtained as

$$\delta = \frac{1}{n}\ln\frac{x_0}{x_n}$$

To determine the number of cycles elapsed for 50 percent reduction in amplitude, we obtain the following relation from the preceding equation:

$$\delta \cong 2\pi\zeta = \frac{1}{n}\ln 2 = \frac{0.693}{n}$$

$$n\zeta = \frac{0.693}{2\pi} = 0.110$$

The last equation is that of a rectangular hyperbola and is plotted in Fig. 2.7-3.

2.8 COULOMB DAMPING

Coulomb damping results from the sliding of two dry surfaces. The damping force is equal to the product of the normal force and the coefficient of friction μ and is assumed to be independent of the velocity, once the motion is initiated. Since the sign of the damping force is always opposite to that of the velocity, the differential equation of motion for each sign is valid only for half-cycle intervals.

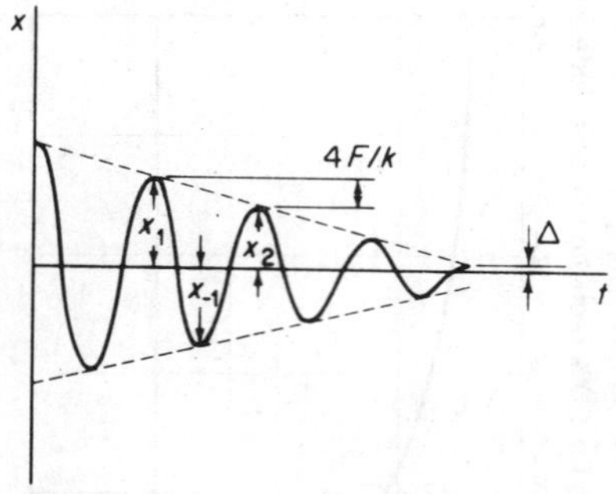

Figure 2.8-1. Free vibration with Coulomb damping.

To determine the decay of amplitude, we resort to the work-energy principle of equating the work done to the change in kinetic energy. Choosing a half-cycle starting at the extreme position with velocity equal to zero and the amplitude equal to X_1, the change in the kinetic energy is zero and the work done on m is also zero.

$$\tfrac{1}{2}k(X_1^2 - X_{-1}^2) - F_d(X_1 + X_{-1}) = 0$$

or

$$\tfrac{1}{2}k(X_1 - X_{-1}) = F_d$$

where X_{-1} is the amplitude after the half-cycle as shown in Fig. 2.8-1.

Repeating this procedure for the next half-cycle, a further decrease in amplitude of $2F_d/k$ will be found, so that the decay in amplitude per cycle is a constant and equal to

$$X_1 - X_2 = \frac{4F_d}{k} \tag{2.8-1}$$

The motion will cease, however, when the amplitude becomes less than Δ, at which position the spring force is insufficient to overcome the static friction force, which is generally greater than the kinetic friction force. It can also be shown that the frequency of oscillation is $\omega_\mu = \sqrt{k/m}$, which is the same as that of the undamped system.

Figure 2.8-1 shows the free vibration of a system with Coulomb damping. It should be noted that the amplitudes decay linearly with time.

Table of Spring Stiffness.

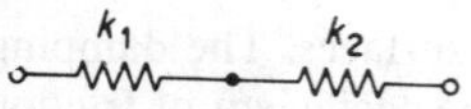

$$k = \frac{1}{1/k_1 + 1/k_2}$$

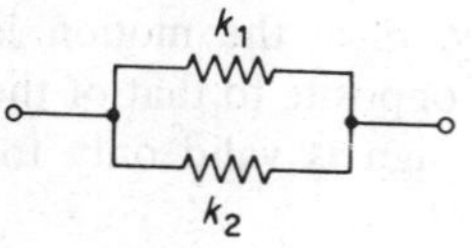

$$k = k_1 + k_2$$

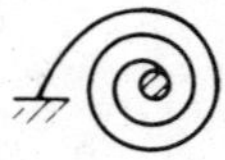

$k = \dfrac{EI}{l}$, I = moment of inertia of cross-sectional area; l = total length

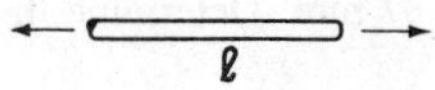

$k = \dfrac{EA}{l}$, A = cross-sectional area

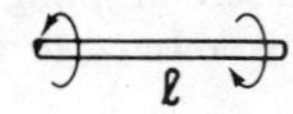

$k = \dfrac{GJ}{l}$, J = torsion constant of cross section

$k = \dfrac{Gd^4}{64nR^3}$, n = number of turns

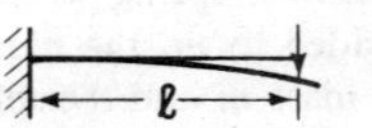

$k = \dfrac{3EI}{l^3}$, k at position of load

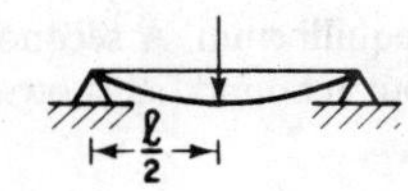

$k = \dfrac{48EI}{l^3}$

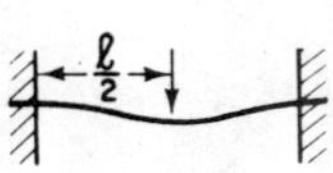

$k = \dfrac{192EI}{l^3}$

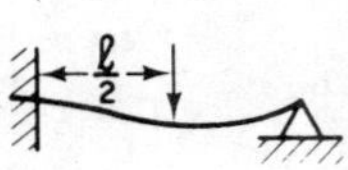

$k = \dfrac{768EI}{7l^3}$

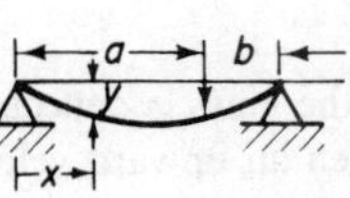

$k = \dfrac{3EIl}{a^2b^2}$, $y_x = \dfrac{Pbx}{6EIl}(l^2 - x^2 - b^2)$

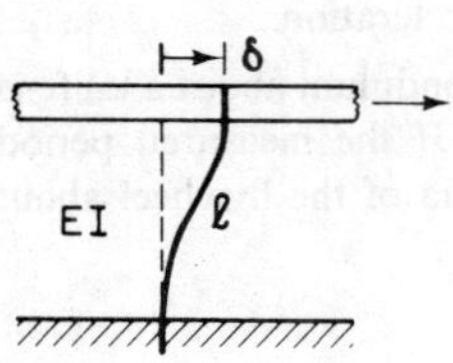

$k = \dfrac{12EI}{l^3}$

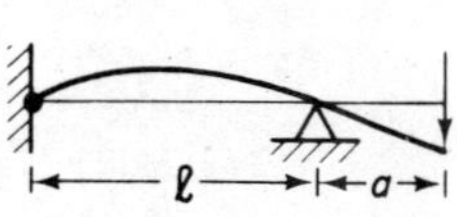

$k = \dfrac{3EI}{(l + a)a^2}$

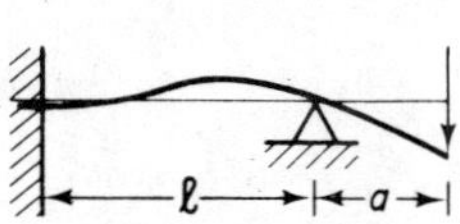

$k = \dfrac{24EI}{a^2(3l + 8a)}$

PROBLEMS

2-1 A 0.453-kg mass attached to a light spring elongates it 7.87 mm. Determine the natural frequency of the system.

2-2 A spring-mass system k_1, m, has a natural frequency of f_1. If a second spring k_2 is added in series with the first spring, the natural frequency is lowered to $\frac{1}{2}f_1$. Determine k_2 in terms of k_1.

2-3 A 4.53-kg mass attached to the lower end of a spring whose upper end is fixed vibrates with a natural period of 0.45 s. Determine the natural period when a 2.26-kg mass is attached to the midpoint of the same spring with the upper and lower ends fixed.

2-4 An unknown mass m kg attached to the end of an unknown spring k has a natural frequency of 94 cpm. When a 0.453-kg mass is added to m, the natural frequency is lowered to 76.7 cpm. Determine the unknown mass m and the spring constant k N/m.

2-5 A mass m_1 hangs from a spring k (N/m) and is in static equilibrium. A second mass m_2 drops through a height h and sticks to m_1 without rebound, as shown in Fig. P2-5. Determine the subsequent motion.

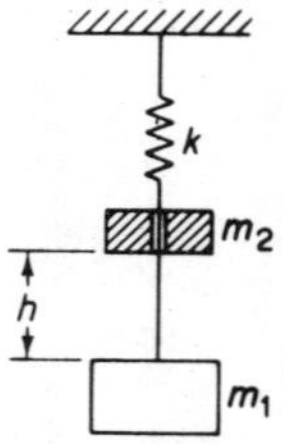

Figure P2-5.

2-6 The ratio k/m of a spring-mass system is given as 4.0. If the mass is deflected 2 cm down, measured from its equilibrium position, and given an upward velocity of 8 cm/s, determine its amplitude and maximum acceleration.

2-7 A flywheel weighing 70 lb was allowed to swing as a pendulum about a knife-edge at the inner side of the rim as shown in Fig. P2-7 If the measured period of oscillation was 1.22 s, determine the moment of inertia of the flywheel about its geometric axis.

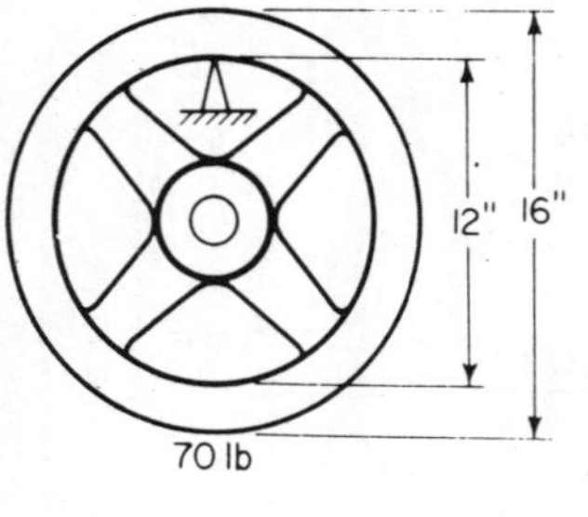

Figure P2-7.

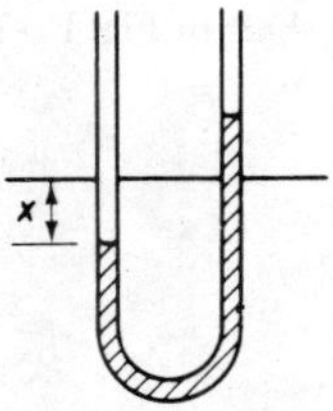

Figure P2-21.

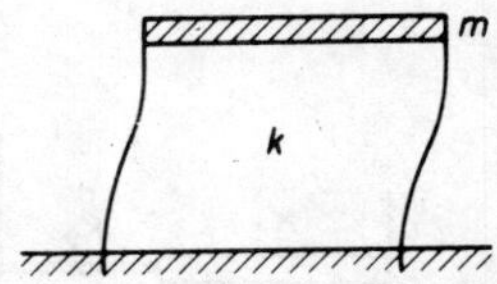

Figure P2-22.

2-22 Figure P2-22 shows a simplified model of a single-story building. The columns are assumed to be rigidly imbedded at the ends. Determine its natural period τ. Refer to the table of stiffness at the end of the chapter.

2-23 Determine the effective mass of the columns of Prob. 2-22 assuming the deflection to be

$$y = \frac{1}{2}y_{\max}\left(1 - \cos\frac{\pi x}{l}\right)$$

2-24 Determine the effective mass at point n for the system shown in Fig. P2-24 and its natural frequency.

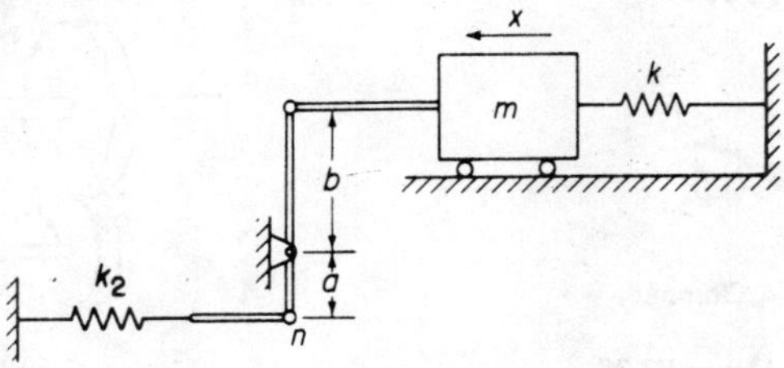

Figure P2-24.

2-25 A uniform cantilever beam of total mass ml has a concentrated mass M at its free end. Determine the effective mass of the beam to be added to M assuming the deflection to be that of a massless beam with a concentrated force at the end, and write the equation for its fundamental frequency.

2-26 Determine the effective mass of the rocket engine shown in Fig. P2-26 to be added to the actuator mass m_1.

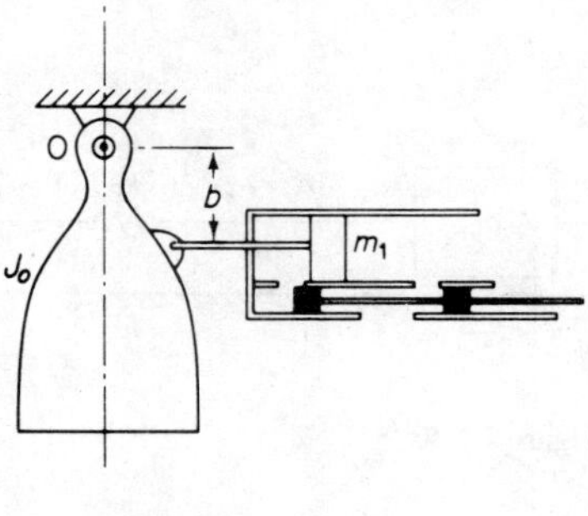

Figure P2-26.

2-27 Determine the effective rotational stiffness of the shaft in Fig. P2-27 and calculate its natural period.

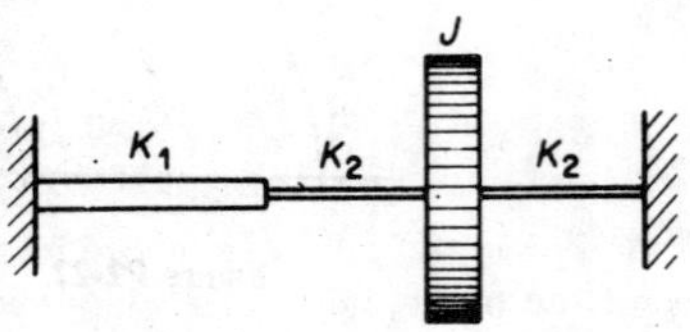

Figure P2-27.

2-28 For purposes of analysis, it is desired to reduce the system of Fig. P2-28 to a simple linear spring-mass system of effective mass m_{eff} and effective stiffness k_{eff}. Determine m_{eff} and k_{eff} in terms of the given quantities.

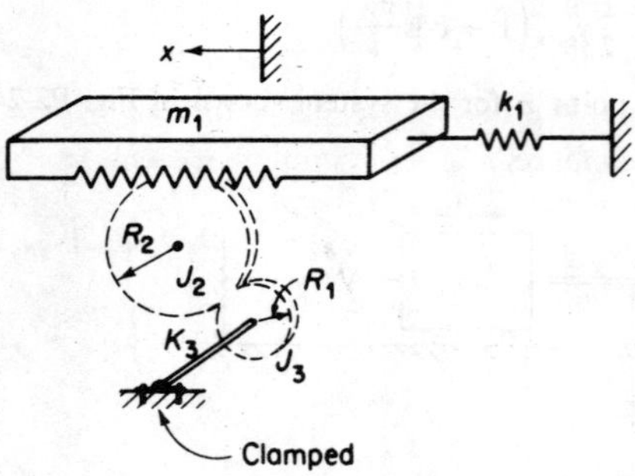

Figure P2-28.

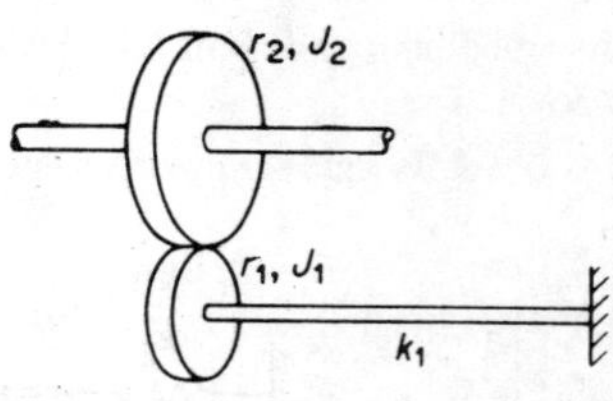

Figure P2-29.

2-29 Determine the effective mass moment of inertia for shaft 1 in the system shown in Fig. P2-29.

2-30 Determine the kinetic energy of the system shown in Fig. P2-30 in terms of $\dot{x}$. Determine the stiffness at m_0 and write the expression for the natural frequency.

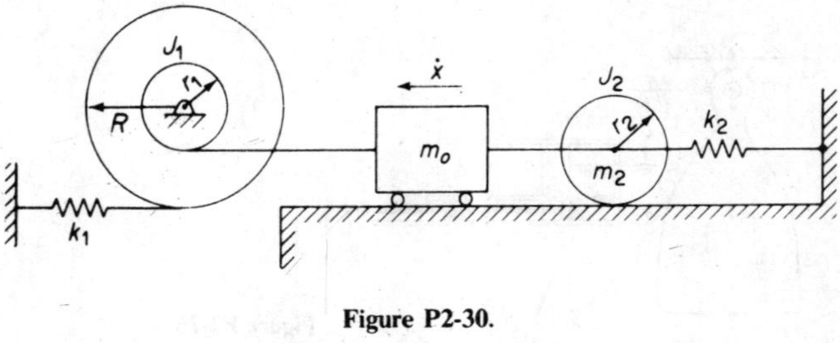

Figure P2-30.

2-31 Tachometers are a reed-type of frequency-measuring instrument consisting of small cantilever beams with weights attached at the ends. When the frequency of vibration corresponds to the natural frequency of one of the reeds, it will vibrate, thereby indicating the frequency. How large a weight must be placed on the end of a reed made of spring steel 0.1016 cm thick, 0.635 cm wide, and 8.890 cm long for a natural frequency of 20 cps?

2-32 A mass of 0.907 kg is attached to the end of a spring with a stiffness of 7.0 N/cm. Determine the critical damping coefficient.

2-33 To calibrate a dashpot, the velocity of the plunger was measured when a given force was applied to it. If a $\frac{1}{2}$-lb weight produced a constant velocity of 1.20 in./s, determine the damping factor ζ when used with the system of Prob. 2-32.

2-34 A vibrating system is started under the following initial conditions: $x = 0$, $\dot{x} = v_0$. Determine the equation of motion when (a) $\zeta = 2.0$, (b) $\zeta = 0.50$, (c) $\zeta = 1.0$. Plot nondimensional curves for the three cases with $\omega_n t$ as abscissa and $x\omega_n/v_0$ as ordinate.

2-35 A vibrating system consisting of a mass of 2.267 kg and a spring of stiffness 17.5 N/cm is viscously damped such that the ratio of any two consecutive amplitudes is 1.00 and 0.98. Determine (a) the natural frequency of the damped system, (b) the logarithmic decrement, (c) the damping factor, and (d) the damping coefficient.

2-36 A vibrating system consists of a mass of 4.534 kg, a spring of stiffness 35.0 N/cm, and a dashpot with a damping coefficient of 0.1243 N/cm/s. Find (a) the damping factor, (b) the logarithmic decrement, and (c) the ratio of any two consecutive amplitudes.

2-37 A vibrating system has the following constants: $m = 17.5$ kg, $k = 70.0$ N/cm, and $c = 0.70$ N/cm/s. Determine (a) the damping factor, (b) the natural frequency of damped oscillation, (c) the logarithmic decrement, and (d) the ratio of any two consecutive amplitudes.

2-38 Set up the differential equation of motion for the system shown in Fig. P2-38. Determine the expression for (a) the critical damping coefficient, and (b) the natural frequency of damped oscillation.

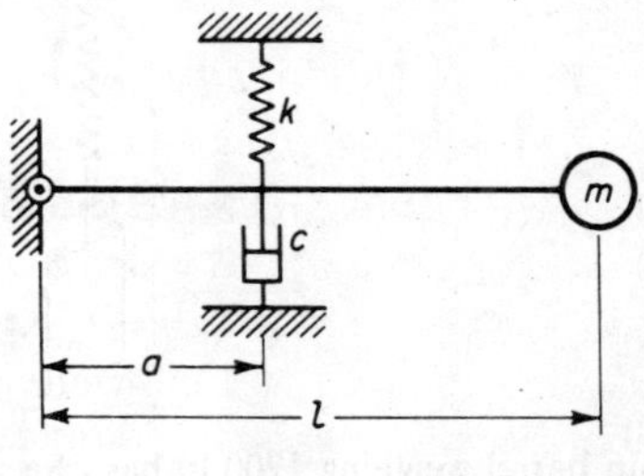

Figure P2-38.

2-39 Write the differential equation of motion for the system shown in Fig. P2-39 and determine the natural frequency of damped oscillation and the critical damping coefficient.

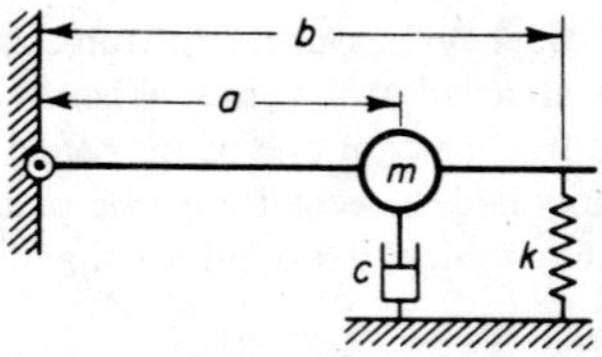

Figure P2-39.

2-40 A spring-mass system with viscous damping is displaced from the equilibrium position and released. If the amplitude diminished by 5% each cycle, what fraction of the critical damping does the system have?

2-41 A rigid uniform bar of mass m and length l is pinned at O and supported by a spring and viscous damper as shown in Fig. P2-41. Measuring θ from the static equilibrium position, determine (a) the equation for small θ (the moment of inertia of the bar about 0 is $ml^2/3$), (b) the equation for the undamped natural frequency, and (c) the expression for critical damping. Use virtual work.

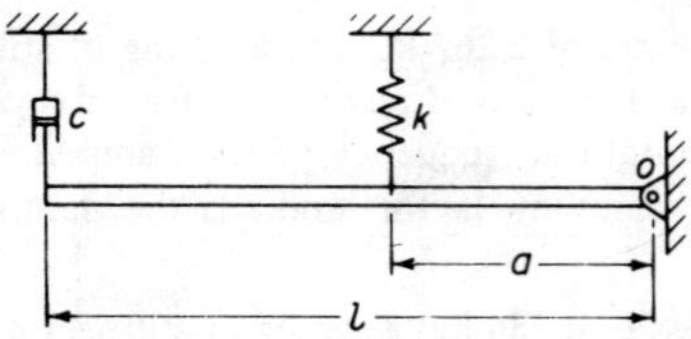

Figure P2-41.

2-42 A thin plate of area A and weight W is attached to the end of a spring and is allowed to oscillate in a viscous fluid as shown in Fig. P2-42. If τ_1 is the natural period of undamped oscillation (that is, with the system oscillating in air), and τ_2 the damped period with the plate immersed in the fluid, show that

$$\mu = \frac{2\pi W}{gA\tau_1\tau_2}\sqrt{\tau_2^2 - \tau_1^2}$$

where the damping force on the plate is $F_d = \mu 2Av$, $2A$ is the total surface area of the plate, and v is its velocity.

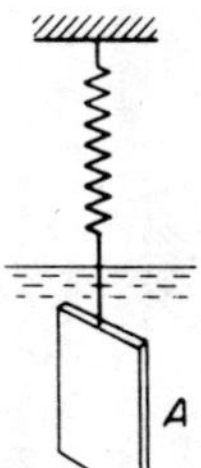

Figure P2-42.

2-43 A gun barrel weighing 1200 lb has a recoil spring of stiffness 20,000 lb/ft. If the barrel recoils 4 ft on firing, determine (a) the initial recoil velocity of the barrel, (b) the critical damping coefficient of a dashpot which is engaged at the end of the recoil stroke, and (c) the time required for the barrel to return to a position 2 in. from its initial position.

2-44 A piston of mass 4.53 kg is traveling in a tube with a velocity of 15.24 m/s and engages a spring and damper as shown in Fig. P2-44. Determine the maximum displacement of the piston after engaging the spring-damper. How many seconds does it take?

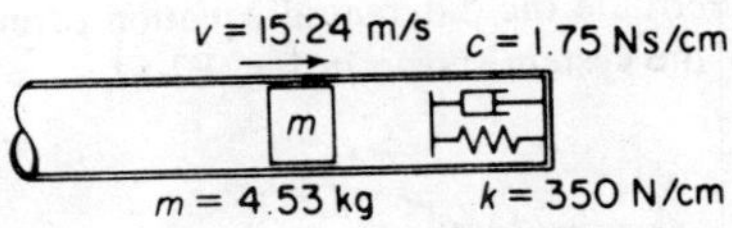

Figure P2-44.

2-45 A shock absorber is to be designed so that its overshoot is 10% of the initial displacement when released. Determine ζ_1. If ζ is made equal to $\frac{1}{2}\zeta_1$, what will be the overshoot?

2-46 Determine the equation of motion for Probs. 2-38 and 2-39 using virtual work.

2-47 Determine the effective stiffness of the springs shown in Fig. P2-47.

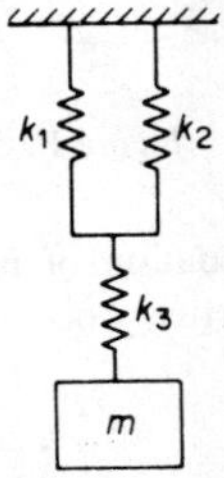

Figure P2-47.

2-48 Determine the flexibility of a simply supported uniform beam of length L at a point $\frac{1}{3}L$ from the end.

2-49 Determine the effective stiffness of the system shown in Fig. P2-49, in terms of the displacement x.

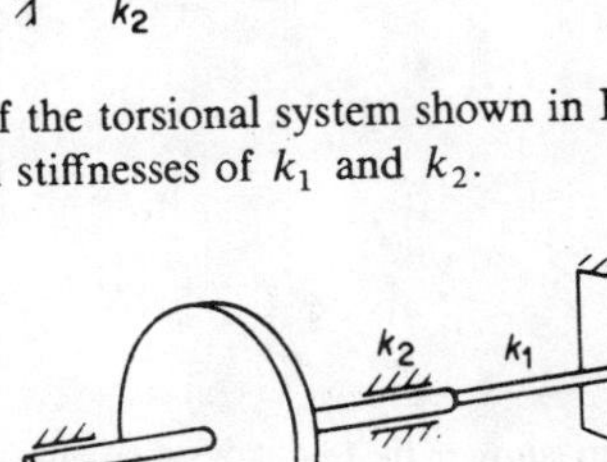

Figure P2-49.

2-50 Determine the effective stiffness of the torsional system shown in Fig. P2-50. The two shafts in series have torsional stiffnesses of k_1 and k_2.

Figure P2-50.

2-51 A spring-mass system m, k, is started with an initial displacement of unity and an initial velocity of zero. Plot ln X versus n where X is amplitude at cycle n for (a) viscous damping with $\zeta = 0.05$, and (b) Coulomb damping with damping force $F_d = 0.05\, k$. When will the two amplitudes be equal?

2-52 Determine the differential equation of motion and establish the critical damping for the system shown in Fig. P2-52.

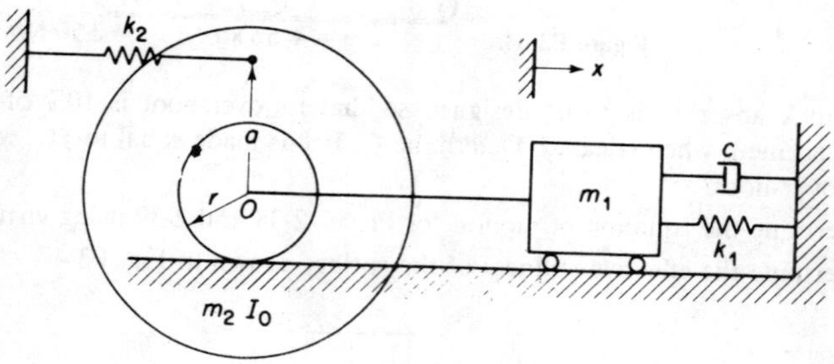

Figure P2-52.

2-53 Determine the differential equation of motion for free vibration of the system shown in Fig. P2-53, using virtual work.

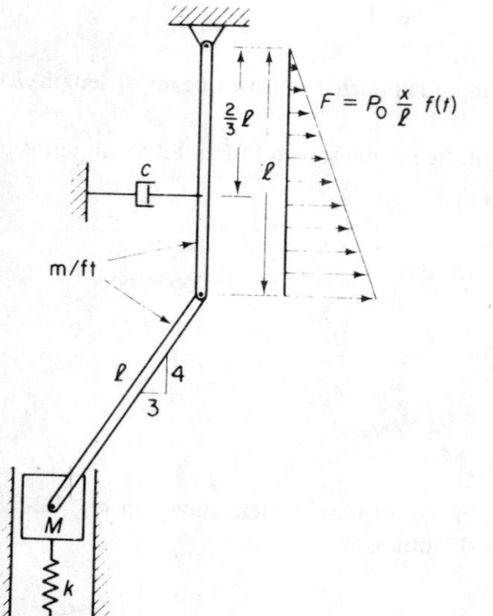

Figure P2-53.

2-54 The system shown in Fig. P2-54 has two rigid uniform beams of length l and mass per unit length m, hinged at the middle and resting on rollers at the test stand. The hinge is restrained from rotation by a torsional spring K and supports a mass M held up by another spring k to a position where the bars are horizontal. Determine the equation of motion using virtual work.

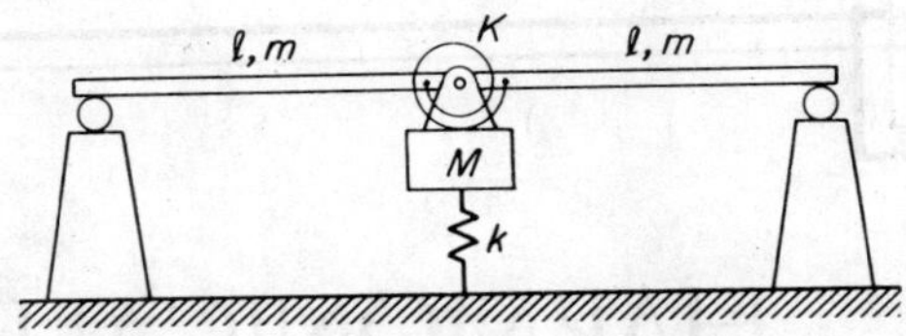

Figure P2-54.

2-55 Two uniform stiff bars are hinged at the middle and constrained by a spring, as shown in Fig. P2-55. Using virtual work, set up the equation of motion for its free vibration.

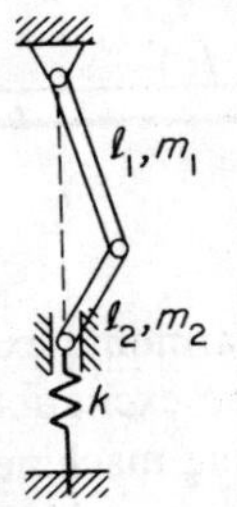

Figure P2-55.

3

HARMONICALLY EXCITED VIBRATION

When a system is subjected to harmonic excitation, it is forced to vibrate at the same frequency as that of the excitation. Common sources of harmonic excitation are unbalance in rotating machines, forces produced by reciprocating machines, or the motion of the machine itself. These excitations may be undesirable for equipment whose operation may be disturbed or for the safety of the structure if large vibration amplitudes develop. Resonance is to be avoided in most cases, and to prevent large amplitudes from developing, dampers and absorbers are often used. Discussion of their behavior is of importance for their intelligent use. Finally, the theory of vibration measuring instruments is presented as a tool for vibration analysis.

3.1 FORCED HARMONIC VIBRATION

Harmonic excitation is often encountered in engineering systems. It is commonly produced by the unbalance in rotating machinery. Although pure harmonic excitation is less likely to occur than periodic or other types of excitation, understanding the behavior of a system undergoing harmonic excitation is essential in order to comprehend how the system will respond to more general types of excitation. Harmonic excitation may be in the form of a force or displacement of some point in the system.

We will first consider a single-DOF system with viscous damping, excited by a harmonic force $F_0 \sin \omega t$ as shown in Fig. 3.1-1. Its differential

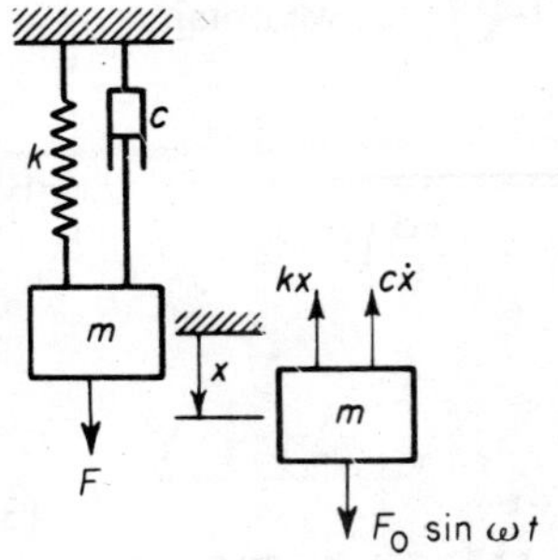

Figure 3.1-1. Viscously damped system with harmonic excitation.

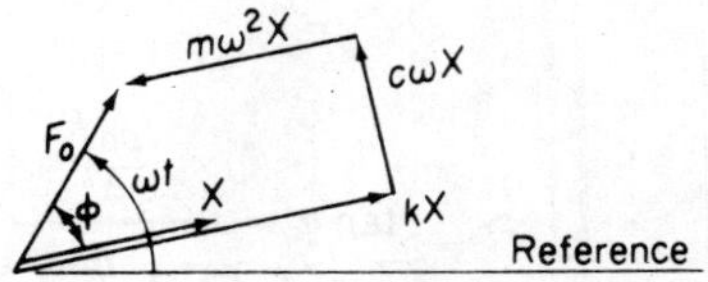

Figure 3.1-2. Vector relationship for forced vibration with damping.

equation of motion is found from the free-body diagram to be

$$m\ddot{x} + c\dot{x} + kx = F_0 \sin \omega t \tag{3.1-1}$$

The solution to this equation consists of two parts, the *complementary function*, which is the solution of the homogeneous equation, and the *particular integral*. The complementary function, in this case, is a damped free vibration that was discussed in Chapter 2.

The particular solution to the preceding equation is a steady-state oscillation of the same frequency ω as that of the excitation. We can assume the particular solution to be of the form

$$x = X \sin(\omega t - \phi) \tag{3.1-2}$$

where X is the amplitude of oscillation and ϕ is the phase of the displacement with respect to the exciting force.

The amplitude and phase in the above equation are found by substituting Eq. (3.1-2) into the differential equation Eq. (3.1-1). Remembering that in harmonic motion the phases of the velocity and acceleration are ahead of the displacement by 90° and 180°, respectively, the terms of the differential equation can also be displayed graphically as in Fig. 3.1-2. It is easily seen from this diagram that

$$X = \frac{F_0}{\sqrt{(k - m\omega^2)^2 + (c\omega)^2}} \tag{3.1-3}$$

and

$$\phi = \tan^{-1} \frac{c\omega}{k - m\omega^2} \tag{3.1-4}$$

We now express Eqs. (3.1-3) and (3.1-4) in nondimensional form that enables a concise graphical p ation of these results. Dividing the numerator and denominator of Eqs. (3.1-3) and (3.1-4) by k, we obtain

$$X = \frac{\frac{F_0}{k}}{\sqrt{\left(1 - \frac{m\omega^2}{k}\right)^2 + \left(\frac{c\omega}{k}\right)^2}} \tag{3.1-5}$$

and

$$\tan\phi = \frac{\frac{c\omega}{k}}{1 - \frac{m\omega^2}{k}} \tag{3.1-6}$$

These equations may be further expressed in terms of the following quantities:

$$\omega_n = \sqrt{\frac{k}{m}} = \text{natural frequency of undamped oscillation}$$

$$c_c = 2m\omega_n = \text{critical damping}$$

$$\zeta = \frac{c}{c_c} = \text{damping factor}$$

$$\frac{c\omega}{k} = \frac{c}{c_c}\frac{c_c\omega}{k} = 2\zeta\frac{\omega}{\omega_n}$$

The nondimensional expressions for the amplitude and phase then become

$$\frac{Xk}{F_0} = \frac{1}{\sqrt{\left[1 - \left(\frac{\omega}{\omega_n}\right)^2\right]^2 + \left[2\zeta\left(\frac{\omega}{\omega_n}\right)\right]^2}} \tag{3.1-7}$$

and

$$\tan\phi = \frac{2\zeta\left(\frac{\omega}{\omega_n}\right)}{1 - \left(\frac{\omega}{\omega_n}\right)^2} \tag{3.1-8}$$

These equations indicate that the nondimensional amplitude Xk/F_0 and the phase ϕ are functions only of the frequency ratio ω/ω_n and the damping factor ζ and can be plotted as shown in Fig. 3.1-3. These curves show that the damping factor has a large influence on the amplitude and phase angle in the frequency region near resonance. Further understanding of the behavior of the system may be obtained by studying the force diagram corresponding to Fig. 3.1-2 in the regions ω/ω_n small, $\omega/\omega_n = 1$, and ω/ω_n large.

For small values of $\omega/\omega_n \ll 1$, both the inertia and damping forces are small, which results in a small phase angle ϕ. The magnitude of the impressed force is then nearly equal to the spring force as shown in Fig. 3.1-4(a).

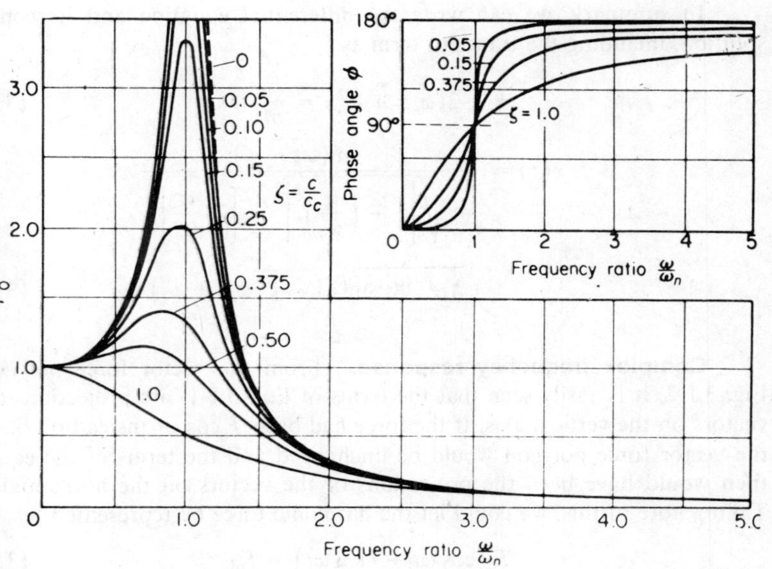

Figure 3.1-3. Plot of Equations (3.1-7) and (3.1-8).

For $\omega/\omega_n = 1.0$, the phase angle is 90° and the force diagram appears as in Fig. 3.1-4(b). The inertia force, which is now larger, is balanced by the spring force, whereas the impressed force overcomes the damping force. The amplitude at resonance can be found, either from Eqs. (3.1-5) or (3.1-7) or from Fig. 3.1-4(b), to be

$$X = \frac{F_0}{c\omega_n} = \frac{F_0}{2\zeta k} \tag{3.1-9}$$

At large values of $\omega/\omega_n \gg 1$, ϕ approaches 180°, and the impressed force is expended almost entirely in overcoming the large inertia force as shown in Fig. 3.1-4(c).

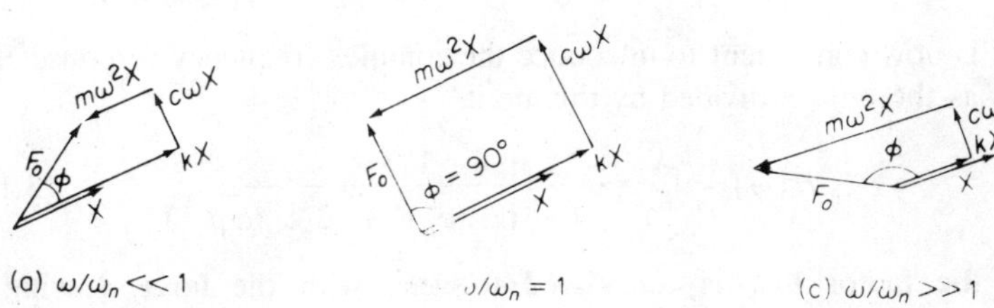

Figure 3.1-4. ationship in forced vibration.

In summary, we can write the differential equation and its complete solution, including the transient term as

$$\ddot{x} + 2\zeta\omega_n\dot{x} + \omega_n^2 x = \frac{F_0}{m}\sin\omega t \tag{3.1-10}$$

$$x(t) = \frac{F_0}{k}\frac{\sin(\omega t - \phi)}{\sqrt{\left[1 - \left(\frac{\omega}{\omega_n}\right)^2\right]^2 + \left[2\zeta\frac{\omega}{\omega_n}\right]^2}} + X_1 e^{-\zeta\omega_n t}\sin\left(\sqrt{1-\zeta^2}\,\omega_n t + \phi_1\right) \tag{3.1-11}$$

Complex frequency response. From the vector force polygon of Fig. 3.1-2, it is easily seen that the terms of Eq. (3.1-1) are projections of the vectors on the vertical axis. If the force had been $F_0\cos\omega t$ instead of $F_0\sin\omega t$, the vector force polygon would be unchanged and the terms of the equation then would have been the projections of the vectors on the horizontal axis. Taking note of this, we could let the harmonic force be represented by

$$F_0(\cos\omega t + i\sin\omega t) = F_0 e^{i\omega t} \tag{3.1-12}$$

This would be equivalent to multiplying the quantities along the vertical axis by $i = \sqrt{-1}$ and using complex vectors. The displacement can then be written as

$$x = Xe^{i(\omega t - \phi)} = (Xe^{-i\phi})e^{i\omega t} = \bar{X}e^{i\omega t} \tag{3.1-13}$$

where $\bar{X}$ is a complex displacement vector:

$$\bar{X} = Xe^{-i\phi} \tag{3.1-14}$$

Substituting into the differential equation and canceling from each side of the equation gives the results

$$(-\omega^2 m + ic\omega + k)\bar{X} = F_0$$

and

$$\bar{X} = \frac{F_0}{(k - \omega^2 m) + i(c\omega)} = \frac{F_0/k}{1 - (\omega/\omega_n)^2 + i(2\zeta\omega/\omega_n)} \tag{3.1-15}$$

It is now convenient to introduce the complex frequency response $H(\omega)$ defined as the output divided by the input,

$$H(\omega) = \frac{\bar{X}}{F_0} = \frac{1/k}{1 - (\omega/\omega_n)^2 + i2\zeta\omega/\omega_n} \tag{3.1-16}$$

(Often the factor $1/k$ is considered together with the force, leaving the frequency response a nondimensional quantity.) Thus $H(\omega)$ depends only on the frequency ratio and the damping factor.

The real and imaginary parts of $H(\omega)$ can be identified by multiplying and dividing Eq. (3.1-16) by the complex conjugate of the denominator. The result is

$$H(\omega) = \frac{1-(\omega/\omega_n)^2}{\left[1-(\omega/\omega_n)^2\right]^2+\left[2\zeta\omega/\omega_n\right]^2} - i\frac{2\zeta\omega/\omega_n}{\left[1-(\omega/\omega_n)^2\right]^2+\left[2\zeta\omega/\omega_n\right]^2} \tag{3.1-17}$$

This equation shows that at resonance, the real part is zero and the response is given by the imaginary part, which is

$$H(\omega) = -i\frac{1}{2\zeta} \tag{3.1-18}$$

It is easily seen that the phase angle is

$$\tan\phi = \frac{2\zeta\omega/\omega_n}{1-(\omega/\omega_n)^2}$$

3.2 ROTATING UNBALANCE

Unbalance in rotating machines is a common source of vibration excitation. We consider here a spring-mass system constrained to move in the vertical direction and excited by a rotating machine that is unbalanced, as shown in Fig. 3.2-1 The unbalance is represented by an eccentric mass m with eccentricity e which is rotating with angular velocity ω. Letting x be the displacement of the nonrotating mass $(M-m)$ from the static equilibrium position, the displacement of m is

$$x + e\sin\omega t$$

The equation of motion is then

$$(M-m)\ddot{x} + m\frac{d^2}{dt^2}(x+e\sin\omega t) = -kx - c\dot{x}$$

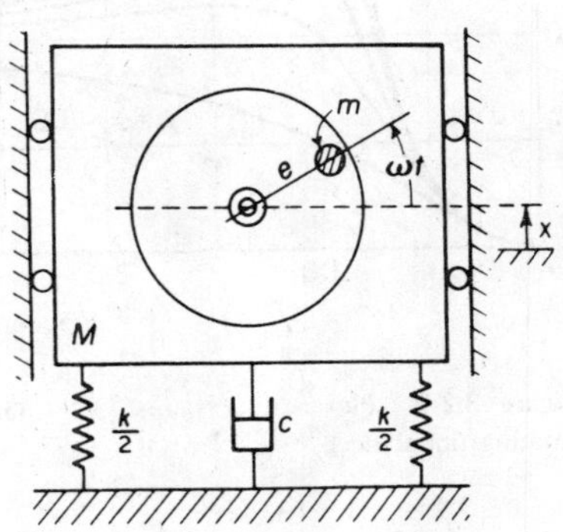

Figure 3.2-1. Harmonic disturbing force resulting from rotating unbalance.

which can be rearranged to

$$M\ddot{x} + c\dot{x} + kx = (me\omega^2)\sin \omega t \tag{3.2-1}$$

It is evident, then, that the above equation is identical to Eq. (3.1-1), where F_0 is replaced by $me\omega^2$, and hence the steady-state solution of the previous section can be replaced by

$$X = \frac{me\omega^2}{\sqrt{(k - M\omega^2)^2 + (c\omega)^2}} \tag{3.2-2}$$

and

$$\tan \phi = \frac{c\omega}{k - M\omega^2} \tag{3.2-3}$$

These can be further reduced to nondimensional form

$$\frac{M}{m}\frac{X}{e} = \frac{\left(\frac{\omega}{\omega_n}\right)^2}{\sqrt{\left[1 - \left(\frac{\omega}{\omega_n}\right)^2\right]^2 + \left[2\zeta\frac{\omega}{\omega_n}\right]^2}} \tag{3.2-4}$$

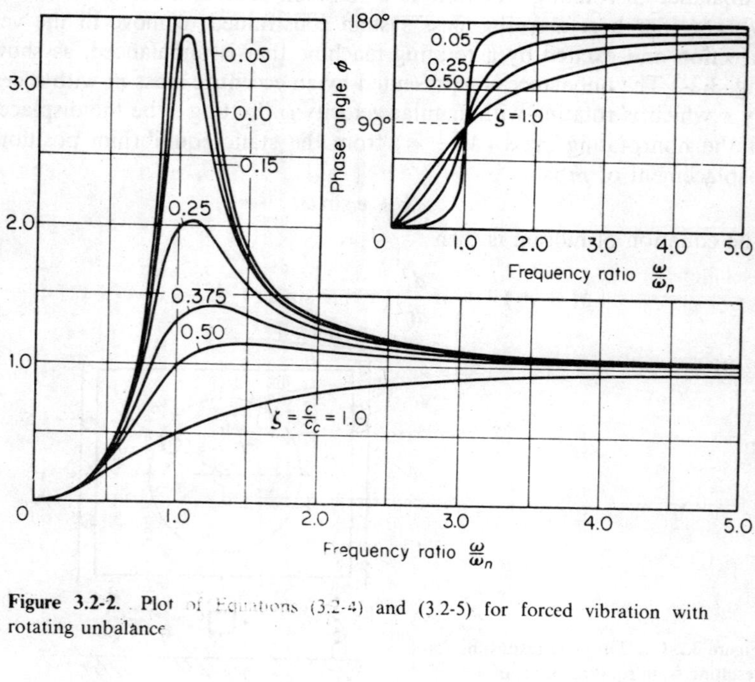

Figure 3.2-2. Plot of Equations (3.2-4) and (3.2-5) for forced vibration with rotating unbalance.

and

$$\tan\phi = \frac{2\zeta\left(\frac{\omega}{\omega_n}\right)}{1-\left(\frac{\omega}{\omega_n}\right)^2} \tag{3.2-5}$$

and presented graphically as in Fig. 3.2-2. The complete solution is given by

$$x(t) = X_1 e^{-\zeta\omega_n t}\sin\left(\sqrt{1-\zeta^2}\,\omega_n t + \phi_1\right) + \frac{me\omega^2}{\sqrt{(k-M\omega^2)^2+(c\omega)^2}}\sin(\omega t - \phi) \tag{3.2-6}$$

Example 3.2-1

A counterrotating eccentric weight exciter is used to produce forced oscillation of a spring-supported mass, as shown in Fig. 3.2-3. By varying the speed of rotation, a resonant amplitude of 0.60 cm was recorded. When the speed of rotation was increased considerably beyond the resonant frequency, the amplitude appeared to approach a fixed value of 0.08 cm. Determine the damping factor of the system.

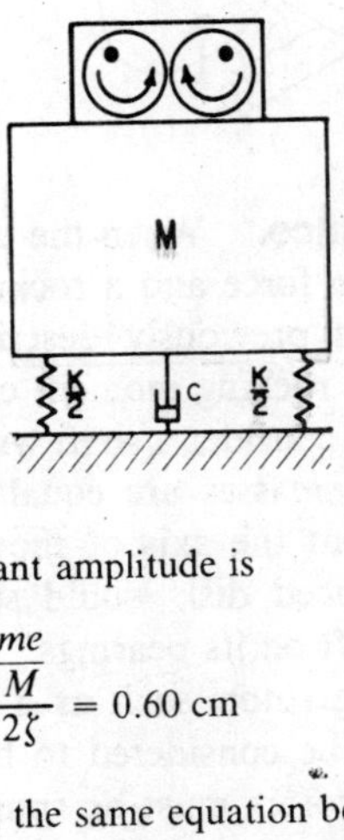

Figure 3.2-3.

Solution: From Eq. (3.2-4), the resonant amplitude is

$$X = \frac{\frac{me}{M}}{2\zeta} = 0.60 \text{ cm}$$

When ω is very much greater than ω_n, the same equation becomes

$$X = \frac{me}{M} = 0.08 \text{ cm}$$

Solving the two equations simultaneously, the damping factor of the system is

$$\zeta = \frac{0.08}{2 \times 0.60} = 0.0666$$

3.3 ROTOR UNBALANCE

In Sec. 3.2 the system was idealized to a spring-mass-damper unit with a rotating unbalance acting in a single plane. It is more likely that the unbalance in a rotating wheel or rotor is distributed in several planes. We wish now to distinguish between two types of rotating unbalance.

Static unbalance. When the unbalanced masses all lie in a single plane, as in the case of a thin rotor disk, the resultant unbalance is a single radial force. As shown in Fig. 3.3-1, such unbalance can be detected by a static test in which the wheel-axle assembly is placed on a pair of horizontal rails. The wheel will roll to a position where the heavy point is directly below the axle. Since such unbalance can be detected without spinning the wheel, it is called *static unbalance*.

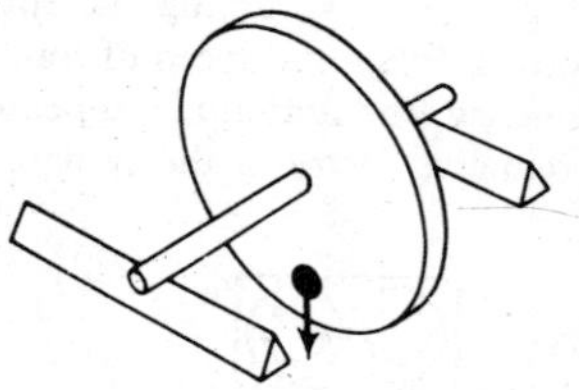

Figure 3.3-1. System with static unbalance.

Dynamic unbalance. When the unbalance appears in more than one plane, the resultant is a force and a rocking moment, which is referred to as *dynamic unbalance*. As previously described, a static test may detect the resultant force, but the rocking moment cannot be detected without spinning the rotor. For example, consider a shaft with two disks, as shown in Fig. 3.3-2. If the two unbalanced masses are equal and 180° apart, the rotor will be statically balanced about the axis of the shaft. However, when the rotor is spinning, each unbalanced disk would set up a rotating centrifugal force, tending to rock the shaft on its bearings.

In general, a long rotor, such as a motor armature or an automobile engine crankshaft, can be considered to be a series of thin disks, each with some unbalance. Such rotors must be spun in order to detect the unbalance. Machines to detect and correct the rotor unbalance are called *balancing machines*. Essentially the balancing machine consists of supporting bearings which are spring mounted so as to detect the unbalanced forces by their motion, as shown in Fig. 3.3-3. Knowing the amplitude of each bearing and their relative phase, it is possible to determine the unbalance of the rotor and correct for them. The problem is that of two DOF, since both translation and angular motion of the shaft take place simultaneously.

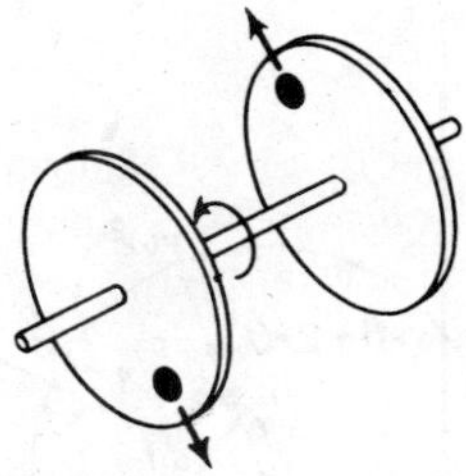

Figure 3.3-2. System with dynamic unbalance.

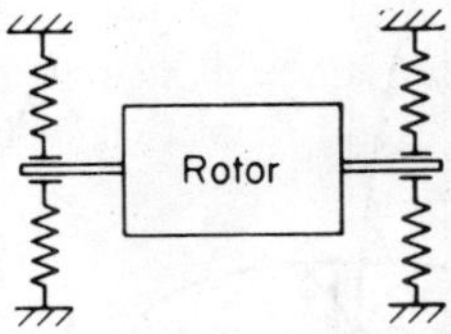

Figure 3.3-3. A rotor balancing machine.

3.4 WHIRLING OF ROTATING SHAFTS

Rotating shafts tend to bow out at certain speeds and whirl in a complicated manner. *Whirling* is defined as the rotation of the plane made by the bent shaft and the line of centers of the bearings. The phenomenon results from such various causes as mass unbalance, hysteresis damping in the shaft, gyroscopic forces, fluid friction in bearings, and so on. The whirling of the shaft may take place in the same or opposite direction as that of the rotation of the shaft and the whirling speed may or may not be equal to the rotation speed.

We will consider here a single disk of mass m symmetrically located on a shaft supported by two bearings as shown in Fig. 3.4-1. The center of mass G of the disk is at a distance e (eccentricity) from the geometric center S of the disk. The center line of the bearings intersects the plane of the disk at O, and the shaft center is deflected by $r = OS$.

We will always assume the shaft (i.e., the line $e = SG$) to be rotating at a constant speed ω and in the general case the line $r = OS$ to be whirling at speed $\dot{\theta}$ which is not equal to ω. For the equation of motion, we can develop the acceleration of the mass center as follows;

$$\mathbf{a}_G = \mathbf{a}_S + \mathbf{a}_{G/S} \tag{3.4-1}$$

where $\mathbf{a}_S$ is the acceleration of S and $\mathbf{a}_{G/S}$ is the acceleration of G with respect to S. The latter term is directed from G to S, since ω is constant. Resolving $\mathbf{a}_G$ in the radial and tangential directions, we have

$$\mathbf{a}_G = \left[(\ddot{r} - r\dot{\theta}^2) - e\omega^2\cos(\omega t - \theta)\right]\mathbf{i} + \left[(r\ddot{\theta} + 2\dot{r}\dot{\theta}) - e\omega^2\sin(\omega t - \theta)\right]\mathbf{j} \tag{3.4-2}$$

Aside from the restoring force of the shaft, we will assume a viscous damping force to be acting at S. The equations of motion resolved [illegible] radial and

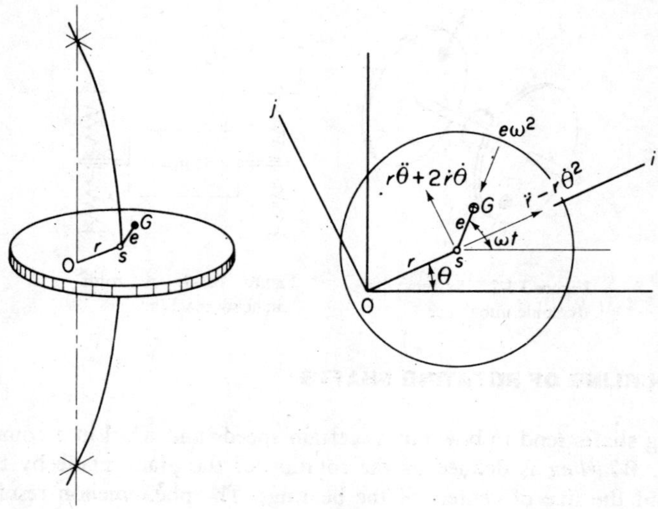

Figure 3.4-1.

tangential directions then become

$$-kr - c\dot{r} = m\left[\ddot{r} - r\dot{\theta}^2 - e\omega^2\cos(\omega t - \theta)\right]$$

$$-cr\dot{\theta} = m\left[r\ddot{\theta} + 2\dot{r}\dot{\theta} - e\omega^2\sin(\omega t - \theta)\right]$$

which can be rearranged to

$$\ddot{r} + \frac{c}{m}\dot{r} + \left(\frac{k}{m} - \dot{\theta}^2\right)r = e\omega^2\cos(\omega t - \theta) \tag{3.4-3}$$

$$r\ddot{\theta} + \left(\frac{c}{m}r + 2\dot{r}\right)\dot{\theta} = e\omega^2\sin(\omega t - \theta) \tag{3.4-4}$$

The general case of whirl as described by the above equations comes under the classification of self-excited motion where the exciting forces inducing the motion are controlled by the motion itself. Since the variables in these equations are r and θ, the problem is that of two DOF. However, in the steady-state synchronous whirl, where $\dot{\theta} = \omega$ and $\ddot{\theta} = \ddot{r} = \dot{r} = 0$, the problem reduces to that of one DOF.

Synchronous whirl. For the synchronous whirl, the whirling speed $\dot{\theta}$ is equal to the rotation speed ω, which we have assumed to be constant. Thus we have

$$\dot{\theta} = \omega$$

and on integrating we obtain

$$\theta = \omega t - \phi$$

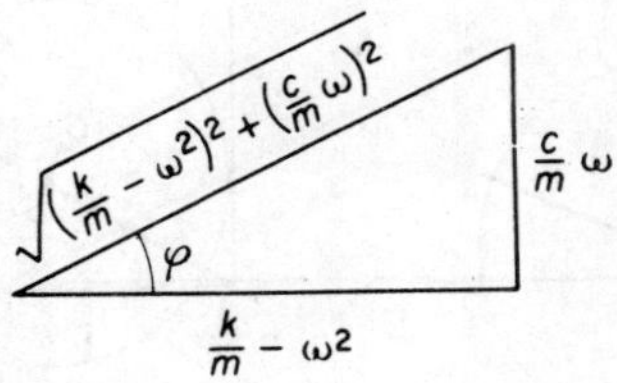

Figure 3.4-2.

where ϕ is the phase angle between e and r, which is now a constant as shown in Fig. 3.4-1. With $\ddot{\theta} = \ddot{r} = \dot{r} = 0$, Eqs. (3.4-3) and (3.4-4) reduce to

$$\left(\frac{k}{m} - \omega^2\right) r = e\omega^2 \cos\phi \qquad \frac{c}{m}\omega r = e\omega^2 \sin\phi \tag{3.4-5}$$

Dividing, we obtain the following equation for the phase angle:

$$\tan\phi = \frac{\dfrac{c}{m}\omega}{\left(\dfrac{k}{m} - \omega^2\right)} = \frac{2\zeta\dfrac{\omega}{\omega_n}}{1 - \left(\dfrac{\omega}{\omega_n}\right)^2} \tag{3.4-6}$$

where $\omega_n = \sqrt{k/m}$ is the critical speed and $\zeta = c/c_{cr}$. Noting from the vector triangle of Fig. 3.4-2 that

$$\cos\phi = \frac{\dfrac{k}{m} - \omega^2}{\sqrt{\left(\dfrac{k}{m} - \omega^2\right)^2 + \left(\dfrac{c}{m}\omega\right)^2}}$$

and substituting into the first of Eq. (3.4-5) gives the amplitude equation

$$r = \frac{me\omega^2}{\sqrt{(k - m\omega^2)^2 + (c\omega)^2}} = \frac{e\left(\dfrac{\omega}{\omega_n}\right)^2}{\sqrt{\left[1 - \left(\dfrac{\omega}{\omega_n}\right)^2\right]^2 + \left[2\zeta\dfrac{\omega}{\omega_n}\right]^2}} \tag{3.4-7}$$

These equations indicate that the eccentricity line $e = SG$ leads the displacement line $r = OS$ by the phase angle ϕ, which depends on the amount of damping and the rotation speed ratio ω/ω_n. When the rotation speed coincides with the critical speed $\omega_n = \sqrt{k/m}$, or the natural frequency of the shaft in lateral vibration, a condition of resonance is encountered in which the amplitude is restrained only by the damping. Figure 3.4-3 shows the disk-shaft system under three different speed conditions. At very high speeds $\omega \gg \omega_n$ the center of mass G tends to approach the fixed point O and the shaft center S rotates about it in a circle of radius e.

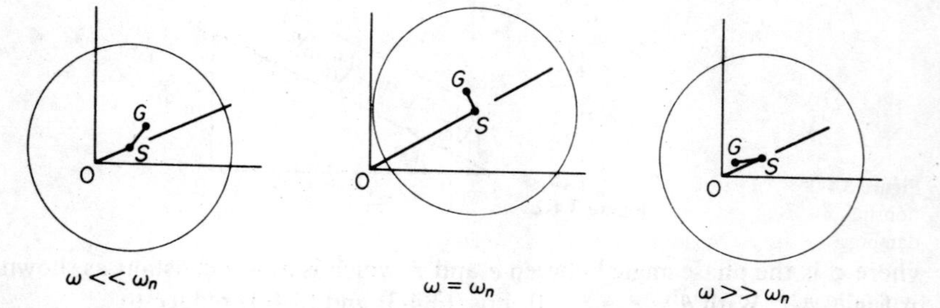

Figure 3.4-3.

It should be noted that the equations for synchronous whirl appear to be the same as those of Sec. 3.2. This is not surprising, since in both cases the exciting force is rotating and equal to $me\omega^2$. However, in Sec. 3.2 the unbalance was in terms of the small unbalanced mass m, whereas in this section the unbalance is defined in terms of the total mass m with eccentricity e. Thus Fig. 3.2-2 is applicable to this problem with the ordinate equal to r/e instead of MX/me.

Example 3.4-1

Turbines operating above the critical speed must run through dangerous speed at resonance each time they are started or stopped. Assuming the critical speed ω_n to be reached with amplitude r_0, determine the equation for the amplitude buildup with time. Assume zero damping.

Solution: We will assume synchronous whirl as before, which makes $\dot{\theta} = \omega =$ constant and $\ddot{\theta} = 0$. However $\ddot{r}$ and $\dot{r}$ terms must be retained unless shown to be zero. With $c = 0$ for the undamped case. the general equations of motion reduce to

$$\begin{aligned} \ddot{r} + \left(\frac{k}{m} - \omega^2\right) r &= e\omega^2 \cos\phi \\ 2\dot{r}\omega &= e\omega^2 \sin\phi \end{aligned} \tag{a}$$

The solution of the second equation with initial deflection equal to r_0 is

$$r = \frac{e\omega}{2} t \sin\phi + r_0 \tag{b}$$

Differentiating this equation twice we find that $\ddot{r} = 0$; so the first equation with the above solution for r becomes

$$\left(\frac{k}{m} - \omega^2\right)\left(\frac{e\omega}{2} t \sin\phi + r_0\right) = e\omega^2 \cos\phi \tag{c}$$

Since the right side of this equation is constant, it is satisfied only if the coefficient of t is zero

$$\left(\frac{k}{m} - \omega^2\right) \sin\phi = 0 \tag{d}$$

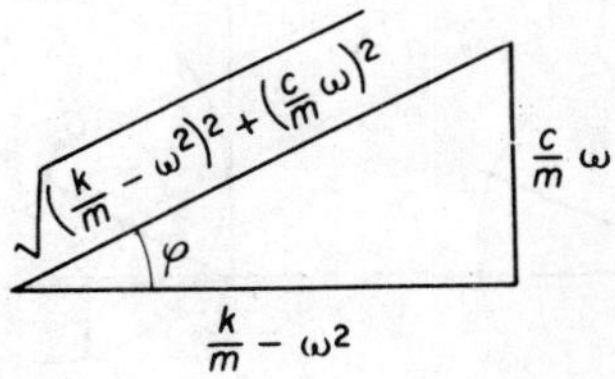

Figure 3.4-2.

where ϕ is the phase angle between e and r, which is now a constant as shown in Fig. 3.4-1. With $\ddot{\theta} = \ddot{r} = \dot{r} = 0$, Eqs. (3.4-3) and (3.4-4) reduce to

$$\left(\frac{k}{m} - \omega^2\right) r = e\omega^2 \cos\phi$$
$$\frac{c}{m}\omega r = e\omega^2 \sin\phi \tag{3.4-5}$$

Dividing, we obtain the following equation for the phase angle:

$$\tan\phi = \frac{\frac{c}{m}\omega}{\left(\frac{k}{m} - \omega^2\right)} = \frac{2\zeta\frac{\omega}{\omega_n}}{1 - \left(\frac{\omega}{\omega_n}\right)^2} \tag{3.4-6}$$

where $\omega_n = \sqrt{k/m}$ is the critical speed and $\zeta = c/c_{cr}$. Noting from the vector triangle of Fig. 3.4-2 that

$$\cos\phi = \frac{\frac{k}{m} - \omega^2}{\sqrt{\left(\frac{k}{m} - \omega^2\right)^2 + \left(\frac{c}{m}\omega\right)^2}}$$

and substituting into the first of Eq. (3.4-5) gives the amplitude equation

$$r = \frac{me\omega^2}{\sqrt{(k - m\omega^2)^2 + (c\omega)^2}} = \frac{e\left(\frac{\omega}{\omega_n}\right)^2}{\sqrt{\left[1 - \left(\frac{\omega}{\omega_n}\right)^2\right]^2 + \left[2\zeta\frac{\omega}{\omega_n}\right]^2}} \tag{3.4-7}$$

These equations indicate that the eccentricity line $e = SG$ leads the displacement line $r = OS$ by the phase angle ϕ, which depends on the amount of damping and the rotation speed ratio ω/ω_n. When the rotation speed coincides with the critical speed $\omega_n = \sqrt{k/m}$, or the natural frequency of the shaft in lateral vibration, a condition of resonance is encountered in which the amplitude is restrained only by the damping. Figure 3.4-3 shows the disk-shaft system under three different speed conditions. At very high speeds $\omega \gg \omega_n$ the center of mass G tends to approach the fixed point O and the shaft center S rotates about it in a circle of radius e.

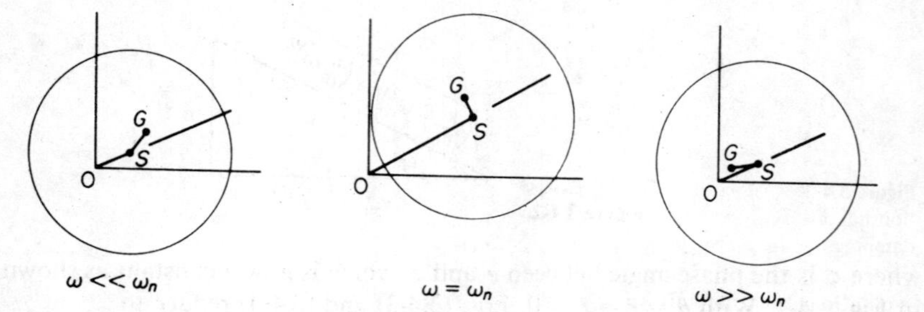

Figure 3.4-3.

It should be noted that the equations for synchronous whirl appear to be the same as those of Sec. 3.2. This is not surprising, since in both cases the exciting force is rotating and equal to $me\omega^2$. However, in Sec. 3.2 the unbalance was in terms of the small unbalanced mass m, whereas in this section the unbalance is defined in terms of the total mass m with eccentricity e. Thus Fig. 3.2-2 is applicable to this problem with the ordinate equal to r/e instead of MX/me.

Example 3.4-1

Turbines operating above the critical speed must run through dangerous speed at resonance each time they are started or stopped. Assuming the critical speed ω_n to be reached with amplitude r_0, determine the equation for the amplitude buildup with time. Assume zero damping.

Solution: We will assume synchronous whirl as before, which makes $\dot{\theta} = \omega =$ constant and $\ddot{\theta} = 0$. However $\ddot{r}$ and $\dot{r}$ terms must be retained unless shown to be zero. With $c = 0$ for the undamped case. the general equations of motion reduce to

$$\ddot{r} + \left(\frac{k}{m} - \omega^2\right) r = e\omega^2 \cos\phi$$

$$2\dot{r}\omega = e\omega^2 \sin\phi \tag{a}$$

The solution of the second equation with initial deflection equal to r_0 is

$$r = \frac{e\omega}{2} t \sin\phi + r_0 \tag{b}$$

Differentiating this equation twice we find that $\ddot{r} = 0$; so the first equation with the above solution for r becomes

$$\left(\frac{k}{m} - \omega^2\right)\left(\frac{e\omega}{2} t \sin\phi + r_0\right) = e\omega^2 \cos\phi \tag{c}$$

Since the right side of this equation is constant, it is satisfied only if the coefficient of t is zero

$$\left(\frac{k}{m} - \omega^2\right) \sin\phi = 0 \tag{d}$$

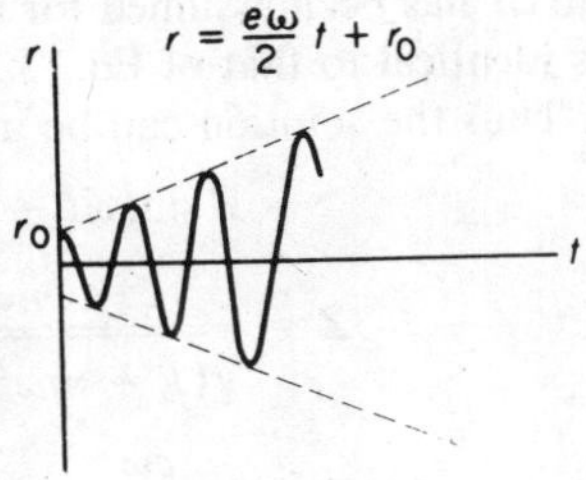

Figure 3.4-4. Amplitude and phase relationship of synchronous whirl with viscous damping.

which leaves the remaining terms

$$\left(\frac{k}{m} - \omega^2\right) r_0 = e\omega^2 \cos\phi \tag{e}$$

With $\omega = \sqrt{k/m}$ the first equation is satisfied, but the second equation is satisfied only if $\cos\phi = 0$, or $\phi = \pi/2$. Thus we have shown that at $\omega = \sqrt{k/m}$, or at resonance, the phase angle is $\pi/2$ as before for the damped case, and the amplitude builds up linearly according to the equation shown in Fig. 3.4-4.

3.5 SUPPORT MOTION

In many cases the dynamical system is excited by the motion of the support point, as shown in Fig. 3.5-1. We let y be the harmonic displacement of the support point and measure the displacement x of the mass m from an inertial reference.

In the displaced position the unbalanced forces are due to the damper and the springs, and the differential equation of motion becomes

$$m\ddot{x} = -k(x - y) - c(\dot{x} - \dot{y}) \tag{3.5-1}$$

Making the substitution

$$z = x - y \tag{3.5-2}$$

Eq. (3.5-1) becomes

$$\begin{aligned} m\ddot{z} + c\dot{z} + kz &= -m\ddot{y} \\ &= m\omega^2 Y \sin\omega t \end{aligned} \tag{3.5-3}$$

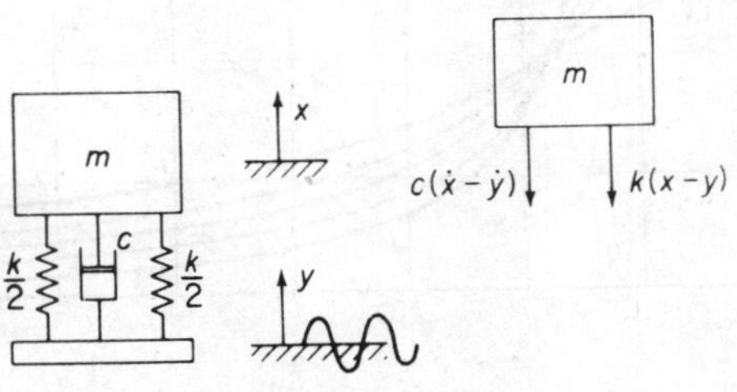

Figure 3.5-1. System excited by motion of support point.

where $y = Y \sin \omega t$ has been assumed for the motion of the base. The form of this equation is identical to that of Eq. (3.2-1), where z replaces x and $m\omega^2 Y$ replaces $me\omega^2$. Thus the solution can be immediately written as

$$z = Z \sin(\omega t - \phi)$$

$$Z = \frac{m\omega^2 Y}{\sqrt{(k - m\omega^2)^2 + (c\omega)^2}} \tag{3.5-4}$$

$$\tan \phi = \frac{c\omega}{k - m\omega^2} \tag{3.5-5}$$

and the curves of Fig. 3.2-2 are applicable with the appropriate change in the ordinate.

If the absolute motion x of the mass is desired, we can solve for $x = z + y$. Using the exponential form of harmonic motion gives

$$\begin{aligned} y &= Ye^{i\omega t} \\ z &= Ze^{i(\omega t - \phi)} = (Ze^{-i\phi})e^{i\omega t} \\ x &= Xe^{i(\omega t - \psi)} = (Xe^{-i\psi})e^{i\omega t} \end{aligned} \tag{3.5-6}$$

Substituting into Eq. (3.5-3), we obtain

$$Ze^{-i\phi} = \frac{m\omega^2 Y}{k - m\omega^2 + i\omega c}$$

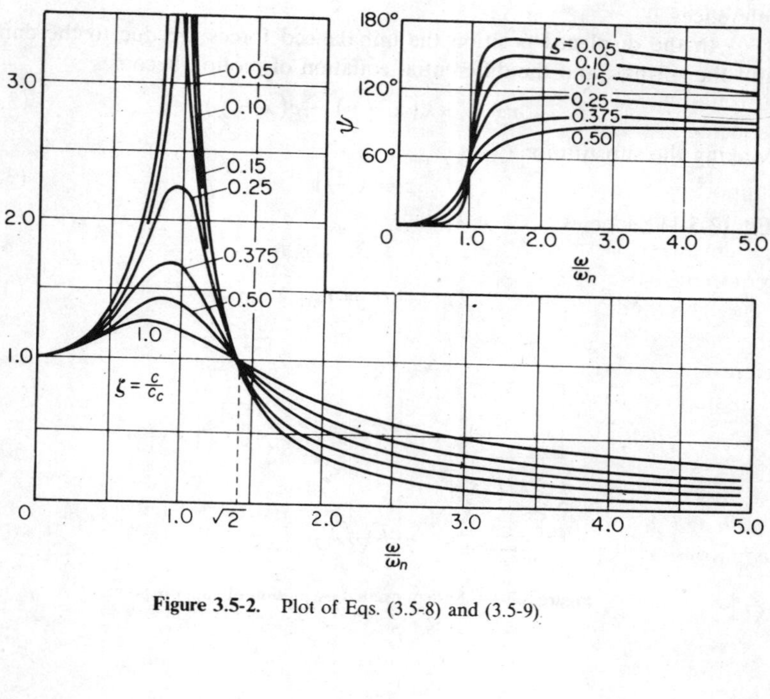

Figure 3.5-2. Plot of Eqs. (3.5-8) and (3.5-9).

and

$$x = (Ze^{-i\phi} + Y)e^{i\omega t}$$
$$= \left(\frac{k + i\omega c}{k - m\omega^2 + i\omega c}\right) Ye^{i\omega t} \tag{3.5-7}$$

The steady-state amplitude and phase from this equation are

$$\left|\frac{X}{Y}\right| = \sqrt{\frac{k^2 + (\omega c)^2}{(k - m\omega^2)^2 + (c\omega)^2}} \tag{3.5-8}$$

and

$$\tan\psi = \frac{mc\omega^3}{k(k - m\omega^2) + (\omega c)^2} \tag{3.5-9}$$

which are plotted in Fig. 3.5-2. It should be observed that the amplitude curves for different damping all have the same value of $|X/Y| = 1.0$ at the frequency $\omega/\omega_n = \sqrt{2}$.

3.6 VIBRATION ISOLATION

Vibratory forces generated by machines and other causes are often unavoidable; however, their effects on a dynamical system can be minimized by proper isolator design. An isolation system attempts either to protect a delicate object from excessive vibration transmitted to it from its supporting structure or to prevent vibratory forces generated by machines from being transmitted to its surroundings. The basic problem is the same for these two objectives, that of reducing the transmitted force.

Figure 3.5-2 for $|X/Y|$ shows that the motion transmitted from the supporting structure to the mass m is less than 1 when the ratio ω/ω_n is greater than $\sqrt{2}$. This indicates that the natural frequency ω_n of the supported system must be small compared to that of the disturbing frequency ω. This requirement can be met by using a soft spring.

The other problem of reducing the force transmitted by the machine to the supporting structure has the same requirement. The force to be isolated is transmitted through the spring and damper, as shown in Fig. 3.6-1. Its equation is

$$F_T = \sqrt{(kX)^2 + (c\omega X)^2} = kX\sqrt{1 + \left(\frac{2\zeta\omega}{\omega_n}\right)^2} \tag{3.6-1}$$

With the disturbing force equal to $F_0 \sin\omega t$, the value of X in the preceding equation is

$$X = \frac{F_0/k}{\sqrt{\left[1 - (\omega/\omega_n)^2\right]^2 + [2\zeta\omega/\omega_n]^2}} \tag{3.6-1a}$$

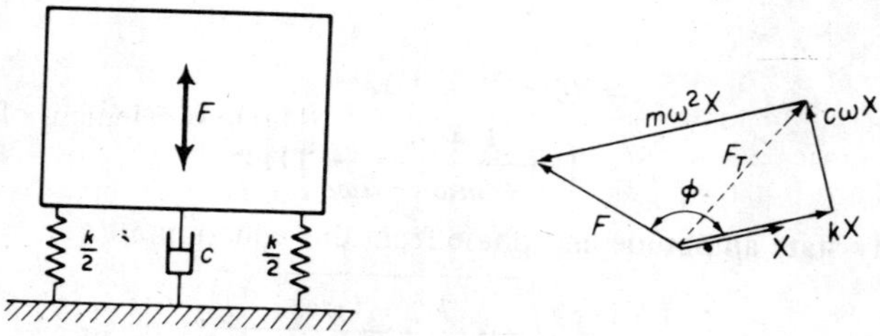

Figure 3.6-1. Disturbing force transmitted through springs and damper.

The transmissibility TR, defined as the ratio of the transmitted force to that of the disturbing force, is then

$$TR = \left|\frac{F_T}{F_0}\right| = \sqrt{\frac{1 + (2\zeta\omega/\omega_n)^2}{\left[1 - (\omega/\omega_n)^2\right]^2 + [2\zeta\omega/\omega_n]^2}} \tag{3.6-2}$$

Comparison of the preceding equation with Eq. (3.5-8) shows that

$$TR = \left|\frac{F_T}{F_0}\right| = \left|\frac{X}{Y}\right|$$

When the damping is negligible, the transmissibility equation reduces to

$$TR = \frac{1}{(\omega/\omega_n)^2 - 1} \tag{3.6-3}$$

where it is understood that the value of ω/ω_n to be used is always greater than $\sqrt{2}$. On further replacing ω_n by Δ/g, where g is the acceleration of gravity and Δ is the statical deflection, Equation (3.6-3) may be expressed as

$$TR = \frac{1}{(2\pi f)^2 \Delta/g - 1}$$

To reduce the amplitude X of the isolated mass m without changing TR, m is often mounted on a large mass M, as shown in Fig. 3.6-2. The stiffness k must then be increased to keep the ratio $k/(m + M)$ constant. The amplitude X is, however, reduced because k appears in the denominator of Eq. (3.6-1a).

Since in the general problem, the mass to be isolated may have six DOF (three translation, and three rotation), the designer of the isolation system must use his or her intuition and ingenuity. The results of the single-DOF analysis should, however, serve as useful guide.

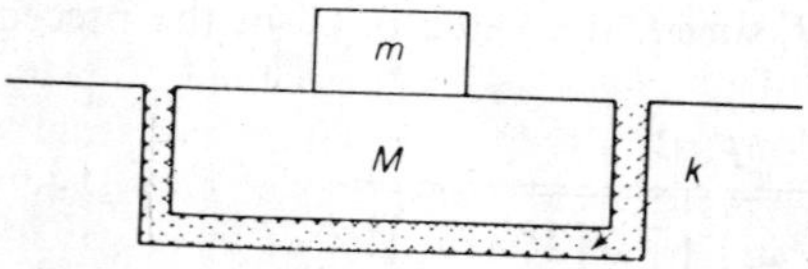

Figure 3.6-2.

and

$$x = (Ze^{-i\phi} + Y)e^{i\omega t}$$
$$= \left(\frac{k + i\omega c}{k - m\omega^2 + i\omega c}\right)Ye^{i\omega t} \tag{3.5-7}$$

The steady-state amplitude and phase from this equation are

$$\left|\frac{X}{Y}\right| = \sqrt{\frac{k^2 + (\omega c)^2}{(k - m\omega^2)^2 + (c\omega)^2}} \tag{3.5-8}$$

and

$$\tan\psi = \frac{mc\omega^3}{k(k - m\omega^2) + (\omega c)^2} \tag{3.5-9}$$

which are plotted in Fig. 3.5-2. It should be observed that the amplitude curves for different damping all have the same value of $|X/Y| = 1.0$ at the frequency $\omega/\omega_n = \sqrt{2}$.

3.6 VIBRATION ISOLATION

Vibratory forces generated by machines and other causes are often unavoidable; however, their effects on a dynamical system can be minimized by proper isolator design. An isolation system attempts either to protect a delicate object from excessive vibration transmitted to it from its supporting structure or to prevent vibratory forces generated by machines from being transmitted to its surroundings. The basic problem is the same for these two objectives, that of reducing the transmitted force.

Figure 3.5-2 for $|X/Y|$ shows that the motion transmitted from the supporting structure to the mass m is less than 1 when the ratio ω/ω_n is greater than $\sqrt{2}$. This indicates that the natural frequency ω_n of the supported system must be small compared to that of the disturbing frequency ω. This requirement can be met by using a soft spring.

The other problem of reducing the force transmitted by the machine to the supporting structure has the same requirement. The force to be isolated is transmitted through the spring and damper, as shown in Fig. 3.6-1. Its equation is

$$F_T = \sqrt{(kX)^2 + (c\omega X)^2} = kX\sqrt{1 + \left(\frac{2\zeta\omega}{\omega_n}\right)^2} \tag{3.6-1}$$

With the disturbing force equal to $F_0 \sin\omega t$, the value of X in the preceding equation is

$$X = \frac{F_0/k}{\sqrt{[1 - (\omega/\omega_n)^2]^2 + [2\zeta\omega/\omega_n]^2}} \tag{3.6-1a}$$

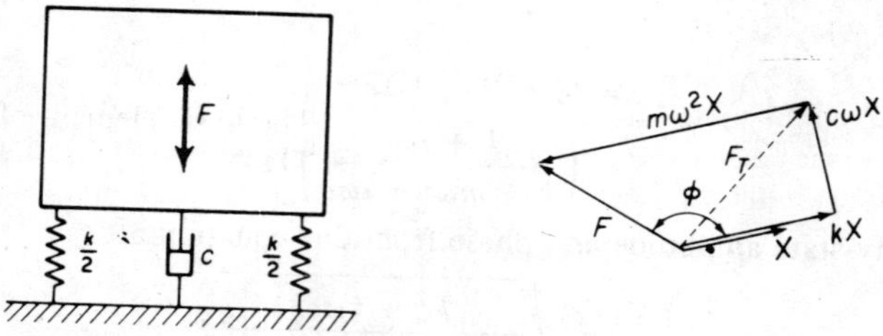

Figure 3.6-1. Disturbing force transmitted through springs and damper.

The transmissibility TR, defined as the ratio of the transmitted force to that of the disturbing force, is then

$$TR = \left|\frac{F_T}{F_0}\right| = \sqrt{\frac{1 + (2\zeta\omega/\omega_n)^2}{[1 - (\omega/\omega_n)^2]^2 + [2\zeta\omega/\omega_n]^2}} \tag{3.6-2}$$

Comparison of the preceding equation with Eq. (3.5-8) shows that

$$TR = \left|\frac{F_T}{F_0}\right| = \left|\frac{X}{Y}\right|$$

When the damping is negligible, the transmissibility equation reduces to

$$TR = \frac{1}{(\omega/\omega_n)^2 - 1} \tag{3.6-3}$$

where it is understood that the value of ω/ω_n to be used is always greater than $\sqrt{2}$. On further replacing ω_n by Δ/g, where g is the acceleration of gravity and Δ is the statical deflection, Equation (3.6-3) may be expressed as

$$TR = \frac{1}{(2\pi f)^2 \Delta/g - 1}$$

To reduce the amplitude X of the isolated mass m without changing TR, m is often mounted on a large mass M, as shown in Fig. 3.6-2. The stiffness k must then be increased to keep the ratio $k/(m + M)$ constant. The amplitude X is, however, reduced because k appears in the denominator of Eq. (3.6-1a).

Since in the general problem, the mass to be isolated may have six DOF, (three translation, and three rotation), the designer of the isolation system must use his or her intuition and ingenuity. The results of the single-DOF analysis should, however, serve as useful guide.

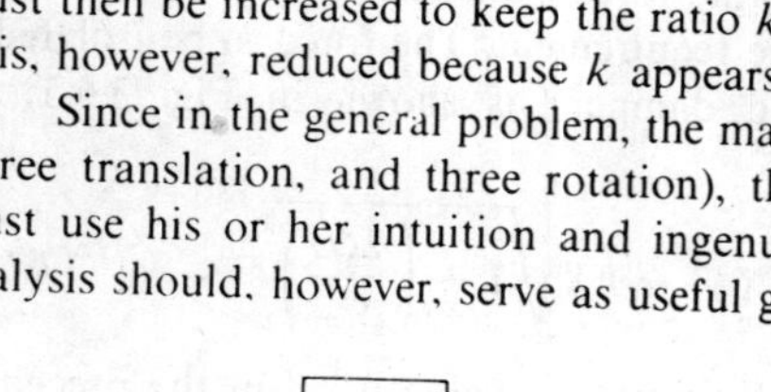

Figure 3.6-2.

Example 3.6-1

A machine of 100 kg mass is supported on springs of total stiffness 700 kN/m and has an unbalanced rotating element, which results in a disturbing force of 350 N at a speed of 3000 rev/min. Assuming a damping factor of $\zeta = 0.20$, determine (a) its amplitude of motion due to the unbalance, (b) the transmissibility, and (c) the transmitted force.

Solution: The statical deflection of the system is

$$\frac{100 \times 9.81}{700 \times 10^3} = 1.401 \times 10^{-3} \text{ m} = 1.401 \text{ mm}$$

and its natural frequency is found to be

$$f_n = \frac{1}{2\pi}\sqrt{\frac{9.81}{1.401 \times 10^{-3}}} = 13.32 \text{ Hz}$$

(a) Substituting into Eq. (3.1-5), the amplitude of vibration is

$$X = \frac{\left(\dfrac{350}{700 \times 10^3}\right)}{\sqrt{\left[1 - \left(\dfrac{50}{13.32}\right)^2\right]^2 + \left[2 \times 0.20 \times \dfrac{50}{13.32}\right]^2}}$$

$$= 3.79 \times 10^{-5} \text{ m}$$
$$= 0.0379 \text{ mm}$$

(b) The transmissibility from Eq. (3.6-2) is

$$TR = \frac{\sqrt{1 + \left(2 \times 0.20 \times \dfrac{50}{13.32}\right)^2}}{\sqrt{\left[1 - \left(\dfrac{50}{13.32}\right)^2\right]^2 + \left[2 \times 0.20 \times \dfrac{50}{13.32}\right]^2}} = 0.137$$

(c) The transmitted force is the disturbing force multiplied by the transmissibility.

$$F_{TR} = 350 \times 0.137 = 47.89 \text{ N}$$

3.7 ENERGY DISSIPATED BY DAMPING

Damping is present in all oscillatory systems. Its effect is to remove energy from the system. Energy in a vibrating system is either dissipated into heat or radiated away. Dissipation of energy into heat can be experienced simply by bending a piece of metal back and forth a number of times. We are all aware of the sound which is radiated from an object given a sharp blow. When a buoy is made to bob up and down in the water, waves radiate out and away from it, thereby resulting in its loss of energy.

In vibration analysis, we are generally concerned with damping in terms of system response. The loss of energy from the oscillatory system results in the decay of amplitude of free vibration. In steady-state forced vibration, the loss of energy is balanced by the energy which is supplied by the excitation.

A vibrating system may encounter many different types of damping forces, from internal molecular friction to sliding friction and fluid resistance. Generally their mathematical description is quite complicated and not suitable for vibration analysis. Thus simplified damping models have been developed that in many cases are found to be adequate in evaluating the system response. For example, we have already used the viscous damping model, designated by the dashpot, which leads to manageable mathematical solutions.

Energy dissipation is usually determined under conditions of cyclic oscillations. Depending on the type of damping present, the force-displacement relationship when plotted may differ greatly. In all cases, however, the force-displacement curve will enclose an area, referred to as the *hysteresis loop*, that is proportional to the energy lost per cycle. The energy lost per cycle due to a damping force F_d is computed from the general equation

$$W_d = \oint F_d \, dx \tag{3.7-1}$$

In general, W_d depends on many factors, such as temperature, frequency, or amplitude.

We consider in this section the simplest case of energy dissipation, that of a spring-mass system with viscous damping. The damping force in this case is $F_d = c\dot{x}$. With the steady-state displacement and velocity

$$x = X \sin(\omega t - \phi)$$

$$\dot{x} = \omega X \cos(\omega t - \phi)$$

the energy dissipated per cycle, from Eq. (3.7-1), becomes

$$W_d = \oint c\dot{x} \, dx = \oint c\dot{x}^2 \, dt$$

$$= c\omega^2 X^2 \int_0^{2\pi/\omega} \cos^2(\omega t - \phi) \, dt = \pi c \omega X^2 \tag{3.7-2}$$

Of particular interest is the energy dissipated in forced vibration at resonance. Substituting $\omega_n = \sqrt{k/m}$ and $c = 2\zeta\sqrt{km}$, the preceding equation at resonance becomes

$$W_d = 2\zeta\pi k X^2 \tag{3.7-3}$$

The energy dissipated per cycle by the damping force can be represented graphically as follows. Writing the velocity in the form

$$\dot{x} = \omega X \cos(\omega t - \phi) = \pm \omega X \sqrt{1 - \sin^2(\omega T - \phi)}$$

$$= \pm \omega \sqrt{X^2 - x^2}$$

the damping force becomes

$$F_d = c\dot{x} = \pm c\omega \sqrt{X^2 - x^2} \tag{3.7-4}$$

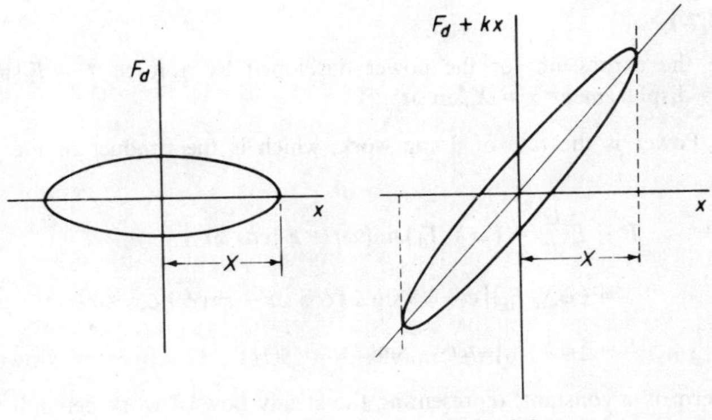

Figure 3.7-1. Energy dissipated by viscous damping.

Rearranging the above equation to

$$\left(\frac{F_d}{c\omega X}\right)^2 + \left(\frac{x}{X}\right)^2 = 1 \tag{3.7-5}$$

we recognize it as that of an ellipse with F_d and x plotted along the vertical and horizontal axes, as shown in Fig. 3.7-1(a). The energy dissipated per cycle is then given by the area enclosed by the ellipse. If we add to F_d the force kx of the lossless spring, the hysteresis loop is rotated as shown in Fig. 3.7-1(b). This representation then conforms to the *Voigt model*, which consists of a dashpot in parallel with a spring.

Damping properties of materials are listed in many different ways depending on the technical areas to which they are applied. Of these we list two relative energy units that have wide usage. First of these is *specific damping capacity*, defined as the energy loss per cycle W_d divided by the peak potential energy U.

$$\frac{W_d}{U} \tag{3.7-6}$$

The second quantity is the *loss coefficient*, defined as the ratio of damping energy loss per radian $W_d/2\pi$ divided by the peak potential or strain energy U.

$$\eta = \frac{W_d}{2\pi U} \tag{3.7-7}$$

For the case of linear damping where the energy loss is proportional to the square of the strain or amplitude, the hysteresis curve is an ellipse. When the damping loss is not a quadratic function of the strain or amplitude, the hysteresis curve is no longer an ellipse.

Example 3.7-1

Determine the expression for the power developed by a force $F = F_0 \sin(\omega t + \phi)$ acting on a displacement $x = X_0 \sin \omega t$.

Solution: Power is the rate of doing work, which is the product of the force and velocity.

$$P = F\frac{dx}{dt} = (\omega X_0 F_0)\sin(\omega t + \phi)\cos \omega t$$

$$= (\omega X_0 F_0)[\cos \phi \cdot \sin \omega t \cos \omega t + \sin \phi \cdot \cos^2 \omega t]$$

$$= \tfrac{1}{2}\omega X_0 F_0[\sin \phi + \sin(2\omega t + \phi)]$$

The first term is a constant, representing the steady flow of work per unit time. The second term is a sine wave of twice the frequency that represents the fluctuating component of power, the average value of which is zero over any interval of time that is a multiple of the period.

Example 3.7-2

A force $F = 10 \sin \pi t$ N acts on a displacement of $x = 2\sin(\pi t - \pi/6)$ m. Determine (a) the work done during the first 6 s; (b) the work done during the first $\frac{1}{2}$ s.

Solution: Rewriting Eq. (3.7-1) as $W = \int F\dot{x}\, dt$ and substituting $F = F_0 \sin \omega t$ and $x = X \sin(\omega t - \phi)$ gives the work done per cycle of

$$W = \pi F_0 X \sin \phi$$

For the force and displacement given in this problem, $F_0 = 10$ N, $X = 2$ m, $\phi = \pi/6$, and the period $\tau = 2$ s. Thus in the 6 s specified in (a), three complete cycles take place, and the work done is

$$W = 3(\pi F_0 X \sin \phi) = 3\pi \times 10 \times 2 \times \sin 30° = 94.2 \text{ N} \cdot \text{m}$$

The work done in part (b) is determined by integrating the expression for work between the limits 0 and $\frac{1}{2}$ s.

$$W = \omega F_0 X_0\left[\cos 30° \int_0^{1/2} \sin \pi t \cos \pi t\, dt + \sin 30° \int_0^{1/2} \sin^2 \pi t\, dt\right]$$

$$= \pi \times 10 \times 2\left[-\frac{0.866}{4\pi}\cos 2\pi t + 0.50\left(\frac{t}{2} - \frac{\sin 2\pi t}{4\pi}\right)\right]_0^{1/2}$$

$$= 16.51 \text{ N} \cdot \text{m}$$

3.8 EQUIVALENT VISCOUS DAMPING

The primary influence of damping on oscillatory systems is that of limiting the amplitude of response at resonance. As seen from the response curves of Fig. 3.1-3, damping has little influence on the response in the frequency regions away from resonance.

sidebands), where X is $0.707X_{\text{res}}$. These points are also referred to as the *half-power points* and are shown in Fig. 3.10-1.

Letting $X = 0.707X_{\text{res}}$ and squaring Eq. (3.1-7), we obtain

$$\frac{1}{2}\left(\frac{1}{2\zeta}\right)^2 = \frac{1}{\left[1 - \left(\frac{\omega}{\omega_n}\right)^2\right]^2 + \left[2\zeta\left(\frac{\omega}{\omega_n}\right)\right]^2}$$

or

$$\left(\frac{\omega}{\omega_n}\right)^4 - 2(1 - 2\zeta^2)\left(\frac{\omega}{\omega_n}\right)^2 + (1 - 8\zeta^2) = 0 \tag{3.10-1}$$

Solving for $(\omega/\omega_n)^2$ we have

$$\left(\frac{\omega}{\omega_n}\right)^2 = (1 - 2\zeta^2) \pm 2\zeta\sqrt{1 - \zeta^2} \tag{3.10-2}$$

Assuming $\zeta \ll 1$ and neglecting higher order terms of ζ, we arrive at the result

$$\left(\frac{\omega}{\omega_n}\right)^2 = 1 \pm 2\zeta \tag{3.10-3}$$

Letting the two frequencies corresponding to the roots of Eq. (3.10-3) be ω_1 and ω_2, we obtain

$$4\zeta = \frac{\omega_2^2 - \omega_1^2}{\omega_n^2} \cong 2\left(\frac{\omega_2 - \omega_1}{\omega_n}\right)$$

The quantity Q is then defined as

$$Q = \frac{\omega_n}{\omega_2 - \omega_1} = \frac{f_n}{f_2 - f_1} = \frac{1}{2\zeta} \tag{3.10-4}$$

Here, again, equivalent damping can be used to define Q for systems with other forms of damping. Thus for structural damping, Q is equal to

$$Q = \frac{1}{\gamma} \tag{3.10-5}$$

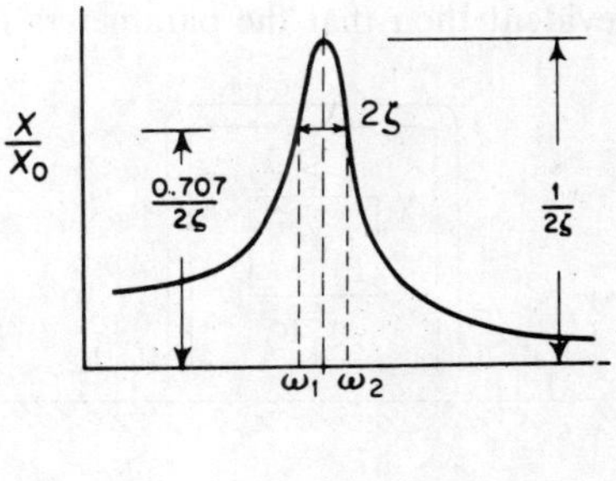

Figure 3.10-1.

3.11 VIBRATION MEASURING INSTRUMENTS

The basic element of many vibration measuring instruments is the seismic unit of Fig. 3.11-1. Depending on the frequency range utilized, displacement, velocity, or acceleration is indicated by the relative motion of the suspended mass with respect to the case.

To determine the behavior of such instruments we consider the equation of motion of m, which is

$$m\ddot{x} = -c(\dot{x} - \dot{y}) - k(x - y) \tag{3.11-1}$$

where x and y are the displacement of the seismic mass and the vibrating body, respectively, both measured with respect to an inertial reference. Letting the relative displacement of the mass m and the case attached to the vibrating body be

$$z = (x - y) \tag{3.11-2}$$

and assuming sinusoidal motion $y = Y \sin \omega t$ of the vibrating body, we obtain the equation

$$m\ddot{z} + c\dot{z} + kz = m\omega^2 Y \sin \omega t \tag{3.11-3}$$

This equation is identical in form to Eq. (3.2-1) with z and $m\omega^2 Y$ replacing x and $me\omega^2$, respectively. The steady-state solution $z = Z \sin(\omega t - \phi)$ is then available from inspection to be

$$Z = \frac{m\omega^2 Y}{\sqrt{(k - m\omega^2)^2 + (c\omega)^2}} = \frac{Y\left(\dfrac{\omega}{\omega_n}\right)^2}{\sqrt{\left[1 - \left(\dfrac{\omega}{\omega_n}\right)^2\right]^2 + \left[2\zeta\dfrac{\omega}{\omega_n}\right]^2}} \tag{3.11-4}$$

and

$$\tan \phi = \frac{\omega c}{k - m\omega^2} = \frac{2\zeta\dfrac{\omega}{\omega_n}}{1 - \left(\dfrac{\omega}{\omega_n}\right)^2} \tag{3.11-5}$$

It is evident then that the parameters involved are the frequency ratio ω/ω_n

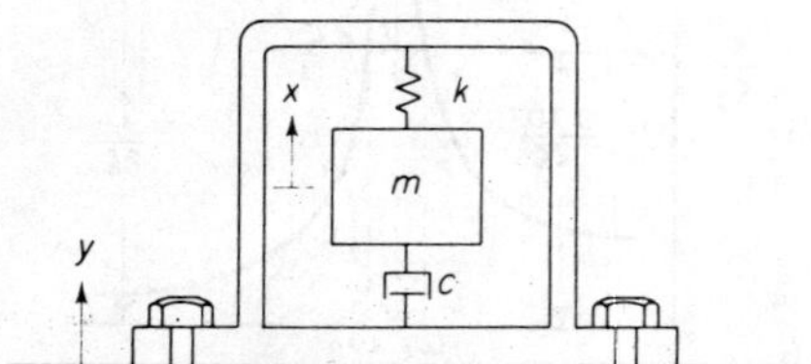

Figure 3.11-1.

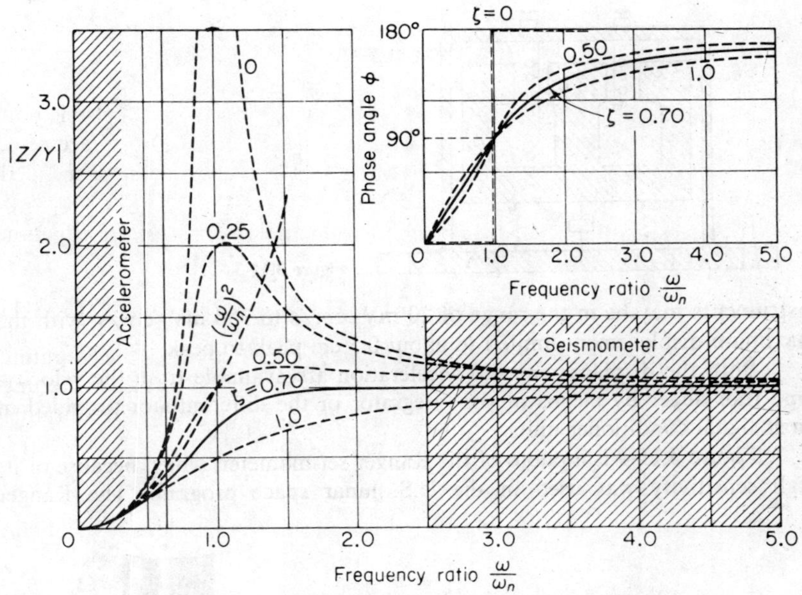

Figure 3.11-2. Response of a vibration-measuring instrument.

and the damping factor ζ. Figure 3.11-2 shows a plot of these equations and is identical to Fig. 3.3-2 except that Z/Y replaces MX/me. The type of instrument is determined by the useful range of frequencies with respect to the natural frequency ω_n of the instrument.

Seismometer — instrument with low natural frequency. When the natural frequency ω_n of the instrument is low in comparison to the vibration frequency ω to be measured, the ratio ω/ω_n approaches a large number, and the relative displacement Z approaches Y regardless of the value of damping ζ, as indicated in Fig. 3.11-2. The mass m then remains stationary while the supporting case moves with the vibrating body. Such instruments are called *seismometers*.

One of the disadvantages of the seismometer is its large size. Since $Z = Y$, the relative motion of the seismic mass must be of the same order of magnitude as that of the vibration to be measured.

The relative motion z is usually converted to an electric voltage by making the seismic mass a magnet moving relative to coils fixed in the case, as shown in Fig. 3.11-3. Since the voltage generated is proportional to the rate of cutting of the magnetic field, the output of the instrument will be proportional to the velocity of the vibrating body. Such instruments are called *velometers*. A typical instrument of this kind may have a natural frequency between 1 Hz to 5 Hz and a useful frequency range of 10 Hz to 2000 Hz. The sensitivity of such

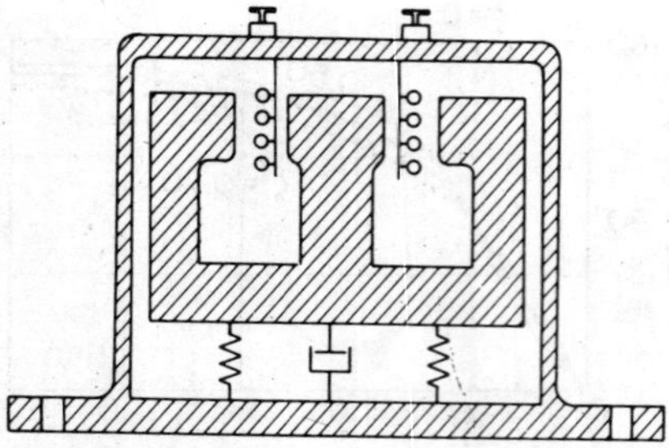

Figure 3.11-3.

instruments may be in the range of 20 mV/cm/s to 350 mV/cm/s, with the maximum displacement limited to about 0.5 cm peak to peak.

Both the displacement and acceleration are available from the velocity-type transducer by means of the integrator or the differentiator provided in most signal conditioner units.

Figure 3.11-4 is a photo of the Ranger seismometer, which because of its high sensitivity was used in the U.S. lunar space program. The Ranger

Figure 3.11-4. Ranger seismometer. (*Courtesy of Kinemetrics, Inc., Pasadena, California.*)

seismometer incorporates a velocity-type transducer with the permanent magnet as the seismic mass. Its natural frequency is nominally 1 Hz with a mass travel of ± 1 mm. Its size is 15 cm in diameter and it weighs 11 lb.

Accelerometer — instrument with high natural frequency. When the natural frequency of the instrument is high compared to that of the vibration to be measured, the instrument indicates acceleration. Examination of Eq. (3.11-4) shows that the factor

$$\sqrt{\left[1-\left(\frac{\omega}{\omega_n}\right)^2\right]^2+\left[2\zeta\frac{\omega}{\omega_n}\right]^2}$$

approaches unity for $\omega/\omega_n \to 0$, so that

$$Z=\frac{\omega^2 Y}{\omega_n^2}=\frac{\text{(acceleration)}}{\omega_n^2} \tag{3.11-6}$$

Thus Z becomes proportional to the acceleration of the motion to be measured with a factor $1/\omega_n^2$. The useful range of the accelerometer can be seen from Fig. 3.11-5 which is a magnified plot of

$$\frac{1}{\sqrt{\left[1-\left(\frac{\omega}{\omega_n}\right)^2\right]^2+\left[2\zeta\frac{\omega}{\omega_n}\right]^2}}$$

for various values of damping ζ. The diagram shows that the useful frequency range of the undamped accelerometer is somewhat limited. However, with $\zeta = 0.7$ the useful frequency range is $0 \le \omega/\omega_n \le 0.20$ with a maximum error less than 0.01 percent. Thus an instrument with a natural frequency of 100 Hz

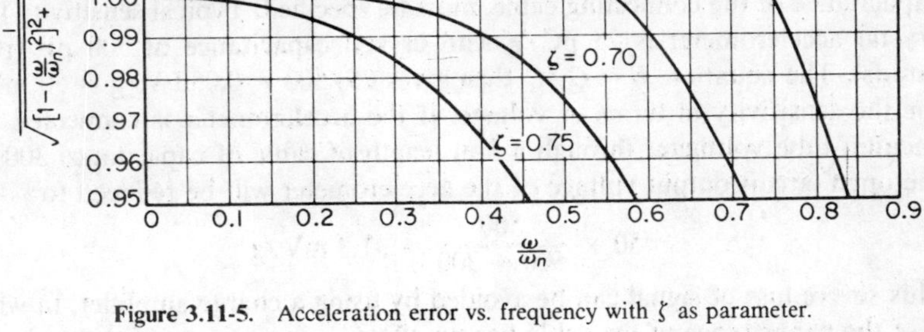

Figure 3.11-5. Acceleration error vs. frequency with ζ as parameter.

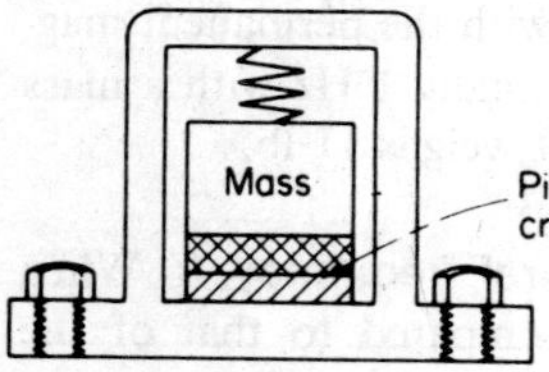

Figure 3.11-6.

has a useful frequency range between 0 Hz to 20 Hz with negligible error. Electromagnetic-type accelerometers generally utilize damping around $\zeta = 0.7$, which not only extends the useful frequency range but also prevents phase distortion for complex waves, as will be shown later. On the other hand, very high natural frequency instruments, such as the piezoelectric crystal accelerometers, have almost zero damping and operate without distortion up to frequencies of $0.06f_n$.

Several different accelerometers are in use today. The seismic mass accelerometer is often used for low-frequency vibration, and the supporting springs may be four electric strain gage wires connected in a bridge circuit. A more accurate variation of this accelerometer is one in which the seismic mass is servo-controlled to have zero relative displacement; the force necessary to accomplish this becomes a measure of the acceleration. Both of these instruments require an external source of electric power.

The piezoelectric properties of crystals like quartz or barium titanate are utilized in accelerometers for higher frequency measurements. The crystals are mounted so that under acceleration they are either compressed or bent to generate an electric charge. Figure 3.11-6 shows one such arrangement. The natural frequency of such accelerometers can be made very high, in the 50,000-Hz range, which enables acceleration measurements to be made up to 3000 Hz. The size of the crystal accelerometer is very small, approximately 1 cm in diameter and height, and it is remarkably rugged and can stand shocks as high as 10,000 g's.

The sensitivity of the crystal accelerometer is given either in terms of charge (picocoulombs = pC = 10^{-12} Coulombs) per g, or in terms of voltage (millivolts = mV = 10^{-3} V) per g. Since the voltage is related to the charge by the equation $E = Q/C$, the capacitance of the crystal, including the shunt capacitance of the connecting cable, must be specified. Typical sensitivity for a crystal accelerometer is 25 pC/g with crystal capacitance of 500 pF (picofarads). The equation $E = Q/C$ then gives $25/500 = 0.050$ V/g = 50 mV/g for the sensitivity in terms of voltage. If the accelerometer is connected to a vacuum tube voltmeter through a 3-m length of cable of capacitance 300 pF, the open circuit output voltage of the accelerometer will be reduced to

$$50 \times \frac{500}{500 + 300} = 31.3 \text{ mV/g}$$

This severe loss of signal can be avoided by using a charge amplifier, in which case the capacitance of the cable has no effect.

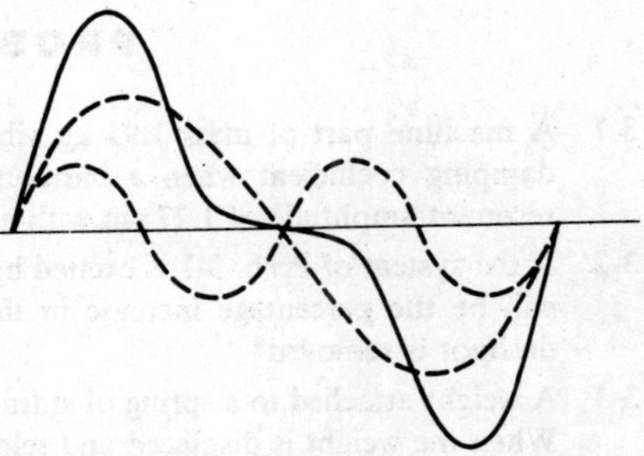

Figure 3.11-7.

Phase distortion. To reproduce a complex wave such as the one shown in Fig. 3.11-7 without changing its shape, the phase of all harmonic components must remain unchanged with respect to the fundamental. This requires that the phase angle be zero or that all the harmonic components must be shifted equally. The first case of zero phase shift corresponds to $\zeta = 0$ for $\omega/\omega_n < 1$. The second case of equal timewise shift of all harmonics is nearly satisfied for $\zeta = 0.70$ for $\omega/\omega_n < 1$. As shown in Fig. 3.11-2, when $\zeta = 0.70$, the phase for $\omega/\omega_n < 1$ can be expressed by the equation

$$\phi \cong \frac{\pi}{2}\frac{\omega}{\omega_n}$$

Thus for $\zeta = 0$, or 0.70, the phase distortion is practically eliminated.

Example 3.11

Investigate the output of an accelerometer with damping $\zeta = 0.70$ when used to measure a periodic motion with the displacement given by the equation

$$y = Y_1 \sin \omega_1 t + Y_2 \sin \omega_2 t$$

Solution: For $\zeta = 0.70$, $\phi \cong \pi/2 \times \omega/\omega_n$, so that $\phi_1 = \pi/2 \times \omega_1/\omega_n$ and $\phi_2 = \pi/2 \times \omega_2/\omega_n$. The output of the accelerometer is then

$$z = Z_1 \sin(\omega_1 t - \phi_1) + Z_2 \sin(\omega_2 t - \phi_2)$$

Substituting for Z_1 and Z_2 from Eq. (3.12-6), the output of the instrument is

$$z = \frac{1}{\omega_n^2}\left[\omega_1^2 Y_1 \sin \omega_1\left(t - \frac{\pi}{2\omega_n}\right) + \omega_2^2 Y_2 \sin \omega_2\left(t - \frac{\pi}{2\omega_n}\right)\right]$$

Since the time function $(t - \pi/2\omega_n)$ in both terms is equal, the shift of both components along the time axis is equal. Thus the instrument faithfully reproduces the acceleration $\ddot{y}$ without distortion. It is obvious that if ϕ_1 and ϕ_2 are both zero, we again obtain zero phase distortion.

PROBLEMS

3-1 A machine part of mass 1.95 kg vibrates in a viscous medium. Determine the damping coefficient when a harmonic exciting force of 24.46 N results in a resonant amplitude of 1.27 cm with a period of 0.20 s.

3-2 If the system of Prob. 3-1 is excited by a harmonic force of frequency 4 cps, what will be the percentage increase in the amplitude of forced vibration when the dashpot is removed?

3-3 A weight attached to a spring of stiffness 525 N/m has a viscous damping device. When the weight is displaced and released, the period of vibration is found to be 1.80 sec, and the ratio of consecutive amplitudes is 4.2 to 1.0. Determine the amplitude and phase when a force $F = 2\cos 3t$ acts on the system.

3-4 Show that for the dampled spring-mass system, the peak amplitude occurs at a frequency ratio given by the expression

$$\left(\frac{\omega}{\omega_n}\right)_p = \sqrt{1 - 2\zeta^2}$$

3-5 A spring-mass is excited by a force $F_0 \sin \omega t$. At resonance the amplitude is measured to be 0.58 cm. At 0.80 resonant frequency, the amplitude is measured to be 0.46 cm. Determine the damping factor ζ of the system.

3-6 Plot the real and imaginary parts of Eq. 3.1-17 for $\zeta = 0.01$ and 0.02

3-7 For the system shown in Fig. P3-7, set up the equation of motion and solve for the steady-state amplitude and phase angle by using complex algebra.

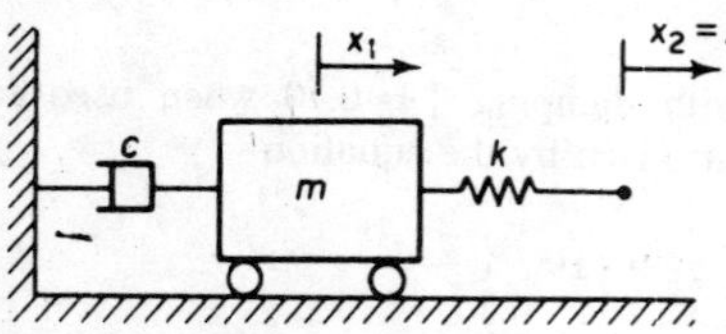

Figure P3-7.

3-8 Shown in Fig. P3-8 is a cylinder of mass m connected to a spring of stiffness k excited through viscous friction c to a piston with motion $y = A \sin \omega t$. Determine the amplitude of the cylinder motion and its phase with respect to the piston.

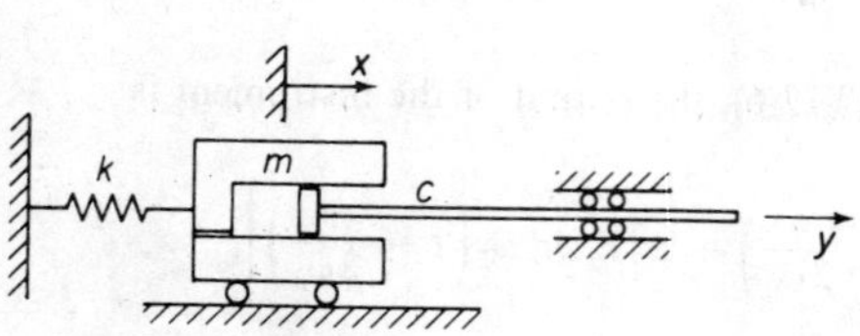

Figure P3-8.

3-9 A counterrotating eccentric mass exciter shown in Fig. P3-9 is used to determine the vibrational characteristics of a structure of mass 181.4 kg. At a speed of 900 rpm, a stroboscope shows the eccentric masses to be at the top at the instant the structure is moving upward through its static equilibrium position, and the corresponding amplitude is 21.6 mm. If the unbalance of each wheel of the exciter is 0.0921 kg m, determine (a) the natural frequency of the structure, (b) the damping factor of the structure, (c) the amplitude at 1200 rpm, and (d) the angular position of the eccentrics at the instant the structure is moving upward through its equilibrium position.

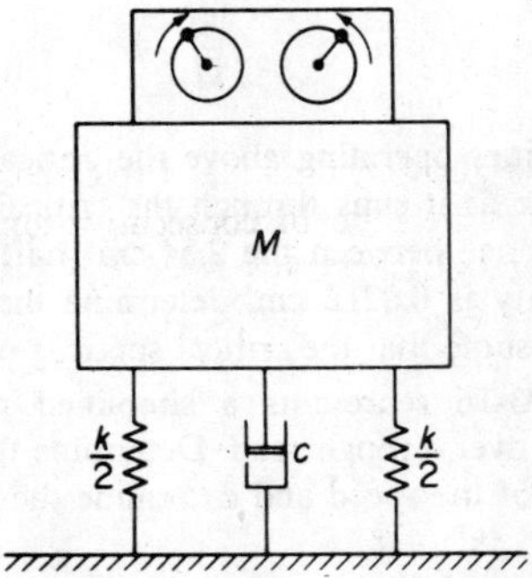

Figure P3-9.

3-10 Solve Eq. 3.2-1 for the complex amplitude, i.e., let $(me\omega^2)\sin \omega t = \overline{F}e^{i\omega t}$ and $x = Xe^{i(\omega t - \phi)} = (Xe^{-i\phi})e^{i\omega t} = \overline{X}e^{i\omega t}$.

3-11 A balanced wheel supported on springs, as shown in Fig. P3-11, is rotating at 1200 rpm. If a bolt weighing 15 g and located 5 cm from center suddenly comes loose and flies off, determine the buildup of vibration if the natural frequency of the system is 18 cps with damping of $\zeta = 0.10$.

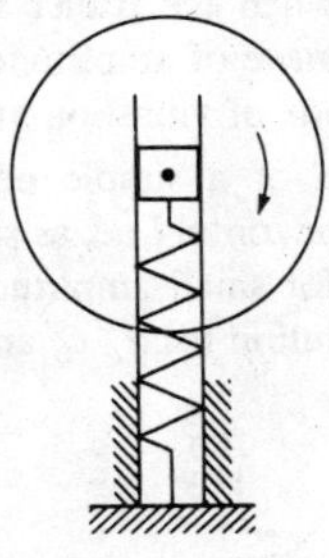

Figure P3-11.

3-12 A solid disk of weight 10 lb is keyed to the center of a $\frac{1}{2}$-in. steel shaft 2 ft between bearings. Determine the lowest critical speed. (Assume shaft to be simply supported at the bearings.)

3-13 Convert all units in Prob. 3-12 to the SI system and recalculate the lowest critical speed.

3-14 The rotor of a turbine 13.6 kg in mass is supported at the midspan of a shaft with bearings 0.4064 m apart, as shown in Fig. P3-14. The rotor is known to have an unbalance of 0.2879 kg · cm. Determine the forces exerted on the bearings at a speed of 6000 rpm if the diameter of the steel shaft is 2.54 cm. Compare this result with that of the same rotor mounted on a steel shaft of diameter [illegible]905 cm. (Assume the shaft to be simply supported at the bearings.)

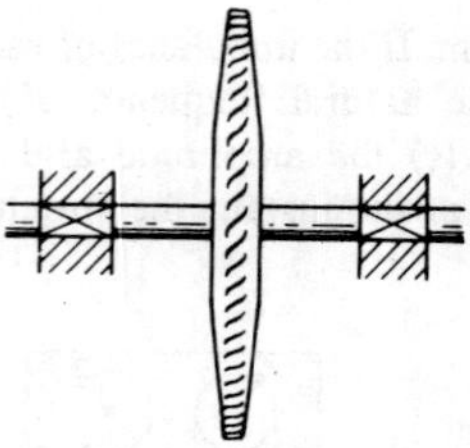

Figure P3-14.

3-15 For turbines operating above the critical speed, stops are provided to limit the amplitude as it runs through the critical speed. In the turbine of Prob. 3-14, if the clearance between the 2.54-cm shaft and the stops is 0.0508 cm, and if the eccentricity is 0.0212 cm, determine the time required for the shaft to hit the stops. Assume that the critical speed is reached with zero amplitude.

3-16 Figure P3-16 represents a simplified diagram of a spring-supported vehicle traveling over a rough road. Determine the equation for the amplitude of W as a function of the speed and determine the most unfavorable speed.

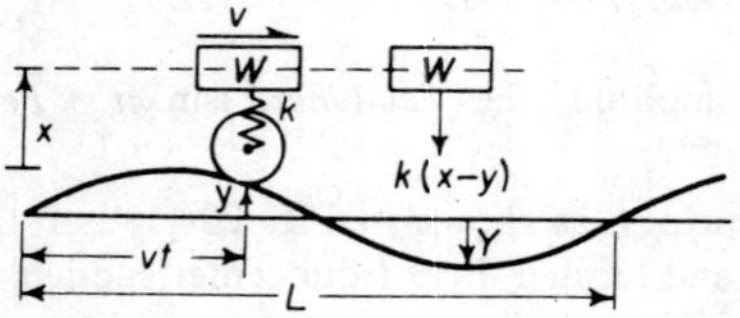

Figure P3-16.

3-17 The springs of an automobile trailer are compressed 10.16 cm under its weight. Find the critical speed when the trailer is traveling over a road with a profile approximated by a sine wave of amplitude 7.62 cm and wave length of 14.63 m. What will be the amplitude of vibration at 64.4 km/h? (Neglect damping.)

3-18 The point of suspension of a simple pendulum is given a harmonic motion $x_0 = X_0 \sin \omega t$ along a horizontal line, as shown in Fig. P3-18. Write the differential equation of motion for small amplitude of oscillation, using the coordinates shown. Determine the solution for x/x_0 and show that when $\omega = \sqrt{2}\,\omega_n$, the node

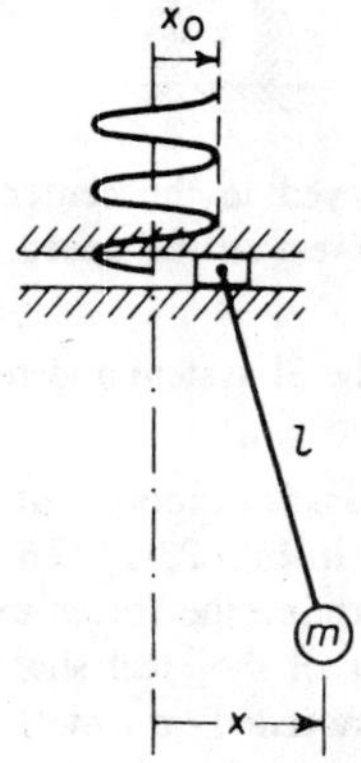

Figure P3-18.

is found at the midpoint of l. Show that in general the distance h from the mass to the node is given by the relation $h = l(\omega_n/\omega)^2$, where $\omega_n = \sqrt{g/l}$.

3-19 Derive Eqs. (3.5-8) and (3.5-9) for the amplitude and phase by letting $y = Y \sin \omega t$ and $x = X \sin(\omega t - \phi)$ in the differential equation (3.5-1).

3-20 An aircraft radio weighing 106.75 N is to be isolated from engine vibrations ranging in frequencies from 1600 cpm to 2200 cpm. What statical deflection must the isolators have for 85% isolation?

3-21 A refrigerator unit weighing 65 lb is to be supported by three springs of stiffness k lb/in. each. If the unit operates at 580 rpm, what should be the value of the spring constant k if only 10% of the shaking force of the unit is to be transmitted to the supporting structure?

3-22 An industrial machine of mass 453.4 kg is supported on springs with a statical deflection of 0.508 cm. If the machine has a rotating unbalance of 0.2303 kg · m, determine (a) the force transmitted to the floor at 1200 rpm and (b) the dynamical amplitude at this speed. (Assume damping to be negligible.)

3-23 If the machine of Prob. 3-22 is mounted on a large concrete block of mass 1136 kg and the stiffness of the springs or pads under the block is increased so that the statical deflection is still 0.508 cm, what will be the dynamical amplitude?

3-24 An electric motor of mass 68 kg is mounted on an isolator block of mass 1200 kg and the natural frequency of the total assembly is 160 cpm with a damping factor of $\zeta = 0.10$ (see Fig. P3-24). If there is an unbalance in the motor that results in a harmonic force of $F = 100 \sin 31.4t$, determine the amplitude of vibration of the block and the force transmitted to the floor.

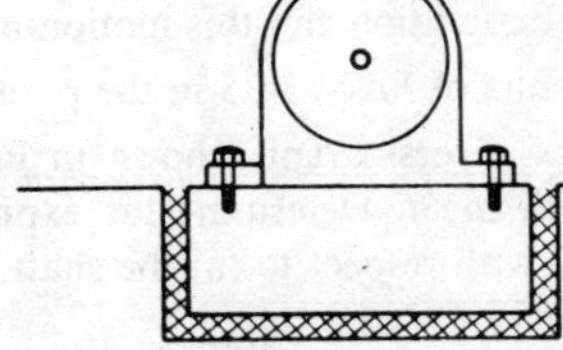

Figure P3-24.

3-25 A sensitive instrument with mass 113 kg is to be installed at a location where the acceleration is 15.24 cm/s^2 at a frequency of 20 Hz. It is proposed to mount the instrument on a rubber pad with the following properties: $k = 2802$ N/cm and $\zeta = 0.10$. What acceleration is transmitted to the instrument?

3-26 If the instrument of Prob. 3-25 can only tolerate an acceleration of 2.03 cm/s^2, suggest a solution assuming that the same rubber pad is the only isolator available. Give numerical values to substantiate your solution.

3-27 For the system shown in Fig. P3-27, verify that the transmissibility $TR = |x/y|$ is

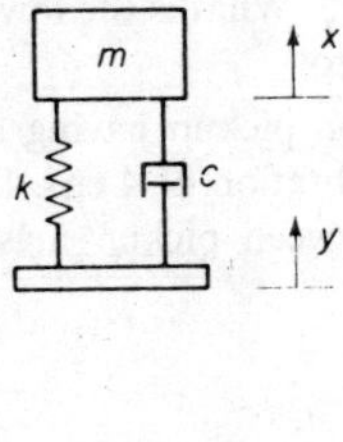

Figure P3-27.

the same as that for force. Plot the transmissibility in decibels, $20 \log|TR|$ versus ω/ω_n between $\omega/\omega_n = 1.50$ to 10 with $\zeta = 0.02, 0.04, \ldots, 0.10$.

3-28 Show that the energy dissipated per cycle for viscous friction can be expressed by

$$W_d = \frac{\pi F_0^2}{k} \frac{2\zeta(\omega/\omega_n)}{\left[1 - (\omega/\omega_n)^2\right]^2 + \left[2\zeta(\omega/\omega_n)\right]^2}$$

3-29 Show that for viscous damping, the loss factor η is independent of the amplitude and proportional to the frequency.

3-30 Express the equation for the free vibration of a single-DOF system in terms of the loss factor η at resonance.

3-31 Show that τ_n/τ_d plotted against ζ is a quarter circle where τ_d = damped natural period and τ_n = undamped natural period.

3-32 For small damping, the energy dissipated per cycle divided by the peak potential energy is equal to 2δ and also to $1/Q$. [See Eq. (3.7-6).] For viscous damping show that

$$\delta = \frac{\pi c \omega_n}{k}$$

3-33 In general, the energy loss per cycle is a function of both amplitude and frequency. State under what condition the logarithmic decrement δ is independent of the amplitude.

3-34 Coulomb damping between dry surfaces is a constant D always opposed to the motion. Determine the equivalent viscous damping.

3-35 Using the result of Prob. 3-34, determine the amplitude of motion of a spring mass system with Coulomb damping when excited by a harmonic force $F_0 \sin \omega t$. Under what condition can this motion be maintained?

3-36 Plot the results of Prob. 3-35 in the permissible range.

3-37 The shaft of a torsiograph, shown in Fig. P3-37, undergoes harmonic torsional oscillation $\theta_0 \sin \omega t$. Determine the expression for the relative amplitude of the outer wheel with respect to (a) the shaft, (b) a fixed reference.

Figure P3-37.

3-38 A commercial-type vibration pickup has a natural frequency of 4.75 cps and a damping factor $\zeta = 0.65$. What is the lowest frequency that can be measured with (a) 1% error, (b) 2% error?

3-39 An undamped vibration pickup having a natural frequency of 1 cps is used to measure a harmonic vibration of 4 cps. If the amplitude indicated by the pickup (relative amplitude between pickup mass and frame) is 0.052 cm, what is the correct amplitude?

3-40 A manufacturer of vibration measuring instruments gives the following specifications for one of its vibration pickups:
Frequency range: Velocity response flat from 10 cps to 1000 cps.
Sensitivity: 0.096 V/cm/s, both volts and velocity in rms values.
Amplitude range: Almost no lower limit to maximum stroke between stops of 0.60 in.

(a) This instrument was used to measure the vibration of a machine with a known frequency of 30 cps. If a reading of 0.024 V is indicated, determine the rms amplitude.

(b) Could this instrument be used to measure the vibration of a machine with known frequency of 12 cps and double amplitude of 0.80 cm? Give reasons.

3-41 A vibration pickup has a sensitivity of 40 mV/cm/s between $f = 10$ Hz to 2000 Hz. If 1 g acceleration is maintained over this frequency range, what will be the output voltage at (a) 10 Hz and (b) 2000 Hz?

3-42 Using the equations of harmonic motion, obtain the relationship for the velocity versus frequency applicable to the velocity pickup.

3-43 A vibration pickup has a sensitivity of 20 mV/cm/s. Assuming that 3 mV (rms) is the accuracy limit of the instrument, determine the upper frequency limit of the instrument for 1g excitation. What voltage would be generated at 200 Hz?

3-44 The sensitivity of a certain crystal accelerometer is given as 18 pC/g, with its capacitance equal to 450 pF. It is used with a vacuum tube voltmeter with connecting cable 5 m long with a capacitance of 50 pF/m. Determine its voltage output per g.

3-45 Specific damping capacity W_d/U is defined as the energy loss per cycle W_d divided by the peak potential energy $U = \frac{1}{2}kX^2$. Show that this quantity is equal to

$$\frac{W_d}{U} = 4\pi\zeta\left(\frac{\omega}{\omega_n}\right)$$

where $\zeta = c/c_{cr}$.

3-46 Logarithmic decrement δ for small damping is equal to $\delta \cong 2\pi\zeta$. Show that δ is related to the specific damping capacity by the equation

$$\frac{W_d}{U} = 2\delta\left(\frac{\omega}{\omega_n}\right)$$

3-47 For a system with hysteresis damping, show that the structural damping factor γ is equal to the loss factor at resonance.

3-48 For viscous damping, the complex frequency response can be written as

$$H(r) = \frac{1}{(1-r^2) + i(2\zeta r)}$$

where $r = \omega/\omega_n$ and $\zeta = c/c_{cr}$. Show that the plot of $H = x + iy$ leads to the equation

$$x^2 + \left(y + \frac{1}{4\zeta r}\right)^2 = \left(\frac{1}{4\zeta r}\right)^2$$

which cannot be a circle since the center and the radius depend on the frequency ratio.

4

TRANSIENT VIBRATION

When a dynamical system is excited by a suddenly applied nonperiodic excitation $F(t)$, the response to such excitation is called *transient response*, since steady-state oscillations are generally not produced. Such oscillations take place at the natural frequencies of the system with the amplitude varying in a manner dependent on the type of excitation.

We first study the response of a spring-mass system to an impulse excitation because this case is important in the understanding of the more general problem of transients.

4.1 IMPULSE EXCITATION

Impulse is the time integral of the force, and we designate it by the notation $\hat{F}$

$$\hat{F} = \int F(t)\, dt \tag{4.1-1}$$

We frequently encounter a force of very large magnitude, which acts for a very short time but with a time integral which is finite. Such forces are called *impulsive*.

Figure 4.1-1 shows an impulsive force of magnitude F/ϵ with a time duration of ϵ. As ϵ approaches zero, such forces tend to become infinite; however, the impulse defined by its time integral is $\hat{F}$, [illegible] considered to be finite. When $\hat{F}$ is equal to unity, such force in th[illegible]ase $\epsilon \to 0$ is called the *uni[illegible]pulse* or the *delta function*. A de[illegible] at $t = \xi$ is

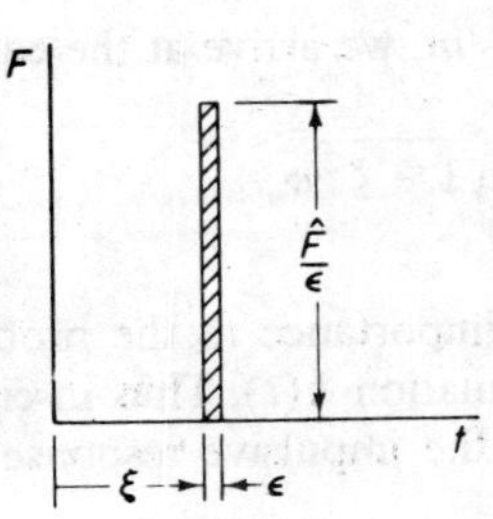

Figure 4.1-1.

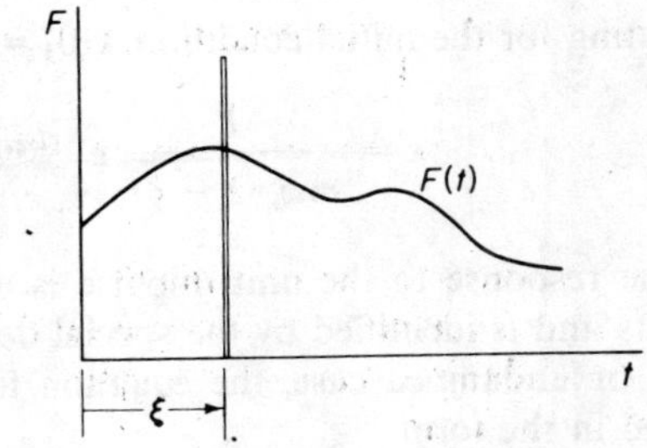

Figure 4.1-2.

identified by the symbol $\delta(t - \xi)$ and has the following properties:

$$\begin{aligned} \delta(t-\xi) &= 0 \qquad \text{for all } t \neq \xi \\ &= \text{greater than any assumed value for } t = \xi \end{aligned} \tag{4.1-2}$$

$$\int_0^\infty \delta(t-\xi)\,dt = 1.0 \qquad 0 < \xi < \infty$$

If $\delta(t - \xi)$ is multiplied by any time function $f(t)$, as shown in Fig. 4.1-2, the product will be zero everywhere except at $t = \xi$, and its time integral will be

$$\int_0^\infty f(t)\delta(t-\xi)\,dt = f(\xi) \qquad 0 < \xi < \infty \tag{4.1-3}$$

Since $F\,dt = m\,dv$, the impulse $\hat{F}$ acting on the mass will result in a sudden change in its velocity equal to $\hat{F}/m$ without an appreciable change in its displacement. Under free vibration we found that the undamped spring-mass system with initial conditions $x(0)$ and $\dot{x}(0)$ behaved according to the equation

$$x = \frac{\dot{x}(0)}{\omega_n}\sin\omega_n t + x(0)\cos\omega_n t$$

Hence the response of a spring-mass system initially at rest and excited by an impulse $\hat{F}$ is

$$x = \frac{\hat{F}}{m\omega_n}\sin\omega_n t = \hat{F}h(t) \tag{4.1-4}$$

where

$$h(t) = \frac{1}{m\omega_n}\sin\omega_n t \tag{4.1-5}$$

is the response to a unit impulse.

When damping is present we can start with the free vibration equation, Eq. (2.6-16), with $x(0) = 0$

$$x = \frac{\dot{x}(0)e^{-\zeta\omega_n t}}{\omega_n\sqrt{1-\zeta^2}}\sin\sqrt{1-\zeta^2}\,\omega_n t$$

Substituting for the initial condition $\dot{x}(0) = \hat{F}/m$, we arrive at the equation

$$x = \frac{\hat{F}}{m\omega_n\sqrt{1-\zeta^2}} e^{-\zeta\omega_n t} \sin\sqrt{1-\zeta^2}\,\omega_n t \tag{4.1-6}$$

The response to the unit impulse is of importance to the problems of transients and is identified by the special designation $h(t)$. Thus in either the damped or undamped case, the equation for the impulsive response can be expressed in the form

$$x = \hat{F}h(t) \tag{4.1-7}$$

where the right side of the equation is given by either Eq. (4.1-4) or (4.1-6).

4.2 ARBITRARY EXCITATION

Having the response $h(t)$ to a unit impulse excitation, it is possible to establish the equation for the response of the system excited by an arbitrary force $f(t)$. For this development, we consider the arbitrary force to be a series of impulses as shown in Fig. 4.2-1. If we examine one of the impulses (shown crosshatched) at time $t = \xi$, its strength is

$$\hat{F} = f(\xi)\Delta\xi$$

and its contribution to the response at time t is dependent upon the elapsed

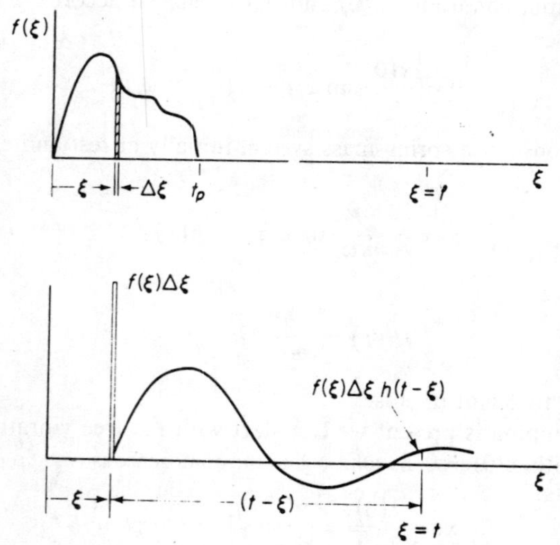

Figure 4.2-1.

time $(t - \xi)$, or

$$f(\xi)\Delta\xi\, h(t-\xi)$$

where $h(t - \xi)$ is the response to a unit impulse started at $t = \xi$. Since the systen we are considering is linear, the principle of superposition holds. Thus, by combining all such contributions, the response to the arbitrary excitation $f(t)$ is represented by the integral

$$x(t) = \int_0^t f(\xi)h(t-\xi)\,d\xi \tag{4.2-1}$$

The above integral is called the *Convolution integral* or is sometimes referred to as the *superposition integral.*

Example 4.2-1

Determine the response of a single-DOF system to the step excitation shown in Fig. 4.2-2.

Figure 4.2-2. Step function excitation.

Solution: Considering the undamped system, we have

$$h(t) = \frac{1}{m\omega_n}\sin\omega_n t$$

Substituting into Eq. (4.2-1) the response of the undamped system is

$$x(t) = \frac{F_0}{m\omega_n}\int_0^t \sin\omega_n(t-\xi)\,d\xi$$

$$= \frac{F_0}{k}(1 - \cos\omega_n t)$$

This result indicates that the peak response to the step excitation of magnitude F_0 is equal to twice the statical deflection.

For a damped system the procedure can be repeated with

$$h(t) = \frac{e^{-\zeta\omega_n t}}{m\omega_n\sqrt{1-\zeta^2}}\sin\sqrt{1-\zeta^2}\,\omega_n t$$

or, alternatively, we can simply consider the differential equation

$$\ddot{x} + 2\zeta\omega_n\dot{x} + \omega_n^2 x = \frac{F}{m}$$

whose solution is the sum of the solutions to the homogeneous equation and that of the particular solution, which for this case is $F_0/m\omega_n^2$. Thus the equation

$$x(t) = Xe^{-\zeta\omega_n t}\sin\left(\sqrt{1-\zeta^2}\,\omega_n t - \phi\right) + \frac{F_0}{m\omega_n^2}$$

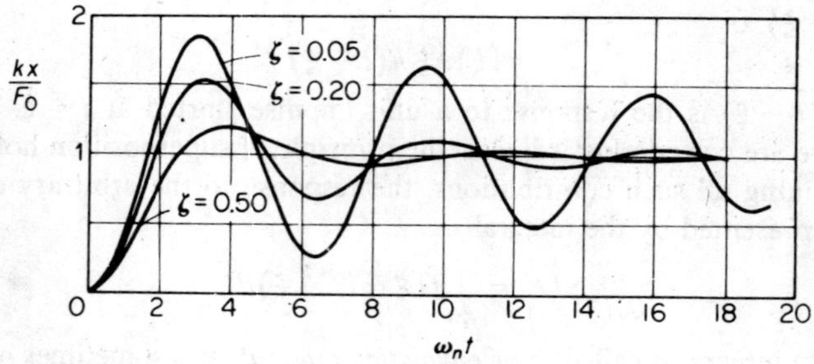

Figure 4.2-3. Response to a unit step function.

fitted to the initial conditions of $x(0) = \dot{x}(0) = 0$ will result in the solution, which is given as

$$x = \frac{F_0}{k}\left[1 - \frac{e^{-\zeta\omega_n t}}{\sqrt{1-\zeta^2}}\cos\left(\sqrt{1-\zeta^2}\,\omega_n t - \psi\right)\right]$$

where

$$\tan\psi = \frac{\zeta}{\sqrt{1-\zeta^2}}$$

Figure 4.2-3 shows a plot of xk/F_0 versus $\omega_n t$ with ζ as parameter, and it is evident that the peak response is less than $2F_0/k$ when damping is present.

Base excitation. Often the support of the dynamical system is subjected to a sudden movement specified by its displacement, velocity, or acceleration. The equation of motion can then be expressed in terms of the relative displacement $z = x - y$ as follows

$$\ddot{z} + 2\zeta\omega_n\dot{z} + \omega_n^2 z = -\ddot{y} \tag{4.2-4}$$

and hence all of the results for the force-excited system apply to the base-excited system for z when the term F/m is replaced by $-\ddot{y}$ or the negative of the base acceleration.

For an undamped system initially at rest, the solution for the relative displacement becomes

$$z = -\frac{1}{\omega_n}\int_0^t \ddot{y}(\xi)\sin\omega_n(t-\xi)\,d\xi \tag{4.2-5}$$

Example 4.2-2

Consider an undamped spring-mass system where the motion of the base is specified by a velocity pulse of the form

$$\dot{y}(t) = v_0 e^{-t/t_0}u(t)$$

where $u(t)$ is a unit step function. The velocity together with its time rate of change is shown in Fig. 4.2-4.

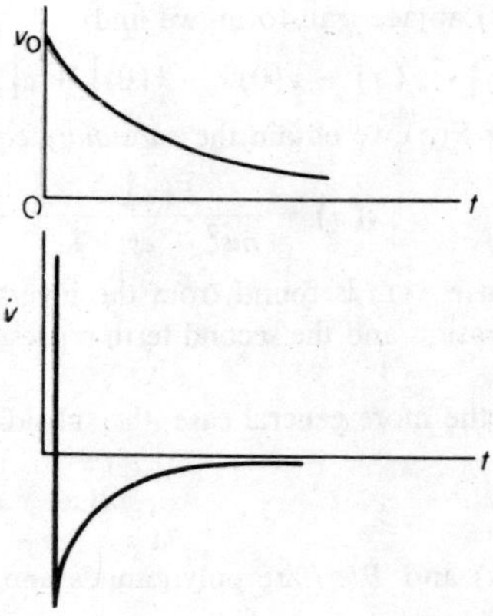

Figure 4.2-4.

Solution: The velocity pulse at $t = 0$ has a sudden jump from zero to v_0, and its rate of change (or acceleration) is infinite. Differentiating $\dot{y}(t)$ and recognizing that $(d/dt)u(t) = \delta(t)$, a delta function at the origin, we obtain

$$\ddot{y} = v_0 e^{-t/t_0}\delta(t) - \frac{v_0}{t_0}e^{-t/t_0}u(t)$$

Substituting $\ddot{y}$ into Eq. (4.2-5) the result is

$$\begin{aligned} z(t) &= -\frac{v_0}{\omega_n}\int_0^t\left[\delta(\xi) - \frac{1}{t_0}e^{-\xi/t_0}u(\xi)\right]\sin\omega_n(t-\xi)\,d\xi \\ &= -\frac{v_0}{\omega_n}\int_0^t\delta(\xi)\sin\omega_n(t-\xi)\,d\xi + \frac{v_0}{\omega_n t_0}\int_0^t e^{-\xi/t_0}\sin\omega_n(t-\xi)\,d\xi \\ &= \frac{v_0 t_0}{1+(\omega_n t_0)^2}\left[e^{-t/t_0} - \omega_n t_0 \sin\omega_n t - \cos\omega_n t\right] \end{aligned}$$

4.3 LAPLACE TRANSFORM FORMULATION

The Laplace transform method of solving the differential equation provides a complete solution, yielding both transient and forced vibration. For those unfamiliar with this method, a brief presentation of the Laplace transform theory is given in the appendix. In this section we illustrate its use by some simple examples.

Example 4.3-1

Formulate the Laplace transform solution of a viscously damped spring-mass system with initial conditions $x(0)$ and $\dot{x}(0)$.

Solution: The equation of motion of the system excited by an arbitrary force $F(t)$ is

$$m\ddot{x} + c\dot{x} + kx = F(t)$$

Taking its Laplace transform, we find

$$m\left[s^2\bar{x}(s) - x(0)s - \dot{x}(0)\right] + c\left[s\bar{x}(s) - x(0)\right] + k\bar{x}(s) = \bar{F}(s)$$

Solving for $\bar{x}(s)$ we obtain the *subsidiary equation*

$$\bar{x}(s) = \frac{\bar{F}(s)}{ms^2 + cs + k} + \frac{(ms + c)x(0) + m\dot{x}(0)}{ms^2 + cs + k} \tag{a}$$

The response $x(t)$ is found from the inverse of Eq. (a); the first term represents the forced vibration and the second term represents the transient solution due to the initial conditions.

For the more general case, the subsidiary equation can be written in the form

$$\bar{x}(s) = \frac{A(s)}{B(s)} \tag{b}$$

where $A(s)$ and $B(s)$ are polynomials and $B(s)$, in general, is of higher order than $A(s)$.

If only the forced solution is considered, we can define the *impedance transform* as

$$\frac{\bar{F}(s)}{\bar{x}(s)} = z(s) = ms^2 + cs + k \tag{c}$$

Its reciprocal is the admittance transform

$$H(s) = \frac{1}{z(s)} \tag{d}$$

Frequently a block diagram is used to denote input and output, as shown in Fig. 4.3-1. The admittance transform $H(s)$ then can also be considered as the *system transfer function*, defined as the ratio in the subsidiary plane of the output over the input with all initial conditions equal to zero.

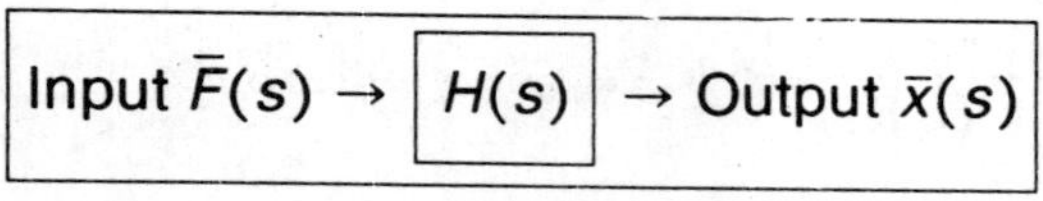

Figure 4.3-1. Block diagram.

Example 4.3-2

The question of how far a body can be dropped without incurring damage is of frequent interest. Such considerations are of paramount importance in the landing of airplanes or the cushioning of packaged articles. In this example we shall discuss some of the elementary aspects of this problem by idealizing the mechanical system in terms of linear spring-mass components.

Consider the spring-mass system of Fig. 4.3-2 dropped through a height h. If x is measured from the position of m at the instant $t = 0$ when the spring first contacts the floor, the differential equation of motion for m applicable as long as the spring remains in contact with the floor is

$$m\ddot{x} + kx = mg \tag{a}$$

Taking the Laplace transform of this equation with the initial conditions $x(0) = 0$ and

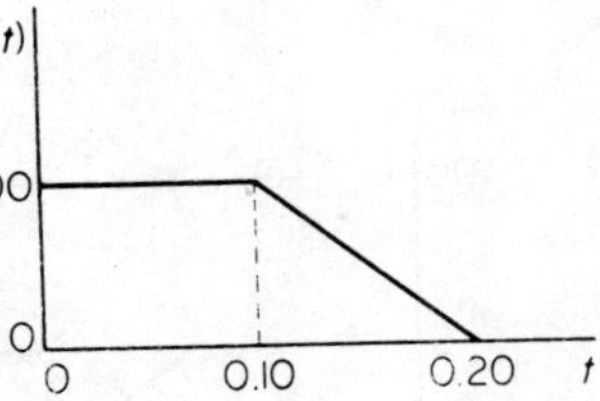

Figure 4.5-2.

with initial conditions

$$x_1 = \dot{x}_1 = 0$$

and forcing function as shown in Fig. 4.5-2.

Solution: The natural period of the system is first found as

$$\omega = \frac{2\pi}{\tau} = \sqrt{\frac{2000}{4}} = 22.36 \text{ rad/s}$$

$$\tau = \frac{2\pi}{22.36} = 0.281 \text{ s}$$

According to the rule $h \leq \tau/10$ and for convenience for representing $F(t)$, we will choose $h = 0.020$ s.

From the differential equation we have

$$\ddot{x} = f(x, t) = \frac{1}{4}F(t) - 500x$$

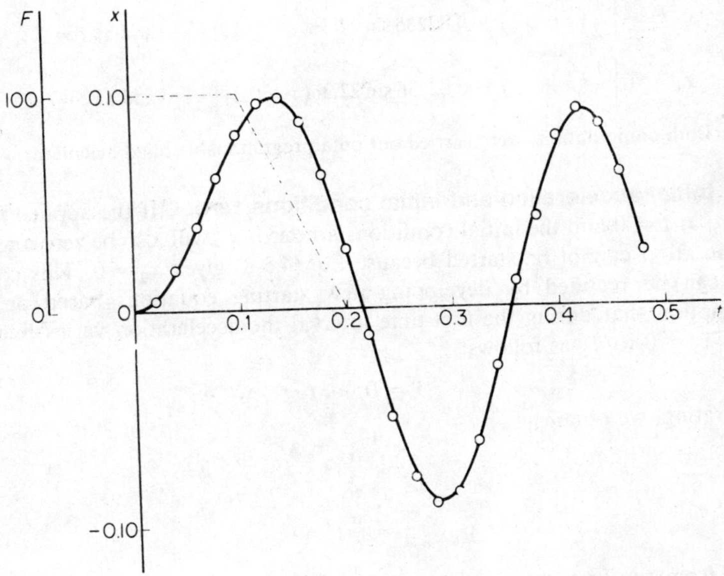

Figure 4.5-3.

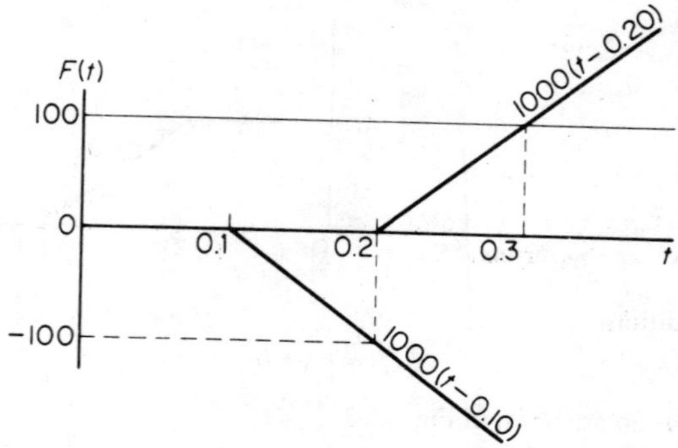

Figure 4.5-4.

Eq. (4.5-8) gives $x_2 = \frac{1}{2}(25)(0.02)^2 = 0.005$. x_3 is then found from Eq. (4.5-7).

$$x_3 = 0.005 - 0 + (0.02)^2(25 - 500 \times 0.005) = 0.0190$$

The following values of x_4, x_5, etc. are now available from Eq. (4.5-7).

Figure 4.5-3 shows the computed values compared with the exact solution. The latter was obtained by the superposition of the solutions for the step function and the ramp function in the following manner. Figure 4.5-4 shows the superposition of forces. The equations to be superimposed for the exact solution are

$$x_1 = 0.05(1 - \cos 22.36t) \qquad 0 \leq t \leq 0.1$$

$$x_2 = -\left[\frac{1}{2}(t - 0.1) - 0.02236 \sin 22.36(t - 0.10)\right] \qquad \text{add at } t = 0.1$$

$$x_3 = +\left[\frac{1}{2}(t - 0.2) - 0.02236 \sin 22.36(t - 0.2)\right] \qquad \text{add at } t = 0.2$$

Both computations were carried out on a programmable hand calculator.

Initial acceleration and initial conditions zero. If the applied force is zero at $t = 0$ and the initial conditions are zero, $\ddot{x}_1$ will also be zero and the computation cannot be started because Eq. (4.5-8) gives $x_2 = 0$. This condition can be rectified by developing new starting equations based on the assumption that during the first-time interval the acceleration varies linearly from $\ddot{x}_1 = 0$ to $\ddot{x}_2$ as follows:

$$\ddot{x} = 0 + \alpha t$$

Integrating, we obtain

$$\dot{x} = \frac{\alpha}{2}t^2$$

$$x = \frac{\alpha}{6}t^3$$

Since from the first equation, $\ddot{x}_2 = \alpha h$, where $h = \Delta t$, the second and third equations become

$$\dot{x}_2 = \frac{h}{2}\ddot{x}_2 \tag{4.5-9}$$

$$x_2 = \frac{h^2}{6}\ddot{x}_2 \tag{4.5-10}$$

Substituting these equations into the differential equation at time $t_2 = h$ enables one to solve for $\ddot{x}_2$ and x_2. Example 4.5-2 illustrates the situation encountered here.

Example 4.5-2

Use the digital computer to solve the problem of a spring-mass system excited by a triangular pulse. The differential equation of motion and the initial conditions are given as

$$0.5\ddot{x} + 8\pi^2 x = F(t)$$

$$x_1 = \dot{x}_1 = 0$$

The triangular force is defined in Fig. 4.5-5.

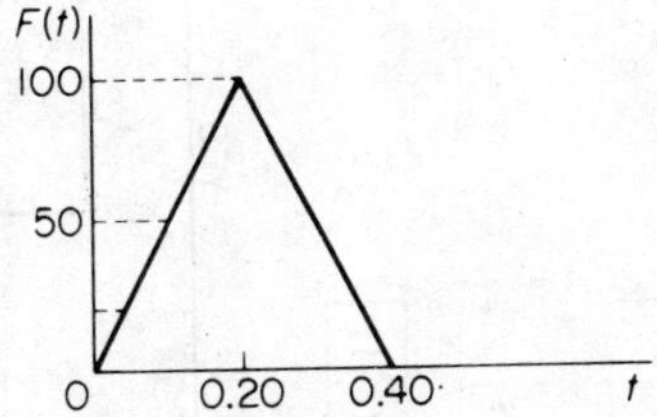

Figure 4.5-5.

Solution: The natural period of the system is

$$\tau = \frac{2\pi}{\omega} = \frac{2\pi}{4\pi} = 0.50$$

The time increment will be chosen as $h = 0.05$, and the differential equation is reorganized as

$$\ddot{x} = f(x, t) = 2F(t) - 16\pi^2 x$$

This equation is to be solved together with the recurrence equation, Eq. (4.5-7),

$$x_{i+1} = 2x_i - x_{i-1} + h^2 f(x, t)$$

Since the force and the acceleration are zero at $t = 0$, it is necessary to start the computational process with Eqs. (4.5-9) and (4.5-10) and the differential equation.

$$x_2 = \tfrac{1}{6}\ddot{x}_2(0.05)^2 = 0.000417\ddot{x}_2$$

$$\ddot{x}_2 = 2F(0.05) - 16\pi^2 x_2 = 50 - 158x_2$$

Their simultaneous solution leads to

$$x_2 = \frac{(0.05)^2 F(0.05)}{3 + 8\pi^2(0.05)^2} = 0.0195$$

$$\ddot{x}_2 = 46.91$$

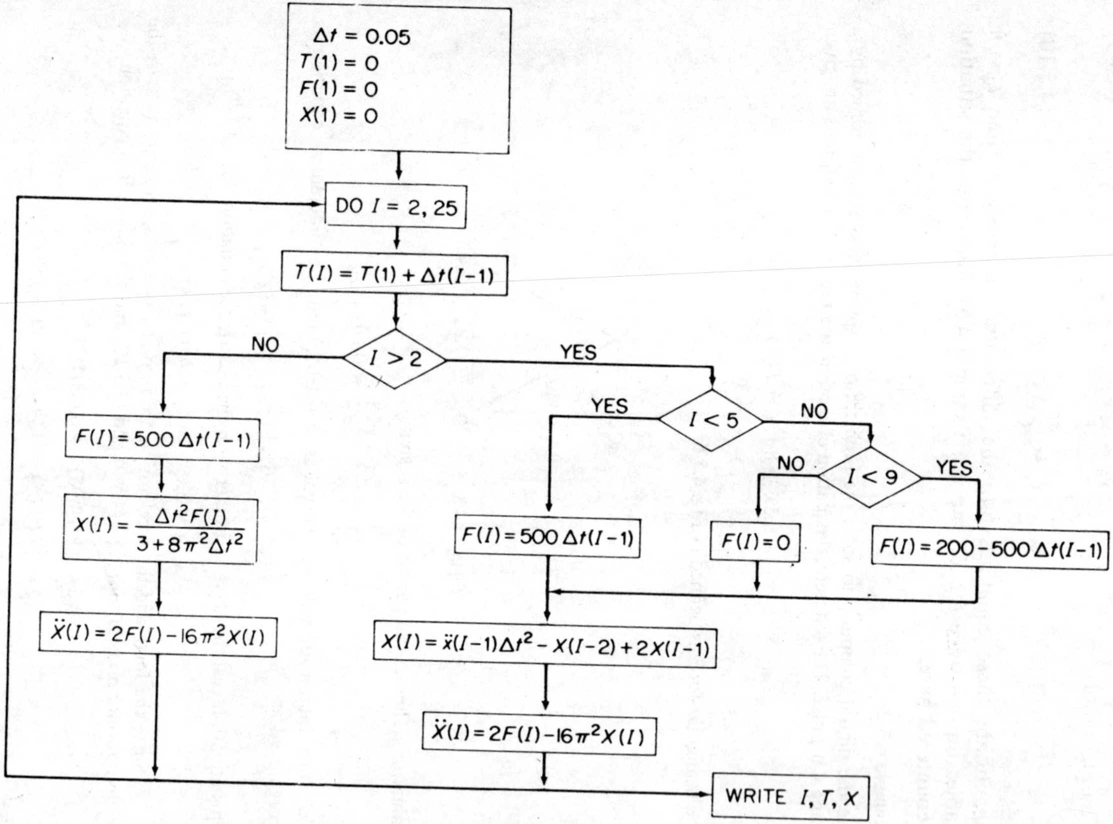
Δt = 0.05
T(1) = 0
F(1) = 0
X(1) = 0
DO I = 2, 25
T(I) = T(1) + Δt(I−1)
I > 2
NO
YES
F(I) = 500 Δt(I−1)
X(I) = Δt²F(I) / (3+8π²Δt²)
Ẍ(I) = 2F(I) − 16π²X(I)
I < 5
YES
NO
I < 9
NO
YES
F(I) = 500 Δt(I−1)
F(I) = 0
F(I) = 200 − 500 Δt(I−1)
X(I) = ẍ(I−1)Δt² − X(I−2) + 2X(I−1)
Ẍ(I) = 2F(I) − 16π²X(I)
WRITE I, T, X

Figure 4.5-6.

J	TIME	DISPL	ACCLRTN	FORCE
1	0.0	0.0	0.0	0.0
2	0.0500	0.020	46.91	25.00
3	0.1000	0.156	75.31	50.00
4	0.1500	0.481	73.97	75.00
5	0.2000	0.992	43.44	100.00
6	0.2500	1.610	-104.25	75.00
7	0.3000	1.968	-210.78	50.00
8	0.3500	1.799	-234.10	25.00
9	0.4000	1.045	-165.01	0.00
10	0.4500	-0.122	19.22	0.0
11	0.5000	-1.240	195.86	0.0
12	0.5500	-1.869	295.19	0.0
13	0.6000	-1.760	277.98	0.0
14	0.6500	-0.957	151.04	0.0
15	0.7000	0.225	-35.52	0.0
16	0.7500	1.318	-208.06	0.0
17	0.8000	1.890	-298.47	0.0
18	0.8500	1.717	-271.05	0.0
19	0.9000	0.865	-136.64	0.0
20	0.9500	-0.328	51.72	0.0
21	1.0000	-1.391	219.66	0.0
22	1.0500	-1.906	300.89	0.0
23	1.1000	-1.668	263.33	0.0
24	1.1500	-0.772	121.83	0.0
25	1.2000	0.429	-67.77	0.0

Figure 4.5-7.

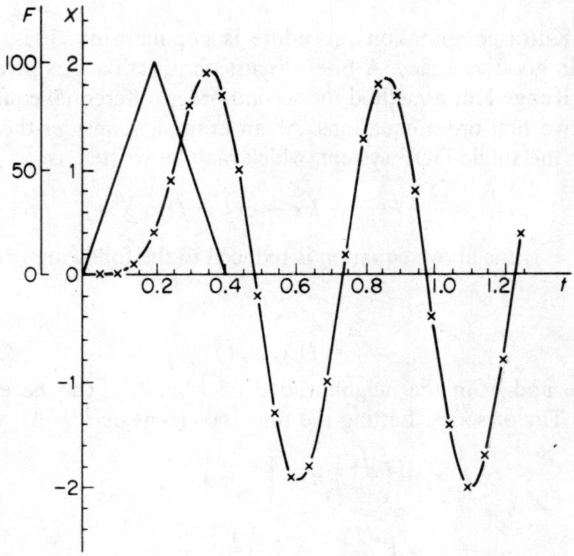

Figure 4.5-8.

The flow diagram for the computation is shown in Fig. 4.5-6. With $h = 0.05$, the time duration for the force must be divided into regions $I = 1$ to 5, $I = 6$ to 9 and $I > 9$. The index I controls the computation path on the diagram.

Shown in Figs. 4.5-7 and 4.5-8 are the computer solution and results printed on a machine. A smaller Δt would have resulted in a smoother plot.

Damped system. When damping is present, the differential equation contains an additional term $\dot{x}_i$ and Eq. (4.5-7) is replaced by

$$x_{i+1} = 2x_i - x_{i-1} + h^2 f(x_i, \dot{x}_i, t_i) \qquad i \geq 2 \tag{4.5-7'}$$

We now need to calculate the velocity at each step as well as the displacement.

Considering again the first three terms of the Taylor series, Eq. (4.5-4), we see that x_2 is available from the expansion of x_{i+1} with $i = 1$

$$x_2 = x_1 + \dot{x}_1 h + \frac{h^2}{2} f(x_1, \dot{x}_1, t_1)$$

The quantity $\dot{x}_2$ is found from the second equation for x_{i-1} with $i = 2$

$$x_1 = x_2 - \dot{x}_2 h + \frac{h^2}{2} f(x_1, \dot{x}_2, t_2)$$

With these results, x_3 can be calculated from Eq. (4.5-7′). The procedure is thus repeated for other values of x_i and $\dot{x}_i$ using the Taylor series.

4.6 RUNGE-KUTTA METHOD *(Method 2)*

The Runge-Kutta computation procedure is popular since it is self-starting and results in good accuracy. A brief discussion of its basis is presented here.

In the Runge-Kutta method the second-order differential equation is first reduced to two first-order equations. As an example, consider the differential equation for the single-DOF system, which may be written as

$$\ddot{x} = \frac{1}{m}[F(t) - kx - c\dot{x}] = f(x, \dot{x}, t) \tag{4.6-1}$$

By letting $\dot{x} = y$, the above equation is reduced to the following two first-order equations:

$$\begin{aligned} \dot{x} &= y \\ \dot{y} &= f(x, y, t) \end{aligned} \tag{4.6-2}$$

Both x and y in the neighborhood of x_i and y_i can be expressed in terms of the Taylor series. Letting the time increment be $h = \Delta t$, we have

$$\begin{aligned} x &= x_i + \left(\frac{dx}{dt}\right)_i h + \left(\frac{d^2x}{dt^2}\right)_i \frac{h^2}{2} + \cdots \\ y &= y_i + \left(\frac{dy}{dt}\right)_i h + \left(\frac{d^2y}{dt^2}\right)_i \frac{h^2}{2} + \cdots \end{aligned} \tag{4.6-3}$$

Instead of using these expressions, it is possible to replace the first derivative by an average slope and ignore higher-order derivatives

$$\begin{aligned} x &= x_i + \left(\frac{dx}{dt}\right)_{iav} h \\ y &= y_i + \left(\frac{dy}{dt}\right)_{iav} h \end{aligned} \tag{4.6-4}$$

If we used Simpson's rule, the average slope in the interval h becomes, i.e.

$$\left(\frac{dy}{dt}\right)_{iav} = \frac{1}{6}\left[\left(\frac{dy}{dt}\right)_{t_i} + 4\left(\frac{dy}{dt}\right)_{t_i+h/2} + \left(\frac{dy}{dt}\right)_{t_i+h}\right]$$

The Runge-Kutta method is very similar to the preceding computations, except that the center term of the given equation is split into two terms and four values of t, x, y, and f are computed for each point i as follows:

t	x	$y = \dot{x}$	$f = \dot{y} = \ddot{x}$
$T_1 = t_i$	$X_1 = x_i$	$Y_1 = y_i$	$F_1 = f(T_1, X_1, Y_1)$
$T_2 = t_i + \frac{h}{2}$	$X_2 = x_i + Y_1\frac{h}{2}$	$Y_2 = y_i + F_1\frac{h}{2}$	$F_2 = f(T_2, X_2, Y_2)$
$T_3 = t_i + \frac{h}{2}$	$X_3 = x_i + Y_2\frac{h}{2}$	$Y_3 = y_i + F_2\frac{h}{2}$	$F_3 = f(T_3, X_3, Y_3)$
$T_4 = t_i + h$	$X_4 = x_i + Y_3 h$	$Y_4 = y_i + F_3 h$	$F_4 = f(T_4, X_4, Y_4)$

These quantities are then used in the following recurrence formula:

$$x_{i+1} = x_i + \frac{h}{6}[Y_1 + 2Y_2 + 2Y_3 + Y_4] \tag{4.6-5}$$

$$y_{i+1} = y_i + \frac{h}{6}[F_1 + 2F_2 + 2F_3 + F_4] \tag{4.6-6}$$

where it is recognized that the four values of Y divided by 6 represent an average slope dx/dt and the four values of F divided by 6 results in an average of dy/dt as defined by Eqs. (4.6-4).

Example 4.6-1

Solve Example 4.5-1 by the Runge-Kutta method.

Solution: The differential equation of motion is

$$\ddot{x} = \frac{1}{4}F(t) - 500x$$

Let $y = \dot{x}$; then

$$\dot{y} = f(x, t) = \frac{1}{4}F(t) - 500x$$

With $h = 0.02$, the following table is calculated:

	t	x	$y = \dot{x}$	f
$t_1 =$	0	0	0	25
	0.01	0	0.25	25
	0.01	0.0025	0.25	23.75
$t_2 =$	0.02	0.0050	0.475	22.50

The calculation for x_2 and y_2 follows.

$$x_2 = 0 + \frac{0.02}{6}(0 + 0.50 + 0.50 + 0.475) = 0.00491667$$

$$y_2 = 0 + \frac{0.02}{6}(25 + 50 + 47.50 + 22.50) = 0.4833333$$

To continue to point 3, we repeat the above table:

$t_2 =$	0.02	0.00491667	0.4833333	22.541665
	0.03	0.0097500	0.70874997	20.12500
	0.03	0.01200417	0.6845833	18.997915
$t_3 =$	0.04	0.01860834	0.8632913	15.695830

We then calculate x_3 and y_3

$$
\begin{aligned}
x_3 &= 0.00491667 \\
&\quad + \frac{0.02}{6}(0.483333 + 1.4174999 + 1.3691666 + 0.8632913) \\
&= 0.00491667 + 0.01377764 = 0.01869431 \\
y_3 &= 0.483333 + 0.38827775 = 0.87161075
\end{aligned}
$$

To complete the calculation, the problem was programmed on a digital computer, and the results showed excellent accuracy. Table 4.6-1 gives the numerical values for the central difference and the Runge-Kutta methods compared with the analytical solution. We see that the Runge-Kutta method gives greater accuracy than the central difference method.

TABLE 4.6-1

COMPARISON OF METHODS FOR PROBLEM 4.5-1

Time t	*Exact Solution*	*Central Difference*	*Runge-Kutta*
0	0	0	0
0.02	0.00492	0.00500	0.00492
0.04	0.01870	0.01900	0.01869
0.06	0.03864	0.03920	0.03862
0.08	0.06082	0.06159	0.06076
0.10	0.08086	0.08167	0.08083
0.12	0.09451	0.09541	0.09447
0.14	0.09743	0.09807	0.09741
0.16	0.08710	0.08712	0.08709
0.18	0.06356	0.06274	0.06359
0.20	0.02949	0.02782	0.02956
0.22	−0.01005	−0.01267	−0.00955
0.24	−0.04761	−0.05063	−0.04750
0.26	−0.07581	−0.07846	−0.07571
0.28	−0.08910	−0.09059	−0.08903
0.30	−0.08486	−0.08461	−0.08485
0.32	−0.06393	−0.06171	−0.06400
0.34	−0.03043	−0.02646	−0.03056
0.36	0.00906	0.01407	0.00887
0.38	0.04677	0.05180	0.04656
0.40	0.07528	0.07916	0.07509
0.42	0.08898	0.09069	0.08886
0.44	0.08518	0.08409	0.08516
0.46	0.06463	0.06066	0.06473
0.48	0.03136	0.02511	0.03157

Although the Runge-Kutta method does not require the evaluation of derivatives beyond the first, its higher accuracy is achieved by four evaluations of the first derivatives to obtain agreement with the Taylor series solution through terms of order h^4. Moreover, the versatility of the Runge-Kutta method is evident in the fact that by replacing the variable by a vector, the same method is applicable to a system of differential equations. For example, the first-order equation of one variable is

$$\dot{x} = f(x, t)$$

For two variables, x and y, as in this problem, we can let $z = \begin{Bmatrix} x \\ y \end{Bmatrix}$ and write the two first-order equations as

$$\begin{Bmatrix} \dot{x} \\ \dot{y} \end{Bmatrix} = \begin{Bmatrix} y \\ f(x, y, t) \end{Bmatrix} = F(x, y, t)$$

or

$$\dot{z} = F(x, y, t)$$

Thus the above vector equation is identical in form to the equation in one variable and can be treated in the same manner.

REFERENCES

[1] CRANDALL, S. H. "Engineering Analysis," *A Survey of Numerical Procedures*. New York: McGraw-Hill Book Company, 1956.

[2] RALSTON, A., AND WILF, H. S. *Mathematical Methods for Digital Computers*, Vols. I and II. New York: John Wiley & Sons, Inc., 1968.

[3] SALVADORI, M. G., AND BARON, M. L. *Numerical Methods in Engineering*. Englewood Cliffs, N.J.: Prentice-Hall, Inc., 1952.

[4] BATHE, K-J., AND WILSON, E. L. *Numerical Methods in Finite Element Analysis*. Englewood Cliffs, N.J.: Prentice-Hall, Inc., 1976.

[5] MEIROVITCH, L. *Computational Methods in Structural Dynamics*. Rockville, Maryland: Sijthoff & Noordhoff, 1980.

PROBLEMS

4-1 Show that the time t_p corresponding to the peak response for the impulsively excited spring-mass system is given by the equation

$$\tan \sqrt{1 - \zeta^2}\, \omega_n t_p = \sqrt{1 - \zeta^2} / \zeta$$

4-2 Determine the peak displacement for the impulsively excited spring-mass system, and show that it can be expressed in the form

$$\frac{x_{\text{peak}} \sqrt{km}}{\hat{F}} = \exp\left(- \frac{\zeta}{\sqrt{1 - \zeta^2}} \tan^{-1} \frac{\sqrt{1 - \zeta^2}}{\zeta} \right)$$

Plot this result as a function of ζ.

4-3 Show that the time t_p corresponding to the peak response of the damped spring-mass system excited by a step force F_0 is $\omega_n t_p = \pi/\sqrt{1-\zeta^2}$

4-4 For the system of Prob. 4-3, show that the peak response is equal to

$$\left(\frac{xk}{F_0}\right)_{\max} = 1 + \exp\left(-\frac{\zeta\pi}{\sqrt{1-\zeta^2}}\right)$$

4-5 A rectangular pulse of height F_0 and duration t_0 is applied to an undamped spring-mass system. Considering the pulse to be the sum of two step pulses, as shown in Fig. P4-5, determine its response for $t > t_0$ by the superposition of the undamped solutions.

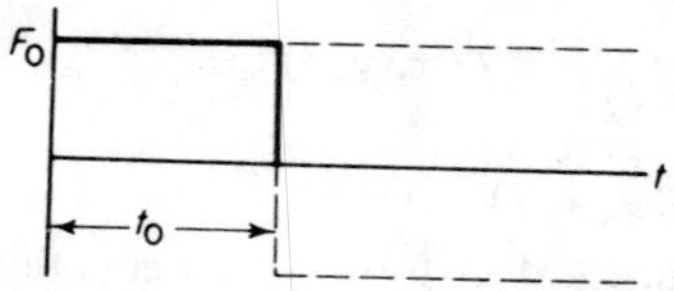

Figure P4-5.

4-6 If an arbitrary force $f(t)$ is applied to an undamped oscillator which has initial conditions other than zero, show that the solution must be in the form

$$x(t) = x_0 \cos \omega_n t + \frac{v_0}{\omega_n} \sin \omega_n t + \frac{1}{m\omega_n}\int_0^t f(\xi) \sin \omega_n (t-\xi)\, d\xi$$

4-7 Show that the response to a unit step function, designated by $g(t)$, is related to the impulsive response $h(t)$ by the equation $h(t) = \dot{g}(t)$.

4-8 Show that the convolution integral can also be written in terms of $g(t)$ as

$$x(t) = f(0)\, g(t) + \int_0^t \dot{f}(\xi)\, g(t-\xi)\, d\xi$$

where $g(t)$ is the response to a unit step function.

4-9 In Sec. 4.3 the subsidiary equation for the viscously damped spring-mass system was given by Eq. (a). Evaluate the second term due to initial conditions by the inverse transforms.

4-10 An undamped spring-mass system is given a base excitation of $\dot{y}(t) = 20(1 - 5t)$. If the natural frequency of the system is $\omega_n = 10 \text{ s}^{-1}$, determine the maximum relative displacement.

4-11 A sinusoidal pulse is considered to be the superposition of two sine waves as shown in Fig. P4-11. Show that its solution is

$$\left(\frac{xk}{F_0}\right) = \frac{1}{(\tau/2t_1 - 2t_1/\tau)}\left(\sin\frac{2\pi t}{\tau} - \frac{2t_1}{\tau}\sin\frac{\pi t}{t_1}\right), \qquad t < t_1$$

$$\left(\frac{xk}{F_0}\right) = \frac{1}{(\tau/2t_1 - 2t_1/\tau)}\left[\left(\sin\frac{2\pi t}{\tau} - \frac{2t_1}{\tau}\sin\frac{\pi t}{t_1}\right) + \left(\sin 2\pi\frac{t-t_1}{\tau} - \frac{2t_1}{\tau}\sin\pi\frac{t-t_1}{t_1}\right)\right], \qquad t > t_1$$

where $\tau = 2\pi/\omega$.

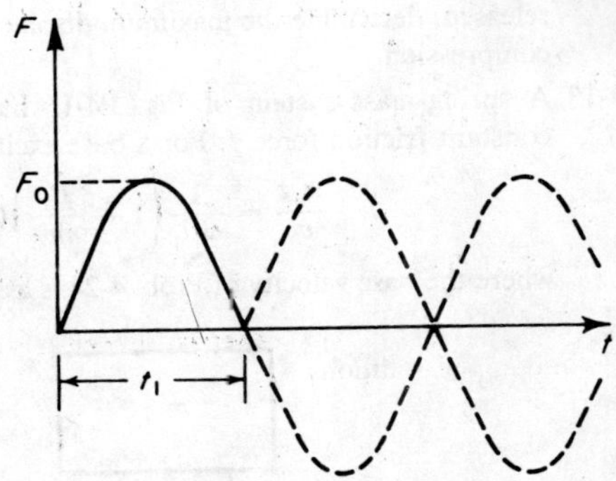

Figure P4-11.

4-12 For the triangular pulse shown in Fig. P4-12 show that the response is

$$x = \frac{2F_0}{k}\left(\frac{t}{t_1} - \frac{\tau}{2\pi t_1}\sin 2\pi\frac{t}{\tau}\right), \qquad 0 < t < \tfrac{1}{2}t_1$$

$$x = \frac{2F_0}{k}\left\{1 - \frac{t}{t_1} + \frac{\tau}{2\pi t_1}\left[2\sin\frac{2\pi}{\tau}\left(t - \frac{1}{2}t_1\right) - \sin 2\pi\frac{t}{\tau}\right]\right\}, \qquad \tfrac{1}{2}t_1 < t < t_1$$

$$x = \frac{2F_0}{k}\left\{\frac{\tau}{2\pi t_1}\left[2\sin\frac{2\pi}{\tau}\left(t - \frac{1}{2}t_1\right) - \sin\frac{2\pi}{\tau}(t - t_1) - \sin 2\pi\frac{t}{\tau}\right]\right\}, \qquad t > t_1$$

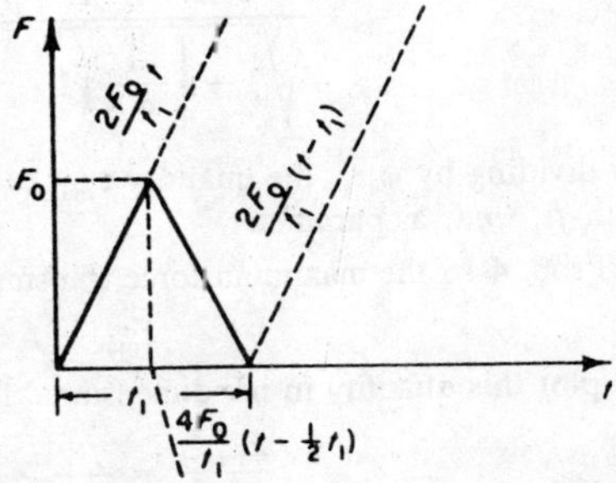

Figure P4-12.

4-13 A spring-mass system slides down a smooth 30° inclined plane, as shown in Fig. P4-13. Determine the time elapsed from first contact of the spring until it breaks contact again.

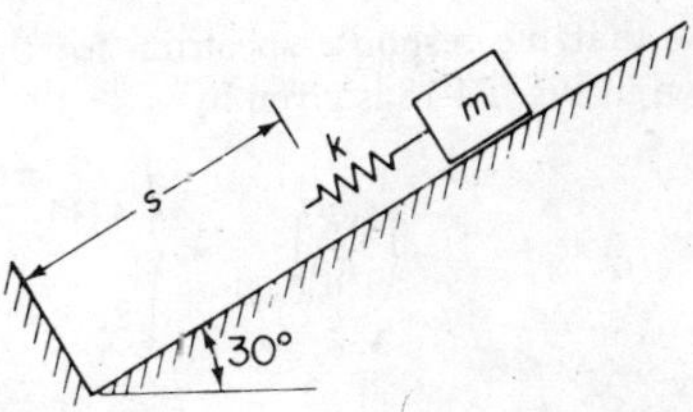

Figure P4-13.

4-14 A 38.6-lb weight is supported on several springs whose combined stiffness is 6.40 lb/in. If the system is lifted so that the bottom of the springs are just free and released, determine the maximum displacement of m, and the time for maximum compression.

4-15 A spring-mass system of Fig. P4-15 has a Coulomb damper, which exerts a constant friction force f. For a base excitation, show that the solution is

$$\frac{\omega_n z}{v_0} = \frac{1}{\omega_n t_1}\left(1 - \frac{f t_1}{m v_0}\right)(1 - \cos\omega_n t) - \sin\omega_n t$$

where the base velocity of Prob. 4-24 is assumed.

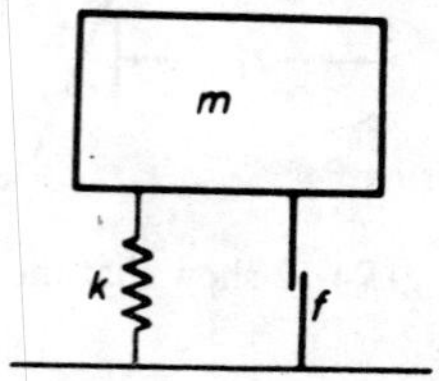

Figure P4-15.

4-16 Show that the peak response for Prob. 4-15 is

$$\frac{\omega_n z_{max}}{v_0} = \frac{1}{\omega_n t_1}\left(1 - \frac{f t_1}{m v_0}\right)\left\{1 - \frac{\frac{1}{\omega_n t_1}\left(1 - \frac{f t_1}{m v_0}\right)}{\sqrt{1 + \left[\frac{1}{\omega_n t_1}\left(1 - \frac{f t_1}{m v_0}\right)\right]^2}}\right\} - \frac{1}{\sqrt{1 + \left[\frac{1}{\omega_n t_1}\left(1 - \frac{f t_1}{m v_0}\right)\right]^2}}$$

By dividing by $\omega_n t_1$, the quantity $z_{max}/v_0 t_1$ can be plotted as a function of $\omega_n t_1$ with $f t_1/m v_0$ as parameter.

4-17 In Prob. 4-16 the maximum force transmitted to m is

$$F_{max} = f + |k z_{max}|$$

To plot this quantity in nondimensional form, multiply by $t_1/m v_0$ to obtain

$$\frac{F_{max} t_1}{m v_0} = \frac{f t_1}{m v_0} + (\omega_n t_1)^2\left(\frac{z_{max}}{v_0 t_1}\right)$$

which again can be plotted as a function of ωt_1 with parameter $f t_1/m v_0$. Plot $|\omega_n z_{max}/v_0|$ and $|z_{max}/v_0 t_1|$ as function of $\omega_n t_1$ for $f t_1/m v_0$ equal to 0, 0.20, and 1.0.

4-18 Show that the response spectrum for the rectangular pulse of time duration t_0 shown in Fig. P4-18 is given by

$$\left(\frac{xk}{F_0}\right)_{max} = \begin{cases} 2\sin\dfrac{\pi t_0}{\tau}, & \dfrac{t_0}{\tau} < 0.50 \\ 2, & \dfrac{t_0}{\tau} > 0.50 \end{cases}$$

where

$$\tau = \frac{2\pi}{\omega_n}$$

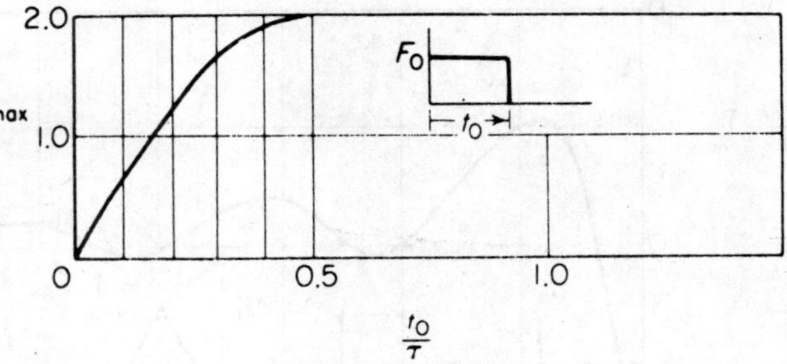

Figure P4-18.

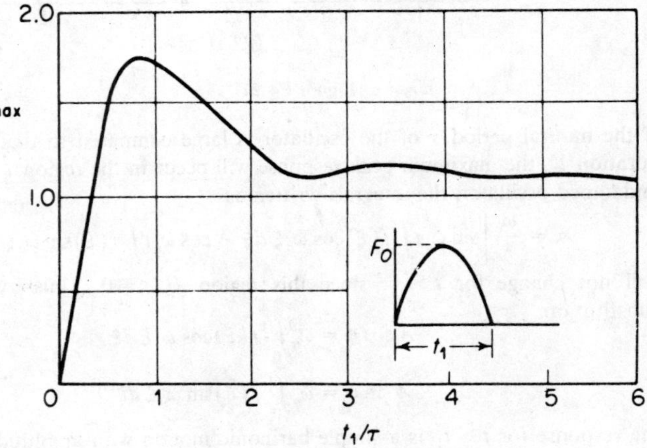

Figure P4-19.

4-19 Shown in Fig. P4-19 is the response spectrum for the sine pulse. Show that for small values of t_1/τ the peak response occurs in the region $t > t_1$. Determine t_p/t_1 when $t_1/\tau = \frac{1}{2}$.

4-20 An undamped spring-mass system with $w = 16.1$ lb has a natural period of [illegible] s. It is subjected to an impulse of 2.0 lb · s, which has a triangular shape with time duration of 0.40 s. Determine the maximum displacement of the mass.

4-21 For a triangular pulse of duration t_1, show that when $t_1/\tau = \frac{1}{2}$, the peak response occurs at $t = t_1$, which can be established from the equation

$$2\cos\frac{2\pi t_1}{\tau}\left(\frac{t_p}{t_1} - 0.5\right) - \cos 2\pi\frac{t_1}{\tau}\left(\frac{t_p}{t_1} - 1\right) - \cos\frac{2\pi t_1}{\tau}\frac{t_p}{t_1} = 0$$

found by differentiating the equation for the displacement for $t > t_1$. The response spectrum for the triangular pulse is shown in Fig. P4-21.

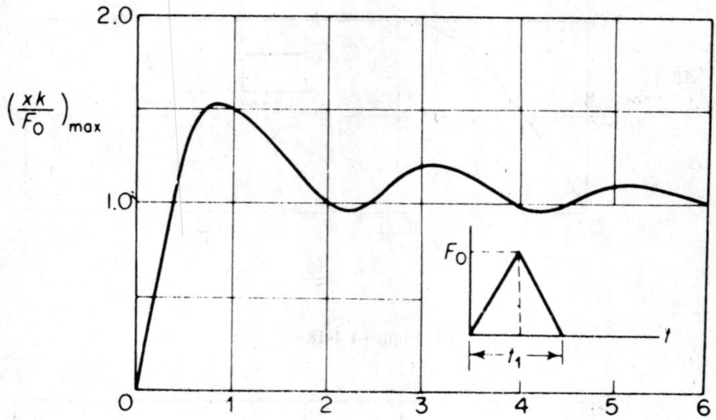

Figure P4-21.

4-22 If the natural period τ of the oscillator is large compared to that of the pulse duration t_1, the maximum peak response will occur in the region $t > t_1$. For the undamped oscillator, the integrals written as

$$x = \frac{\omega_n}{k}\left[\sin \omega_n t \int_0^t f(\xi)\cos \omega_n \xi \, d\xi - \cos \omega_n t \int_0^t f(\xi)\sin \omega_n \xi \, d\xi\right]$$

will not change for $t > t_1$, since this region $f(t) = 0$. Thus by making the substitution

$$A \cos \phi = \omega_n \int_0^{t_1} f(\xi)\cos \omega_n \xi \, d\xi$$

$$A \sin \phi = \omega_n \int_0^{t_1} f(\xi)\sin \omega_n \xi \, d\xi$$

the response for $t > t_1$ is a simple harmonic motion with amplitude A. Discuss the nature of the response spectrum for this case.

4-23 An undamped spring-mass system, m, k, is given a force excitation $F(t)$ as shown in Fig. P4-23. Show that for $t < t_0$

$$\frac{kx(t)}{F_0} = \frac{1}{\omega_n t_0}(\omega_n t - \sin \omega_n t)$$

and for $t > t_0$

$$\frac{kx(t)}{F_0} = \frac{1}{\omega_n t_0}[\sin \omega_n (t - t_0) - \sin \omega_n t] + \cos \omega_n (t - t_0)$$

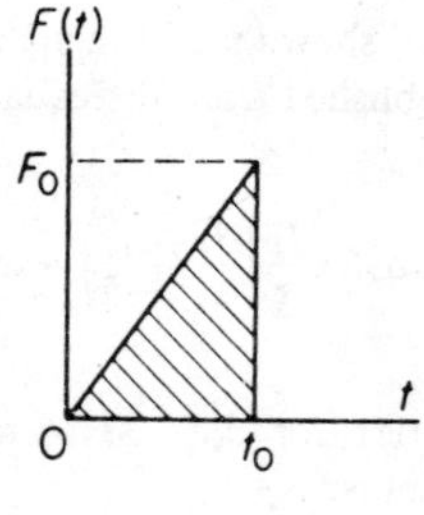

Figure P4-23.

4-24 The base of an undamped spring-mass system, m, k, is given a velocity pulse as shown in Fig. P4-24. Show that if the peak occurs at $t < t_1$, the response spectrum is given by the equation

$$\frac{\omega_n z_{max}}{v_0} = \frac{1}{\omega_n t_1} - \frac{1}{\omega_n t_1\sqrt{1 + (\omega_n t_1)^2}} - \frac{\omega_n t_1}{\sqrt{1 + (\omega_n t_1)^2}}$$

Plot this result.

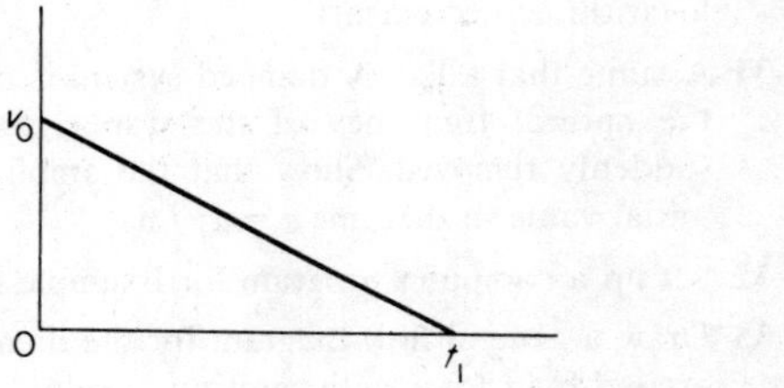

Figure P4-24.

4-25 In Prob. 4-24 if $t > t_1$ show that the solution is

$$\frac{\omega_n z}{v_0} = -\sin \omega_n t + \frac{1}{\omega_n t_1}[\cos \omega_n (t - t_1) - \cos \omega_n t]$$

4-26 Determine the time response for Prob. 4-10 using numerical integration

4-27 Determine the time response for Prob. 4-20 using numerical integration.

4-28 Figure P4-28 shows the response spectra for the undamped spring-mass system under two different base velocity excitations. Solve the problem for the base

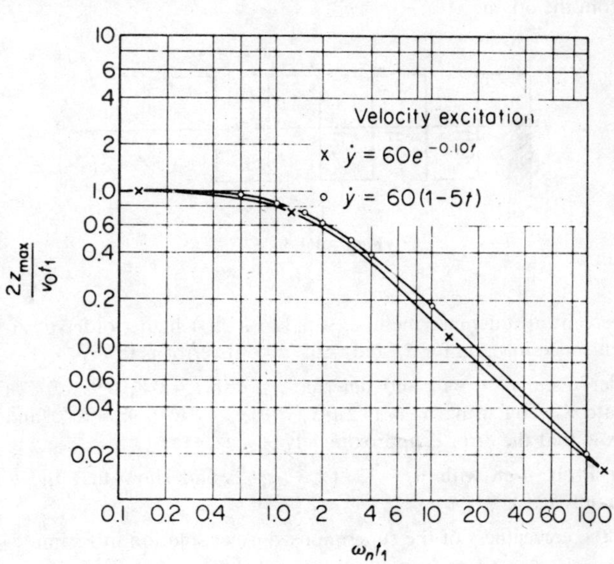

Figure P4-28.

velocity excitation of $\dot{y}(t) = 60e^{-0.10t}$ and verify a few of the points on the spectra.

4-29 A spring-mass system with viscous damping is initially at rest with zero displacement. If the system is activated by a harmonic force of frequency $\omega = \omega_n = \sqrt{k/m}$, determine the equation for its motion.

4-30 In Prob. 4-29 show that with small damping the amplitude will build up to a value $(1 - e^{-1})$ times the steady-state value in the time $t = 1/f_1\delta$. (δ = logarithmic decrement).

4-31 Assume that a lightly damped system is driven by a force $F_0 \sin \omega_n t$ where ω_n is the natural frequency of the system. Determine the equation if the force is suddenly removed. Show that the amplitude decays to a value e^{-1} times the initial value in the time $t = 1/f_n\delta$.

4-32 Set up a computer program for Example 4.5-1.

4-33 Draw a general flow diagram for the damped system with zero initial conditions excited by a force with zero initial value.

4-34 Draw a flow diagram for the damped system excited by base motion $y(t)$ with initial conditions $x(0) = X_1$ and $\dot{x}(0) = V_1$.

4-35 Write a Fortran program for Prob. 4-34, where the base motion is a half sine wave.

4-36 Determine the response of an undamped spring-mass system to the alternating square wave of force shown in Fig. P4-36 by superimposing the solution to the step function and matching the displacement and velocity at each transition time. Plot the result and show that the peaks of the response will increase as straight lines from the origin.

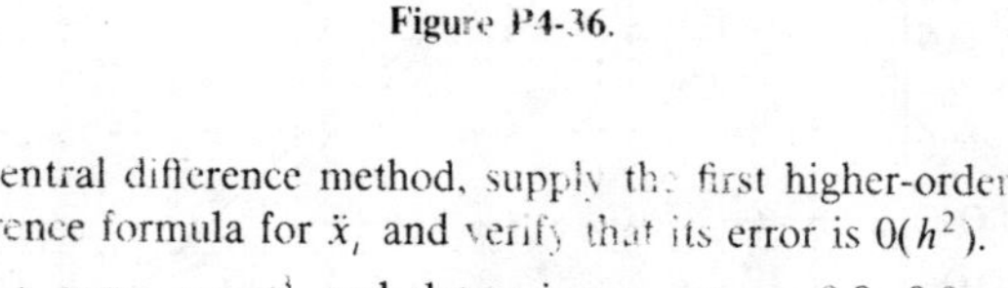

Figure P4-36.

4-37 For the central difference method, supply the first higher-order term left out in the recurrence formula for $\ddot{x}_i$ and verify that its error is $0(h^2)$.

4-38 Consider a curve $x = t^3$ and determine x_i at $t = 0.8, 0.9, 1.0, 1.1$, and 1.2. Calculate $\dot{x}_{1.0}$ by using $\dot{x}_i = 1/2h(x_{i+1} - x_{i-1})$ with $h = 0.20$ and $h = 0.10$, and show that the error is approximately $0(h^2)$

4-39 Repeat Prob. 4-38 with $\dot{x}_i = 1/h(x_i - x_{i-1})$ and show that the error is approximately $0(h)$.

4-40 Verify the correctness of the superimposed exact solution in Example 4.5-1.

4-41 Calculate the problem in Example 4.5-2 by using the Runge-Kutta method.

4-42 A large box of weight W resting on a barge is to be hoisted by a crane, as shown in Fig. P4-42. Assuming the stiffness of the crane boom to be k_c, determine the equation of motion if the extended point of the boom is given a displacement $x = Vt$. Use the method of Laplace transformation.

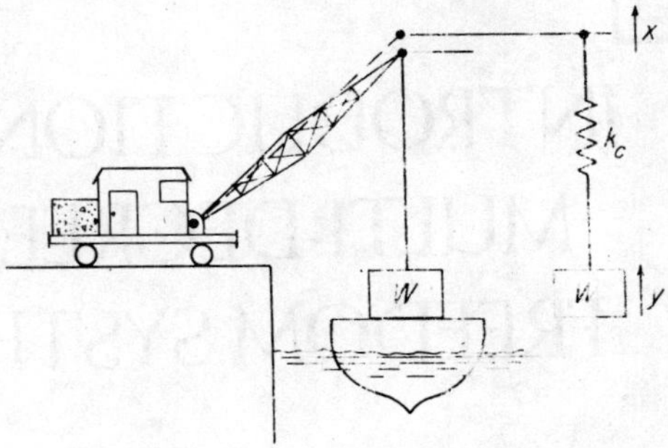

Figure P4-42.

5

INTRODUCTION TO MULTI-DEGREE OF FREEDOM SYSTEMS

A multi-DOF system is one that requires two or more coordinates to describe its motion. These coordinates are called generalized coordinates when they are independent of each other and equal in number to the degrees of freedom of the system.

The N-DOF system differs from the single-DOF system in that it has N natural frequencies, and for each of the natural frequencies there corresponds a natural state of vibration with a displacement configuration known as the *normal mode*. Mathematical terms related to these quantities are known as *eigenvalues* and *eigenvectors*. They are established from the N simultaneous equations of motion of the system and possess certain dynamic properties associated with the system.

Normal mode vibrations are free vibrations that depend only on the mass and stiffness of the system and how they are distributed. These vibrations are important not only in establishing the spectrum of resonance but also in the analysis of the behavior of the system under initial conditions or force excitation in terms of normal mode solutions. Again, damping is generally omitted except when its concern is of importance in limiting the amplitude of vibration.

Numerical results are easily obtained for the two-DOF system, but this is generally not the case for systems of higher DOF for which computers are necessary. Matrix methods are es[illegible]ial for these problems. They provide compact notation and an organi[illegible]cedure for their analyses and soluti[illegible]

We start first with the fr[illegible]ion of simple systems and deter[illegible] their natural frequencies and [illegible]ponding normal modes. To p[illegible] normal mode vibrations the [illegible]t be given specific initial con[illegible]

For the more general initial conditions, the resulting free vibration may contain several normal modes simultaneously. The problem then is that of finding how much of each normal mode is present in the free vibration.

As in the single-DOF system, forced harmonic vibration of multi-DOF systems will take place at the frequency of the excitation. When the excitation frequency coincides with one of the natural frequencies of the system, a condition of resonance will be encountered with large amplitudes. It is also possible to express the response of forced vibration in terms of normal mode summation.

Finally, the numerical computational procedure of the previous chapter is adapted to a two-DOF system as an illustration.

5.1 NORMAL MODE VIBRATION

Consider the undamped system of Figure 5.1-1. Using coordinates x_1 and x_2 measured from inertial reference, the differential equations of motion for the system become

$$\begin{aligned} m\ddot{x}_1 &= -k(x_1 - x_2) - kx_1 \\ 2m\ddot{x}_2 &= k(x_1 - x_2) - kx_2 \end{aligned} \tag{5.1-1}$$

We now define a normal mode oscillation as one in which each mass undergoes harmonic motion of the same frequency, passing simultaneously through the equilibrium position. For such motion we can let

$$\begin{aligned} x_1 &= A_1 e^{i\omega t} \\ x_2 &= A_2 e^{i\omega t} \end{aligned} \tag{5.1-2}$$

Substituting these into the differential equations gives

$$\begin{aligned} (2k - \omega^2 m) A_1 - kA_2 &= 0 \\ -kA_1 + (2k - 2\omega^2 m) A_2 &= 0 \end{aligned} \tag{5.1-3}$$

which are satisfied for any A_1 and A_2 if the following determinant is zero:

$$\begin{vmatrix} (2k - \omega^2 m) & -k \\ -k & (2k - 2\omega^2 m) \end{vmatrix} = 0 \tag{5.1-4}$$

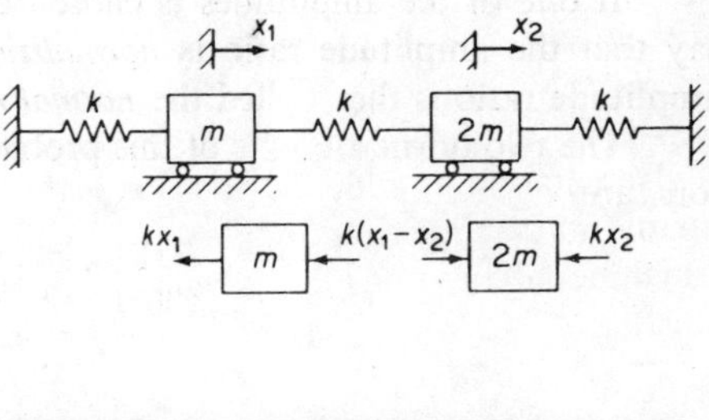

Figure 5.1-1.

Letting $\omega^2 = \lambda$, the above determinant leads to the *characteristic equation*

$$\lambda^2 - \left(3\frac{k}{m}\right)\lambda + \frac{3}{2}\left(\frac{k}{m}\right)^2 = 0 \tag{5.1-5}$$

The two roots λ_1 and λ_2 of this equation are the eigenvalues of the system

$$\lambda_1 = \left(\frac{3}{2} - \frac{1}{2}\sqrt{3}\right)\frac{k}{m} = 0.634\frac{k}{m}$$

and

$$\lambda_2 = \left(\frac{3}{2} + \frac{1}{2}\sqrt{3}\right)\frac{k}{m} = 2.366\frac{k}{m}$$

and the *natural frequencies* of the system are found to be

$$\omega_1 = \lambda_1^{1/2} = \sqrt{0.6 \text{[illegible]}} \tag{5.1-6}$$

and

$$\omega_2 = \lambda_2^{1/2} = \sqrt{2.3 \text{[illegible]}}$$

From Eq. 5.1-3 two expressions for the ratio of the amplitudes are found:

$$\frac{A_1}{A_2} = \frac{k}{2k - \omega^2 m} = \frac{2k - \omega^2 m}{\text{[illegible]}} \tag{5.1-7}$$

Substitution of the natural frequencies in either of these equations lead to the ratio of the amplitudes. For $\omega_1^2 = 0.634k/m$ [illegible]

$$\left(\frac{A_1}{A_2}\right)^{(1)} = \frac{k}{2k - \omega_1^2 m} = \frac{1}{2 - \text{[illegible]}} = \text{[illegible]}731 \tag{5.1-8}$$

which is the amplitude ratio corresponding [illegible] natural frequency.

Similarly, using $\omega_2^2 = 2.366k/m$, we obtain

$$\left(\frac{A_1}{A_2}\right)^{(2)} = \frac{k}{2k - \omega_2^2 m} = \frac{1}{2 - 2.\text{[illegible]}} = \text{[illegible]} \tag{5.1-9}$$

for the amplitude ratio corresponding to the second natural frequency. Eq. 5.1-7 enables us to find only the ratio of the amplitudes and not their absolute values, which are arbitrary.

If one of the amplitudes is chosen equal to 1 or any other number, we say that the amplitude ratio is *normalized* to that number. The normalized amplitude ratio is then called the *normal mode* and is designated by $\phi_i(x)$.

The two normal modes of this problem, which we can now call eigenvectors, are

$$\text{[illegible]} \begin{Bmatrix} \text{[illegible]}731 \\ \text{[illegible]}00 \end{Bmatrix} \qquad \phi_2(x) = \begin{Bmatrix} -2.73 \\ 1.00 \end{Bmatrix}$$

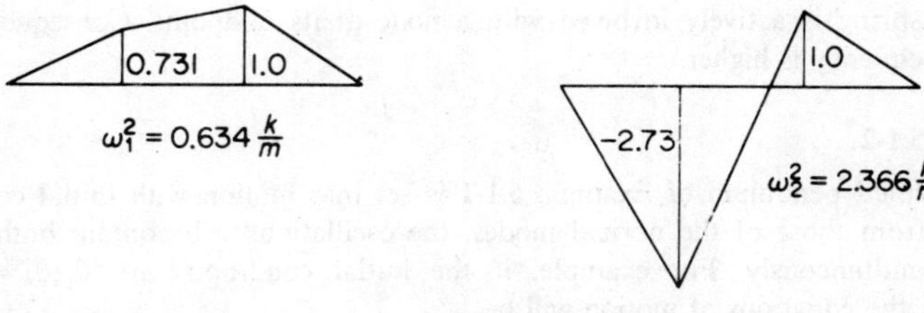

Figure 5.1-2. Normal modes of the system shown in Figure 5.1-1.

These normal modes are displayed graphically in Fig. 5.1-2. In the first normal mode, the two masses move in phase; in the second mode the two masses move in opposition, or out of phase with each other.

Example 5.1-1

In Fig. 5.1-3 the two pendulums are coupled by means of a weak spring k, which is unstrained when the two pendulum rods are in the vertical position. Determine the normal mode vibrations.

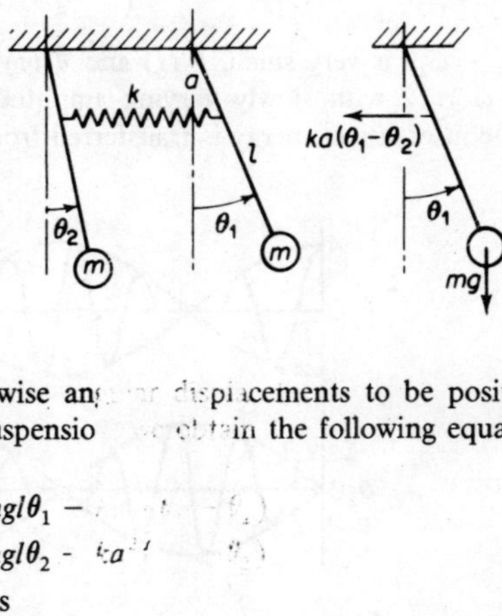

Figure 5.1-3. Coupled pendulum.

Solution: Assuming the counterclockwise angular displacements to be positive and taking moments about the points of suspension, we obtain the following equations of motion for small oscillations

$$ml^2\ddot{\theta}_1 = -mgl\theta_1 - ka^2(\theta_1 - \theta_2)$$
$$ml^2\ddot{\theta}_2 = -mgl\theta_2 + ka^2(\theta_1 - \theta_2)$$

Assuming the normal mode solutions as

$$\theta_1 = A_1 \cos \omega t$$

$$\theta_2 = A_2 \cos \omega t$$

the natural frequencies and mode shapes are found to be

$$\omega_1 = \sqrt{\frac{g}{l}} \qquad \omega_2 = \sqrt{\frac{g}{l} + 2\frac{k}{m}\frac{a^2}{l^2}}$$

$$\left(\frac{A_1}{A_2}\right)^{(1)} = 1.0 \qquad \left(\frac{A_1}{A_2}\right)^{(2)} = -1.0$$

Thus in the first mode the two pendulums move in phase and the spring remains unstretched. In the second mode the two pendulums move in opposition and the coupling spring is actively involved with a node at its midpoint. Consequently, the natural frequency is higher.

Example 5.1-2

If the coupled pendulum of Example 5.1-1 is set into motion with initial conditions differing from those of the normal modes, the oscillations will contain both normal modes simultaneously. For example, if the initial conditions are $\theta_1(0) = A$ and $\theta_2(0) = 0$, the equations of motion will be

$$\theta_1(t) = \tfrac{1}{2}A\cos\omega_1 t + \tfrac{1}{2}A\cos\omega_2 t$$

$$\theta_2(t) = \tfrac{1}{2}A\cos\omega_1 t - \tfrac{1}{2}A\cos\omega_2 t$$

Consider the case where the coupling spring is very weak, and show that a beating phenomena takes place between the two pendulums.

Solution: The preceding equations can be rewritten as follows:

$$\theta_1(t) = A\cos\left(\frac{\omega_1 - \omega_2}{2}\right)t\cos\left(\frac{\omega_1 + \omega_2}{2}\right)t$$

$$\theta_2(t) = -A\sin\left(\frac{\omega_1 - \omega_2}{2}\right)t\sin\left(\frac{\omega_1 + \omega_2}{2}\right)t$$

Since $(\omega_1 - \omega_2)$ is very small, $\theta_1(t)$ and $\theta_2(t)$ will behave like $\cos(\omega_1 + \omega_2)t/2$ and $\sin(\omega_1 + \omega_2)t/2$ with slowly varying amplitudes as shown in Fig. 5.1-4. Since the system is conservative, energy is transferred from one pendulum to the other.

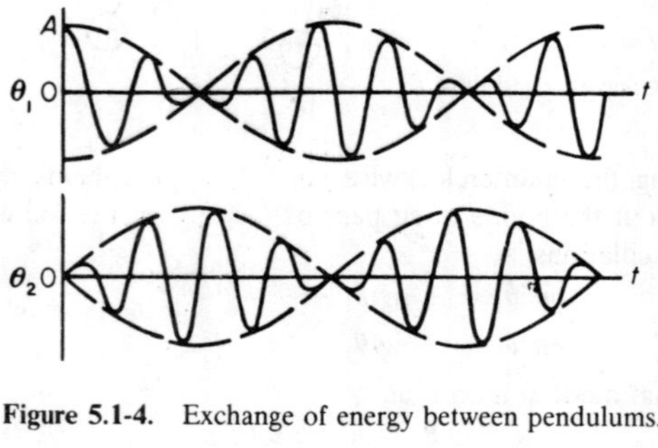

Figure 5.1-4. Exchange of energy between pendulums.

Example 5.1-3

Determine the free vibration of the system shown in Fig. 5.1-1 when the initial conditions are $x_1(0) = 5$, $x_2(0) = 1$, and $\dot{x}_1(0) = \dot{x}_2(0) = 0$. The natural frequencies and normal modes of the system have previously been given as:

$$\omega_1 = \sqrt{0.634\frac{k}{m}} \qquad \omega_2 = \sqrt{2.366\frac{k}{m}}$$

$$\phi_1(x) = \begin{Bmatrix} 0.731 \\ 1.000 \end{Bmatrix} \qquad \phi_2(x) = \begin{Bmatrix} -2.731 \\ 1.000 \end{Bmatrix}$$

Solution: Any free vibration can be considered to be the superposition of its normal modes. For each of these modes, the time solution can be expressed as

$$\begin{Bmatrix} x_1 \\ x_2 \end{Bmatrix}_1 = \begin{Bmatrix} 0.731 \\ 1.000 \end{Bmatrix} \sin \omega_1 t \qquad \begin{Bmatrix} x_1 \\ x_2 \end{Bmatrix}_2 = \begin{Bmatrix} -2.731 \\ 1.000 \end{Bmatrix} \sin \omega_2 t$$

The general solution for the free vibration can then be written as

$$\begin{Bmatrix} x_1 \\ x_2 \end{Bmatrix} = A \begin{Bmatrix} 0.731 \\ 1.000 \end{Bmatrix} \sin(\omega_1 t + \psi_1) + B \begin{Bmatrix} -2.731 \\ 1.000 \end{Bmatrix} \sin(\omega_2 t + \psi_2)$$

where A and B allow different amounts of each mode and ψ_1 and ψ_2 allow the two modes' different phases or starting values. Fitting the above equation to the four initial conditions, we obtain

$$\begin{Bmatrix} x_1(0) \\ x_2(0) \end{Bmatrix} = \begin{Bmatrix} 5 \\ 1 \end{Bmatrix} = A \begin{Bmatrix} 0.731 \\ 1.000 \end{Bmatrix} \sin \psi_1 + B \begin{Bmatrix} -2.731 \\ 1.000 \end{Bmatrix} \sin \psi_2$$

$$\begin{Bmatrix} \dot{x}_1(0) \\ \dot{x}_2(0) \end{Bmatrix} = \begin{Bmatrix} 0 \\ 0 \end{Bmatrix} = \omega_1 A \begin{Bmatrix} 0.731 \\ 1.000 \end{Bmatrix} \cos \psi_1 + \omega_2 B \begin{Bmatrix} -2.731 \\ 1.000 \end{Bmatrix} \cos \psi_2$$

For the second set of equations to be satisfied, $\cos \psi_1 = \cos \psi_2 = 0$, so that $\psi_1 = \psi_2 = 90°$. Substituting the 90° into the first set of equations, we obtain

$$\begin{Bmatrix} 5 \\ 1 \end{Bmatrix} = A \begin{Bmatrix} 0.731 \\ 1.000 \end{Bmatrix} + B \begin{Bmatrix} -2.731 \\ 1.000 \end{Bmatrix}$$

Solving for A and B, $A = 2.233$ and $B = -1.233$. The equation for the free vibration is then

$$\begin{Bmatrix} x_1 \\ x_2 \end{Bmatrix} = 2.233 \begin{Bmatrix} 0.731 \\ 1.000 \end{Bmatrix} \cos \omega_1 t - 1.233 \begin{Bmatrix} 2.731 \\ 1.000 \end{Bmatrix} \cos \omega_2 t$$

5.2 COORDINATE COUPLING

The differential equations of motion for the two-DOF system are in general *coupled*, in that both coordinates appear in each equation. In the most general case the two equations for the undamped system have the form

$$\begin{aligned} m_{11}\ddot{x}_1 + m_{12}\ddot{x}_2 + k_{11}x_1 + k_{12}x_2 &= 0 \\ m_{21}\ddot{x}_1 + m_{22}\ddot{x}_2 + k_{21}x_1 + k_{22}x_2 &= 0 \end{aligned} \tag{5.2-1}$$

These equations can be expressed in matrix form (see Appendix C) as

$$\begin{bmatrix} m_{11} & m_{12} \\ m_{21} & m_{22} \end{bmatrix} \begin{Bmatrix} \ddot{x}_1 \\ \ddot{x}_2 \end{Bmatrix} + \begin{bmatrix} k_{11} & k_{12} \\ k_{21} & k_{22} \end{bmatrix} \begin{Bmatrix} x_1 \\ x_2 \end{Bmatrix} = \begin{Bmatrix} 0 \\ 0 \end{Bmatrix} \tag{5.2-2}$$

which immediately reveals the type of coupling present. Mass or *dynamical coupling* exists if the mass matrix is nondiagonal, whereas stiffness or *static coupling* exists if the stiffness matrix is nondiagonal.

It is also possible to establish the type of coupling from the expressions for the kinetic and potential energies. Cross products of coordinates in either expression denote coupling, dynamic or static, depending on whether they are found in T or U. The choice of coordinates establishes the type of coupling, and both dynamic and static coupling may be present.

It is possible to find a coordinate system which has neither form of coupling. The two equations are then decoupled and each equation may be solved independently of the other. Such coordinates are called *principal coordinates* (also called *normal coordinates*).

Although it is always possible to decouple the equations of motion for the undamped system, this is not always the case for a damped system. The following matrix equations show a system which has zero dynamic and static coupling, but the coordinates are coupled by the damping matrix.

$$\begin{bmatrix} m_{11} & 0 \\ 0 & m_{22} \end{bmatrix} \begin{Bmatrix} \ddot{x}_1 \\ \ddot{x}_2 \end{Bmatrix} + \begin{bmatrix} c_{11} & c_{12} \\ c_{21} & c_{22} \end{bmatrix} \begin{Bmatrix} \dot{x}_1 \\ \dot{x}_2 \end{Bmatrix} + \begin{bmatrix} k_{11} & 0 \\ 0 & k_{22} \end{bmatrix} \begin{Bmatrix} x_1 \\ x_2 \end{Bmatrix} = \begin{Bmatrix} 0 \\ 0 \end{Bmatrix} \qquad (5.2\text{-}3)$$

If in the above equation $c_{12} = c_{21} = 0$, then the damping is said to be *proportional* (proportional to the stiffness or mass matrix), and the system equations become uncoupled.

Example 5.2-1

Figure 5.2-1 shows a rigid bar with its center of mass not coinciding with its geometric center, i.e., $l_1 \neq l_2$, and supported by two springs, k_1, k_2. It represents a two-DOF system, since two coordinates are necessary to describe its motion. The choice of the coordinates will define the type of coupling that can be immediately determined from the mass and stiffness matrices. Mass or *dynamical coupling* exists if the mass matrix is nondiagonal, whereas stiffness or *static coupling* exists if the stiffness matrix is nondiagonal. It is also possible to have both forms of coupling.

Static coupling. Choosing coordinates x and θ, shown in Fig. 5.2-2, where x is the linear displacement of the center of mass, the system will have

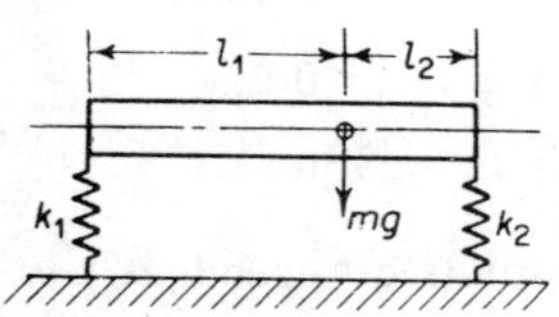

Figure 5.2-1.

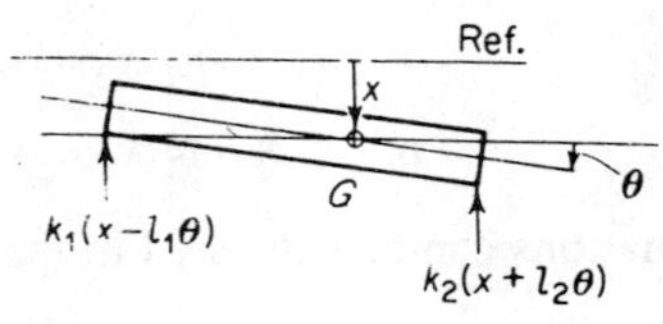

Figure 5.2-2. Coordinates leading to static coupling.

static coupling as shown by the matrix equation

$$\begin{bmatrix} m & 0 \\ 0 & J \end{bmatrix}\begin{Bmatrix} \ddot{x} \\ \ddot{\theta} \end{Bmatrix} + \begin{bmatrix} (k_1 + k_2) & (k_2 l_2 - k_1 l_1) \\ (k_2 l_2 - k_1 l_1) & (k_1 l_1^2 + k_2 l_2^2) \end{bmatrix}\begin{Bmatrix} x \\ \theta \end{Bmatrix} = \begin{Bmatrix} 0 \\ 0 \end{Bmatrix}$$

If $k_1 l_1 = k_2 l_2$, the coupling disappears, and we obtain uncoupled x and θ vibrations.

Dynamic coupling. There is some point C along the bar where a force applied normal to the bar produces pure translation; i.e., $k_1 l_3 = k_2 l_4$. (See Fig. 5.2-3.) The equations of motion in terms of x_c and θ can be shown to be

$$\begin{bmatrix} m & me \\ me & J_c \end{bmatrix}\begin{Bmatrix} \ddot{x}_c \\ \ddot{\theta} \end{Bmatrix} + \begin{bmatrix} (k_1 + k_2) & 0 \\ 0 & (k_1 l_3^2 + k_2 l_4^2) \end{bmatrix}\begin{Bmatrix} x_c \\ \theta \end{Bmatrix} = \begin{Bmatrix} 0 \\ 0 \end{Bmatrix}$$

which shows that the coordinates chosen eliminated the static coupling and introduced dynamic coupling.

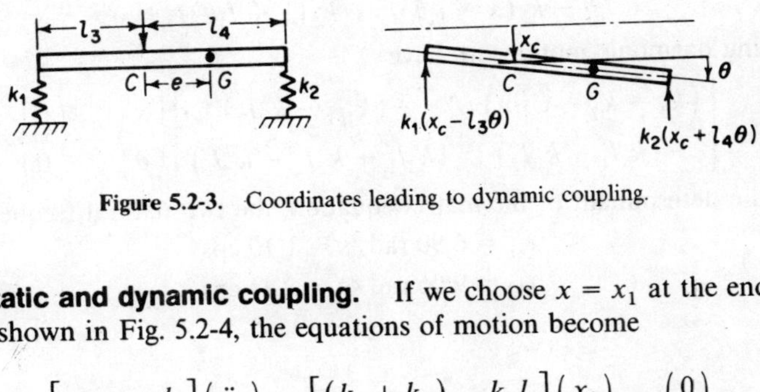

Figure 5.2-3. Coordinates leading to dynamic coupling.

Static and dynamic coupling. If we choose $x = x_1$ at the end of the bar, as shown in Fig. 5.2-4, the equations of motion become

$$\begin{bmatrix} m & ml_1 \\ ml_1 & J_1 \end{bmatrix}\begin{Bmatrix} \ddot{x}_1 \\ \ddot{\theta} \end{Bmatrix} + \begin{bmatrix} (k_1 + k_2) & k_2 l \\ k_2 l & k_2 l^2 \end{bmatrix}\begin{Bmatrix} x_1 \\ \theta \end{Bmatrix} = \begin{Bmatrix} 0 \\ 0 \end{Bmatrix}$$

and both static and dynamic coupling are now present.

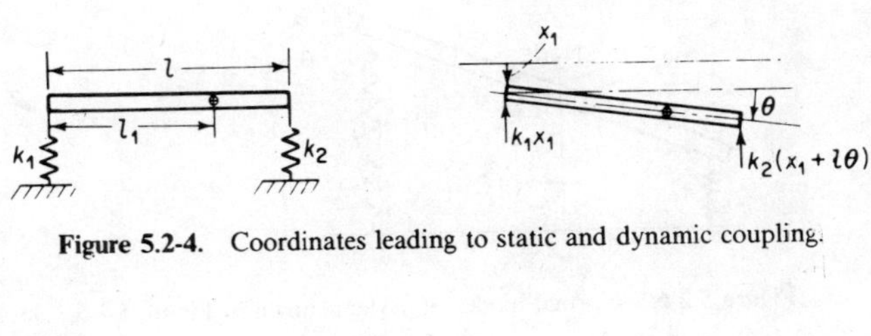

Figure 5.2-4. Coordinates leading to static and dynamic coupling.

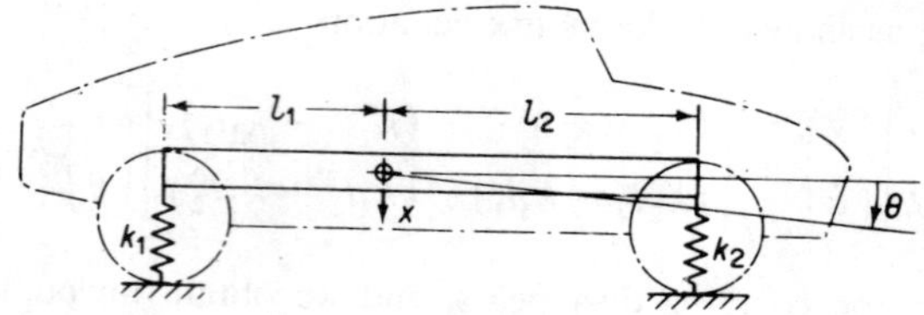

Figure 5.2-5.

Example 5.2-2

Determine the normal modes of vibration of an automobile simulated by the simplified two-DOF system with the following numerical values (See Fig. 5.2-5):

$$W = 3220 \text{ lb} \qquad l_1 = 4.5 \text{ ft} \qquad k_1 = 2400 \text{ lb/ft}$$
$$J_c = \frac{W}{g} r^2 \qquad l_2 = 5.5 \text{ ft} \qquad k_2 = 2600 \text{ lb/ft}$$
$$r = 4 \text{ ft} \qquad l = 10 \text{ ft}$$

The equations of motion indicate static coupling.

$$m\ddot{x} + k_1(x - l_1\theta) + k_2(x + l_2\theta) = 0$$
$$J_c\ddot{\theta} - k_1(x - l_1\theta)l_1 + k_2(x + l_2\theta)l_2 = 0$$

Assuming harmonic motion, we have

$$\begin{bmatrix} (k_1 + k_2 - \omega^2 m) & -(k_1 l_1 - k_2 l_2) \\ -(k_1 l_1 - k_2 l_2) & (k_1 l_1^2 + k_2 l_2^2 - \omega^2 J_c) \end{bmatrix} \begin{Bmatrix} x \\ \theta \end{Bmatrix} = \begin{Bmatrix} 0 \\ 0 \end{Bmatrix}$$

From the determinant of the matrix equation, the two natural frequencies are

$$\omega_1 = 6.90 \text{ rad/s} = 1.10 \text{ cps}$$
$$\omega_2 = 9.06 \text{ rad/s} = 1.44 \text{ cps}$$

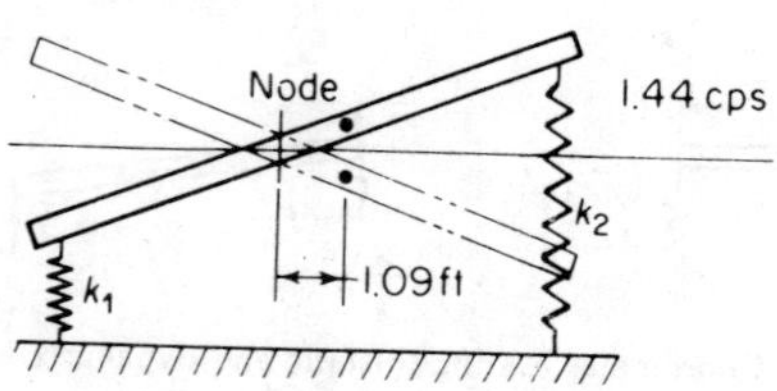

Figure 5.2-6. Normal modes of system shown in Figure 5.2-5.

The amplitude ratios for the two frequencies are

$$\left(\frac{x}{\theta}\right)_{\omega_1} = -14.6 \text{ ft/rad} = -3.06 \text{ in./deg}$$

$$\left(\frac{x}{\theta}\right)_{\omega_2} = 1.09 \text{ ft/rad} = 0.288 \text{ in./deg}$$

The mode shapes are illustrated by the diagrams of Fig. 5.2-6.

5.3 FORCED HARMONIC VIBRATION

Consider here a system excited by a harmonic force $F_1 \sin \omega t$ expressed by the matrix equation

$$\begin{bmatrix} m_1 & 0 \\ 0 & m_2 \end{bmatrix}\begin{Bmatrix} \ddot{x}_1 \\ \ddot{x}_2 \end{Bmatrix} + \begin{bmatrix} k_{11} & k_{12} \\ k_{21} & k_{22} \end{bmatrix}\begin{Bmatrix} x_2 \\ x_2 \end{Bmatrix} = \begin{Bmatrix} F_1 \\ 0 \end{Bmatrix} \sin \omega t \tag{5.3-1}$$

Since the system is undamped, the solution can be assumed as

$$\begin{Bmatrix} x_1 \\ x_2 \end{Bmatrix} = \begin{Bmatrix} X_1 \\ X_2 \end{Bmatrix} \sin \omega t$$

Substituting this into the differential equation, we obtain

$$\begin{bmatrix} (k_{11} - m_1\omega^2) & k_{12} \\ k_{21} & (k_{22} - m_2\omega^2) \end{bmatrix}\begin{Bmatrix} X_1 \\ X_2 \end{Bmatrix} = \begin{Bmatrix} F_1 \\ 0 \end{Bmatrix} \tag{5.3-2}$$

or, in simpler notation,

$$[Z(\omega)]\begin{Bmatrix} X_1 \\ X_2 \end{Bmatrix} = \begin{Bmatrix} F_1 \\ 0 \end{Bmatrix}$$

Premultiplying by $[Z(\omega)]^{-1}$ we obtain (see Appendix C)

$$\begin{Bmatrix} X_1 \\ X_2 \end{Bmatrix} = [Z(\omega)]^{-1}\begin{Bmatrix} F_1 \\ 0 \end{Bmatrix} = \frac{\operatorname{adj}[Z(\omega)]\begin{Bmatrix} F_1 \\ 0 \end{Bmatrix}}{|Z(\omega)|} \tag{5.3-3}$$

Referring to Eq. (5.3-2), the determinant $|Z(\omega)|$ can be expressed as

$$|Z(\omega)| = m_1 m_2(\omega_1^2 - \omega^2)(\omega_2^2 - \omega^2) \tag{5.3-4}$$

where ω_1 and ω_2 are the normal mode frequencies. Thus Eq. (5.3-3) becomes

$$\begin{Bmatrix} X_1 \\ X_2 \end{Bmatrix} = \frac{1}{|Z(\omega)|}\begin{bmatrix} (k_{22} - m_2\omega^2) & -k_{12} \\ -k_{21} & (k_{11} - m_1\omega^2) \end{bmatrix}\begin{Bmatrix} F_1 \\ 0 \end{Bmatrix} \tag{5.3-5}$$

or

$$\begin{aligned} X_1 &= \frac{(k_{22} - m_2\omega^2) F_1}{m_1 m_2(\omega_1^2 - \omega^2)(\omega_2^2 - \omega^2)} \\ X_2 &= \frac{-k_{12} F_1}{m_1 m_2(\omega_1^2 - \omega^2)(\omega_2^2 - \omega^2)} \end{aligned} \tag{5.3-6}$$

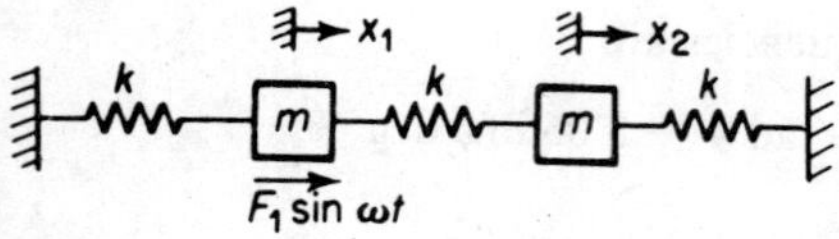

Figure 5.3-1.

Example 5.3-1

Apply Eqs. (5.3-6) to the system shown in Fig. 5.3-1 when m_1 is excited by the force $F_1 \sin \omega t$. Plot its frequency response curve.

The equation of motion for the system is

$$\begin{bmatrix} m & 0 \\ 0 & m \end{bmatrix} \begin{Bmatrix} \ddot{x}_1 \\ \ddot{x}_2 \end{Bmatrix} + \begin{bmatrix} 2k & -k \\ -k & 2k \end{bmatrix} \begin{Bmatrix} x_1 \\ x_2 \end{Bmatrix} = \begin{Bmatrix} F_1 \\ 0 \end{Bmatrix} \sin \omega t$$

Thus we have $k_{11} = k_{22} = 2k$ and $k_{12} = k_{21} = -k$. Equation 5.3-6 then becomes

$$X_1 = \frac{(2k - m\omega^2) F_1}{m^2(\omega_1^2 - \omega^2)(\omega_2^2 - \omega^2)}$$

$$X_2 = \frac{kF_1}{m^2(\omega_1^2 - \omega^2)(\omega_2^2 - \omega^2)}$$

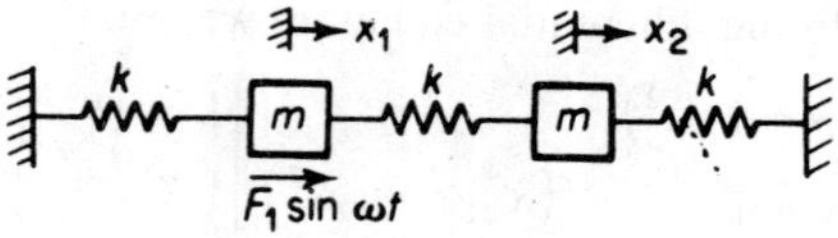

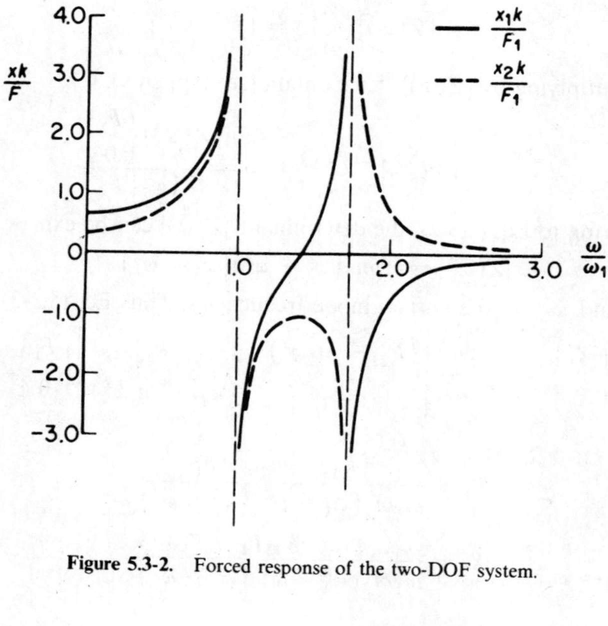

Figure 5.3-2. Forced response of the two-DOF system.

where $\omega_1^2 = k/m$ and $\omega_2^2 = 3k/m$ are obtained from the determinant of the matrix equation. When plotted these results appear as in Fig. 5.3-2.

Example 5.3-2

Express the equations for X_1 and X_2 in Ex. 5.3-1 as the sum of the normal modes.

Solution: Consider X_1 and expand the equation in terms of partial fractions.

$$\frac{(2k - m\omega^2)F_1}{m^2(\omega_1^2 - \omega^2)(\omega_2^2 - \omega^2)} = \frac{C_1}{(\omega_1^2 - \omega^2)} + \frac{C_2}{(\omega_2^2 - \omega^2)}$$

To solve for C_1, multiply by $(\omega_1^2 - \omega^2)$ and let $\omega = \omega_1$

$$C_1 = \frac{(2k - m\omega_1^2)F_1}{m^2(\omega_2^2 - \omega_1^2)} = \frac{F_1}{2m}$$

Similarly, C_2 is evaluated by multiplying by $(\omega_2^2 - \omega^2)$ and letting $\omega = \omega_2$

$$C_2 = \frac{(2k - m\omega_2^2)F_1}{m^2(\omega_1^2 - \omega_2^2)} = \frac{F_1}{2m}$$

An alternative form of X_1 is then

$$X_1 = \frac{F_1}{2m}\left[\frac{1}{\omega_1^2 - \omega^2} + \frac{1}{\omega_2^2 - \omega^2}\right]$$

$$= \frac{F_1}{2k}\left[\frac{1}{1 - (\omega/\omega_1)^2} + \frac{1}{3 - (\omega/\omega_1)^2}\right]$$

Treating X_2 in the same manner, its equation is

$$X_2 = \frac{F_1}{2k}\left[\frac{1}{1 - (\omega/\omega_1)^2} - \frac{1}{3 - (\omega/\omega_1)^2}\right]$$

The amplitudes X_1 and X_2 are now expressed as the sum of normal modes, their time solution being

$$x_1 = X_1 \sin \omega t$$

$$x_2 = X_2 \sin \omega t$$

5.4 DIGITAL COMPUTATION

The finite difference method of Sec. 4.5 can easily be extended to the solution of systems with two DOF. The procedure is illustrated by the following problem which is programmed and solved by the digital computer.

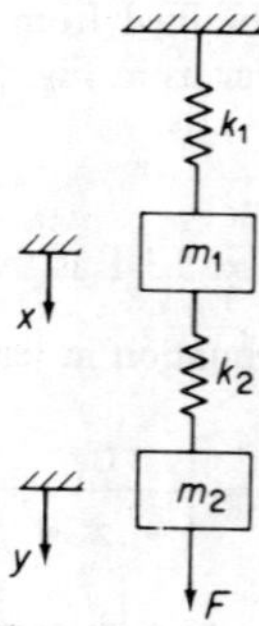

Figure 5.4-1.

The system to be solved is shown in Fig. 5.4-1. To avoid confusion with subscripts, we let the displacements be x and y.

$$k_1 = 36 \text{ kN/m}$$

$$k_2 = 36 \text{ kN/m}$$

$$m_1 = 100 \text{ kg}$$

$$m_2 = 25 \text{ kg}$$

$$F = \begin{cases} 4000 \text{ N}, & t > 0 \\ 0, & t < 0 \end{cases}$$

Initial conditions:

$$x = \dot{x} = y = \dot{y} = 0$$

The equations of motion are

$$100\ddot{x} = -36{,}000x + 36{,}000(y - x)$$

$$25\ddot{y} = -36{,}000(y - x) + F$$

which can be rearranged to

$$\ddot{x} = -720x + 360y$$

$$\ddot{y} = 1440(x - y) + 160$$

These equations are to be solved together with the recurrence equations of Sec. 4.5.

$$x_{i+1} = \ddot{x}_i \Delta t^2 + 2x_i - x_{i-1}$$

$$y_{i+1} = \ddot{y}_i \Delta t^2 + 2y_i - y_{i-1}$$

Calculations for the natural periods of the system reveal that they do not differ substantially. They are $\tau_1 = 0.3803$ and $\tau_2 = 0.1462$ s. We therefore arbitrarily choose a value of $\Delta t = 0.01$ s. which is smaller than $\tau_2/10$.

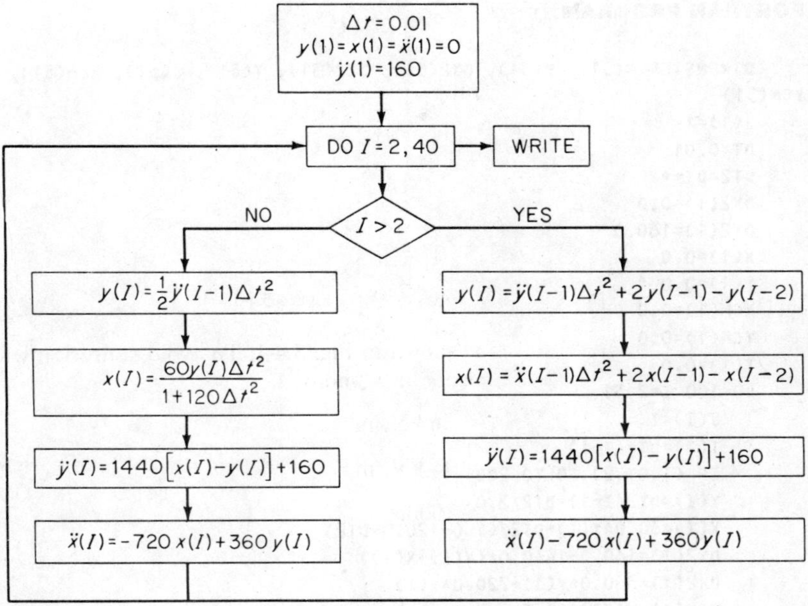

Figure 5.4-2. Flow diagram for computation.

To start the computation, note that the initial accelerations are $\ddot{x}_1 = 0$ and $\ddot{y}_1 = 160$, so that the starting equation, Eq. (4.5-8), can be used only for y.

$$y_2 = \tfrac{1}{2}\ddot{y}_1 \Delta t^2$$

For the calculation of x_2, the special starting equation, Eq. (4.5-10), must be used together with the differential equations

$$x_2 = \tfrac{1}{6}\ddot{x}_2 \Delta t^2$$

$$\ddot{x}_2 = -720x_2 + 360y_2$$

Eliminating $\ddot{x}_2$ gives the following equation for x_2:

$$x_2 = \frac{60y_2\Delta t^2}{1 + 120\Delta t^2}$$

The flow diagram for the computation is shown in Fig. 5.4-2. The Fortran program with the computed results and plot are presented in the following section.

FORTRAN PROGRAM

```
      DIMENSION X(51), Y(51), DX2(51), DY2(51), T(51), J(51), XCM(51),
YCM(51)
      J(1)=1
      DT=0.01
      DT2=DT**2
      DX2(1)=0.0
      DY2(1)=160.0
      X(1)=0.0
      Y(1)=0.0
      XCM(1)=0.0
      YCM(1)=0.0
      T(1)=0.0
      DO 100 I=2,51
         J(I)=I
         T(I)=DT*(I-1)
         IF (I.GT.2) GO TO 200
         Y(I)=DY2(I-1)*DT2/2.0
         X(I)=60.0*Y(I)*DT2/(1.0+120.0*DT2)
         DY2(I)=160.0-1440.0*(Y(I)-X(I))
         DX2(I)=360.0*Y(I)-720.0*X(I)
         YCM(I)=Y(I)*100.0
         XCM(I)=X(I)*100.0
         GO TO 100
200      Y(I)=DY2(I-1)*DT2+2.0*Y(I-1)-Y(I-2)
         X(I)=DX2(I-1)*DT2+2.0*X(I-1)-X(I-2)
         DY2(I)=160.0-1440.0*(Y(I)-X(I))
         DX2(I)=360.0*Y(I)-720.0*X(I)
         YCM(I)=Y(I)*100.0
         XCM(I)=X(I)*100.0
100   CONTINUE
      WRITE (*,300)
300   FORMAT (1X," J  TIME  DISPL. X,cm  DISPL.  Y,cm ",/)
      WRITE (*,400)  (J(I), T(I), XCM(I), YCM(I), I=1,51)
400   FORMAT (1X, I4, F10.4, 2F14.6)
      STOP
      END
```

J	TIME	DISPL. X, cm	DISPL. Y, cm
1	0.	0.	0.
2	0.0100	0.004743	0.800000
3	0.0200	0.037945	3.085483
4	0.0300	0.179492	6.532120
5	0.0400	0.543272	10.663978
6	0.0500	1.251839	14.938455

	TIME	DISPL. X, cm	DISPL. Y, cm
7	0.0600	2.408059	18.842060
8	0.0700	4.069212	21.979170
9	0.0800	6.228632	24.137247
10	0.0900	8.808531	25.316486
11	0.1000	11.665609	25.718578
12	0.1100	14.608634	25.697044
13	0.1200	17.424929	25.678774
14	0.1300	19.911066	26.071951
15	0.1400	21.902195	27.177958
16	0.1500	23.294771	29.124250
17	0.1600	24.058596	31.831100
18	0.1700	24.236122	35.018711
19	0.1800	23.929321	38.253628
20	0.1900	23.276741	41.025845
21	0.2000	22.425163	42.842190
22	0.2100	21.501293	43.318481
23	0.2200	20.588797	42.253105
24	0.2300	19.715019	39.668064
25	0.2400	18.849812	35.809784
26	0.2500	17.916569	31.109264
27	0.2600	16.813269	26.109001
28	0.2700	15.439338	21.370152
29	0.2800	13.723100	17.377266
30	0.2900	11.644379	14.458179
31	0.3000	9.247756	12.733907
32	0.3100	6.643715	12.107629
33	0.3200	3.997202	12.294546
34	0.3300	1.505494	12.886645
35	0.3400	-0.630690	13.439858
36	0.3500	-2.237630	13.566913
37	0.3600	-3.195051	13.018113
38	0.3700	-3.453777	11.734618
39	0.3800	-3.041385	9.863994
40	0.3900	-2.054909	7.734994
41	0.4000	-0.642020	5.796249
42	0.4100	1.025759	4.530393
43	0.4200	2.782778	4.359870
44	0.4300	4.496392	5.562246
45	0.4400	6.086507	8.211138
46	0.4500	7.533995	12.154083
47	0.4600	8.876582	17.031736
48	0.4700	10.193198	22.335049
49	0.4800	11.579966	27.489933
50	0.4900	13.122614	31.953781
51	0.5000	14.870770	35.305943

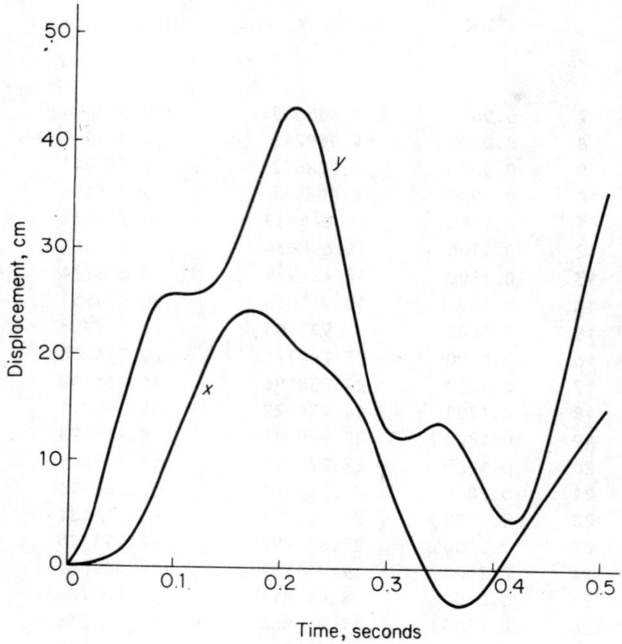

Figure 5.4-3.

5.5 VIBRATION ABSORBER

A spring-mass system k_2, m_2, Fig. 5.5-1, tuned to the frequency of the exciting force such that $\omega^2 = k_2/m_2$, will act as a vibration absorber and reduce the motion of the main mass m_1 to zero. Making the substitution

$$\omega_{11}^2 = \frac{k_1}{m_1} \qquad \omega_{22}^2 = \frac{k_2}{m_2}$$

and assuming the motion to be harmonic, the equation for the amplitude X_1 can be shown to be equal to

$$\frac{X_1 k_1}{F_0} = \frac{\left[1 - \left(\frac{\omega}{\omega_{22}}\right)^2\right]}{\left[1 + \frac{k_2}{k_1} - \left(\frac{\omega}{\omega_{11}}\right)^2\right]\left[1 - \left(\frac{\omega}{\omega_{22}}\right)^2\right] - \frac{k_2}{k_1}} \tag{5.5-1}$$

Figure 5.5-2 shows a plot of this equation with $\mu = m_2/m_1$ as parameter. Note that $k_2/k_1 = \mu(\omega_{22}/\omega_{11})^2$. Since the system is one of two DOF, two natural frequencies exist. These are shown against μ in Fig. 5.5-3.

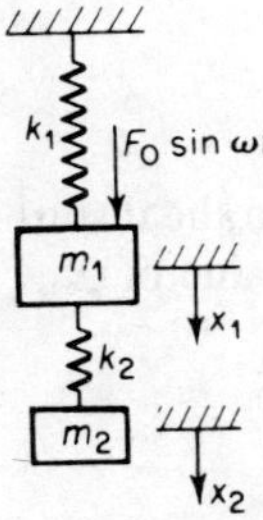

Figure 5.5-1. Vibration absorber.

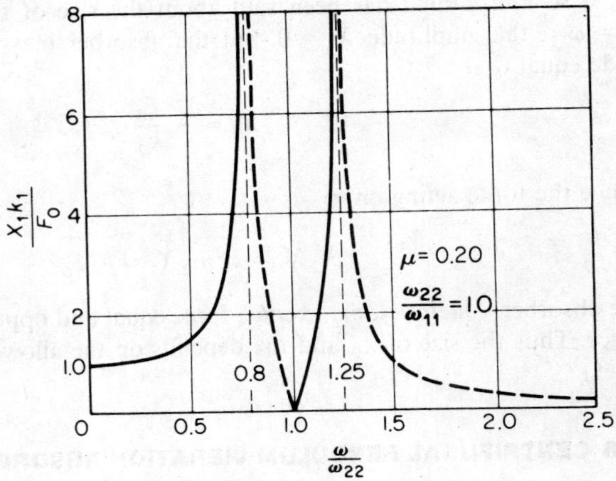

Figure 5.5-2. Response versus frequency.

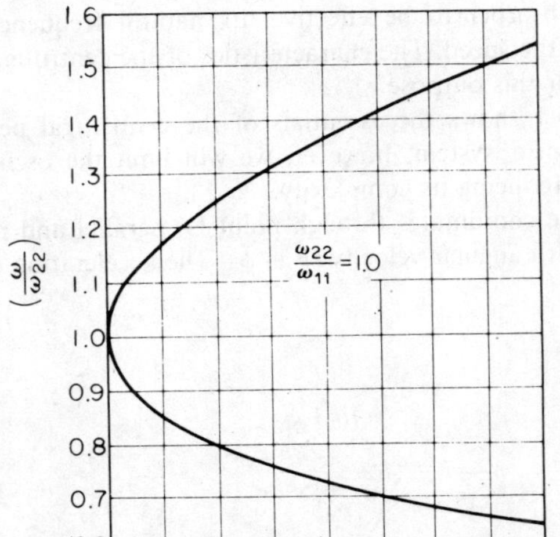

Figure 5.5-3. Natural frequencies versus $\mu = m_2/m_1$.

So far nothing has been said about the size of the absorber mass. At $\omega = \omega_{22}$, the amplitude $X_1 = 0$, but the absorber mass undergoes an amplitude equal to

$$X_2 = -\frac{F_0}{k_2} \tag{5.5-2}$$

Since the force acting on m_2 is

$$k_2 X_2 = \omega^2 m_2 X_2 = -F_0$$

the absorber system k_2, m_2 exerts a force equal and opposite to the disturbing force. Thus the size of k_2 and m_2 depends on the allowable value of X_2.

5.6 CENTRIFUGAL PENDULUM VIBRATION ABSORBER

The vibration absorber of Sec. 5.5 is only effective at one frequency, $\omega = \omega_{22}$. Also, with resonant frequencies on each side of ω_{22}, the usefulness of the spring-mass absorber is narrowly limited.

For a rotating system such as the automobile engine, the exciting torques are proportional to the rotational speed n, which may vary over a wide range. Thus for the absorber to be effective, its natural frequency must also be proportional to the speed. The characteristics of the centrifugal pendulum are ideally suited for this purpose.

Figure 5.6-1 shows the essentials of the centrifugal pendulum. It is a two-DOF nonlinear system; however, we will limit the oscillations to small angles, thereby reducing its complexity.

Placing the coordinates through point O′ parallel and normal to r, the line r rotates with angular velocity $(\dot{\theta} + \dot{\phi})$. The acceleration of m is equal to

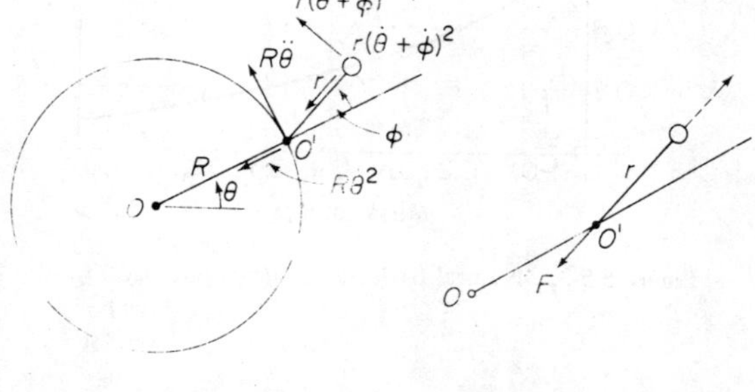

Figure 5.6-1. Centrifugal pendulum.

the vector sum of the acceleration of O′ and the acceleration of m relative to O′

$$a_m = \left[R\ddot{\theta}\sin\phi - R\dot{\theta}^2\cos\phi - r(\dot{\theta}+\dot{\phi})^2\right]i + \left[R\ddot{\theta}\cos\phi + R\dot{\theta}^2\sin\phi + r(\ddot{\theta}+\ddot{\phi})\right]j \tag{5.6-1}$$

Since the moment about O′ is zero, we have, from the j-component of a_m,

$$M_{0'} = m\left[R\ddot{\theta}\cos\phi + R\dot{\theta}^2\sin\phi + r(\ddot{\theta}+\ddot{\phi})\right]r = 0 \tag{5.6-2}$$

Assuming ϕ to be small, we let $\cos\phi = 1$ and $\sin\phi = \phi$ and arrive at the equation for the pendulum:

$$\ddot{\phi} + \left(\frac{R}{r}\dot{\theta}^2\right)\phi = -\left(\frac{R+r}{r}\right)\ddot{\theta} \tag{5.6-3}$$

If we assume the motion of the wheel to be a steady rotation n plus a small sinusoidal oscillation of frequency ω, we can write

$$\begin{aligned}\theta &= nt + \theta_0\sin\omega t\\ \dot{\theta} &= n + \omega\theta_0\cos\omega t \cong n\\ \ddot{\theta} &= -\omega^2\theta_0\sin\omega t\end{aligned} \tag{5.6-4}$$

Then Eq. (5.6-3) becomes

$$\ddot{\phi} + \left(\frac{R}{r}n^2\right)\phi = \left(\frac{R+r}{r}\right)\omega^2\theta_0\sin\omega t \tag{5.6-3'}$$

and we recognize the natural frequency of the pendulum to be

$$\omega_n = n\sqrt{\frac{R}{r}} \tag{5.6-5}$$

and its steady-state solution to be

$$\phi = \frac{(R+r)/r}{-\omega^2 + Rn^2/r}\omega^2\theta_0\sin\omega t \tag{5.6-6}$$

The same pendulum in a gravity field would have a natural frequency of $\sqrt{g/r}$, so it can be concluded that for the contrifugal pendulum the gravity field is replaced by the centrifugal field Rn^2.

We next consider the torque exerted by the pendulum on the wheel. With the j-component of a_m equal to zero, the pendulum force is a tension along r, given by m times the i-component of a_m. Recognizing that the major term of ma_m is $-(R+r)n^2$, the torque exerted by the pendulum on the wheel is

$$T = -m(R+r)n^2R\phi \tag{5.6-7}$$

Substituting for ϕ from Eq. 5.6-6 into the last equation, we obtain

$$T = -\frac{m(R+r)^2Rn^2/r}{Rn^2/r - \omega^2}\omega^2\theta_0\sin\omega t = -\left[\frac{m(R+r)^2}{1 - r\omega^2/Rn^2}\right]\ddot{\theta}$$

Since we can write the torque equation as $T = J_{\text{eff}}\ddot{\theta}$, the pendulum behaves like a wheel of rotational inertia:

$$J_{\text{eff}} = -\frac{m(R + r)^2}{1 - r\omega^2/Rn^2} \tag{5.6-8}$$

which can become infinite at its natural frequency.

This poses some difficulties in the design of the pendulum. For example, to suppress a disturbing torque of frequency equal to four times the rotational speed n, the pendulum must meet the requirement $\omega^2 = (4n)^2 = n^2R/r$, or $r/R = \frac{1}{16}$. Such a short effective pendulum has been made possible by the Chilton bifilar design (see Prob. 5-40).

5.7 VIBRATION DAMPER

In contrast to the vibration absorber, where the exciting force is opposed by the absorber, energy is dissipated by the vibration damper. Figure 5.7-1 represents a friction type of vibration damper, commonly known as the Lanchester damper, which has found practical use in torsional systems such as gas and diesel engines in limiting the amplitudes of vibration at critical speeds. The damper consists of two flywheels a free to rotate on the shaft and driven only by means of the friction rings b when the normal pressure is maintained by the spring-loaded bolts c.

When properly adjusted, the flywheels rotate with the shaft for small oscillations. However, when the torsional oscillations of the shaft tend to become large, the flywheels will not follow the shaft because of their large inertia, and energy is dissipated by friction due to the relative motion. The

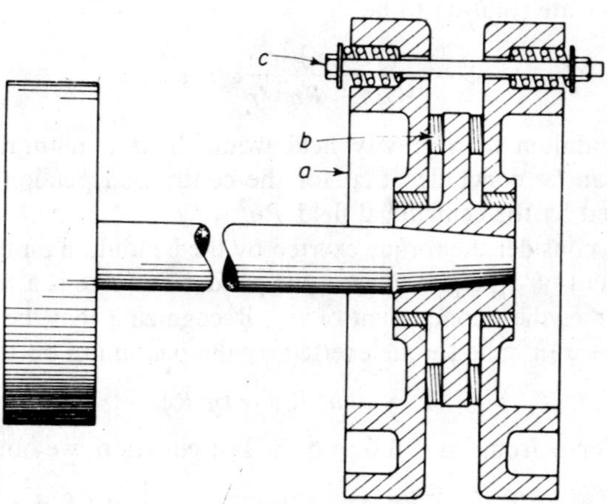

Figure 5.7-1. Torsional vibration damper.

dissipation of energy thus limits the amplitude of oscillation, thereby preventing high torsional stresses in the shaft.

In spite of the simplicity of the torsional damper the mathematical analysis for its behavior is rather complicated. For instance, the flywheels may slip continuously, for part of the cycle, or not at all, depending on the pressure exerted by the spring bolts. If the pressure on the friction ring is either too great for slipping or zero, no energy is dissipated, and the damper becomes ineffective. Maximum energy dissipation takes place at some intermediate pressure, resulting in optimum damper effectiveness.

Obviously, the damper should be placed in a position where the amplitude of oscillation is the greatest. This position generally is found on the side of the shaft away from the main flywheel, since the node is usually near the largest mass.

The untuned viscous vibration damper. In this section we discuss another interesting application of a vibration damper, which has found practical use in suppressing the torsional vibrations of automobile engines. In a rotating system such as an automobile engine, the disturbing frequencies for torsional oscillations are proportional to the rotational speed. However, there is generally more than one such frequency, and the centrifugal pendulum has the disadvantage that several pendulums tuned to the order number of the disturbance must be used. In contrast to the centrifugal pendulum, the untuned viscous torsional damper is effective over a wide operating range. It consists of a free rotational mass within a cylindrical cavity filled with viscous fluid, as shown in Fig. 5.7-2. Such a system is generally incorporated into the end pulley of a crankshaft which drives the fan belt, and is often referred to as the Houdaille damper.

We can examine the untuned viscous damper as a two-DOF system by considering the crankshaft, to which it is attached, as being fixed at one end with the damper at the other end. With the torsional stiffness of the shaft equal to K in.· lb/rad, the damper can be considered to be excited by a harmonic torque $M_0 e^{i\omega t}$. The damper torque results from the viscosity of the fluid within the pulley cavity, and we will assume it to be proportional to the relative rotational speed between the pulley and the free mass. Thus the two

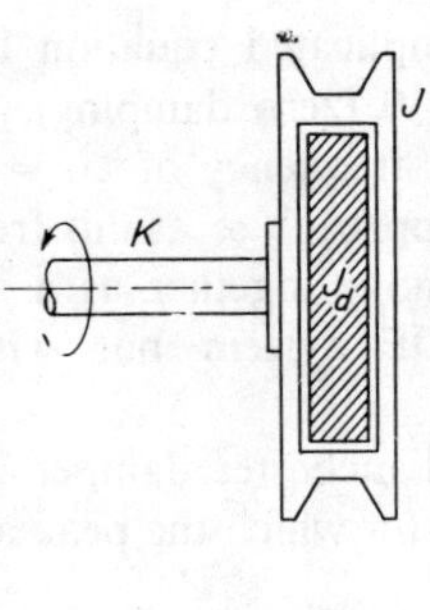

Figure 5.7-2. Untuned viscous damper.

equations of motion for the pulley and the free mass are

$$\begin{aligned} J\ddot{\theta} + K\theta + c(\dot{\theta} - \dot{\phi}) &= M_0 e^{i\omega t} \\ J_d\ddot{\phi} - c(\dot{\theta} - \dot{\phi}) &= 0 \end{aligned} \tag{5.7-1}$$

Assuming the solution to be in the form

$$\begin{aligned} \theta &= \theta_0 e^{i\omega t} \\ \phi &= \phi_0 e^{i\omega t} \end{aligned} \tag{5.7-2}$$

where θ_0 and ϕ_0 are complex amplitudes, their substitution into the differential equations results in

$$\left[\left(\frac{K}{J} - \omega^2\right) + i\frac{c\omega}{J}\right]\theta_0 - \frac{ic\omega}{J}\phi_0 = \frac{M_0}{J}$$

and

$$\left(-\omega^2 + i\frac{c\omega}{J_d}\right)\phi_0 = \frac{ic\omega}{J_d}\theta_0 \tag{5.7-3}$$

Eliminating ϕ_0 between the two equations, the expression for the amplitude θ_0 of the pulley becomes

$$\frac{\theta_0}{M_0} = \frac{(\omega^2 J_d - ic\omega)}{[\omega^2 J_d(K - J\omega^2)] + ic\omega[\omega^2 J_d - (K - J\omega^2)]} \tag{5.7-4}$$

Letting $\omega_n^2 = K/J$ and $\mu = J_d/J$, the critical damping is

$$c_c = 2J\omega_n, \qquad c = \frac{c}{c_c}2J\omega_n = 2\zeta J\omega_n$$

The amplitude equation then becomes

$$\left|\frac{K\theta_0}{M_0}\right| = \sqrt{\frac{\mu^2(\omega/\omega_n)^2 + 4\zeta^2}{\mu^2(\omega/\omega_n)^2(1 - \omega^2/\omega_n^2)^2 + 4\zeta^2[\mu(\omega/\omega_n)^2 - (1 - \omega^2/\omega_n^2)]^2}} \tag{5.7-5}$$

which indicates that $|K\theta_0/M_0|$ is a function of three parameters, ζ, μ, and (ω/ω_n).

This rather complicated equation lends itself to the following simple interpretation. If $\zeta = 0$ (zero damping), we have an undamped single-DOF system with resonant frequency of $\omega_1 = \sqrt{K/J}$. A plot of $|K\theta_0/M_0|$ versus frequency ratio will approach ∞ at this frequency. If $\zeta = \infty$, the damper mass and the wheel will move together as a single mass, and again we have an undamped single-DOF system but with a lower natural frequency of $\sqrt{K/(J + J_d)}$.

Thus, like the Lanchester damper of the previous section, there is an optimum damping ζ_0 for which the peak amplitude is a minimum as shown in

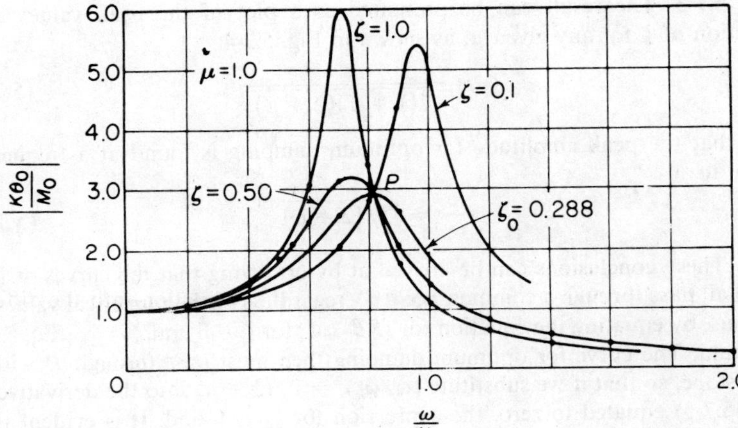

Figure 5.7-3. Response of an untuned viscous damper (all curves pass through P).

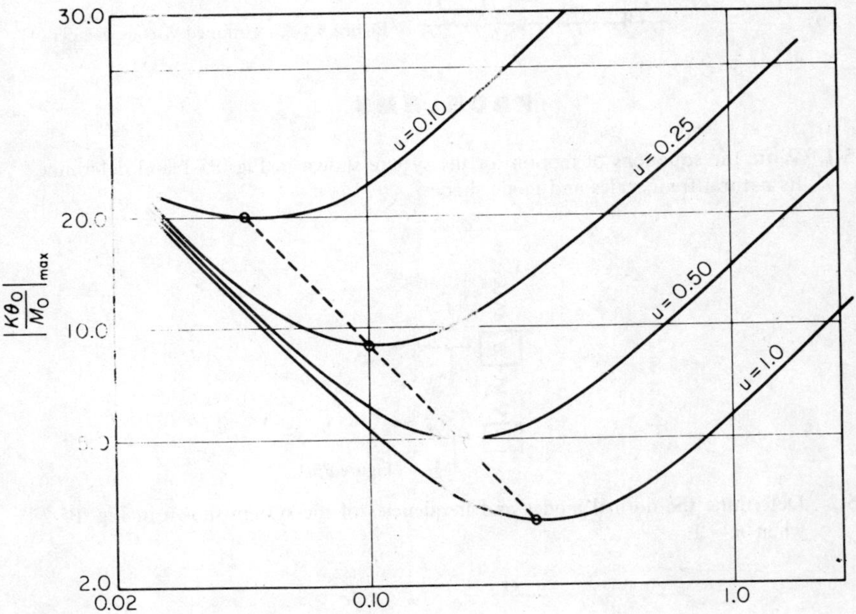

Figure 5.7-4.

Fig. 5.7-3. The result can be presented as a plot of the peak values as a function of ζ for any given μ, as shown in Fig. 5.7-4.

$$\zeta_0 = \frac{\mu}{\sqrt{2(1+\mu)(2+\mu)}} \tag{5.7-6}$$

and that the peak amplitude for optimum damping is found at a frequency equal to

$$\frac{\omega}{\omega_n} = \sqrt{2/(2+\mu)} \tag{5.7-7}$$

These conclusions can be arrived at by observing that the curves of Fig. 5.7-3 all pass through a common point P, regardless of the numerical values of ζ. Thus, by equating the equation for $|K\theta_0/M|$ for $\zeta = 0$ and $\zeta = \infty$, Eq. 5.7-7 is found. The curve for optimum damping then must pass through P with a zero slope, so that if we substitute $(\omega/\omega_n)^2 = 2/(2+\mu)$ into the derivative of Eq. (5.7-5) equated to zero, the expression for ζ_0 is found. It is evident that these conclusions apply also to the linear spring-mass system of Fig. 5.7-5, which is a special case of the damped vibration absorber with the damper spring equal to zero.

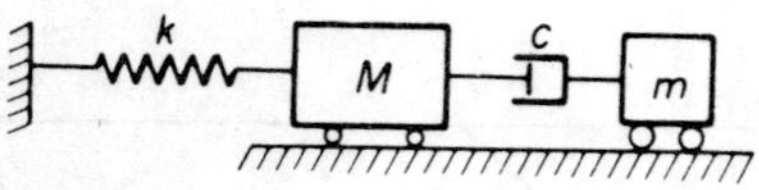

Figure 5.7-5. Untuned viscous damper.

PROBLEMS

5-1 Write the equations of motion for the system shown in Fig. P5-1 and determine its natural frequencies and mode shapes.

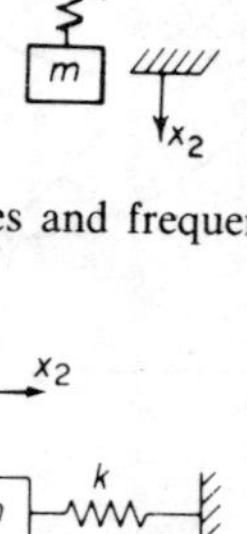

Figure P5-1.

5-2 Determine the normal modes and frequencies of the system shown in Fig. P5-2 when $n = 1$.

Figure P5-2.

5-3 For the system of Prob. 5-2, determine the natural frequencies as a function of n.

5-4 Determine the natural frequencies and mode shapes of the system shown in Fig. P5-4.

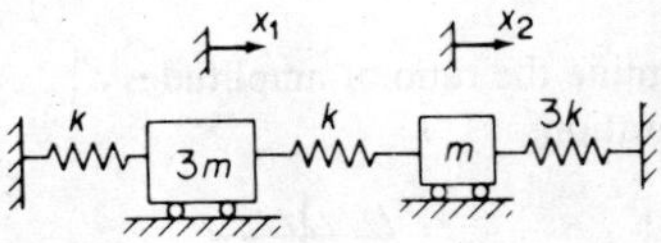

Figure P5-4.

5-5 Determine the normal modes of the torsional system shown in Fig. P5-5 for $K_1 = K_2$ and $J_1 = 2J_2$.

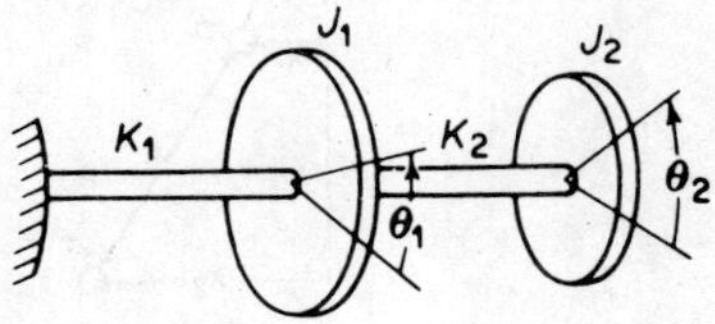

Figure P5-5.

5-6 If $K_1 = 0$ in the torsional system of Prob. 5-5, the system becomes a degenerate two-DOF system with only one natural frequency. Discuss the normal modes of this system as well as a linear spring-mass system equivalent to it. Show that the system can be treated as one of a single DOF by using the coordinate $\phi = (\theta_1 - \theta_2)$.

5-7 Determine the natural frequency of the torsional system shown in Fig. P5-7, and draw the normal mode curve. $G = 11.5 \times 10^6$ psi.

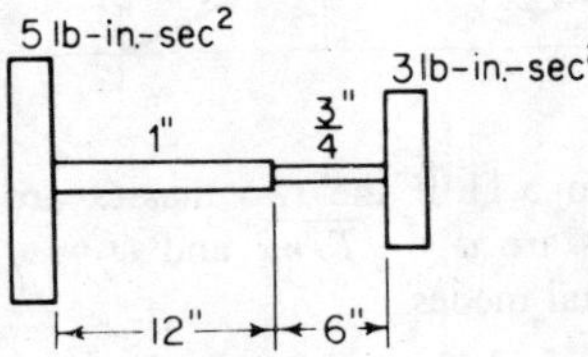

Figure P5-7.

5-8 An electric train made up of two cars each of weight 50,000 lb is connected by couplings of stiffness equal to 16,000 lb/in., as shown in Fig. P5-8. Determine the natural frequency of the system.

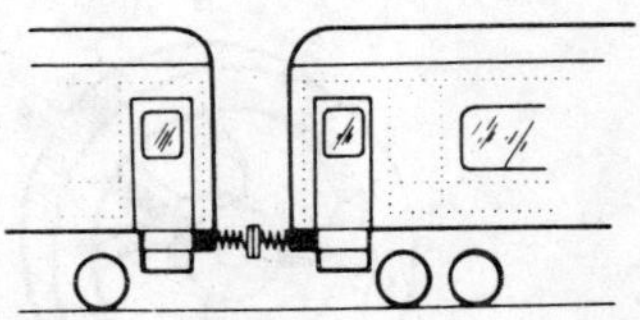

Figure P5-8.

5-9 Assuming small amplitudes, set up the differential equations of motion for the double pendulum using the coordinates shown in Fig. P5-9. Show that the natural frequencies of the system are given by the equation

$$\omega = \sqrt{\frac{g}{l}(2 \pm \sqrt{2})}$$

Determine the ratio of amplitudes x_1/x_2 and locate the nodes for the two modes of vibration.

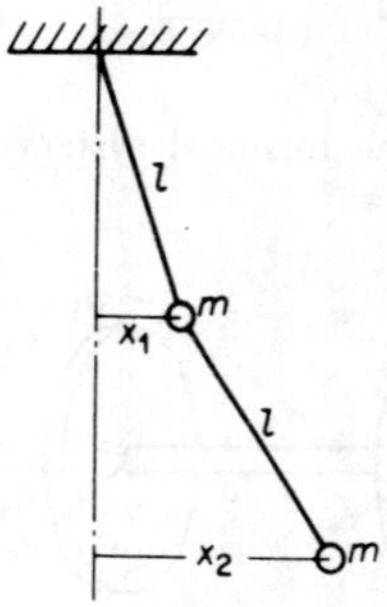

Figure P5-9.

5-10 Set up the equations of motion of the double pendulum in terms of angles θ_1 and θ_2 measured from the vertical.

5-11 Two masses m_1 and m_2 are attached to a light string with tension T, as shown in Fig. P5-11. Assuming that T remains unchanged when the masses are displaced normal to the string, write the equations of motion expressed in matrix form.

Figure P5-11.

5-12 In Problem 5-11 if the two masses are made equal, show that normal mode frequencies are $\omega = \sqrt{T/ml}$ and $\omega_2 = \sqrt{3T/ml}$. Establish the configuration for these normal modes.

5-13 In Problem 5-11 if $m_1 = 2m$, and $m_2 = m$, determine the normal mode frequencies and mode shapes.

5 14 A torsional system shown in Fig. P5-14 is composed of a shaft of stiffness K_1, a hub of radius r and moment of inertia J_1, four leaf springs of stiffness k_2, and an

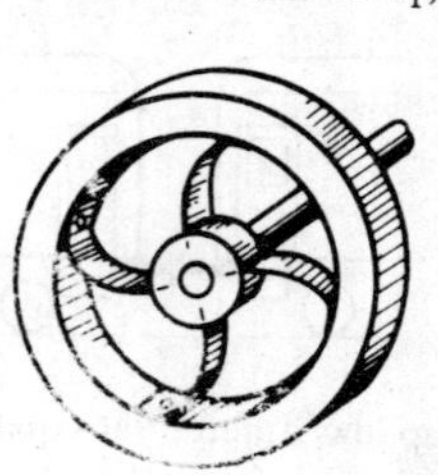

Figure P5-14.

outer wheel of radius R and moment of inertia J_2. Set up the differential equations for torsional oscillation, assuming one end of the shaft to be fixed. Show that the frequency equation reduces to

$$\omega^4 - \left(\omega_{11}^2 + \omega_{22}^2 + \frac{J_2}{J_1}\omega_{22}^2\right)\omega^2 + \omega_{11}^2\omega_{22}^2 = 0$$

where ω_{11} and ω_{22} are uncoupled frequencies given by the expressions

$$\omega_{11}^2 = \frac{K_1}{J_1}, \qquad \omega_{22}^2 = \frac{4k_2R^2}{I}$$

5-15 Two equal pendulums free to rotate about the x-x axis are coupled together by a rubber hose of torsional stiffness k lb · in./rad, as shown in Fig. P5-15. Determine the natural frequencies for the normal modes of vibration, and describe how these motions may be started.
If $l = 19.3$ in., mg = 3.86 lb, and $k = 2.0$ lb · in./rad, determine the beat period for a motion started with $\theta_1 = 0$ and $\theta_2 = \theta_0$. Examine carefully the phase of the motion as the amplitude approaches zero.

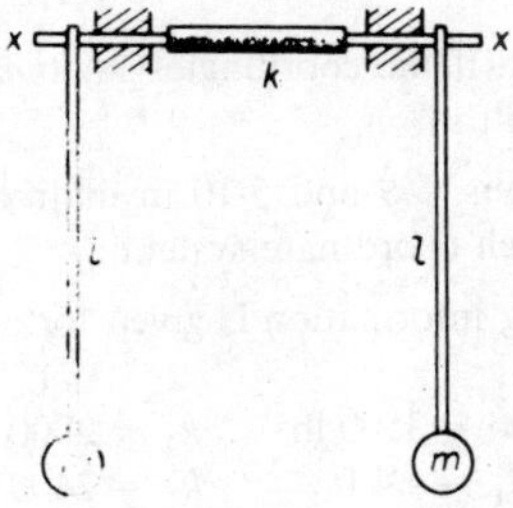

Figure P5-15.

5-16 Determine the equations of motion for the system of Prob. 5-4 when the initial conditions are $x_1(0) = A$, $\dot{x}_1(0) = x_2(0) = \dot{x}_2(0) = 0$.

5-17 The double pendulum of Prob. 5-9 is started with the following initial conditions: $x_1(0) = x_2(0) = X$, $\dot{x}_1(0) = \dot{x}_2(0) = 0$. Determine the equations of motion.

5-18 The lower mass of Prob. 5-1 is given a sharp blow, imparting to it an initial velocity $\dot{x}_2(0) = V$. Determine the equation of motion.

5-19 If the system of Prob. 5-1 is started with initial conditions $x_1(0) = 0$, $x_2(0) = 1.0$, $\dot{x}_1(0) = \dot{x}_2(0) = 0$, show that the equations of motion are

$$x_1(t) = 0.447\cos\omega_1 t - 0.447\cos\omega_2 t$$

$$x_2(t) = 0.722\cos\omega_1 t + 0.27[illegible]\cos\omega_2 t$$

$$\omega_1 = \sqrt{0.382k/m} \qquad \omega_2 = \sqrt{[illegible]18k/m}$$

5-20 Choose coordinates x for the displacement of c and θ clockwise for the rotation of the uniform bar shown in Fig. P5-20, and determine the natural frequencies and mode shapes.

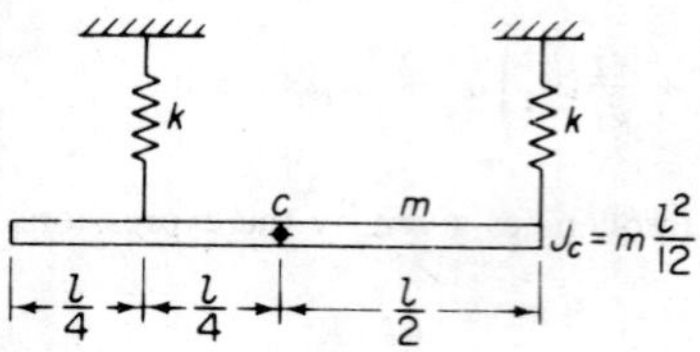

Figure P5-20.

5-21 Set up the matrix equation of motion for the system shown in Fig. P5-21, using coordinates x_1 and x_2 at m and $2m$. Determine the equation for the normal mode frequencies and describe the mode shapes.

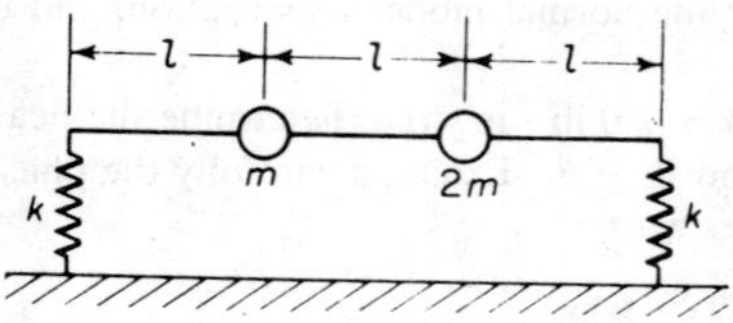

Figure P5-21.

5-22 In Prob. 5-21, if the coordinates x at m and θ are used, what form of coupling will they result in?

5-23 Compare Probs. 5-9 and 5-10 in matrix form and indicate the type of coupling present in each coordinate system.

5-24 The following information is given for a certain automobile shown in Fig. P5-24.

$$W = 3500 \text{ lb} \qquad k_1 = 2000 \text{ lb/ft}$$
$$l_1 = 4.4 \text{ ft} \qquad k_2 = 2400 \text{ lb/ft}$$
$$l_2 = 5.6 \text{ ft} \qquad r = 4 \text{ ft} = \text{radius of gyration about } cg$$

Determine the normal modes of vibration and locate the node for each mode.

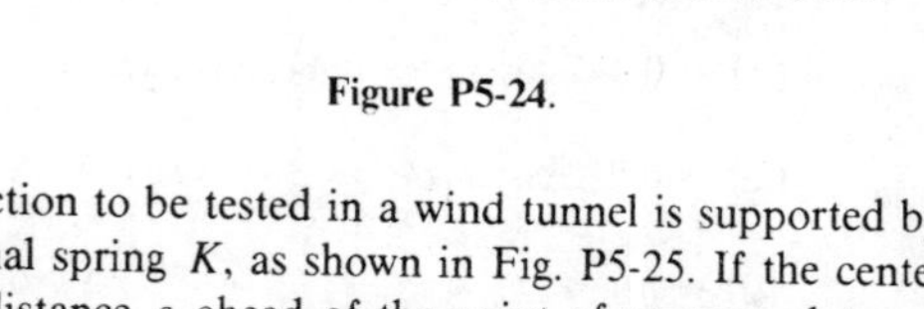

Figure P5-24.

5-25 An airfoil section to be tested in a wind tunnel is supported by a linear spring k and a torsional spring K, as shown in Fig. P5-25. If the center of gravity of the section is a distance e ahead of the point of support, determine the differential equations of motion of the system.

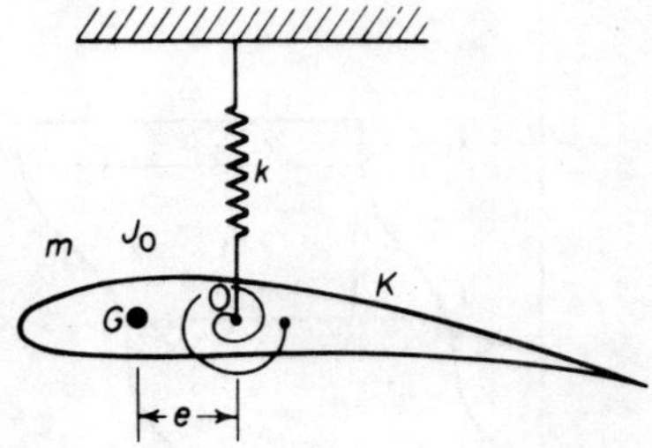

Figure P5-25.

5-26 Determine the natural frequencies and normal modes of the system shown in Fig. P5-26 when

$$gm_1 = 3.86 \text{ lb} \qquad k_1 = 20 \text{ lb/in.}$$
$$gm_2 = 1.93 \text{ lb} \qquad k_2 = 10 \text{ lb/in.}$$

When forced by $F_1 = F_0 \sin \omega t$, determine the equations for the amplitudes and plot them against ω/ω_{11}.

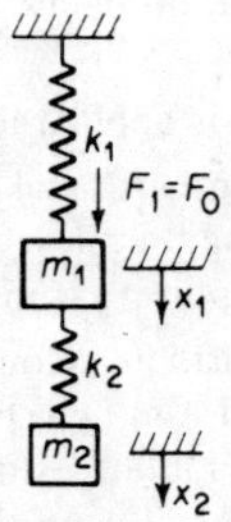

Figure P5-26.

5-27 A rotor is mounted in bearings that are free to move in a single plane, as shown in Fig. P5-27. The rotor is symmetrical about O with total mass M and moment of inertia J_0 about an axis perpendicular to the shaft. If a small unbalance mr acts at an axial distance b from its center O, determine the equations of motion for a rotational speed ω.

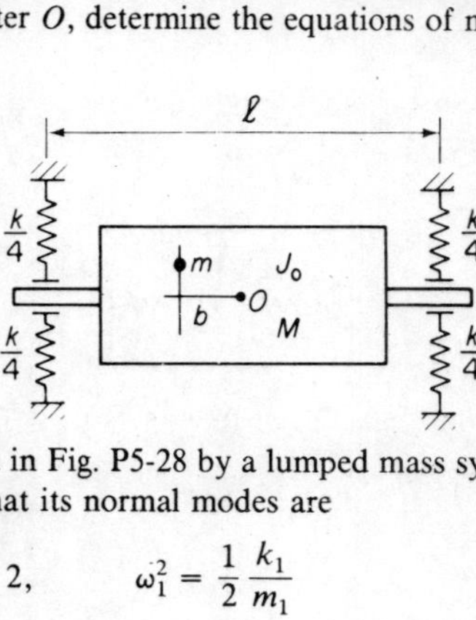

Figure P5-27.

5-28 A two-story building is represented in Fig. P5-28 by a lumped mass system where $m_1 = \frac{1}{2}m_2$ and $k_1 = \frac{1}{2}k_2$. Show that its normal modes are

$$\left(\frac{x_1}{x_2}\right)^{(1)} = 2, \qquad \omega_1^2 = \frac{1}{2}\frac{k_1}{m_1}$$

$$\left(\frac{x_1}{x_2}\right)^{(2)} = -1, \qquad \omega_2^2 = 2\frac{k_1}{m_1}$$

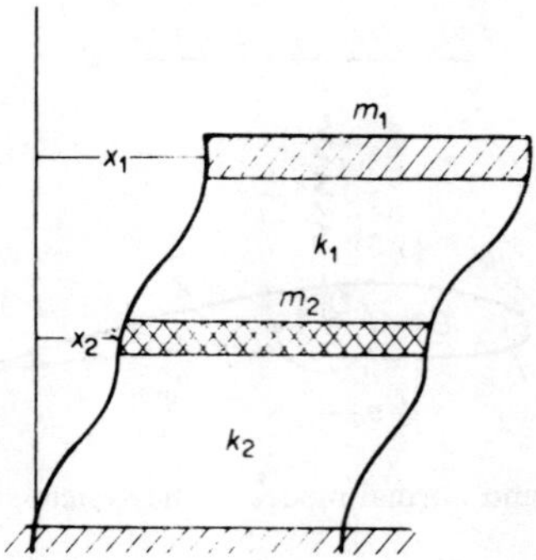

Figure P5-28.

5-29 In Prob. 5-28, if a force is applied to m_1 to deflect it by unity and the system is released from this position, determine the equation of motion of each mass by the normal mode summation method.

5-30 In Prob. 5-29, determine the ratio of the maximum shear in the first and second stories.

5-31 Repeat Prob. 5-29 if the load is applied to m_2, displacing it by unity.

5-32 Assume in Prob. 5-28 that an earthquake causes the ground to oscillate in the horizontal direction according to the equation $x_g = X_g \sin \omega t$. Determine the response of the building and plot it against ω/ω_1.

5-33 To simulate the effect of an earthquake on a rigid building, the base is assumed to be connected to the ground through two springs: K_h for the translational stiffness, and K_r for the rotational stiffness. If the ground is now given a harmonic motion $Y = Y_0 \sin \omega t$, set up the equations of motion in terms of the coordinates shown in Fig. P5-33.

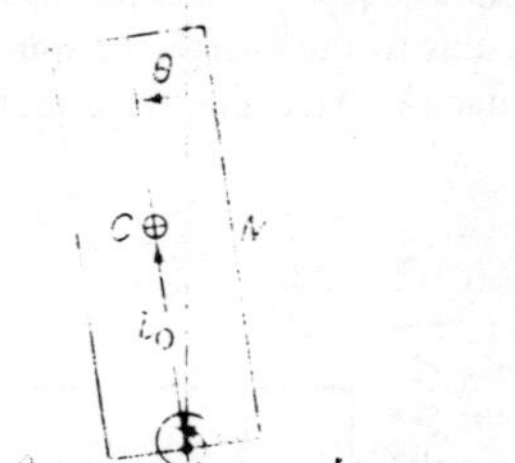

Figure P5-33.

5-34 Solve the equations of Prob. 5-33 by letting

$$\omega_h^2 = \frac{K_h}{M}, \qquad \left(\frac{l}{\rho_c}\right)^2 = 2$$

$$\omega_r^2 = \frac{K_r}{M\rho_c^2}, \qquad \left(\frac{\omega_r}{\omega_h}\right)^2 = 4$$

The first natural frequency and mode shape are

$$\frac{\omega_1}{\omega_h} = 0.734 \quad \text{and} \quad \frac{Y_0}{l_0\theta} = -1.14$$

which indicate a motion that is predominantly translational. Establish the second natural frequency and its mode ($Y_1 = Y_0 - 2l_0\theta_0$ = displacement of top).

5-35 The response and mode configuration for Probs. 5-33 and 5-34 are shown in Fig. P5-35. Verify the mode shapes for several values of the frequency ratio.

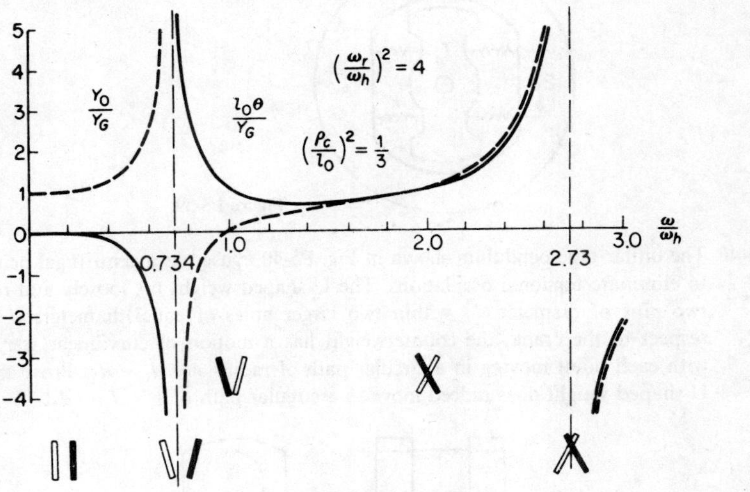

Figure P5-35.

5-36 The expansion joints of a concrete highway are 45 ft apart. These joints cause a series of impulses at equal intervals to affect cars traveling at a constant speed. Determine the speeds at which pitching motion and up-and-down motion are most apt to arise for the automobile of Prob. 5-24.

5-37 For the system shown in Fig. P5-37, W_1 = 200 lb and the absorber weighs W_2 = 50 lb. If W_1 is excited by a 2 lb-in. unbalance rotating at 1800 rpm, determine the proper value of the absorber spring k_2. What will be the amplitude of W_2?

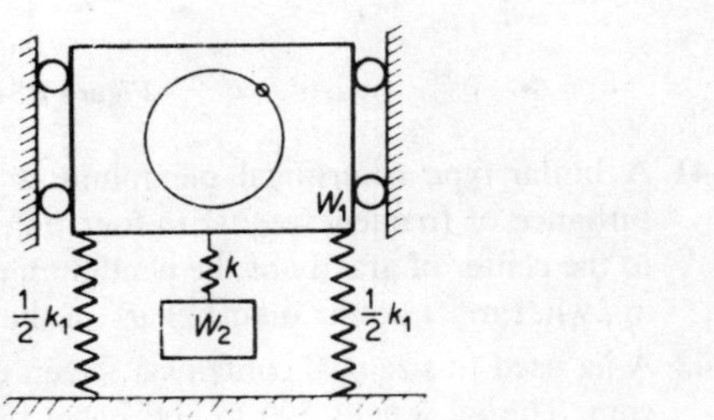

Figure P5-37.

5-38 In Prob. 5-37, if a dashpot c is introduced between W_1 and W_2, determine the amplitude equations by the complex algebra method.

5-39 A flywheel of moment of inertia I has a torsional absorber of moment of inertia I_d free to rotate on the shaft and connected to the flywheel by four springs of stiffness k lb/in., as shown in Fig. P5-39. Set up the differential equations of motion for the system, and discuss the response of the system to an oscillatory torque.

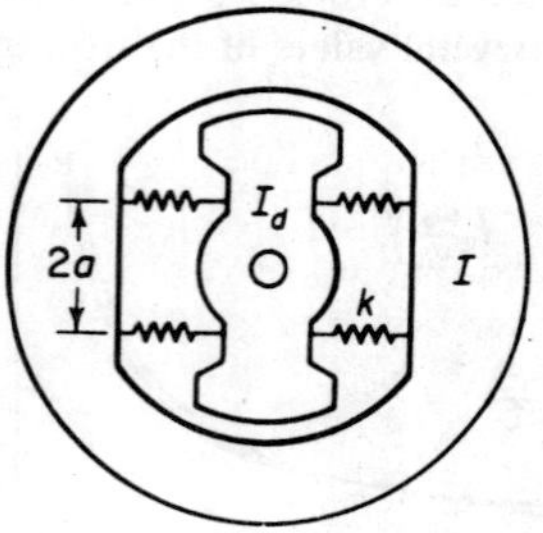

Figure P5-39.

5-40 The bifilar-type pendulum shown in Fig. P5-40 is used as a centrifugal pendulum to eliminate torsional oscillations. The U-shaped weight fits loosely and rolls on two pins of diameter d_2 within two larger holes of equal diameters d_1. With respect to the crank, the counterweight has a motion of curvilinear translation with each point moving in a circular path of radius $r = d_1 - d_2$. Prove that the U-shaped weight does indeed move in a circular path of $r = d_1 - d_2$.

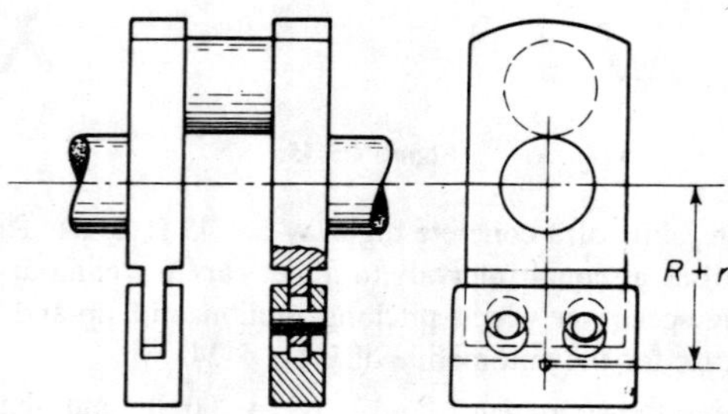

Figure P5-40.

5-41 A bifilar-type centrifugal pendulum is proposed to eliminate a torsional disturbance of frequency equal to four times the rotational speed. If the distance R to the center of gravity of the pendulum mass is made equal to 4.0 in. and $d_1 = \frac{3}{4}$ in., what must be the diameter d_2 of the pins?

5-42 A jig used to size coal contains a screen that reciprocates with a frequency of 600 cpm. The jig weighs 500 lb and has a fundamental frequency of [illegible] m. If an absorber weighing 125 lb is to be installed to eliminate the vibration of the jig frame, determine the absorber spring stiffness. What will be the resulting two natural frequencies of the system?

5-43 In a certain refrigeration plant a section of pipe carrying the refrigerant vibrated violently at a compressor speed of 232 rpm. To eliminate this difficulty it was proposed to clamp a spring-mass system to the pipe to act as an absorber. For a trial test a 2.0-lb absorber tuned to 232 cpm resulted in two natural frequencies of 198 and 272 cpm. If the absorber system is to be designed so that the natural frequencies lie outside the region 160 to 320 cpm, what must be the weight and spring stiffness?

5-44 A type of damper frequently used on automobile crankshafts is shown in Fig. P5-44. J represents a solid disk free to spin on the shaft, and the space between the disk and case is filled with a silicone oil of coefficient of viscosity μ. The damping action results from any relative motion between the two. Derive an equation for the damping torque exerted by the disk on the case due to a relative velocity of ω.

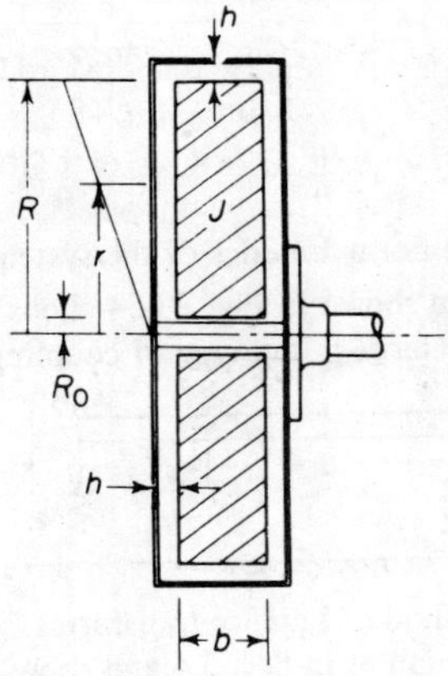

Figure P5-44.

5-45 For the Houdaille viscous damper with mass ratio $\mu = 0.25$, determine the optimum damping ζ_0 and the frequency at which the damper is most effective.

5-46 If the damping for the viscous damper of Prob. 5-45 is equal to $\zeta = 0.10$, determine the peak amplitude as compared to the optimum.

5-47 Establish the relationships given by Eqs. (5.7-7) and (5.7-8).

5-48 Derive the equations of motion for the two masses in Fig. 5.7-5 and follow the parallel development of the untuned torsional vibration damper problem.

5-49 Draw the flow diagram and develop the Fortran program for the computation of the response of the system shown in Prob. 5-4 when the mass $3m$ is excited by a rectangular pulse of magnitude 100 lb and duration $6\pi\sqrt{m/k}$ sec.

5-50 In Prob. 5-28 assume the following data: $k_1 = 4 \times 10^3$ lb/in., $k_2 = 6 \times 10^3$ lb/in., $m_1 = m_2 = 100$. Develop the flow diagram and the Fortran program for the case where the ground is given a displacement $y = 10'' \sin \pi t$ for 4 s.

5-51 Figure P5-51 shows a degenerate 3 DFS. Its characteristic equation yields one zero root and two elastic vibration frequencies. Discuss the physical significance of the fact that three coordinates are required but only two natural frequencies are obtained.

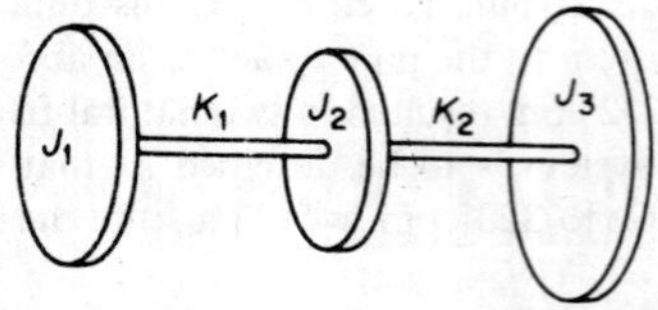

Figure P5-51.

5-52 The two uniform rigid bars shown in Fig. P5-52 are of equal length but of different masses. Determine the equations of motion and the natural frequencies and mode shapes using matrix methods.

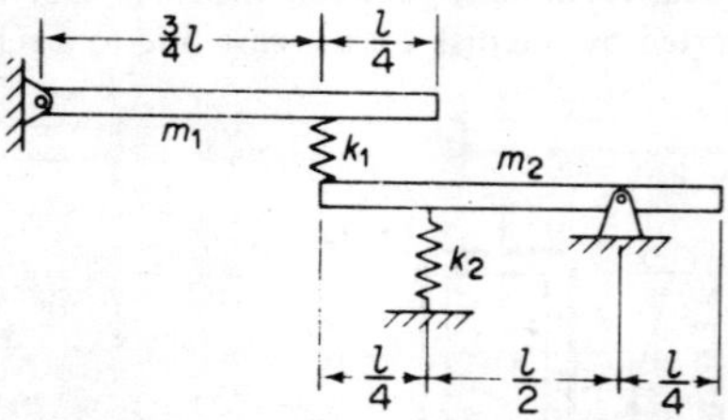

Figure P5-52.

5-53 Show that the normal modes of the system of Prob. 5-51 are orthogonal.

5-54 For the system shown in Fig. P5-54 choose coordinates x_1 and x_2 at the ends of the bar and determine the type of coupling this introduces.

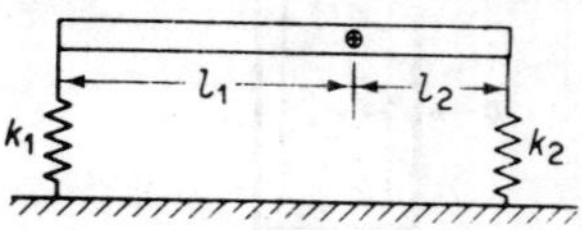

Figure P5-54.

5-55 Using the method of Laplace transforms, solve analytically the problem solved by the digital computer in Sec. 5.4 and show that the solution is

$$x_{cm} = 13.01(1 - \cos \omega_1 t) - 1.90(1 - \cos \omega_2 t)$$
$$y_{cm} = 16.08(1 - \cos \omega_1 t) + 6.14(1 - \cos \omega_2 t)$$

5-56 Consider the free vibration of any two degrees of freedom system with arbitrary initial conditions, and show by examination of the subsidiary equations of Laplace transforms that the solution is the sum of normal modes.

5-57 Determine by the method of Laplace transformation the solution to the forced vibration problem shown in Fig. P5-57. Initial conditions are $x_1(0)$, $\dot{x}_1(0)$, $x_2(0)$ and $\dot{x}_2(0)$.

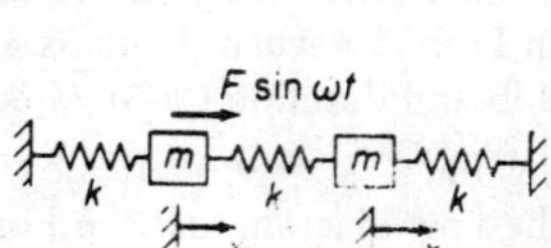

Figure P5-57.

6

PROPERTIES OF VIBRATING SYSTEMS

In this chapter we discuss the various properties of vibrating systems and the matrix techniques applicable to them. These concepts form the basis for the treatment and understanding of the behavior of large systems.

Since the elastic behavior of a structure can be expressed in terms of the stiffness or the flexibility, the equations of motion may be formulated by either the stiffness matrix $[k]$ or the flexibility matrix $[a]$. In the stiffness formulation, the force is expressed in terms of the displacement by the equation

$$\{f\} = [k]\{x\}$$

whereas its inverse

$$\{x\} = [k]^{-1}\{f\} = [a]\{f\}$$

leads to the flexibility approach. The choice as to which approach to adopt depends on the problem. Some problems are more easily pursued on the basis of stiffness, whereas for others the flexibility approach may be desirable.

6.1 FLEXIBILITY MATRIX

The flexibility matrix written in terms of its coefficients a_{ij} is

$$\begin{Bmatrix} x_1 \\ x_2 \\ x_3 \end{Bmatrix} = \begin{bmatrix} a_{11} & a_{12} & a_{13} \\ a_{21} & a_{22} & a_{23} \\ a_{31} & a_{32} & a_{33} \end{bmatrix} \begin{Bmatrix} f_1 \\ f_2 \\ f_3 \end{Bmatrix} \tag{6.1-1}$$

The flexibility influence coefficient a_{ij} is defined as the displacement at i due to a unit force applied at j with all other forces equal to zero. Thus the first column of the above matrix represents the displacements corresponding to $f_1 = 1$ and $f_2 = f_3 = 0$. The second column is equal to the displacements for $f_2 = 1$ and $f_1 = f_3 = 0$, and so on.

Example 6.1-1

Determine the flexibility matrix for the axial displacements of the two spring system shown in Fig. 6.1-1.

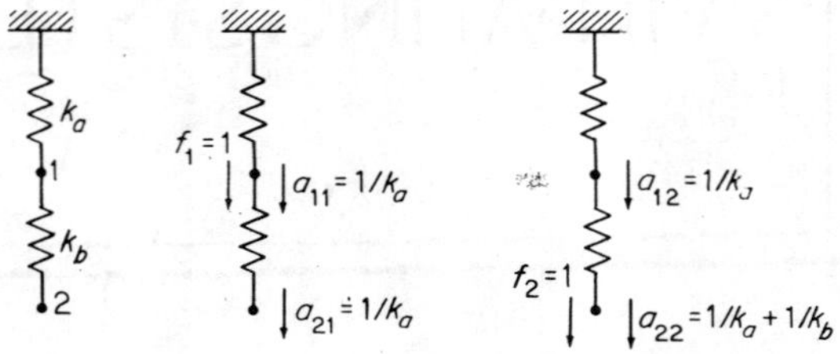

Figure 6.1-1.

Solution: The accompanying figures show a unit force applied at each station with the other force equal to zero. With the unit force applied at 1, only spring (a) is stretched, and station 2 is displaced with spring (b) unstretched. With $f_2 = 1$ and $f_1 = 0$, spring (b) is stretched by $1/k_b$ and the displacement of station 2 is the sum of the stretches in each spring. The flexibility matrix can then be written as

$$\begin{Bmatrix} x_1 \\ x_2 \end{Bmatrix} = \begin{bmatrix} \dfrac{1}{k_a} & \dfrac{1}{k_a} \\ \dfrac{1}{k_a} & \left(\dfrac{1}{k_a} + \dfrac{1}{k_b}\right) \end{bmatrix} \begin{Bmatrix} f_1 \\ f_2 \end{Bmatrix}$$

Example 6.1-2

Determine the flexibility influence coefficients for stations (1), (2), and (3) of the uniform cantilever beam shown in Fig. 6.1-2.

Solution: The influence coefficients may be determined by placing unit loads at (1), (2) and (3) as shown, and calculating the deflections at these points. Using the area moment method,† the deflection at the various stations is equal to the moment of the M/EI area about the position in question. For example, the value of $a_{21} = a_{12}$ is

†Egor P. Popov, *Introduction to Mechanics of Solids* (Englewood Cliffs, N.J.: Prentice-Hall, Inc., 1968), p. 411.

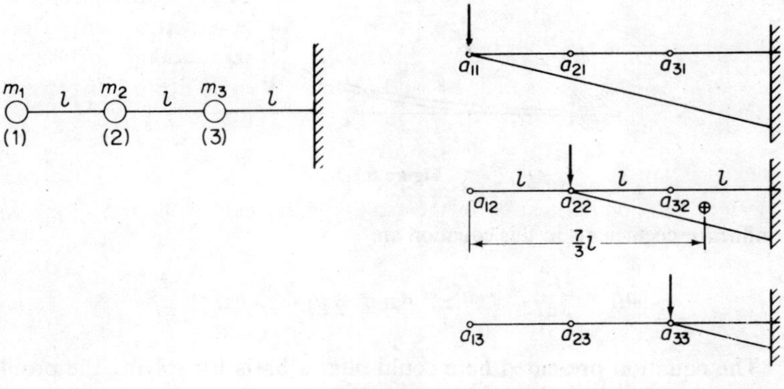

Figure 6.1-2.

found from Fig. 6.1-2 as follows:

$$a_{12} = \frac{1}{EI}\left[\frac{1}{2}(2l)^2 \times \frac{7}{3}l\right] = \frac{14}{3}\frac{l^3}{EI}$$

The other values (determined as before) are

$$a_{11} = \frac{27}{3}\frac{l^3}{EI}, \qquad a_{21} = a_{12} = \frac{14}{3}\frac{l^3}{EI}$$

$$a_{22} = \frac{8}{3}\frac{l^3}{EI}, \qquad a_{23} = a_{32} = \frac{2.5}{3}\frac{l^3}{EI}$$

$$a_{33} = \frac{1}{3}\frac{l^3}{EI}, \qquad a_{13} = a_{31} = \frac{4}{3}\frac{l^3}{EI}$$

The flexibility matrix can now be written as

$$a = \frac{l^3}{3EI}\begin{bmatrix} 27 & 14 & 4 \\ 14 & 8 & 2.5 \\ 4 & 2.5 & 1 \end{bmatrix}$$

and the symmetry about the diagonal should be noted.

Example 6.1-3

The flexibility influence coefficients can be used to set up the equations of a flexible shaft supported by a rigid bearing at one end with a force P and a moment M at the other end, as shown in Fig. 6.1-3.

The deflection and slope at the free end is

$$\begin{aligned} y &= a_{11}P + a_{12}M \\ \theta &= a_{21}P + a_{22}M \end{aligned} \tag{a}$$

which can be expressed by the matrix equation

$$\begin{Bmatrix} y \\ \theta \end{Bmatrix} = \begin{bmatrix} a_{11} & a_{12} \\ a_{21} & a_{22} \end{bmatrix}\begin{Bmatrix} P \\ M \end{Bmatrix} \tag{b}$$

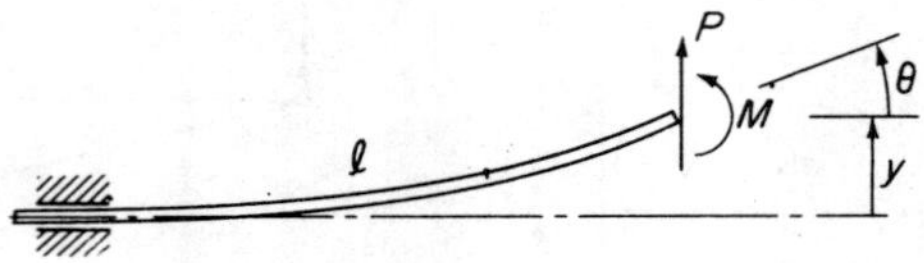

Figure 6.1-3.

The influence coefficients in this equation are

$$a_{11} = \frac{l^3}{3EI}, \qquad a_{12} = a_{21} = \frac{l^2}{2EI}, \qquad a_{22} = \frac{l}{EI} \tag{c}$$

The equation presented here could offer a basis for solving the problem of the gyroscopic whirl of a spinning wheel fixed to the end of an overhanging shaft. P and M in this case would be replaced by the inertia force and the gyroscopic moment of the spinning wheel. By including the flexibility of the supporting bearing, a still more general problem can be examined (see Prob. 6-41).

Reciprocity theorem. The reciprocity theorem states that in a linear system $a_{ij} = a_{ji}$. For the proof of this theorem, we consider the work done by forces f_i and f_j where the order of loading is i followed by j and then by its reverse. Reciprocity results when we recognize that the the work done is independent of the order of loading.

Applying f_i, the work done is $\frac{1}{2}f_i^2 a_{ii}$. Applying f_j, the work done by f_j is $\frac{1}{2}f_j^2 a_{jj}$. However, i undergoes further displacement $a_{ij}f_j$, and the additional work done by f_i becomes $a_{ij}f_j f_i$. Thus the total work done is

$$W = \tfrac{1}{2}f_i^2 a_{ii} + \tfrac{1}{2}f_j^2 a_{jj} + a_{ij}f_j f_i$$

We now reverse the order of loading, in which case the total work done is

$$W = \tfrac{1}{2}f_j^2 a_{jj} + \tfrac{1}{2}f_i^2 a_{ii} + a_{ji}f_i f_j$$

Since the work done in the two cases must be equal, we find that

$$a_{ij} = a_{ji}$$

Example 6.1-4

Figure 6.1-4 shows an overhanging beam with P first applied at 1 and then at 2. In (a) the deflection at 2 is

$$y_2 = a_{21}P$$

In (b) the deflection at 1 is

$$y_1 = a_{12}P$$

Since $a_{12} = a_{21}$, y_1 will equal y_2.

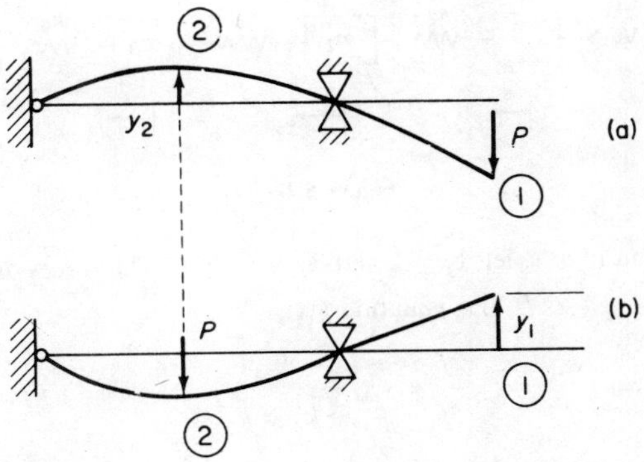

Figure 6.1-4.

6.2 STIFFNESS MATRIX

The stiffness matrix written in terms of its coefficients k_{ij} is

$$\begin{Bmatrix} f_1 \\ f_2 \\ f_3 \end{Bmatrix} = \begin{bmatrix} k_{11} & k_{12} & k_{13} \\ k_{21} & k_{22} & k_{23} \\ k_{31} & k_{32} & k_{33} \end{bmatrix} \begin{Bmatrix} x_1 \\ x_2 \\ x_3 \end{Bmatrix} \tag{6.2-1}$$

The elements of the stiffness matrix have the following interpretation. If $x_1 = 1.0$ and $x_2 = x_3 = 0$, the forces at 1, 2, and 3 that are required to maintain this displacement according to Eq. (6.2-1) are k_{11}, k_{21}, and k_{31} in the first column. Similarly, the forces f_1, f_2, and f_3 required to maintain the displacement configuration $x_1 = 0$, $x_2 = 1.0$, and $x_3 = 0$ are k_{12}, k_{22}, and k_{32} in the second column. Thus the general rule for establishing the stiffness elements of any column is to set the displacement corresponding to that column to unity with all other displacements equal to zero and measure the forces required at each station.

Example 6.2-1

Figure 6.2-1 shows a three-DOF system. Determine the stiffness matrix.

Solution: Let $x_1 = 1.0$ and $x_2 = x_3 = 0$. The forces required at 1, 2, and 3, considering forces to the right as positive, are

$$f_1 = k_1 + k_2 = k_{11}$$
$$f_2 = -k_2 = k_{21}$$
$$f_3 = 0 = k_{31}$$

Repeat with $x_2 = 1$, and $x_1 = x_3 = 0$. The forces are now

$$f_1 = -k_2 = k_{12}$$
$$f_2 = k_2 + k_3 = k_{22}$$
$$f_3 = -k_3 = k_{32}$$

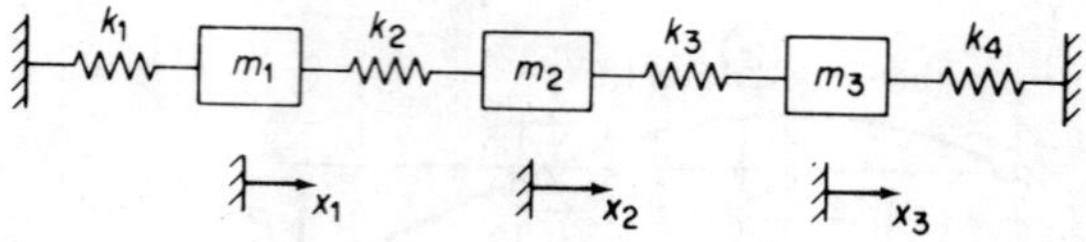

Figure 6.2-1.

For the last column of k's, let $x_3 = 1$ and $x_1 = x_2 = 0$. The forces are

$$f_1 = 0 = k_{13}$$
$$f_2 = -k_3 = k_{23}$$
$$f_3 = k_3 + k_4 = k_{33}$$

The stiffness matrix can now be written as

$$K = \begin{bmatrix} (k_1 + k_2) & -k_2 & 0 \\ -k_2 & (k_2 + k_3) & -k_3 \\ 0 & -k_3 & (k_3 + k_4) \end{bmatrix}$$

Example 6.2-2

Consider the four-story building with rigid floors shown in Fig. 6.2-2. Show diagramatically the significance of the terms of the stiffness matrix.

Solution: The stiffness matrix for the problem is a 4×4 matrix. The elements of the first column are obtained by giving station 1 a unit displacement with the displacement of all other stations equal to zero, as shown in Fig. 6.2-2(b). The forces required for this configuration are the elements of the first column. Similarly, the elements of the second column are the forces necessary to maintain the configuration shown in Fig. 6.2-2(c).

It is evident from these diagrams that $k_{11} = k_{22} = k_{33}$ and that they can be determined from the deflection of a fixed-fixed beam of length $2l$, which is

$$k_{11} = k_{22} = k_{33} = \frac{192\,EI}{(2l)^3} = 24\frac{EI}{l^3}$$

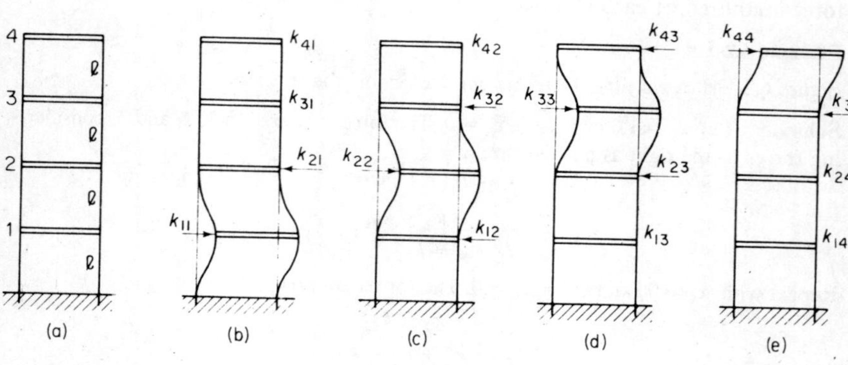

Figure 6.2-2.

The stiffness matrix is then easily found as

$$[k] = \frac{EI}{l^3}\begin{bmatrix} 24 & -12 & 0 & 0 \\ -12 & 24 & -12 & 0 \\ 0 & -12 & 24 & -12 \\ 0 & 0 & -12 & 12 \end{bmatrix}$$

6.3 STIFFNESS OF BEAM ELEMENTS

Structural elements are generally composed of beam elements. If the ends of the element are rigidly connected to the adjoining structure instead of being pinned, the element will act like a beam with moments and lateral forces acting at the joints. For the most part, the relative axial displacements will be small compared to the lateral displacements of the beam and can be assumed to be zero.

TABLE 6.3-1. Stiffness for beam elements.

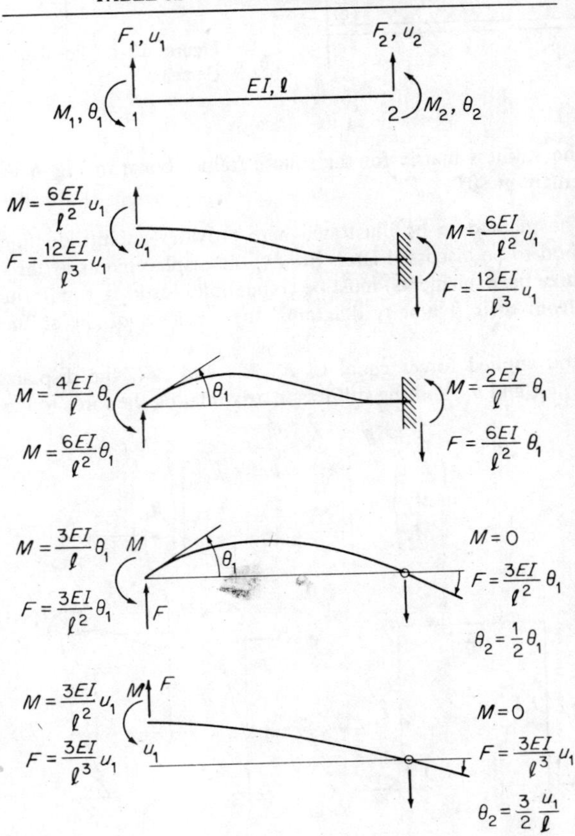

Table 6.3-1 shows the lateral forces and moments of uniform beam elements when each of the end displacements are taken separately. They relate to the following stiffness matrix:

$$\begin{Bmatrix} F_1 \\ M_1 \\ F_2 \\ M_2 \end{Bmatrix} = \begin{bmatrix} k_{11} & k_{12} & k_{13} & k_{14} \\ k_{21} & k_{22} & k_{23} & k_{24} \\ k_{31} & k_{32} & k_{33} & k_{34} \\ k_{41} & k_{42} & k_{43} & k_{44} \end{bmatrix} \begin{Bmatrix} u_1 \\ \theta_1 \\ u_2 \\ \theta_2 \end{Bmatrix}$$

They can be determined by the area-moment method.

The positive sense of these coordinates is arbitrary; however, the notation shown in Fig. 6.3-1 conforms to that generally used in the finite element method.

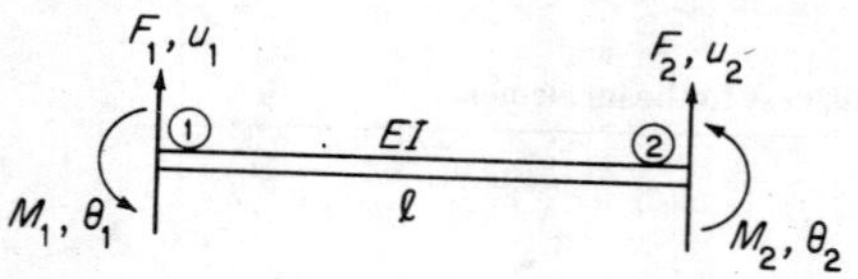

Figure 6.3-1. Positive sense for beam element.

Example 6.3-1

Determine the stiffness matrix for the square frame shown in Fig. 6.3-2. Assume the corners to remain at 90°

Solution: The method to be illustrated here provides an introduction to the finite element method to be discussed later. Briefly, the displacements at the joints (corners joining the three beam elements) must be compatible. Insuring equilibrium of forces at the corners from their free-body diagrams, the elements of the stiffness matrix are found.

With the applied forces equal to F, M_1, and M_2, the displacements of the corners are u, θ_1, and θ_2, and the stiffness matrix relating the force to the displacement is

$$\begin{Bmatrix} F \\ M_1 \\ M_2 \end{Bmatrix} = \begin{bmatrix} k_{11} & k_{12} & k_{13} \\ k_{21} & k_{22} & k_{23} \\ k_{31} & k_{32} & k_{33} \end{bmatrix} \begin{Bmatrix} u \\ \theta_1 \\ \theta_2 \end{Bmatrix}$$

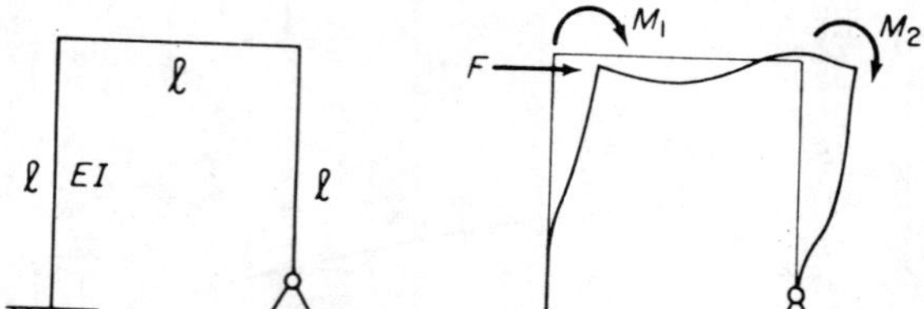

Figure 6.3-2.

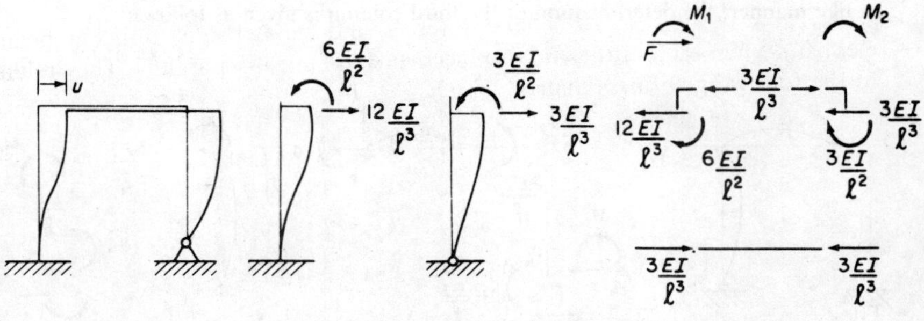

Figure 6.3-3(a).

The first column of the stiffness matrix can be determined by letting $u = 1$ and $\theta_1 = \theta_2 = 0$, which results in the configuration, Fig. 6.3-3(a).

Considering each member and making use of Table 6.3-1, we have

$$\begin{Bmatrix} F \\ M_1 \\ M_2 \end{Bmatrix} = \begin{bmatrix} \dfrac{15EI}{l^3} & 0 & 0 \\ -\dfrac{6EI}{l^2} & 0 & 0 \\ -\dfrac{3EI}{l^2} & 0 & 0 \end{bmatrix} \begin{Bmatrix} u \\ 0 \\ 0 \end{Bmatrix}$$

The second column of the stiffness matrix is determined by letting $u = 0$, $\theta_1 = 1$, and $\theta_2 = 0$, which results in the configuration, Fig. 6.3-3(b).

$$\begin{Bmatrix} F \\ M_1 \\ M_2 \end{Bmatrix} = \begin{bmatrix} 0 & -\dfrac{6EI}{l^2} & 0 \\ 0 & \dfrac{8EI}{l} & 0 \\ 0 & \dfrac{2EI}{l} & 0 \end{bmatrix} \begin{Bmatrix} 0 \\ \theta_1 \\ 0 \end{Bmatrix}$$

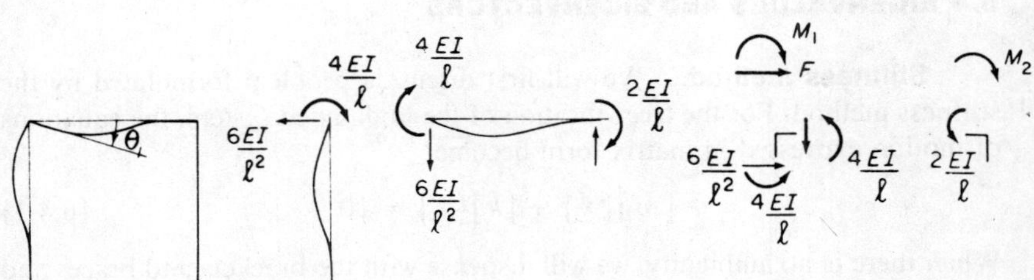

Figure 6.3-3(b).

In like manner, the determination of the third column is given as follows:

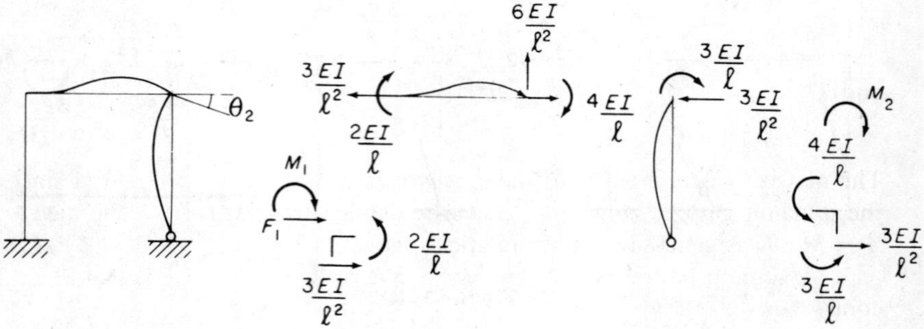

Figure 6.3-3(c).

$$\begin{Bmatrix} F \\ M_1 \\ M_2 \end{Bmatrix} = \begin{bmatrix} 0 & 0 & -\dfrac{3EI}{l^2} \\ 0 & 0 & \dfrac{2EI}{l} \\ 0 & 0 & \dfrac{7EI}{l} \end{bmatrix} \begin{Bmatrix} 0 \\ 0 \\ \theta_2 \end{Bmatrix}$$

The system stiffness is found from the superposition of the preceding three configurations, which becomes

$$\begin{Bmatrix} F \\ M_1 \\ M_2 \end{Bmatrix} = \frac{EI}{l} \begin{bmatrix} \dfrac{15}{l^2} & -\dfrac{6}{l} & -\dfrac{3}{l} \\ -\dfrac{6}{l} & 8 & 2 \\ -\dfrac{3}{l} & 2 & 7 \end{bmatrix} \begin{Bmatrix} u \\ \theta_1 \\ \theta_2 \end{Bmatrix}$$

It should be noted that the matrix is symmetric about the diagonal.

6.4 EIGENVALUES AND EIGENVECTORS

Stiffness method. We will first discuss a problem formulated by the stiffness method. For the free vibration of the undamped system, the equations of motion expressed in matrix form become

$$[M]\{\ddot{x}\} + [k]\{x\} = \{0\} \tag{6.4-1}$$

When there is no ambiguity, we will dispense with the brackets and braces and use capital letters and simply write the matrix equation as

$$M\ddot{X} + KX = 0 \tag{6.4-2}$$

If we premultiply the above equation by M^{-1}, we obtain the following terms:

$$M^{-1}M = I \qquad \text{(a unit matrix)}$$

$$M^{-1}K = A \qquad \text{(a system matrix)}$$

and

$$I\ddot{X} + AX = 0 \tag{6.4-3}$$ †

The matrix A is referred to as the *system matrix*, or the dynamic matrix since the dynamic properties of the system are defined by this matrix. The matrix $A = M^{-1}K$ is generally not symmetric.

Assuming harmonic motion $\ddot{X} = -\lambda X$, where $\lambda = \omega^2$, Eq. (6.4-3) becomes

$$[A - \lambda I]\{X\} = 0 \tag{6.4-4}$$

The *characteristic equation* of the system is the determinant equated to zero, or

$$|A - \lambda I| = 0 \tag{6.4-5}$$

The roots λ_i of the characteristic equation are called *eigenvalues*, and the natural frequencies of the system are determined from them by the relationship

$$\lambda_i = \omega_i^2 \tag{6.4-6}$$

By substituting λ_i into the matrix equation, Eq. (6.4-4), we obtain the corresponding mode shape X_i which is called the *eigenvector*. Thus for an n-degrees of freedom system, there will be n eigenvalues and n eigenvectors.

Adjoint matrix. It is also possible to find the eigenvectors from the adjoint matrix (see Appendix C) of the system. If, for conciseness, we make the abbreviation $B = A - \lambda I$ and start with the definition of the inverse

$$B^{-1} = \frac{1}{|B|}\operatorname{adj} B \tag{6.4-7}$$

we can premultiply by $|B|B$ to obtain

$$|B|I = B \operatorname{adj} B$$

or in terms of the original expression for B

$$|A - \lambda I|I = [A - \lambda I]\operatorname{adj}[A - \lambda I] \tag{6.4-8}$$

If now we let $\lambda = \lambda_i$, an eigenvalue, then the determinant on the left side of

†Some computer algorithms will solve the eigen-problem of Eq. 6.4-2 in terms of the two real symmetric matrices M and K. However the usual computer algorithms are written for the standard form of the equation written as Eq. 6.4-4, where the matrix A must be symmetric. This requires M to be decomposed to $M = Q^TQ$, where $Q = M^{1/2}$. With the transformation $x = Q^{-1}u$, Eq. 6.4-2 can be written as Eq. 6.4-4 with $A = Q^{-T}KQ^{-1}$, a symmetric matrix. (See Meirovitch "Computational Methods in Structural Dynamics" p. 41, Sijthoff & Noordhoff, 1980.)

the equation is zero and we obtain

$$0 = [A - \lambda_i I]\text{adj}[A - \lambda_i I] \tag{6.4-9}$$

The above equation is valid for all λ_i and represents n equations for the n-DOF system. Comparing this equation with Eq. (6.4-4) for the ith mode

$$[A - \lambda_i I]\{X\}_i = 0$$

we recognize that the adjoint matrix, adj$[A - \lambda_i I]$, must consist of columns, each of which is the eigenvector X_i (multiplied by an arbitrary constant). Eigenvalues and eigenvectors can be calculated for any symmetric matrix by standard subroutine programs.

Example 6.4-1

Consider the two-story building shown in Fig. 6.4-1. The equation of motion can be expressed in matrix notation as

$$\begin{bmatrix} 2m & 0 \\ 0 & m \end{bmatrix}\begin{Bmatrix} \ddot{x}_1 \\ \ddot{x}_2 \end{Bmatrix} + \begin{bmatrix} 3k & -k \\ -k & k \end{bmatrix}\begin{Bmatrix} x_1 \\ x_2 \end{Bmatrix} = \begin{Bmatrix} 0 \\ 0 \end{Bmatrix} \tag{a}$$

Premultiplying by the inverse of the mass matrix

$$M^{-1} = \begin{bmatrix} \dfrac{1}{2m} & 0 \\ 0 & \dfrac{1}{m} \end{bmatrix}$$

and letting $\lambda = \omega^2$, Eq. (a) becomes

$$\begin{bmatrix} \left(\dfrac{3}{2}\dfrac{k}{m} - \lambda\right) & -\dfrac{k}{2m} \\ -\dfrac{k}{m} & \left(\dfrac{k}{m} - \lambda\right) \end{bmatrix}\begin{Bmatrix} x_1 \\ x_2 \end{Bmatrix} = \begin{Bmatrix} 0 \\ 0 \end{Bmatrix} \tag{b}$$

The characteristic equation from the determinant of the above matrix is

$$\lambda^2 - \frac{5}{2}\frac{k}{m}\lambda + \left(\frac{k}{m}\right)^2 = 0 \tag{c}$$

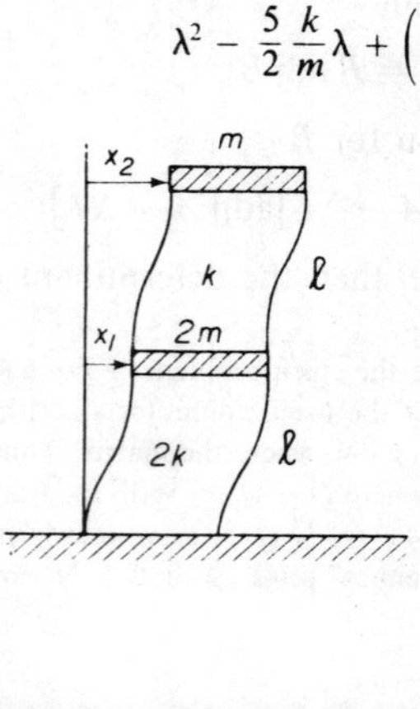

Figure 6.4-1.

from which the eigenvalues are found to be

$$\lambda_1 = \frac{1}{2}\frac{k}{m} \qquad \text{(d)}$$

$$\lambda_2 = 2\frac{k}{m}$$

The eigenvectors can be found from Eq. (b) by substituting the above values of λ. We will, however, illustrate the use of the adjoint matrix in their evaluation.

The adjoint matrix from Eq. (b) is

$$\operatorname{adj}[A - \lambda I] = \begin{bmatrix} \left(\dfrac{k}{m} - \lambda_i\right) & \dfrac{k}{2m} \\ \dfrac{k}{m} & \left(\dfrac{3}{2}\dfrac{k}{m} - \lambda_i\right) \end{bmatrix} \qquad \text{(e)}$$

The two columns of the adjoint matrix are simply the two expressions for x_1/x_2 found from matrix equation (b). Although differing in expression, both columns result in the same numbers when λ_i is substituted.

Substituting $\lambda_i = \frac{1}{2}(k/m)$, we obtain from (e)

$$\begin{bmatrix} 0.50 & 0.50 \\ 1.00 & 1.00 \end{bmatrix} \frac{k}{m}$$

Here each column is already normalized to unity and the first eigenvector is

$$X_1 = \begin{Bmatrix} 0.50 \\ 1.00 \end{Bmatrix}$$

Similarly, when $\lambda_2 = 2(k/m)$ is used, the adjoint matrix gives

$$\begin{bmatrix} -1.0 & 0.5 \\ 1.0 & -0.5 \end{bmatrix} \frac{k}{m}$$

Normalizing to unity, the second eigenvector from either column is

$$X_2 = \begin{Bmatrix} -1.00 \\ 1.00 \end{Bmatrix}$$

The two normal modes are shown in Fig. 6.4-2.

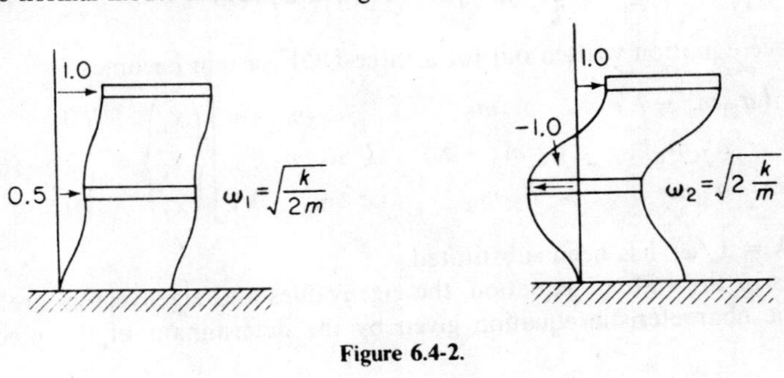

Figure 6.4-2.

Computer solution in standard eigen-problem form—**see footnote p. 165.**

To express Eq. (a) in the standard eigen-problem form, calculate the following quantities.

$$Q = M^{1/2} = \sqrt{m}\begin{bmatrix} \sqrt{2} & 0 \\ 0 & 1 \end{bmatrix}$$

$$Q^{-1} = \frac{1}{\sqrt{m}}\begin{bmatrix} 1/\sqrt{2} & 0 \\ 0 & 1 \end{bmatrix}$$

$$A = Q^{-T}KQ^{-1} = \frac{k}{m}\begin{bmatrix} 0.7071 & 0 \\ 0 & 1 \end{bmatrix}\begin{bmatrix} 3 & -1 \\ -1 & 1 \end{bmatrix}\begin{bmatrix} 0.7071 & 0 \\ 0 & 1 \end{bmatrix}$$

$$= \frac{k}{m}\begin{bmatrix} 1.500 & -0.7071 \\ -0.7071 & 1.000 \end{bmatrix}$$

With the coordinate transformation $x = Q^{-1}u$ and premultiplication by Q^{-T}, the equation of motion becomes

$$\left[\begin{bmatrix} 1.500 & -0.7071 \\ -0.7071 & 1.00 \end{bmatrix} - \lambda\begin{bmatrix} 1 & 0 \\ 0 & 1 \end{bmatrix}\right]\begin{Bmatrix} u_1 \\ u_2 \end{Bmatrix} = \begin{Bmatrix} 0 \\ 0 \end{Bmatrix}$$

where $\lambda = m\omega^2/k$. Computer results are $\lambda = 2.00, 0.50$

eigenvectors in u		*eigenvectors in x*		*normalized*	
−0.816	0.577	−0.577	0.408	−1.0	0.50
0.577	0.816	0.577	0.816	1.0	1.00

Flexibility method. We now formulate the equations of motion in terms of flexibility and make some comparisons with the stiffness method. For the equations of motion, we will again assume the masses to be lumped at each station, in which case the mass matrix is diagonal $\lceil m \rfloor$. Replacing $\{f\}$ by the D'Alembert's inertia force

$$\{f\} = -\lceil m \rfloor\{\ddot{x}\} = \omega^2\lceil m \rfloor\{x\} \tag{6.4-10}$$

Substituting this into the equation for $\{x\}$, we obtain

$$\{x\} = \omega^2[a]\lceil m \rfloor\{x\}$$

Rearranging the equation to

$$\left[[a]\lceil m \rfloor - \frac{1}{\omega^2}[I]\right]\left\{x\right\} = 0 \tag{6.4-11}$$

the above equation written out for a three-DOF system becomes

$$\begin{bmatrix} (a_{11}m_1 - \lambda) & a_{12}m_2 & a_{13}m_3 \\ a_{21}m_1 & (a_{22}m_2 - \lambda) & a_{23}m_3 \\ a_{31}m_1 & a_{32}m_2 & (a_{33}m_3 - \lambda) \end{bmatrix}\begin{Bmatrix} x_1 \\ x_2 \\ x_3 \end{Bmatrix} = \begin{Bmatrix} 0 \\ 0 \\ 0 \end{Bmatrix} \tag{6.4-12}$$

where $\lambda = 1/\omega^2$ has been substituted.

As in the stiffness equation, the eigenvalues and eigenvectors are found from the characteristic equation given by the determinant of the preceding equation. It should be noted here, however, that the eigenvalues λ are the reciprocals of the squares of the natural frequencies, so that the lowest eigenvalue corresponds to the highest natural frequency.

Example 6.4-2

Using the flexibility matrix, determine the equation of motion and the eigenvalues and eigenvectors for the system solved by the stiffness method in Example 6.4-1.

Solution: The stiffness matrix in Ex. 6.4-1 is

$$[k] = k\begin{bmatrix} 3 & -1 \\ -1 & 1 \end{bmatrix}$$

Taking its inverse, the flexibility matrix is

$$[a] = \frac{1}{2k}\begin{bmatrix} 1 & 1 \\ 1 & 3 \end{bmatrix}$$

The dynamic matrix for the problem is

$$[a][m] = \frac{m}{2k}\begin{bmatrix} 1 & 1 \\ 1 & 3 \end{bmatrix}\begin{bmatrix} 2 & 0 \\ 0 & 1 \end{bmatrix} = \frac{m}{2k}\begin{bmatrix} 2 & 1 \\ 2 & 3 \end{bmatrix}$$

and the equation of motion from Eq. (6.4-11) becomes

$$\left[\frac{m}{2k}\begin{bmatrix} 2 & 1 \\ 2 & 3 \end{bmatrix} - \frac{1}{\omega^2}\begin{bmatrix} 1 & 0 \\ 0 & 1 \end{bmatrix}\right]\begin{Bmatrix} x_1 \\ x_2 \end{Bmatrix} = \begin{Bmatrix} 0 \\ 0 \end{Bmatrix}$$

If we let the eigenvalue be $\lambda = 2k/m\omega^2$, its characteristic equation is

$$\begin{vmatrix} (2-\lambda) & 1 \\ 2 & (3-\lambda) \end{vmatrix} = 0$$

or

$$\lambda^2 - 5\lambda + 4 = 0$$

The eigenvalues are then

$$\lambda_2 = \frac{2k}{m\omega_2^2} = 1, \qquad \omega_2^2 = \frac{2k}{m}$$

$$\lambda_1 = \frac{2k}{m\omega_1^2} = 4, \qquad \omega_1^2 = \frac{1}{2}\frac{k}{m}$$

Solving for the eigenvectors, we have

$$\frac{x_1}{x_2} = \frac{(3-\lambda)}{-2} = \begin{cases} 0.5 & \text{for } \lambda_1 \\ -1.0 & \text{for } \lambda_2 \end{cases}$$

or

$$X_1 = \begin{Bmatrix} 0.50 \\ 1.00 \end{Bmatrix}, \qquad X_2 = \begin{Bmatrix} -1.00 \\ 1.00 \end{Bmatrix}$$

Thus the natural frequencies and mode shapes are found to be identical to those determined by the stiffness method.

Example 6.4-3

Using the flexibility matrix for the cantilever beam of Example 6.1-2, determine the matrix equation for the normal mode vibration of the system.

Solution: The term $[a][m]$ from Ex. 6.1-2 is

$$\frac{l^3}{3EI}\begin{bmatrix} 27 & 14 & 4 \\ 14 & 8 & 2.5 \\ 4 & 2.5 & 1 \end{bmatrix}\begin{bmatrix} m_1 & 0 & 0 \\ 0 & m_2 & 0 \\ 0 & 0 & m_3 \end{bmatrix}$$

If we let $\lambda = 3EI/\omega^2 l^3$, the normal mode equation from Eq. (6.4-11) becomes

$$\left[\begin{bmatrix} 27 & 14 & 4 \\ 14 & 8 & 2.5 \\ 4 & 2.5 & 1 \end{bmatrix}\begin{bmatrix} m_1 & 0 & 0 \\ 0 & m_2 & 0 \\ 0 & 0 & m_3 \end{bmatrix} - \lambda \begin{bmatrix} 1 & 0 & 0 \\ 0 & 1 & 0 \\ 0 & 0 & 1 \end{bmatrix}\right]\begin{Bmatrix} x_1 \\ x_2 \\ x_3 \end{Bmatrix} = \begin{Bmatrix} 0 \\ 0 \\ 0 \end{Bmatrix}$$

6.5 ORTHOGONAL PROPERTIES OF THE EIGENVECTORS

The normal modes, or the eigenvectors of the system, can be shown to be *orthogonal* with respect to the mass and stiffness matrices as follows. Let the equation for the ith mode be

$$KX_i = \lambda_i MX_i \tag{6.5-1}$$

Premultiply by the transpose of mode j

$$X_j'KX_i = \lambda_i(X_j'MX_i) \tag{6.5-2}$$

Next, start with the equation for the jth mode and premultiply by X_i' to obtain

$$X_i'KX_j = \lambda_j(X_i'MX_j) \tag{6.5-3}$$

Since K and M are symmetric matrices, the following relationships hold†

$$\begin{aligned} X_j'MX_i &= X_i'MX_j \\ X_j'KX_i &= X_i'KX_j \end{aligned} \tag{6.5-4}$$

Thus, subtracting Eq. (6.5-3) from Eq. (6.5-2), we obtain

$$0 = (\lambda_i - \lambda_j)X_i'MX_j \tag{6.5-5}$$

If $\lambda_i \neq \lambda_j$, the above equation requires that

$$X_i'MX_j = 0 \tag{6.5-6}$$

It is also evident from Eq. (6.5-2) or Eq. (6.5-3) that as a consequence of Eq. (6.5-6)

$$X_i'KX_j = 0 \tag{6.5-7}$$

Equations (6.5-6) and (6.5-7) define the *orthogonal* character of the normal modes.

†See Appendix C.

Finally, if $i = j$, Eq. (6.5-5) is satisfied for any finite value of the products given by Eqs. (6.5-6) or (6.5-7). We therefore let

$$\begin{aligned} X_i' M X_i &= M_i \\ X_i' K X_i &= K_i \end{aligned} \tag{6.5-8}$$

These are called the *generalized mass* and the *generalized stiffness*, respectively.

Example 6.5-1

Verify that the two normal modes of the system considered in Example 6.3-1 are orthogonal.

Solution: The mass matrix and the two normal modes are

$$M = \begin{bmatrix} 2m & 0 \\ 0 & m \end{bmatrix} \qquad X_1 = \begin{Bmatrix} 0.50 \\ 1.00 \end{Bmatrix} \qquad X_2 = \begin{Bmatrix} -1.00 \\ 1.00 \end{Bmatrix}$$

Substituting into Eq. (6.5-6), we have

$$\begin{aligned} X_1' M X_2 &= (0.50 \quad 1.00) \begin{bmatrix} 2m & 0 \\ 0 & m \end{bmatrix} \begin{Bmatrix} -1.00 \\ 1.00 \end{Bmatrix} \\ &= (0.50 \quad 1.00) \begin{Bmatrix} -2m \\ m \end{Bmatrix} = -m + m = 0 \end{aligned}$$

The student should verify that $X_1' K X_2$ also equals zero.

Example 6.5-2

Consider the problem of initiating the free vibration of a system from a specified distribution of the displacement. As previously stated, free vibrations are the superposition of normal modes, and we wish to determine how much of each mode will be present.

Solution: We will first express the displacement at time zero by the equation

$$u(o) = c_1 X_1 + c_2 X_2 + \cdots + c_i X_i + \cdots$$

where X_i are the normal modes and c_i are the coefficients indicating how much of each mode is present. Premultiplying the above equation by $X_i' M$ and taking note of the orthogonal property of X_i, we obtain

$$X_i' M u(0) = 0 + 0 + \cdots + c_i X_i' M X_i + 0 + \cdots$$

The coefficient c_i of any mode is then found as

$$c_i = \frac{X_i' M u(0)}{X_i' M X_i}$$

6.6 REPEATED ROOTS

When repeated roots are found in the characteristic equation, the corresponding eigenvectors are not unique, and a linear combination of such eigenvectors may also satisfy the equation of motion. To illustrate this point, let X_1 and X_2 be eigenvectors belonging to a common eigenvalue λ_0, and X_3 be a third eigenvector belonging to λ_3 that is different from λ_0. We can then write

$$\begin{aligned} AX_1 &= \lambda_0 X_1 \\ AX_2 &= \lambda_0 X_2 \\ AX_3 &= \lambda_3 X_3 \end{aligned} \tag{6.6-1}$$

By multiplying the second equation by a constant b and adding it to the first, we obtain another equation

$$A(X_1 + bX_2) = \lambda_0(X_1 + bX_2) \tag{6.6-2}$$

Thus a new eigenvector $X_{12} = X_1 + bX_2$, which is a linear combination of the first two, also satisfies the basic equation

$$AX_{12} = \lambda_0 X_{12} \tag{6.6-3}$$

and hence no unique mode exists for λ_0.

Any of the modes corresponding to λ_0 must be orthogonal to X_3 if they are to be a normal mode. If all three modes are orthogonal, they are linearly independent and may be combined to describe the free vibration resulting from any initial condition.

Example 6.6-1

Consider the system shown in Fig. 6.6-1 where the connecting bar is rigid and negligible in weight.

The two normal modes of vibration are shown to be translation and rotation, which are orthogonal. The natural frequencies for the two modes, however, are equal

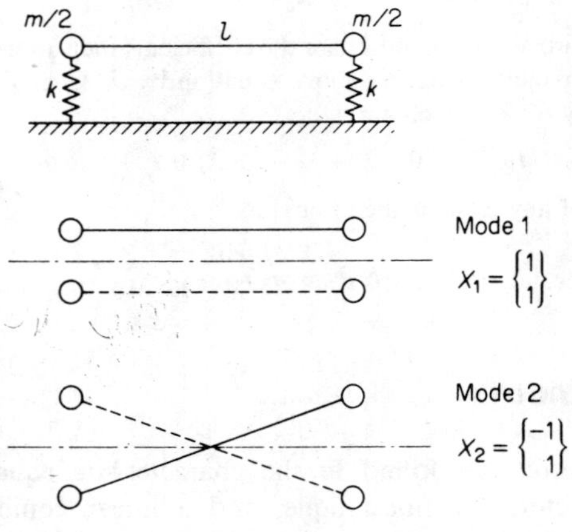

Figure 6.6-1.

and can be calculated to be

$$\omega_n = \sqrt{\frac{2k}{m}}$$

The example illustrates that different eigenvectors may have equal eigenvalues.

6.7 MODAL MATRIX P

It is possible to uncouple the equations of motion of an n-DOF system, provided we know beforehand the normal modes of the system. When the n normal modes (or eigenvectors) are assembled into a square matrix with each normal mode represented by a column, we call it the *modal matrix P*. Thus the modal matrix for a three-DOF system may appear as

$$P = \begin{bmatrix} \begin{Bmatrix} x_1 \\ x_2 \\ x_3 \end{Bmatrix}_1 & \begin{Bmatrix} x_1 \\ x_2 \\ x_3 \end{Bmatrix}_2 & \begin{Bmatrix} x_1 \\ x_2 \\ x_3 \end{Bmatrix}_3 \end{bmatrix} = [X_1 X_2 X_3] \tag{6.7-1}$$

The modal matrix makes it possible to include all the orthogonality relations of Sec. 6.5 into one equation. For this operation we need also the transpose of P, which is

$$P' = \begin{bmatrix} (x_1 x_2 x_3)_1 \\ (x_1 x_2 x_3)_2 \\ (x_1 x_2 x_3)_3 \end{bmatrix} = [X_1 X_2 X_3]' \tag{6.7-2}$$

with each row corresponding to a mode. If we now form the product $P'MP$ or $P'KP$, the result will be a diagonal matrix, since the off-diagonal terms simply express the orthogonality relations which are zero.

As an example consider a two-DOF system. Performing the indicated operation with the modal matrix, we have

$$\begin{aligned} P'MP &= [X_1 \quad X_2]'[M][X_1 \quad X_2] \\ &= \begin{bmatrix} X_1'MX_1 & X_1'MX_2 \\ X_2'MX_1 & X_2'MX_2 \end{bmatrix} \\ &= \begin{bmatrix} M_1 & 0 \\ 0 & M_2 \end{bmatrix} \end{aligned} \tag{6.7-3}$$

In the preceding equation, the off-diagonal terms are zero because of orthogonality, and the diagonal terms are the generalized mass M_i.

It is evident that a similar formulation applies also to the stiffness matrix K that results in the following equation:

$$P'KP = \begin{bmatrix} K_1 & 0 \\ 0 & K_2 \end{bmatrix} \tag{6.7-4}$$

The diagonal terms here are the generalized stiffness K_i.

If each of the columns of the modal matrix P is divided by the square root of the generalized mass M_i, the new matrix is called the *weighted modal matrix* and designated as $\tilde{P}$. It is easily seen that the diagonalization of the mass matrix by the weighted modal matrix results in the unit matrix

$$\tilde{P}'M\tilde{P} = I \tag{6.7-5}$$

Since $M_i^{-1}K_i = \lambda_i$, the stiffness matrix treated similarly by the weighted modal matrix becomes a diagonal matrix of the eigenvalues

$$\tilde{P}'K\tilde{P} = \begin{bmatrix} \lambda_1 & & & 0 \\ & \lambda_2 & & \\ & & \ddots & \\ 0 & & & \lambda_n \end{bmatrix} = \Lambda \tag{6.7-6}$$

To decouple the equations of motion by the use of the P matrix, we consider the following general equation of undamped forced vibration:

$$M\ddot{X} + KX = F \tag{6.7-7}$$

Making the coordinate transformation $X = PY$, the above equation becomes

$$MP\ddot{Y} + KPY = F$$

Next premultiply by the transpose P' to obtain

$$(P'MP)\ddot{Y} + (P'KP)Y = P'F \tag{6.7-8}$$

The products $P'MP$ and $P'KP$ being diagonal matrices, the equations in Y are uncoupled.

Example 6.7-1

Consider the two-story building of Example 6.3-1 excited by a force $F(t)$ at the top. Its equation of motion is

$$m\begin{bmatrix} 2 & 0 \\ 0 & 1 \end{bmatrix}\begin{Bmatrix} \ddot{x}_1 \\ \ddot{x}_2 \end{Bmatrix} + k\begin{bmatrix} 3 & -1 \\ -1 & 1 \end{bmatrix}\begin{Bmatrix} x_1 \\ x_2 \end{Bmatrix} = \begin{Bmatrix} 0 \\ F \end{Bmatrix}$$

The normal modes of the homogeneous equation are

$$X_1 = \begin{Bmatrix} 0.5 \\ 1 \end{Bmatrix}, \qquad X_2 = \begin{Bmatrix} -1 \\ 1 \end{Bmatrix}$$

from which the P matrix is assembled as

$$P = \begin{bmatrix} 0.5 & -1 \\ 1 & 1 \end{bmatrix}$$

Writing out the terms of Eq. (6.7-8), we have

$$m\begin{bmatrix} 0.5 & 1 \\ -1 & 1 \end{bmatrix}\begin{bmatrix} 2 & 0 \\ 0 & 1 \end{bmatrix}\begin{bmatrix} 0.5 & -1 \\ 1 & 1 \end{bmatrix}\begin{Bmatrix} \ddot{y}_1 \\ \ddot{y}_2 \end{Bmatrix}$$

$$+k\begin{bmatrix} 0.5 & 1 \\ -1 & 1 \end{bmatrix}\begin{bmatrix} 3 & -1 \\ -1 & 1 \end{bmatrix}\begin{bmatrix} 0.5 & -1 \\ 1 & 1 \end{bmatrix}\begin{Bmatrix} y_1 \\ y_2 \end{Bmatrix} = \begin{bmatrix} 0.5 & 1 \\ -1 & 1 \end{bmatrix}\begin{Bmatrix} 0 \\ F_2 \end{Bmatrix}$$

or

$$m\begin{bmatrix}1.5 & 0\\0 & 3\end{bmatrix}\begin{Bmatrix}\ddot{y}_1\\\ddot{y}_2\end{Bmatrix} + k\begin{bmatrix}0.75 & 0\\0 & 6\end{bmatrix}\begin{Bmatrix}y_1\\y_2\end{Bmatrix} = \begin{Bmatrix}F_2\\F_2\end{Bmatrix}$$

which are uncoupled.
The solutions for y_1 and y_2 are in the form

$$y_i = y_i(0)\cos\omega_i t + \frac{\dot{y}_i(0)}{\omega_i}\sin\omega_i t + \frac{F_2}{k_i}\frac{\sin\omega t}{1-(\omega/\omega_i)^2}$$

which can be expressed in terms of the original coordinates by the P matrix as

$$\begin{Bmatrix}x_1\\x_2\end{Bmatrix} = \begin{bmatrix}0.5 & -1\\1 & 1\end{bmatrix}\begin{Bmatrix}y_1\\y_2\end{Bmatrix}$$

Example 6.7-2

For Example 6.7-1, determine the generalized mass and the $\tilde{P}$ matrix. Numerically verify Eqs. (6.7-5) and (6.7-6).

Solution: The calculations for the generalized mass are

$$M_1 = (0.5 \quad 1)\begin{bmatrix}2 & 0\\0 & 1\end{bmatrix}\begin{Bmatrix}0.5\\1\end{Bmatrix} = 1.5$$

$$M_2 = (-1 \quad 1)\begin{bmatrix}2 & 0\\0 & 1\end{bmatrix}\begin{Bmatrix}-1\\1\end{Bmatrix} = 3.0$$

Dividing the first column of P by $\sqrt{M_1}$ and the second column by $\sqrt{M_2}$, the $\tilde{P}$ matrix becomes

$$\tilde{P} = \begin{bmatrix}0.4083 & -0.5773\\0.8165 & 0.5773\end{bmatrix}$$

Equations (6.7-5) and (6.7-6) are simply verified by substitution.

6.8 MODAL DAMPING IN FORCED VIBRATION

The equation of motion of an n-DOF system with viscous damping and arbitrary excitation $F(t)$ can be presented in the matrix form

$$M\ddot{X} + C\dot{X} + KX = F \tag{6.8-1}$$

It is generally a set of n coupled equations.

We have found that the solution of the homogeneous undamped equation

$$M\ddot{X} + KX = 0 \tag{6.8-2}$$

leads to the eigenvalues and eigenvectors that describe the normal modes of the system and the modal matrix P or $\tilde{P}$. If we let $X = \tilde{P}Y$ and premultiply Eq. (6.8-1) by $\tilde{P}'$ as in Sec. 6.7, we obtain

$$\tilde{P}'M\tilde{P}\ddot{Y} + \tilde{P}'C\tilde{P}\dot{Y} + \tilde{P}'K\tilde{P}Y = \tilde{P}'F \tag{6.8-3}$$

We have already shown that $\tilde{P}'M\tilde{P}$ and $\tilde{P}'K\tilde{P}$ are diagonal matrices. In general, $\tilde{P}'C\tilde{P}$ is not diagonal and the preceding equation is coupled by the damping matrix.

If C is proportional to M or K, it is evident that $\tilde{P}'C\tilde{P}$ becomes diagonal, in which case we can say that the system has *proportional damping.* Eq. (6.8-3) is then completely uncoupled and its ith equation will have the form

$$\ddot{y}_i + 2\zeta_i\omega_i\dot{y}_i + \omega_i^2 y_i = \tilde{f}_i(t) \tag{6.8-4}$$

Thus instead of n coupled equations we would have n uncoupled equations similar to that of a single-DOF system.

Rayleigh damping. Rayleigh introduced proportional damping in the form

$$C = \alpha M + \beta K \tag{6.8-5}$$

where α and β are constants. The application of the weighted modal matrix $\tilde{P}$ here results in

$$\begin{aligned} \tilde{P}'C\tilde{P} &= \alpha\tilde{P}'M\tilde{P} + \beta\tilde{P}'K\tilde{P} \\ &= \alpha I + \beta\Lambda \end{aligned} \tag{6.8-6†}$$

where I is a unit matrix and Λ is a diagonal matrix of the eigenvalues [see Eq. (6.7-6)].

$$\Lambda = \begin{bmatrix} \omega_1^2 & & & \\ & \omega_2^2 & & \\ & & \ddots & \\ & & & \omega_n^2 \end{bmatrix} \tag{6.8-7}$$

Thus instead of Eq. (6.8-4), we obtain for the ith equation

$$\ddot{y}_i + \left(\alpha + \beta\omega_i^2\right)\dot{y}_i + \omega_i^2 y_i = \tilde{f}_i(t) \tag{6.8-8}$$

and the modal damping can be defined by the equation

$$2\zeta_i\omega_i = \alpha + \beta\omega_i^2 \tag{6.8-9}$$

†It can be shown that $C = \alpha M'' + \beta K''$ can also be diagonalized (see Probs. 6-29 and 6-30)

6.9 NORMAL MODE SUMMATION

The forced vibration equation for the n-DOF system

$$M\ddot{X} + C\dot{X} + KX = F \tag{6.9-1}$$

can be routinely solved by the digital computer. However, for systems of large numbers of degrees of freedom, the computation can be costly. It is possible, however, to cut down the size of the computation (or reduce the degrees of freedom of the system) by a procedure known as the *mode summation method*. Essentially, the displacement of the structure under forced excitation is approximated by the sum of a limited number of normal modes of the system multiplied by generalized coordinates.

For example, consider a 50-story building with 50 DOF. The solution of its undamped homogeneous equation will lead to 50 eigenvalues and 50 eigenvectors that describe the normal modes of the structure. If we know that the excitation of the building centers around the lower frequencies, the higher modes will not be excited and we would be justified in assuming the forced response to be the superposition of only a few of the lower frequency modes; perhaps $\phi_1(x)$, $\phi_2(x)$, and $\phi_3(x)$ may be sufficient. Then the deflection under forced excitation may be written as

$$x_i = \phi_1(x_i)q_1(t) + \phi_2(x_i)q_2(t) + \phi_3(x_i)q_3(t) \tag{6.9-2}$$

or in matrix notation the position of all n floors can be expressed in terms of the modal matrix P composed of only the three modes. (See Fig. 6.9-1.)

$$\begin{Bmatrix} x_1 \\ x_2 \\ \vdots \\ x_n \end{Bmatrix} = \begin{bmatrix} \phi_1(x_1) & \phi_2(x_1) & \phi_3(x_1) \\ \vdots & \vdots & \vdots \\ \phi_1(x_n) & \phi_2(x_n) & \phi_3(x_n) \end{bmatrix} \begin{Bmatrix} q_1 \\ q_2 \\ \vdots \\ q_n \end{Bmatrix} \tag{6.9-3}$$

The use of the limited modal matrix then reduces the system to that equal to the number of modes used. For example, for the 50-story building, each of the matrices such as K is a 50×50 matrix; using three normal modes, P is a 50×3 matrix and the product $P'KP$ becomes

$$P'KP = (3 \times 50)(50 \times 50)(50 \times 3) = (3 \times 3) \text{ matrix}$$

Thus instead of solving the 50 coupled equations represented by Eq. (6.9-1), we need only solve the three by three equations represented by

$$P'MP\ddot{q} + P'CP\dot{q} + P'KPq = P'F \tag{6.9-4}$$

If the damping matrix is assumed to be proportional, the preceding equations become uncoupled, and if the force $F(x, t)$ is separable to $p(x)f(t)$, the three equations take the form

$$\ddot{q}_i + 2\zeta_i\omega_i\dot{q}_i + \omega_i^2 q_i = \Gamma_i f(t) \tag{6.9-5}$$

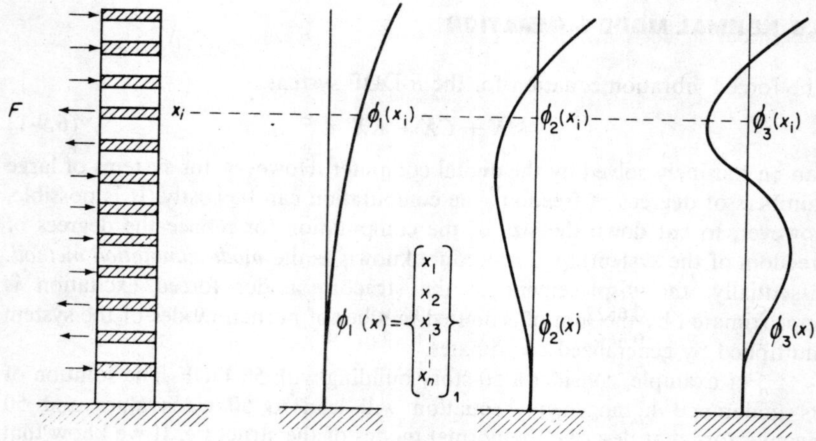

Figure 6.9-1. Building displacement represented by normal modes.

where the term

$$\Gamma_i = \frac{\sum_j \phi_i(x_j)p(x_j)}{\sum_j m_j\phi_i^2(x_j)} \tag{6.9-6}$$

is called the *mode participation factor*.

In many cases we are interested only in the maximum peak value of x_i, in which case the following procedure has been found to give acceptable results. We first find the maximum value of each $q_j(t)$ and combine them in the form

$$|x_i|_{\max} = |\phi_1(x_i)q_{1_{\max}}| + \sqrt{|\phi_2(x_i)q_{2_{\max}}|^2 + |\phi_3(x_i)q_{3_{\max}}|^2} \tag{6.9-7}†$$

Thus the first mode response is supplemented by the square root of the sum of the squares of the peaks for the higher modes. For the above computation, a shock spectrum for the particular excitation can be used to determine $q_{i_{\max}}$. If the predominant excitation is about a higher frequency, the normal modes centering about that frequency may be used.

Example 6.9-1

Consider the 10-story building of equal rigid floors and equal interstory stiffness. If the foundation of the building undergoes horizontal translation $u_0(t)$, determine the response of the building.

Solution: We will assume the normal modes of the building to be known. Given are the first three normal modes, which have been computed from the undamped

†The method is used by the Shock and Vibration groups in various industries and military.

homogeneous equation and are as follows:

Floor	$\omega_1 = 0.1495\sqrt{k/m}$	$\omega_2 = 0.4451\sqrt{k/m}$	$\omega_3 = 0.7307\sqrt{k/m}$
	$\phi_1(x)$	$\phi_2(x)$	$\phi_3(x)$
10	1.0000	1.0000	1.0000
9	0.9777	0.8019	0.4662
8	0.9336	0.4451	−0.3165
7	0.8686	0.0000	−0.9303
6	0.7840	−0.4451	−1.0473
5	0.6822	−0.8019	−0.6052
4	0.5650	−1.0000	1.6010
3	0.4352	−1.0000	0.8398
2	0.2954	−0.8019	1.0711
1	0.1495	−0.4451	0.7307
0	0.0000	0.0000	0.0000

The equation of motion of the building due to ground motion $u_0(t)$ is

$$M\ddot{X} + C\dot{X} + KX = -M1\ddot{u}_0(t)$$

where 1 is a unit vector and X is a 10×1 vector. Using the three given modes, we make the transformation

$$X = Pq$$

where P is a 10×3 matrix and q is a 3×1 vector, i.e.,

$$P = \begin{bmatrix} \phi_1(x_1) & \phi_2(x_1) & \phi_3(x_1) \\ \phi_1(x_2) & \phi_2(x_2) & \phi_3(x_2) \\ \vdots & \vdots & \vdots \\ \vdots & \vdots & \vdots \\ \phi_1(x_{10}) & \phi_2(x_{10}) & \phi_3(x_{10}) \end{bmatrix}, \quad q = \begin{Bmatrix} q_1 \\ q_2 \\ \vdots \\ q_3 \end{Bmatrix}$$

Premultiplying by P', we obtain

$$P'MP\ddot{q} + P'CP\dot{q} + P'KPq = -P'M1u_0(t)$$

and by assuming C to be a proportional damping matrix, the above equation results in three uncoupled equations

$$m_{11}\ddot{q}_1 + c_{11}\dot{q}_1 + k_{11}q_1 = -\ddot{u}_0(t)\sum_{i=1}^{10} m_i\phi_1(x_i)$$

$$m_{22}\ddot{q}_2 + c_{22}\dot{q}_2 + k_{22}q_2 = -\ddot{u}_0(t)\sum_{i=1}^{10} m_i\phi_2(x_i)$$

$$m_{33}\ddot{q}_3 + c_{33}\dot{q}_3 + k_{33}q_3 = -\ddot{u}_0(t)\sum_{i=1}^{10} m_i\phi_3(x_i)$$

where m_{ii}, c_{ii}, and k_{ii} are generalized mass, generalized damping, and generalized stiffness. The $q_j(t)$ are then independently solved from each of the above equations.

The displacement x_i of any floor must be found from the equation $X = Pq$ to be

$$x_i = \phi_1(x_i)q_1(t) + \phi_2(x_i)q_2(t) + \phi_3(x_i)q_3(t)$$

Thus the time solution for any floor is composed of the normal modes used.

From the numerical information supplied on the normal modes, we now determine the numerical values for the first equation, which can be rewritten as

$$\ddot{q}_1 + 2\zeta_1\omega_1\dot{q}_1 + \omega_1^2 q_1 = -\frac{\sum m\phi_1}{\sum m\phi_1^2}\ddot{u}(t)$$

We have, for the first mode,

$$m_{11} = \sum m\phi_1^2 = 5.2803m$$

$$\frac{c_{11}}{m_{11}} = 2\zeta_1\omega_1 = 0.299\sqrt{\frac{k}{m}}\,\zeta_1$$

$$\frac{k_{11}}{m_{11}} = \omega_1^2 = 0.02235\frac{k}{m}$$

$$\sum m\varphi_1 = 6.6912m$$

The equation for the first mode then becomes

$$\ddot{q}_1 + 0.299\sqrt{\frac{k}{m}}\,\zeta_1\dot{q}_1 + 0.02235\frac{k}{m}q_1 = -1.2672\ddot{u}_0(t)$$

Thus given the values for k/m and ζ_1, the above equation can be solved for any $\ddot{u}_0(t)$.

PROBLEMS

6-1 Determine the flexibility matrix for the spring-mass system shown in Fig. P6-1.

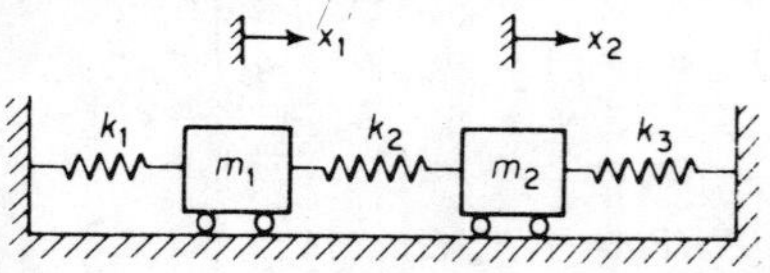

Figure P6-1.

6-2 Three equal springs of stiffness k lb/in. are joined at one end, the other ends being arranged symmetrically at 120° from each other, as shown in Fig. P6-2. Prove that the influence coefficients of the junction in a direction making an angle θ with any spring is independent of θ and equal to $1/1.5k$.

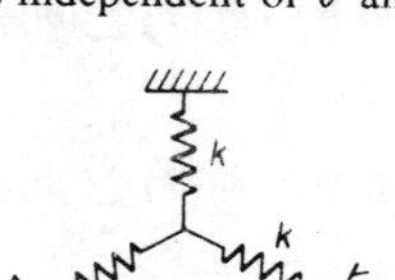

Figure P6-2.

6-3 A simply supported uniform beam of length l is loaded with weights at positions $0.25l$ and $0.6l$. Determine the flexibility influence coefficients for these positions.

6-4 Determine the flexibility matrix for the cantilever beam shown in Fig. P6-4 and calculate the stiffness matrix from its inverse.

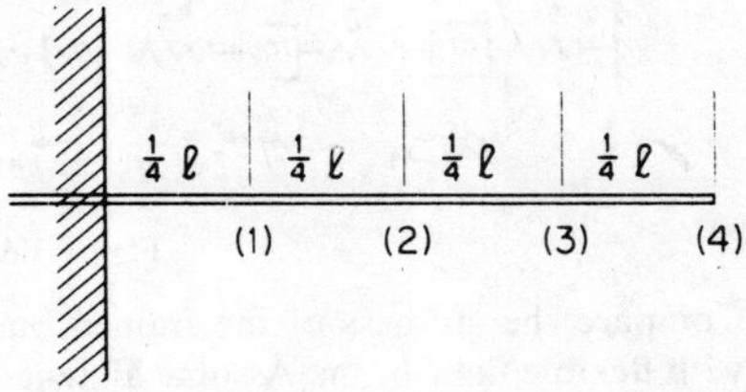

Figure P6-4.

6-5 Determine the influence coefficients for the triple pendulum shown in Fig. P6-5.

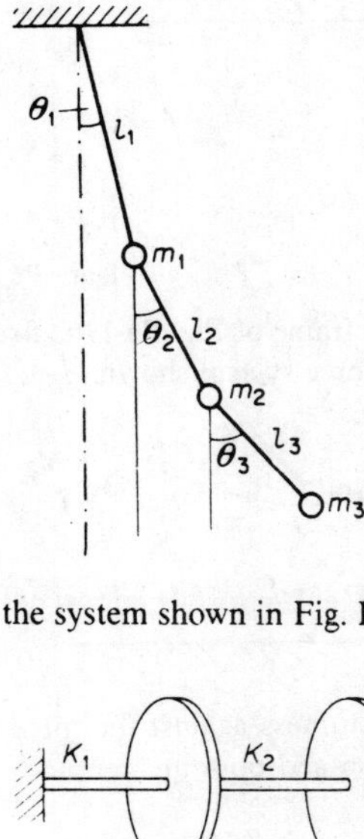

Figure P6-5.

6-6 Determine the stiffness matrix for the system shown in Fig. P6-6 and establish the flexibility matrix by its inverse.

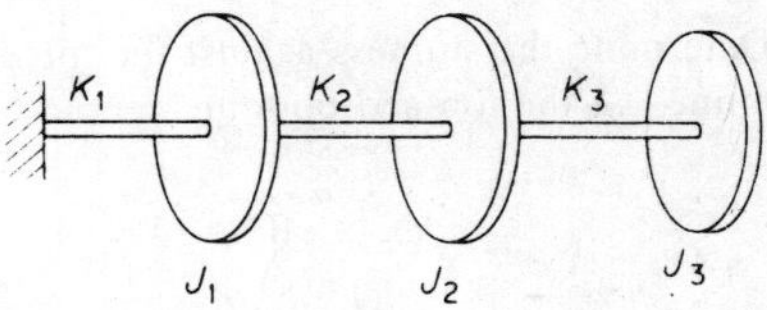

Figure P6-6

6-7 Determine the flexibility matrix for the uniform beam of Fig. P6-7 by using the area-moment method.

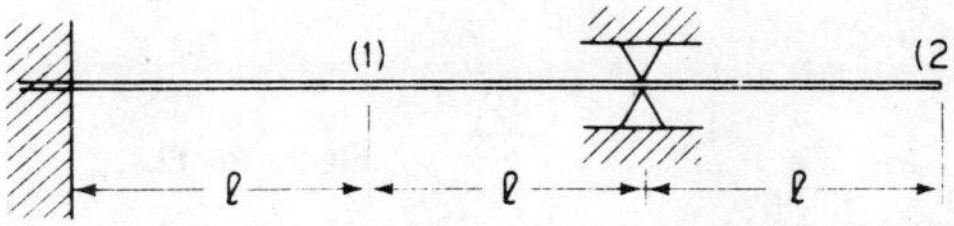

Figure P6-7.

6-8 Determine the flexibility matrix for the four-story building of Fig. 6.2-2 and invert it to arrive at the stiffness matrix given in the text.

6-9 Consider a system with n springs in series as presented in Fig. P6-9 and show that the stiffness matrix is a band matrix along the diagonal.

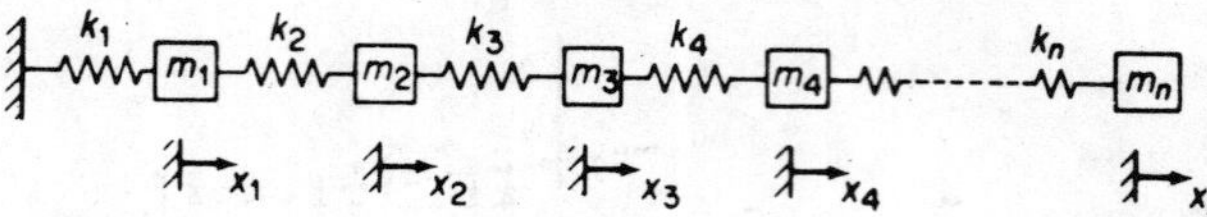

Figure P6-9.

6-10 Compare the stiffness of the framed building with rigid floor beam versus that with flexible floor beam. Assume all length and EI to be equal. If the floor mass is pinned at the corners as shown in Fig. P6-10(b), what is the ratio of the two natural frequencies?

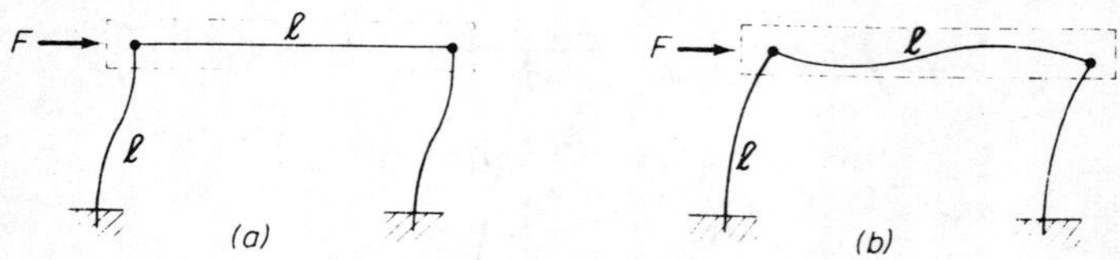

Figure P6-10.

6-11 The rectangular frame of Fig. P6-11 is fixed in the ground. Determine the stiffness matrix for the force system shown.

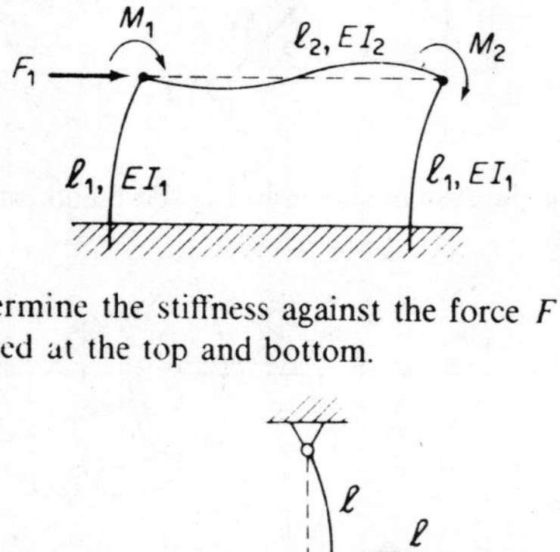

Figure P6-11.

6-12 Determine the stiffness against the force F for the frame of Fig. P6-12, which is pinned at the top and bottom.

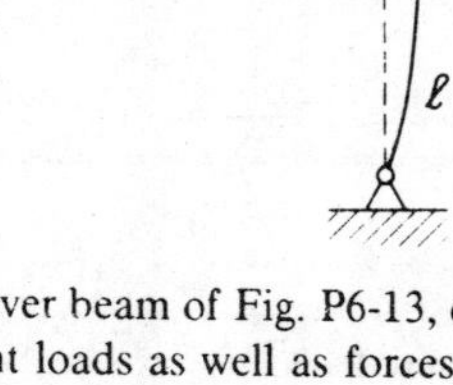

Figure P6-12.

6-13 Using the cantilever beam of Fig. P6-13, demonstrate that the reciprocity theorem holds for moment loads as well as forces.

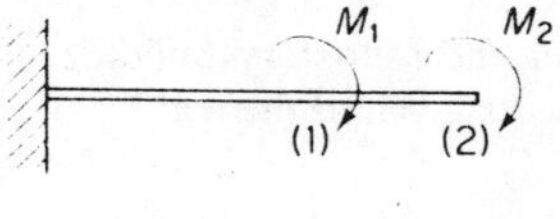

Figure P6-13

6-14 Verify each of the results given in Table 6.3-1 by the area-moment method and superposition.

6-15 Using the adjoint matrix, determine the normal modes of the spring-mass system shown in Fig. P6-15.

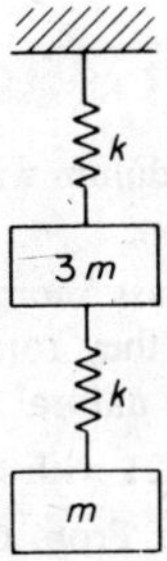

Figure P6-15.

6-16 For the system shown in Fig. P6-16, write the equations of motion in matrix form and determine the normal modes from the adjoint matrix.

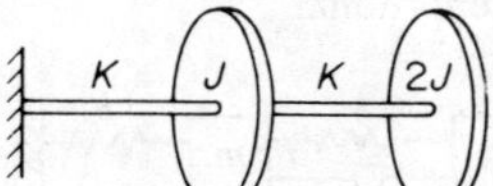

Figure P6-16.

6-17 Determine the modal matrix P and the weighted modal matrix $\tilde{P}$ for the system shown in Fig. P6-17. Show that P or $\tilde{P}$ will diagonalize the stiffness matrix.

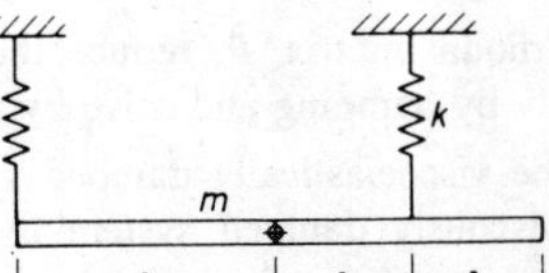

Figure P6-17.

6-18 Determine the flexibility matrix for the spring-mass system of three DOF shown in Fig. P6-18 and write its equation of motion in matrix form.

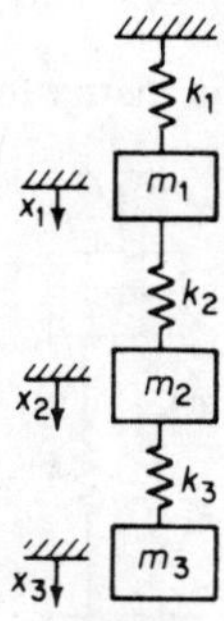

Figure P6-18.

6-19 Determine the modal matrix P and the weighted modal matrix $\tilde{P}$ for the system shown in Fig. P6-19 and diagonalize the stiffness matrix, thereby decoupling the equations.

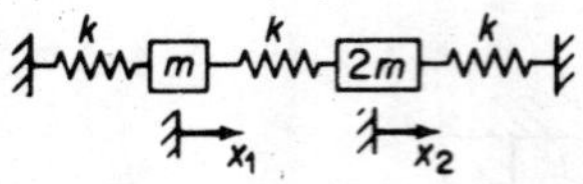

Figure P6-19.

6-20 Determine $\tilde{P}$ for the double pendulum with coordinates θ_1 and θ_2. Show that $\tilde{P}$ decouples the equations of motion.

6-21 If in Prob. 6-11 masses and mass moment of inertia m_1, J_1 and m_2, J_2 are attached to the corners so that they rotate as well as translate, determine the equations of motion and find the natural frequencies and mode shapes.

6-22 Repeat the procedure of Prob. 6-21 with the frame of Fig. P6-12.

6-23 If the lower end of the frame of Prob. 6-12 is rigidly fixed to the ground, the rotation of the corners will differ. Determine its stiffness matrix and determine its matrix equation of motion for m_i, J_i at the corners.

6-24 Determine the damping matrix for the system presented in Fig. P6-24 and show that it is not proportional.

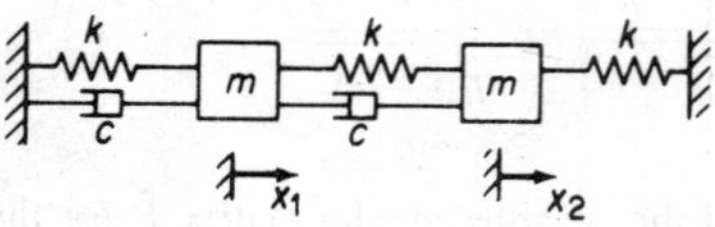

Figure P6-24.

6-25 Using the modal matrix $\tilde{P}$, reduce the system of Prob. 6-24 to one which is coupled only by damping and solve by the Laplace transform method.

6-26 Consider the viscoelastically damped system of Fig. P6-26. The system differs from the viscously damped system by the addition of the spring k_1, which introduces one more coordinate x_1 to the system. The equations of motion for the system in inertial coordinates x and x_1 are

$$m\ddot{x} = -kx - c(\dot{x} - \dot{x}_1) + F$$

$$0 = c(\dot{x} - \dot{x}_1) - k_1 x_1$$

Write the equation of motion in matrix form.

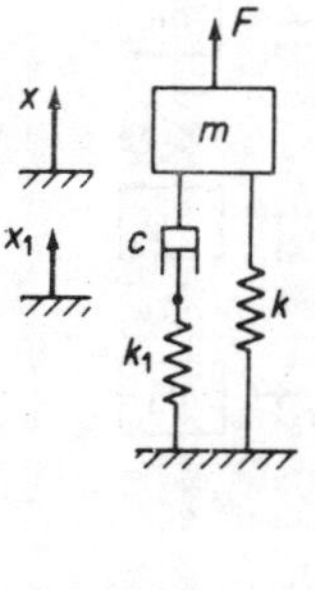

Figure P6-26.

6-27 Show, by comparing the viscoelastic system of Fig. P6-26 to the viscously damped system, that the equivalent viscous damping and the equivalent stiffness are

$$c_{eq} = \frac{c}{1 + \left(\frac{\omega c}{k_1}\right)^2}$$

$$k_{eq} = \frac{k + (k_1 + k)\left(\frac{\omega c}{k_1}\right)^2}{1 + \left(\frac{\omega c}{k_1}\right)^2}$$

6-28 Verify the relationship of Eq. (6.5-7)

$$X_i' K X_j = 0$$

by applying it to Prob. 6-16.

6-29 Starting with the matrix equation

$$K\phi_s = \omega_s^2 M \phi_s$$

premultiply first by KM^{-1} and, using the orthogonality relation $\phi_r' M \phi_s = 0$, show that

$$\phi_r' K M^{-1} K \phi_s = 0$$

Repeat to show that

$$\phi_r' [KM^{-1}]^h K \phi_s = 0$$

for $h = 1, 2, \ldots, n$, where n = number of degrees of freedom of the system.

6-30 In a manner similar to Prob. 6-29, show that

$$\phi_r' [MK^{-1}]^h M \phi_s = 0, \qquad h = 1, 2, \ldots$$

6-31 Evaluate the numerical coefficients for the equations of motion for the second and third modes of Example 6.9-1.

6-32 If the acceleration $\ddot{u}(t)$ of the ground in Example 6.9-1 is a single sine pulse of amplitude a_0 and duration t_1 as shown in Fig. P6-32, determine the maximum q for each mode and the value of x_{max} as given in Sec. 6.9.

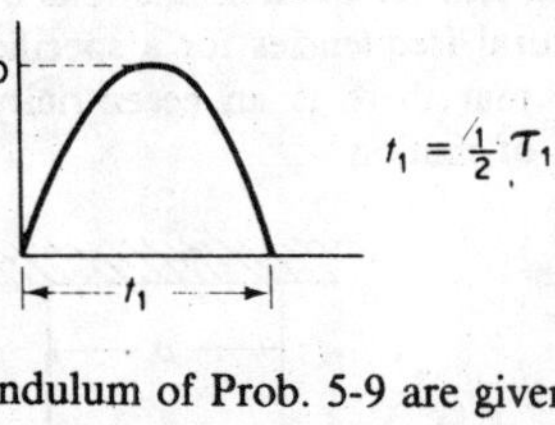

Figure P6-32.

6-33 The normal modes of the double pendulum of Prob. 5-9 are given as

$$\omega_1 = 0.764\sqrt{\frac{g}{l}}, \qquad \omega_2 = 1.850\sqrt{\frac{g}{l}}$$

$$\phi_1 = \begin{Bmatrix} \theta_1 \\ \theta_2 \end{Bmatrix}_{(1)} = \begin{Bmatrix} 0.707 \\ 1.00 \end{Bmatrix}, \qquad \phi_2 = \begin{Bmatrix} \theta_1 \\ \theta_2 \end{Bmatrix}_{(2)} = \begin{Bmatrix} -0.707 \\ 1.00 \end{Bmatrix}$$

If the lower mass is given an impulse $F_0\delta(t)$, determine the response in terms of the normal modes.

6-34 The normal modes of the three mass torsional system of Fig. P6-6 are given for $J_1 = J_2 = J_3$ and $K_1 = K_2 = K_3$.

$$\phi_1 = \begin{Bmatrix} 0.328 \\ 0.591 \\ 0.737 \end{Bmatrix} \qquad \lambda_1 = \frac{J\omega_1^2}{k} = 0.198, \qquad \phi_2 = \begin{Bmatrix} 0.737 \\ 0.328 \\ -0.591 \end{Bmatrix}$$

$$\lambda_2 = 1.555, \qquad \phi_3 = \begin{Bmatrix} 0.591 \\ -0.737 \\ 0.328 \end{Bmatrix} \qquad \lambda_3 = 3.247$$

Determine the equations of motion if a torque $M(t)$ is applied to the free end. If $M(t) = M_0 u(t)$ where $u(t)$ is a unit step function, determine the time solution and the maximum response of the end mass from the shock spectrum.

6-35 Using two normal modes, set up the equations of motion for the five-story building whose foundation stiffness in translation and rotation are k_t and $K_r = \infty$ (see Fig. P6-35).

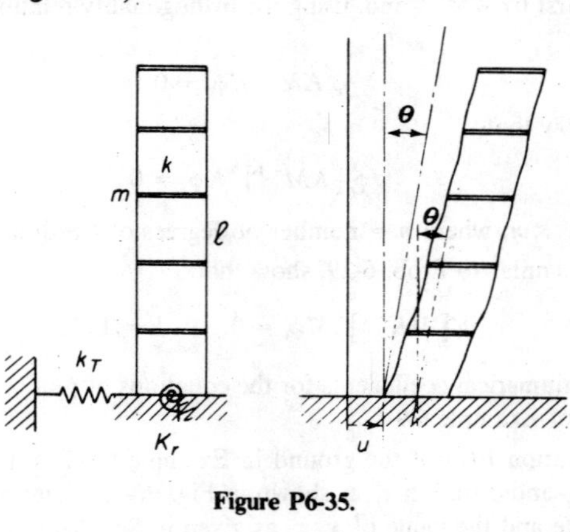

Figure P6-35.

6-36 The lateral and torsional oscillations of the system shown in Fig. P6-36 will have equal natural frequencies for a specific value of a/L. Determine this value and assuming that there is an eccentricity e of mass equal to me, determine the equations of motion.

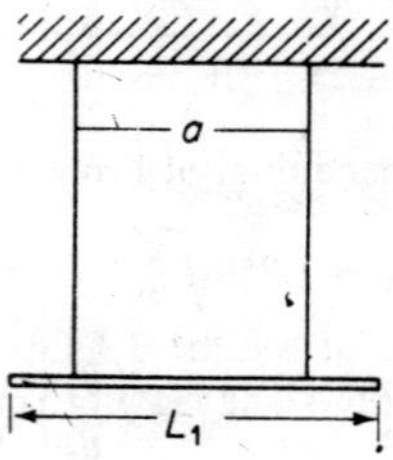

Figure P6-36.

6-37 Assume that a three-story building with rigid floor girders has Rayleigh damping. If the modal dampings for the first and second modes are 0.05% and 0.13%, respectively, determine the modal damping for the third mode.

6-38 The normal modes of a three-DOF system with $m_1 = m_2 = m_3$ and $k_1 = k_2 = k_3$ are given as

$$X_1 = \begin{Bmatrix} 0.737 \\ 0.591 \\ 0.328 \end{Bmatrix}, \qquad X_2 = \begin{Bmatrix} -0.591 \\ 0.328 \\ 0.737 \end{Bmatrix}, \qquad X_3 = \begin{Bmatrix} 0.328 \\ -0.737 \\ 0.591 \end{Bmatrix}$$

Verify the orthogonal properties of these modes.

6-39 The system of Prob. 6-38 is given an initial displacement of

$$X = \begin{Bmatrix} 0.520 \\ -0.100 \\ 0.205 \end{Bmatrix}$$

and released. Determine how much of each mode will be present in the free vibration.

6-40 In general, the free vibration of an undamped system can be represented by the modal sum

$$X(t) = \sum_i A_i X_i \sin \omega_i t + \sum_i B_i X_i \cos \omega_i t$$

If the system is started from zero displacement and an arbitrary distribution of velocity $\dot{X}(0)$, determine the coefficients A_i and B_i.

6-41 Figure P6-41 shows a shaft supported by a bearing, which has translational and rotational flexibility. Show that the left side of the shaft flexibility equation (a) or (b) of Example 6.1-3 should now be replaced by

$$\begin{Bmatrix} \eta \\ \theta - \beta \end{Bmatrix}$$

From the relationship between η, β, y, θ, and the loads P and M, determine the new flexibility equation

$$\begin{Bmatrix} y \\ \theta \end{Bmatrix} = \begin{bmatrix} \bar{a}_{11} & \bar{a}_{12} \\ \bar{a}_{21} & \bar{a}_{22} \end{bmatrix} \begin{Bmatrix} P \\ M \end{Bmatrix}$$

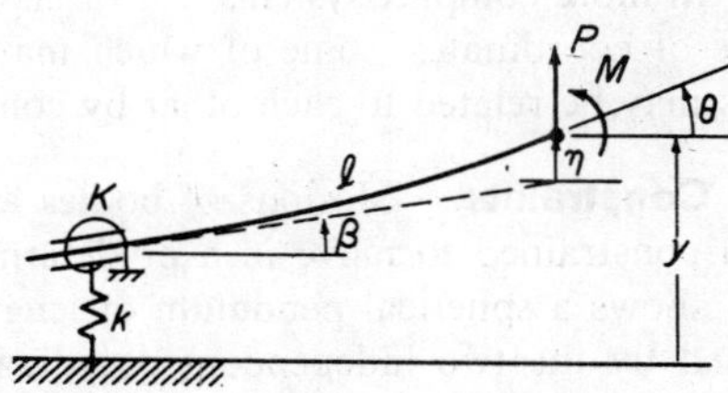

Figure P6-41.

6-42 Set up the matrix equation of motion for the 3-DOF system of Fig. P6-18 in terms of stiffness. Transform it to the standard eigen-problem form where A is symmetric.

7

LAGRANGE'S EQUATION

Joseph L. C. Lagrange (1736–1813) developed a general treatment of dynamical systems formulated from the scalar quantities of kinetic energy T, potential energy U, and work W. Lagrange's equations are in terms of generalized coordinates, and preliminary to discussing these equations, we must have clearly in mind the basic concepts of coordinates and their classification.

7.1 GENERALIZED COORDINATES

In the previous chapters, equations of motion were formulated in terms of physical coordinates. As long as they were independent coordinates, equal in number to the degrees of freedom of the system, they also qualified as generalized coordinates.

In more complex systems it is often convenient to describe the system in terms of coordinates, some of which may not be independent. Such coordinates may be related to each other by constraint equations.

Constraints. Motions of bodies are not always free motions and are often constrained to move in a predetermined manner. As an example, Fig. 7.1-1 shows a spherical pendulum of length l. Its position can be completely defined by the two independent coordinates ψ and ϕ. Hence ψ and ϕ are generalized coordinates, and the spherical pendulum represents a system of two DOF.

The position of the spherical pendulum can also be described by the three rectangular coordinates x, y, z, which exceed the degrees of freedom of

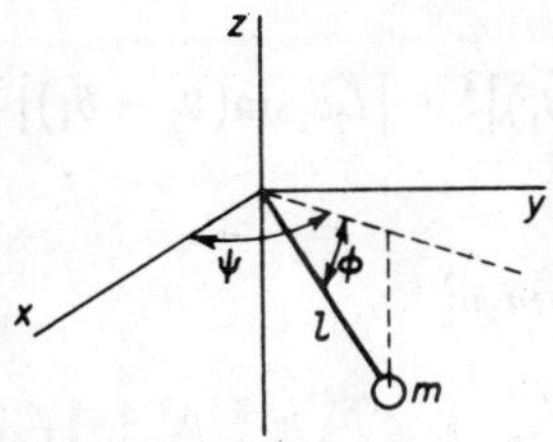

Figure 7.1-1.

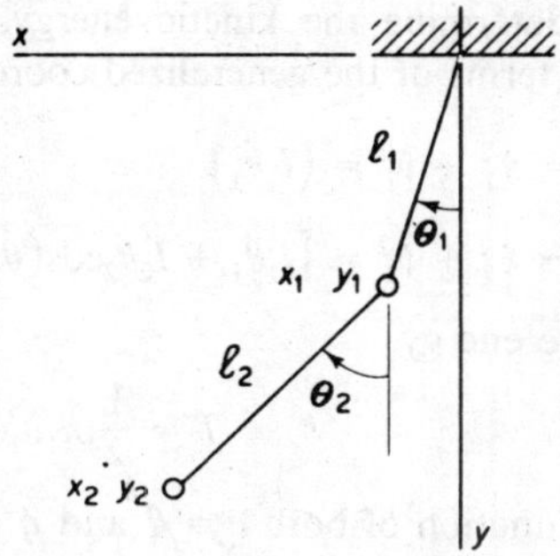

Figure 7.1-2.

the system by one. The coordinates x, y, z are, however, not independent, because they are related by the *constraint equation*

$$x^2 + y^2 + z^2 - l^2 = 0 \tag{7.1-1}$$

One of the coordinates can be eliminated by the above equation, thereby reducing the number of necessary coordinates to two.

The excess coordinates exceeding the number of degrees of freedom of the system are called *superfluous coordinates*, and constraint equations equal in number to the superfluous coordinates are necessary for their elimination. Constraints are called *holonomic* if the excess coordinates can be eliminated through equations of constraint. We will deal only with holonomic systems in this text.

Let us examine now the problem of defining the position of the double pendulum of Fig. 7.1-2. The double pendulum has only two DOF and the angles θ_1 and θ_2 completely define the position of m_1 and m_2. Thus θ_1 and θ_2 are generalized coordinates, i.e., $\theta_1 = q_1$ and $\theta_2 = q_2$.

The position of m_1 and m_2 can also be expressed in rectangular coordinates x, y. However they are related by the constraint equations

$$l_1^2 = x_1^2 + y_1^2$$

$$l_2^2 = (x_2 - x_1)^2 + (y_2 - y_1)^2$$

and hence are not independent. We can express the rectangular coordinates x_i, y_i in terms of the generalized coordinates θ_1 and θ_2

$$x_1 = l_1 \sin\theta_1, \qquad x_2 = l_1 \sin\theta_1 + l_2 \sin\theta_2$$

$$y_1 = l_1 \cos\theta_1, \qquad y_2 = l_1 \cos\theta_1 + l_2 \cos\theta_2$$

and these can also be considered as constraint equations.

To determine the kinetic energy, the squares of the velocity can be written in terms of the generalized coordinates.

$$v_1^2 = \dot{x}_1^2 + \dot{y}_1^2 = \left(l_1\dot{\theta}_1\right)^2$$

$$v_2^2 = \dot{x}_2^2 + \dot{y}_2^2 = \left[l_1\dot{\theta}_1 + l_2\dot{\theta}_2\cos(\theta_2 - \theta_1)\right]^2 + \left[l_2\dot{\theta}_2\sin(\theta_2 - \theta_1)\right]^2$$

The kinetic energy

$$T = \frac{1}{2}m_1v_1^2 + \frac{1}{2}m_2v_2^2$$

is then a function of both $q = \theta$ and $\dot{q} = \dot{\theta}$

$$T = T(q_1, q_2, \ldots, \dot{q}_1, \dot{q}_2, \ldots) \tag{7.1-2}$$

For the potential energy, the reference can be chosen at the level of the support point.

$$U = -m_1(l_1\cos\theta) - m_2(l_1\cos\theta_1 + l_2\cos\theta_2)$$

The potential energy is then seen to be a function only of the generalized coordinates

$$U = U(q_1, q_2, \ldots) \tag{7.1-3}$$

Example 7.1-1

Consider the plane mechanism shown in Fig. 7.1-3, where the members are assumed to be rigid. Describe all possible motions in terms of generalized coordinates.

Solution: As shown in Fig. 7.1-3, the displacements can be obtained by the superposition of two displacements q_1 and q_2. Since q_1 and q_2 are independent, they are generalized coordinates, and the system has two DOF.

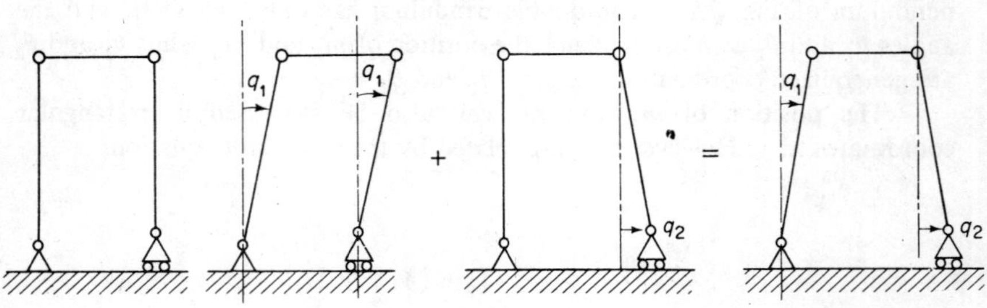

Figure 7.1-3.

Example 7.1-2

The plane frame shown in Fig. 7.1-4 has flexible members. Determine a set of generalized coordinates of the system. Assume that the corners remain at 90°.

Solution: There are two translational modes, q_1 and q_2, and each of the four corners can rotate independently, making a total of six generalized coordinates, $q_1, q_2, \ldots, q_6$

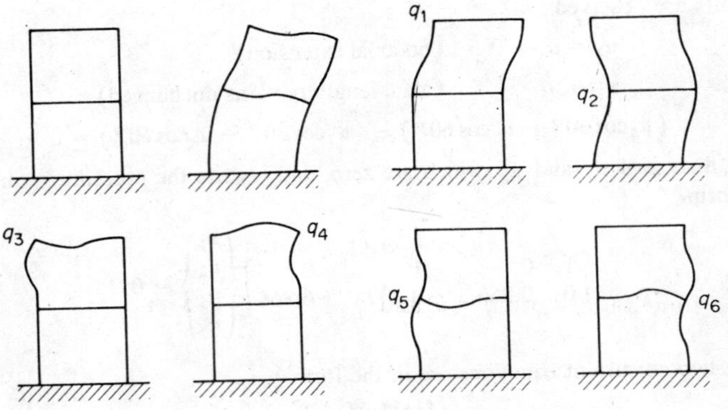

Figure 7.1-4.

Allowing each of these displacements to take place with all others equal to zero, the displacement of the frame can be seen to be the superposition of the six generalized coordinates.

Example 7.1-3

In defining the motion of a framed structure, the number of coordinates chosen often exceeds the number of degrees of freedom of the system so that constraint equations are involved. It is then desirable to express all of the coordinates u in terms of the fewer generalized coordinates q by a matrix equation of the form

$$u = Cq$$

The generalized coordinates q can be chosen arbitrarily from the coordinates u.

As an illustration of this equation we consider the framed structure of Fig. 7.1-5 consisting of four beam elements. We will be concerned only with the displacement of the joints and not the stresses in the members, which would require an added consideration of the distribution of the masses.

In Fig. 7.1-5 we have four element members with three joints that may undergo displacement. Two linear displacements and one rotation may be possible for each joint. We can label them u_1 to u_9. For compatibility of displacement, the following

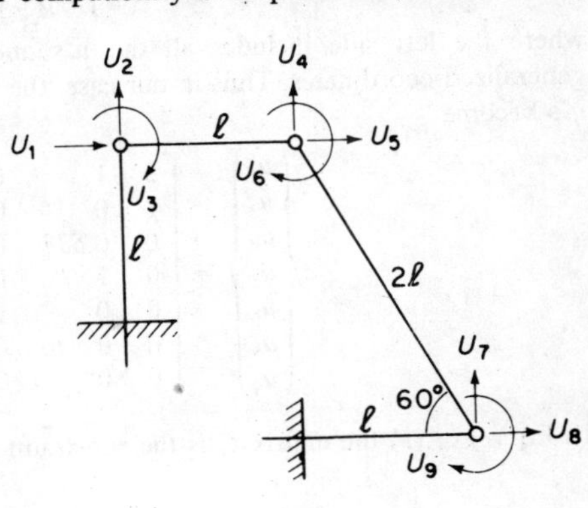

Figure 7.1-5.

constraints are observed

$$u_2 = u_8 = 0 \qquad \text{(no axial extension)}$$
$$u_1 = u_5 \qquad \text{(axial length remains unchanged)}$$
$$(u_4 \cos 30^\circ - u_5 \cos 60^\circ) - (u_7 \cos 30^\circ - u_8 \cos 60^\circ) = 0$$

We now disregard u_2 and u_8, which are zero, and rewrite the preceding equations in matrix form.

$$\begin{bmatrix} 1 & 0 & -1 & 0 \\ 0 & 0.866 & -0.500 & -0.866 \end{bmatrix} \begin{Bmatrix} u_1 \\ u_4 \\ u_5 \\ u_7 \end{Bmatrix} = 0 \tag{a}$$

Thus the two constraint equations are in the form

$$[A]\{u\} = 0 \tag{b}$$

We actually have seven coordinates $(u_1, u_3, u_4, u_5, u_6, u_7, u_9)$ and two constraint equations. Thus the degrees of freedom of the system are $7 - 2 = 5$, indicating that of the seven coordinates, five can be chosen as generalized coordinates q.

Of the four coordinates in the constraint equation, we choose u_5 and u_7 as two of the generalized coordinates and partition Eq. (a) as

$$[a \mid b]\begin{Bmatrix} u \\ \hline q \end{Bmatrix} = [a]\{u\} + [b]\{q\} = 0 \tag{c}$$

Thus the superfluous coordinates u can be expressed in terms of q as

$$\{u\} = -[a]^{-1}[b]\{q\} \tag{d}$$

Applying the preceding procedure to Eq. (a), we have

$$\begin{bmatrix} 1 & 0 \\ 0 & 0.866 \end{bmatrix} \begin{Bmatrix} u_1 \\ u_4 \end{Bmatrix} + \begin{bmatrix} -1 & 0 \\ -0.5 & -0.866 \end{bmatrix} \begin{Bmatrix} u_5 \\ u_7 \end{Bmatrix} = \begin{Bmatrix} 0 \\ 0 \end{Bmatrix}$$

$$\begin{Bmatrix} u_1 \\ u_4 \end{Bmatrix} = \begin{bmatrix} 1 & 0 \\ 0 & \dfrac{1}{0.866} \end{bmatrix} \begin{bmatrix} 1 & 0 \\ 0.5 & 0.866 \end{bmatrix} \begin{Bmatrix} u_5 \\ u_7 \end{Bmatrix} = \begin{bmatrix} 1 & 0 \\ 0.578 & 1 \end{bmatrix} \begin{Bmatrix} u_5 \\ u_7 \end{Bmatrix}$$

By supplying the remaining q_i as identities, all the u's can be expressed in terms of the q's as

$$\{u\} = [C]\{q\} \tag{e}$$

where the left side includes all the u's and the right column contains only the generalized coordinates. Thus in our case, the seven u's expressed in terms of the five q's become

$$\begin{Bmatrix} u_1 \\ u_3 \\ u_4 \\ u_5 \\ u_6 \\ u_7 \\ u_9 \end{Bmatrix} = \begin{bmatrix} 0 & 1 & 0 & 0 & 0 \\ 1 & 0 & 0 & 0 & 0 \\ 0 & 0.578 & 0 & 1 & 0 \\ 0 & 1 & 0 & 0 & 0 \\ 0 & 0 & 1 & 0 & 0 \\ 0 & 0 & 0 & 1 & 0 \\ 0 & 0 & 0 & 0 & 1 \end{bmatrix} \begin{Bmatrix} u_3 \\ u_5 \\ u_6 \\ u_7 \\ u_9 \end{Bmatrix} \tag{f}$$

In Eq. (e) or (f) the matrix C is the constraint matrix relating u to q.

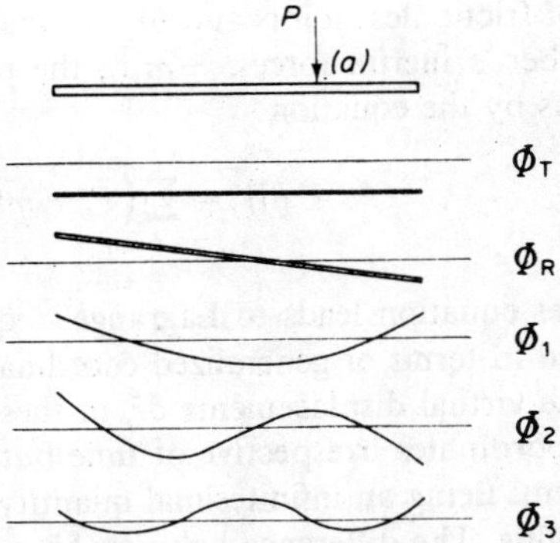

Figure 7.1-6.

Example 7.1-4

In the lumped mass models we treated earlier, n coordinates were assigned to the n masses of the n-DOF system, and each coordinate was independent and qualified as a generalized coordinate. For the flexible continuous body of infinite degrees of freedom, an infinite number of coordinates are required. Such a body can be treated as a system of a finite number of degrees of freedom by considering its deflection to be the sum of its normal modes multiplied by generalized coordinates.

$$y(x,t) = \phi_1(x)q_1(t) + \phi_2(x)q_2(t) + \phi_3(x)q_3(t) + \cdots$$

In many problems only a finite number of normal modes are sufficient, and the series can be terminated at n terms, thereby reducing the problem to that of a system of n DOF. For example, the motion of a slender free-free beam struck by a force P at point (a) can be described in terms of two rigid body motions of translation and rotation plus its normal modes of elastic vibration, as shown in Fig. 7.1-6.

$$y(x,t) = \phi_T q_T + \phi_R q_R + \phi_1(x)q_1 + \phi_2(x)q_2 + \cdots$$

7.2 VIRTUAL WORK

In Chapter 2 the method of virtual work was briefly introduced with examples for single-DOF problems. The advantage of the virtual work method over the vector method is considerably greater for multi-DOF systems. For interconnected bodies of many degrees of freedom, Newton's vector method is burdened with the necessity of accounting for all joint and constraint forces in the free-body diagrams, whereas these forces are excluded in the virtual work method.

In reviewing the method of virtual work, we summarize the virtual work equation as

$$\delta W = \sum_i \bar{F}_i \cdot \delta \bar{r}_i = 0 \tag{7.2-1}$$

where $\bar{F}_i$ are the applied forces excluding all constraint forces and internal forces of frictionless joints and $\delta\bar{r}_i$ are the virtual displacements. By including D'Alembert's inertia forces, $-m_i\ddot{\bar{r}}_i$, the procedure is extended to dynamical problems by the equation

$$\delta W = \sum_i \left(\bar{F}_i - m\ddot{\bar{r}}_i\right) \cdot \delta\bar{r}_i = 0 \tag{7.2-2}$$

This later equation leads to Lagrange's equation when the displacement $\bar{r}_i$ is expressed in terms of generalized coordinates.

The virtual displacements $\delta\bar{r}_i$ in these equations are arbitrary variations of the coordinates irrespective of time but compatible with the constraints of the system. Being an infinitesimal quantity, $\delta\bar{r}_i$ obeys all the rules of differential calculus. The difference between $\delta\bar{r}_i$ and $d\bar{r}_i$ is that $d\bar{r}_i$ takes place in the time dt, whereas $\delta\bar{r}_i$ is an arbitrary number that may be equal to $d\bar{r}_i$ but is assigned instantaneously irrespective of time. Although the virtual displacement $\delta\bar{r}$ is distinguished from $d\bar{r}$, the latter is often substituted for $\delta\bar{r}$ to ensure compatibility of displacement.

Example 7.2-1

We first illustrate the virtual work method for a problem of static equilibrium. Figure 7.2-1 shows a double pendulum with generalized coordinates θ_1 and θ_2. Determine its static equilibrium position when a horizontal force P is applied to m_2.

With the system in its equilibrium position, give θ_2 a virtual displacement $\delta\theta_2$ [Fig. 7.2-1(a)] and write the equation for the virtual work δW of all the applied forces:

$$\delta W = -(m_2 g \sin\theta_2) l\,\delta\theta_2 + (P\cos\theta_2) l\,\delta\theta_2 = 0$$

From the equilibrium position (with $\delta\theta_2 = 0$), give θ_1 a virtual displacement $\delta\theta_1$ as in Fig. 7.2-1(b) and write the equation for δW:

$$\delta W = -(m_1 g \sin\theta_1) l\,\delta\theta_1 - (m_2 g \sin\theta_1) l\,\delta\theta_1 + (P\cos\theta_1) l\,\delta\theta_1 = 0$$

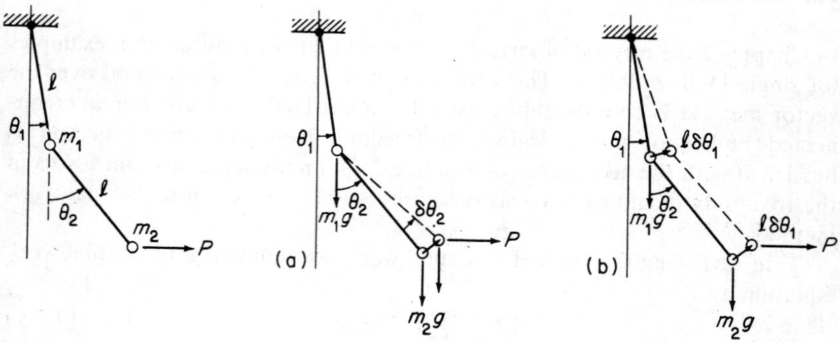

Figure 7.2-1.

These equations lead to the two equilibrium angles, given as

$$\tan \theta_2 = \frac{P}{m_2 g}$$

$$\tan \theta_1 = \frac{P}{(m_1 + m_2) g}$$

Example 7.2-2

Determine the equations of motion for the system shown in Fig. 7.2-2.

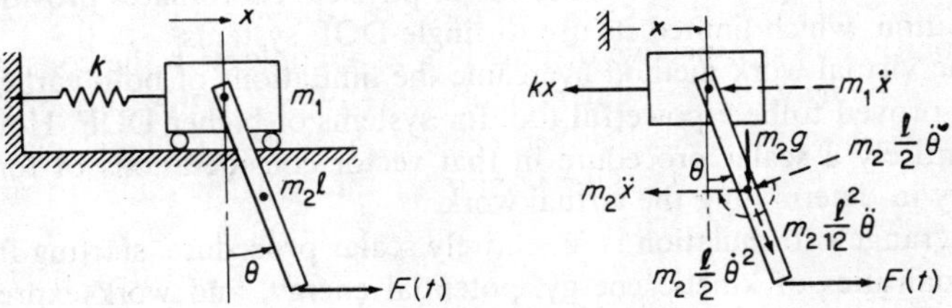

Figure 7.2-2.

Solution: The generalized coordinates for the problem are x and θ. Sketch the system in the displaced position with all the active forces and inertia forces. Giving x a virtual displacement δx, the virtual work equation is

$$\delta W = -[(m_1 + m_2)\ddot{x} + kx]\,\delta x - \left(m_2 \frac{l}{2} \ddot{\theta} \cos \theta\right) \delta x + \left(m_2 \frac{l}{2} \dot{\theta}^2 \sin \theta\right) \delta x + F(t)\,\delta x = 0$$

Since δx is an arbitrary, the preceding equation leads to

$$(m_1 + m_2)\ddot{x} + m_2 \frac{l}{2}(\ddot{\theta} \cos \theta - \dot{\theta}^2 \sin \theta) + kx = F(t)$$

Next, allow a virtual displacement $\delta\theta$. δW is then

$$\delta W = -\left(m_2 \frac{l}{2} \ddot{\theta}\right) \frac{l}{2}\,\delta\theta - \left(m_2 \frac{l^2}{12} \ddot{\theta}\right) \delta\theta - (m_2 g \sin \theta) \frac{l}{2}\,\delta\theta - (m_2 \ddot{x} \cos \theta) \frac{l}{2}\,\delta\theta + [F(t) \cos \theta]\, l\,\delta\theta = 0$$

from which we obtain

$$m_2 \frac{l^2}{3} \ddot{\theta} + m_2 \frac{l}{2} \ddot{x} \cos \theta + m_2 g \frac{l}{2} \sin \theta = F(t)\, l \cos \theta$$

These are nonlinear differential equations, which for small angles simplify to

$$(m_1 + m_2)\ddot{x} + m_2 \frac{l}{2} \ddot{\theta} + kx = F(t)$$

$$m_2 \frac{l^2}{3} \ddot{\theta} + m_2 \frac{l}{2} \ddot{x} + m_2 g \frac{l}{2} \theta = lF(t)$$

7.3 LAGRANGE'S EQUATION

In our previous discussions we were able to formulate the equations of motion by three different methods. Newton's vector method offered a simple approach for systems of a few degrees of freedom. The necessity for the consideration of forces of constraints and free-body diagrams in this method led to algebraic difficulties for systems of higher degrees of freedom.

The energy method overcame the difficulties of the vector method. However, the energy principle in terms of physical coordinates provided only one equation, which limited its use to single-DOF systems.

The virtual work method overcame the limitations of both earlier methods and proved to be a powerful tool for systems of higher DOF. However it is not entirely a scalar procedure in that vector considerations of forces are necessary in determining the virtual work.

Lagrange's formulation is an entirely scalar procedure, starting from the scalar quantities of kinetic energy, potential energy, and work expressed in terms of generalized coordinates. It is presented here as

$$\frac{d}{dt}\left(\frac{\partial T}{\partial \dot{q}_i}\right) - \frac{\partial T}{\partial q_i} + \frac{\partial U}{\partial q_i} = Q_i \tag{7.3-1}$$

The left side of this equation, when summed for all the q_i, is a statement of the principle of conservation of energy and is equivalent to

$$d(T + U) = 0$$

The right side of Lagrange's equation results from dividing the work term in the dynamical relationship† $dT = dW$ into the work done by the potential and nonpotential forces as follows:

$$dT = dW_p + dW_{np}$$

The work of the potential forces was shown earlier to be equal to $dW_p = -dU$, which is included in the left side of Lagrange's equation. The nonpotential work is equal to the work done by the nonpotential forces in a virtual displacement expressed in terms of the generalized coordinates. Thus Lagrange's equation, Eq. (7.3-1), is the q_i component of the energy equation

$$d(T + U) = \delta W_{np}$$

We can write the right side of this equation as

$$\delta W = \sum Q_i \, \delta q_i = Q_1 \, \delta q_1 + Q_2 \, \delta q_2 + \cdots$$

The quantity Q_i is called the *generalized force*. In spite of its name, Q_i may have units other than that of a force; i.e., if δq_i is an angle, Q_i has the units of

† $dW = \bar{F} \cdot d\bar{r} = \bar{F} \cdot \dot{\bar{r}}\, dt, \qquad T = \frac{1}{2} m \dot{\bar{r}} \cdot \dot{\bar{r}}$

$dT = m \ddot{\bar{r}} \cdot \dot{\bar{r}}\, dt = \bar{F} \cdot \dot{\bar{r}}\, dt \qquad \therefore\ dT = dW$

a moment. The only requirement is that the product $Q_i\,\delta q_i$ be in the units of work.

Lagrange's equations are differential equations of motion expressed in terms of generalized coordinates. We present here a brief development for the general form of these equations in terms of the kinetic and potential energies.

Consider first a conservative system where the sum of the kinetic and potential energies is a constant. The differential of the total energy is then zero:

$$d(T+U)=0 \tag{7.3-2}$$

The kinetic energy T is a function of the generalized coordinates q_i and the generalized velocities $\dot{q}_i$, whereas the potential energy U is a function only of q_i.

$$\begin{aligned} T &= T(q_1, q_2, \ldots, q_N, \dot{q}_1, \dot{q}_2, \ldots, \dot{q}_N) \\ U &= U(q_1, q_2, \ldots, q_N) \end{aligned} \tag{7.3-3}$$

The differential of T is

$$dT=\sum_{i=1}^{N}\frac{\partial T}{\partial q_i}\,dq_i+\sum_{i=1}^{N}\frac{\partial T}{\partial \dot{q}_i}\,d\dot{q}_i \tag{7.3-4}$$

To eliminate the second term with $d\dot{q}_i$, we start with the equation for kinetic energy,

$$T=\frac{1}{2}\sum_{i=1}^{N}\sum_{j=1}^{N}m_{ij}\dot{q}_i\dot{q}_j \tag{7.3-5}$$

Differentiating this equation with respect to $\dot{q}_i$, multiplying by $\dot{q}_i$, and summing over i from 1 to N, we obtain a result equal to

$$\sum_{i=1}^{N}\frac{\partial T}{\partial \dot{q}_i}\dot{q}_i=\sum_{i=1}^{N}\sum_{j=1}^{N}m_{ij}\dot{q}_j\dot{q}_i=2T$$

or

$$2T=\sum_{i=1}^{N}\frac{\partial T}{\partial \dot{q}_i}\dot{q}_i \tag{7.3-6}$$

We now form the differential of $2T$ from the preceding equation by using the product rule in calculus:

$$2\,dT=\sum_{i=1}^{N}d\left(\frac{\partial T}{\partial \dot{q}_i}\right)\dot{q}_i+\sum_{i=1}^{N}\frac{\partial T}{\partial \dot{q}_i}\,d\dot{q}_i \tag{7.3-7}$$

Subtracting Eq. (7.3-4) from this equation, the second term with $d\dot{q}_i$ is eliminated. By shifting the scalar quantity dt, the term $d(\partial T/\partial \dot{q}_i)\dot{q}_i$ becomes $d/dt\,(\partial T/\partial \dot{q}_i)\,dq_i$ and the result is

$$dT=\sum_{i=1}^{N}\left[\frac{d}{dt}\left(\frac{\partial T}{\partial \dot{q}_i}\right)-\frac{\partial T}{\partial q_i}\right]dq_i \tag{7.3-8}$$

From Eq. (7.3-3) the differential of U is

$$dU = \sum_{i=1}^{N} \frac{\partial U}{\partial q_i} dq_i \tag{7.3-9}$$

Thus, Eq. (7.3-2) for the invariance of the total energy becomes

$$d(T + U) = \sum_{i=1}^{N} \left[\frac{d}{dt}\left(\frac{\partial T}{\partial \dot{q}_i}\right) - \frac{\partial T}{\partial q_i} + \frac{\partial U}{\partial q_i} \right] dq_i = 0 \tag{7.3-10}$$

Since the N generalized coordinates are independent of one another, the dq_i may assume arbitrary values. Therefore, the above equation is satisfied only if

$$\frac{d}{dt}\left(\frac{\partial T}{\partial \dot{q}_i}\right) - \frac{\partial T}{\partial q_i} + \frac{\partial U}{\partial q_i} = 0, \qquad i = 1, 2, \ldots, N \tag{7.3-11}$$

This is Lagrange's equation for the case in which all forces have a potential U. They can be somewhat modified by introducing the Lagrangian $L = (T - U)$. Since $\partial U/\partial \dot{q}_i = 0$, Eq. (7.3-11) can be written in terms of L as

$$\frac{d}{dt}\left(\frac{\partial L}{\partial \dot{q}_i}\right) - \frac{\partial L}{\partial q_i} = 0, \qquad i = 1, 2, \ldots, N \tag{7.3-12}$$

Nonconservative systems. When the system is also subjected to given forces that do not have a potential, we have instead of Eq. (7.3-2)

$$d(T + U) = \delta W_{np} \tag{7.3-13}$$

where δW_{np} is the work of the nonpotential forces when the system is subjected to an arbitrary infinitesimal displacement. Expressing δW_{np} in terms of the generalized coordinates q_i, we have

$$\delta W_{np} = \sum_{i=1}^{N} Q_i \, \delta q_i \tag{7.3-14}$$

The quantities Q_i are known as the generalized forces associated with the generalized coordinate q_i. Lagrange's equation including nonconservative forces then becomes

$$\frac{d}{dt}\left(\frac{\partial T}{\partial \dot{q}_i}\right) - \frac{\partial T}{\partial q_i} + \frac{\partial U}{\partial q_i} = Q_i, \qquad i = 1, 2, \ldots, N \tag{7.3-15}$$

Example 7.3-1

Using Lagrange's method, set up the equations of motion for the system shown in Fig. 7.3-1.

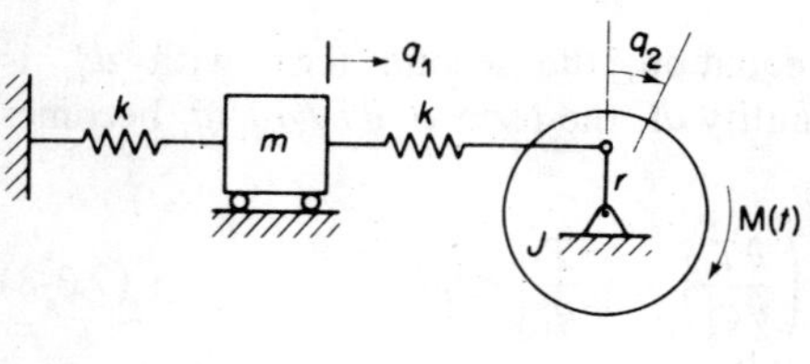

Figure 7.3-1.

Solution: The kinetic and potential energies are

$$T = \tfrac{1}{2}m\dot{q}_1^2 + \tfrac{1}{2}J\dot{q}_2^2$$

$$U = \tfrac{1}{2}kq_1^2 + \tfrac{1}{2}k(rq_2 - q_1)^2$$

and from the work done by the external moment, the generalized force is

$$\delta W = \mathfrak{M}(t)\,\delta q_2 \qquad \therefore\ Q_2 = \mathfrak{M}(t)$$

Substituting into Lagrange's equation, the equations of motion are

$$m\ddot{q}_1 + 2kq_1 - krq_2 = 0$$

$$J\ddot{q}_2 - krq_1 + kr^2q_2 = \mathfrak{M}(t)$$

Example 7.3-2

Write the Lagrangian for the system shown in Fig. 7.3-2.

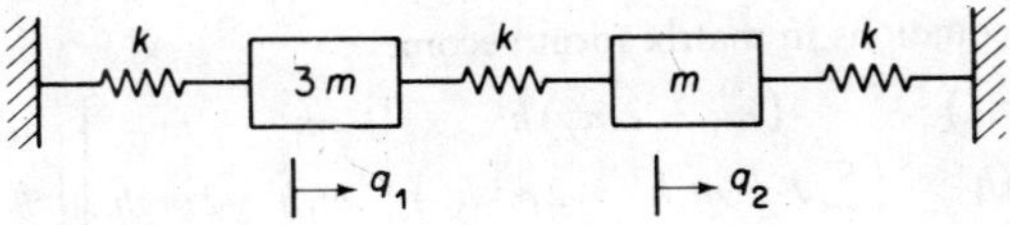

Figure 7.3-2.

Solution:

$$T = \frac{1}{2}m(\dot{q}_1\dot{q}_2)\begin{bmatrix}3 & 0\\ 0 & 1\end{bmatrix}\begin{Bmatrix}\dot{q}_1\\ \dot{q}_2\end{Bmatrix}$$

$$U = \frac{1}{2}k(q_1q_2)\begin{bmatrix}2 & -1\\ -1 & 2\end{bmatrix}\begin{Bmatrix}q_1\\ q_2\end{Bmatrix}$$

$$L = T - U$$

$$= \frac{1}{2}(3m)\dot{q}_1^2 + \frac{1}{2}(m)\dot{q}_2^2 - kq_1^2 - kq_2^2 + kq_1q_2$$

Example 7.3-3

Fig. 7.3-3 shows a simplified model of a two-story building whose foundation is subject to translation and rotation. Determine T and U and the equations of motion.

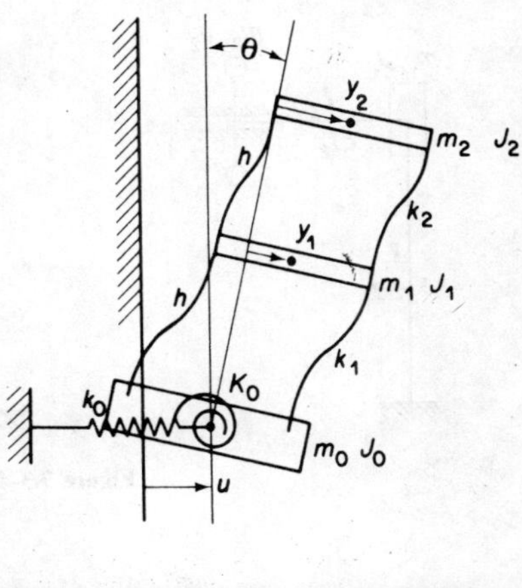

Figure 7.3-3.

Solution: We choose u and θ for the translation and rotation of the foundation and y for the elastic displacement of the floors. The equations for T and U become

$$T = \frac{1}{2}m_0\dot{u}^2 + \frac{1}{2}J_0\dot{\theta}^2 + \frac{1}{2}m_1(\dot{u} + h\dot{\theta} + \dot{y}_1)^2 + \frac{1}{2}J_1\dot{\theta}^2 + \frac{1}{2}m_2(\dot{u} + 2h\dot{\theta} + \dot{y}_2)^2 + \frac{1}{2}J_2\dot{\theta}^2$$

$$U = \frac{1}{2}k_0u^2 + \frac{1}{2}K_0\theta^2 + \frac{1}{2}k_1y_1^2 + \frac{1}{2}k_2(y_2 - y_1)^2$$

where u, θ, y_1, and y_2 are the generalized coordinates. Substituting into Lagrange's equation, we obtain, for example,

$$\frac{\partial T}{\partial \dot{\theta}} = (J_0 + J_1 + J_2)\dot{\theta} + m_1h(\dot{u} + h\dot{\theta} + \dot{y}_1) + m_2 2h(\dot{u} + 2h\dot{\theta} + \dot{y}_2)$$

$$\frac{\partial U}{\partial \theta} = K_0$$

The four equations in matrix form become

$$\left[\begin{array}{cc|cc} (m_0 + m_1 + m_2) & (m_1 + 2m_2)h & m_1 & m_2 \\ (m_1 + 2m_2)h & \left(\sum J + m_1h^2 + 2m_2h^2\right) & m_1h & 2m_2h \\ \hline m_1 & m_1h & m_1 & 0 \\ m_2 & 2m_2h & 0 & m_2 \end{array}\right]\left\{\begin{array}{c} \ddot{u} \\ \ddot{\theta} \\ \hline \ddot{y}_1 \\ \ddot{y}_2 \end{array}\right\} + \left[\begin{array}{cc|cc} k_0 & 0 & 0 & 0 \\ 0 & K_0 & 0 & 0 \\ \hline 0 & 0 & (k_1 + k_2) & -k_2 \\ 0 & 0 & -k_2 & k_2 \end{array}\right]\left\{\begin{array}{c} u \\ \theta \\ \hline y_1 \\ y_2 \end{array}\right\} = \{0\}$$

It should be noted that the equation represented by the upper left corner of the matrices is that of rigid body translation and rotation.

Example 7.3-4

Determine the generalized coordinates for the system shown in Fig. 7.3-4 and evaluate the stiffness and the mass matrices for the equations of motion.

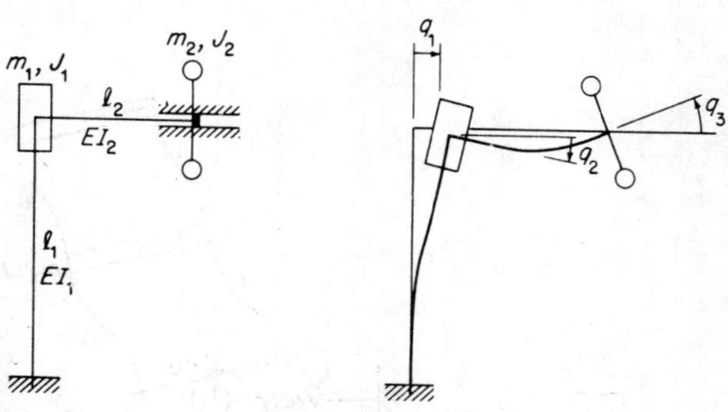

Figure 7.3-4.

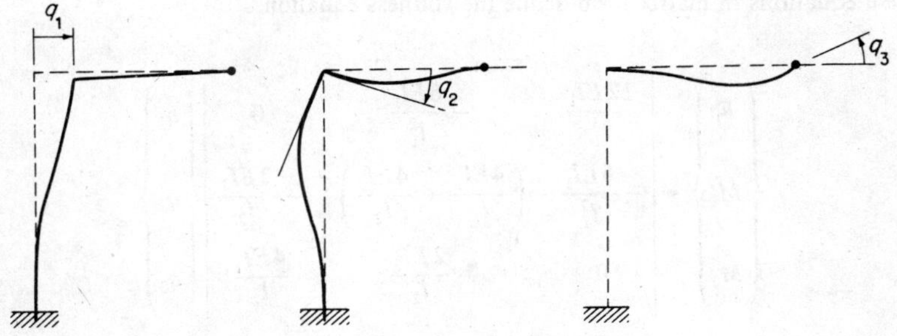

Figure 7.3-4(a).

Solution: As shown in Fig. 7.3-4 (a), three generalized coordinates are required.

The displaced configuration of the frame can be considered to be the superposition of the following shapes with each coordinate given separately.

The forces and moments for the free-body diagrams shown in Fig. 7.3-4 (b) are obtained from Table 6.3-1.

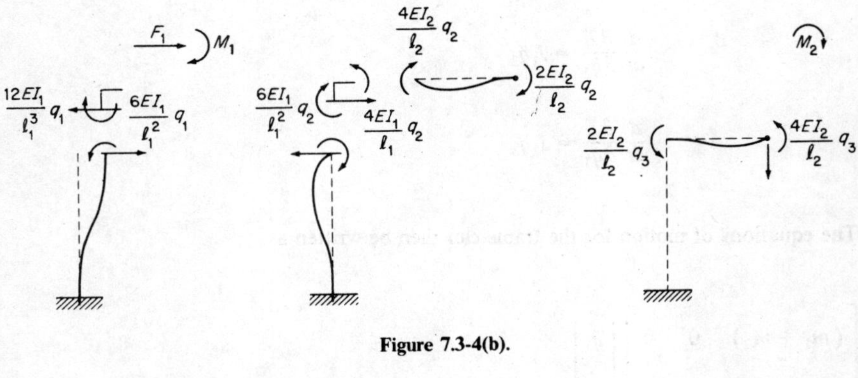

Figure 7.3-4(b).

The equations for the forces and moments necessary to maintain these deflections are obtained from their superposition as follows:

$$F_1 = \frac{12EI_1}{l_1^3}q_1 - \frac{6EI_1}{l_1^2}q_2$$

$$M_1 = -\frac{6EI_1}{l_1^2}q_1 + \left(\frac{4EI_1}{l_1} + \frac{4EI_2}{l_2}\right)q_2 - \frac{2EI_2}{l_2}q_3$$

$$M_2 = -\frac{2EI_2}{l_2}q_2 + \frac{4EI_2}{l_2}q_3$$

These equations in matrix form define the stiffness equation.

$$\begin{Bmatrix} F_1 \\ M_1 \\ M_2 \end{Bmatrix} = \begin{vmatrix} \dfrac{12EI_1}{l_1^3} & -\dfrac{6EI_1}{l_1^2} & 0 \\ -\dfrac{6EI_1}{l_1^2} & \left(\dfrac{4EI_1}{l_1} + \dfrac{4EI_2}{l_2}\right) & -\dfrac{2EI_2}{l_2} \\ 0 & -\dfrac{2EI_2}{l_2} & \dfrac{4EI_1}{l_2} \end{vmatrix} \begin{Bmatrix} q_1 \\ q_2 \\ q_3 \end{Bmatrix}$$

The mass matrix is found from the kinetic energy:

$$T = \tfrac{1}{2}(m_1 + m_2)\dot{q}_1^2 + \tfrac{1}{2}J_1\dot{q}_2^2 + \tfrac{1}{2}J_2\dot{q}_3^2$$

$$\frac{d}{dt}\frac{\partial T}{\partial \dot{q}_1} = (m_1 + m_2)\ddot{q}_1$$

$$\frac{d}{dt}\frac{\partial T}{\partial \dot{q}_2} = J_1\ddot{q}_2$$

$$\frac{d}{dt}\frac{\partial T}{\partial \dot{q}_3} = J_2\ddot{q}_3$$

The equations of motion for the frame can then be written as

$$\begin{bmatrix} (m_1 + m_2) & 0 & 0 \\ 0 & J_1 & 0 \\ 0 & 0 & J_2 \end{bmatrix} \begin{Bmatrix} \ddot{q}_1 \\ \ddot{q}_2 \\ \ddot{q}_3 \end{Bmatrix}$$

$$+ \begin{bmatrix} \dfrac{12EI_1}{l_1^3} & -\dfrac{6EI_1}{l_1^2} & 0 \\ -\dfrac{6EI_1}{l_1^2} & \left(\dfrac{4EI_1}{l_1} + \dfrac{4EI_2}{l_2}\right) & -\dfrac{2EI_2}{l_2} \\ 0 & -\dfrac{2EI_2}{l_2} & \dfrac{4EI_2}{l_2} \end{bmatrix} \begin{Bmatrix} q_1 \\ q_2 \\ q_3 \end{Bmatrix} = \begin{Bmatrix} F_1 \\ M_1 \\ M_2 \end{Bmatrix}$$

7.4 KINETIC ENERGY, POTENTIAL ENERGY, AND GENERALIZED FORCE IN TERMS OF GENERALIZED COORDINATE *q*

In the previous section, the use of Lagrange's equation was demonstrated for simple problems. We now discuss the quantities T, U, and Q from a more general point of view.

Kinetic energy. Representing the system by N particles, the instantaneous position of each particle may be expressed in terms of the N generalized coordinates

$$\mathbf{r}_j = \mathbf{r}_j(q_1 q_2, \ldots, q_N) \tag{7.4-1}$$

The velocity of the jth particle is

$$\mathbf{v}_j = \sum_{i=1}^{N} \frac{\partial \mathbf{r}_j}{\partial q_i} \dot{q}_i \tag{7.4-2}$$

and the kinetic energy of the system becomes

$$T = \frac{1}{2} \sum_{j=1}^{N} m_j \mathbf{v}_j \cdot \mathbf{v}_j = \frac{1}{2} \sum_{i=1}^{N} \sum_{j=1}^{N} \left(\sum_{i=1}^{N} m_i \frac{\partial \mathbf{r}_i}{\partial q_j} \cdot \frac{\partial \mathbf{r}_i}{\partial q_i} \right) \dot{q}_i \dot{q}_j \tag{7.4-3}$$

Defining the *generalized mass* as

$$m_{ij} = \sum_{i=1}^{N} m_i \frac{\partial \mathbf{r}_i}{\partial q_j} \cdot \frac{\partial \mathbf{r}_i}{\partial q_i} \tag{7.4-4}$$

the kinetic energy may be written as

$$\begin{aligned} T &= \frac{1}{2} \sum_{i=1}^{N} \sum_{j=1}^{N} m_{ij} \dot{q}_i \dot{q}_j \\ &= \frac{1}{2} \{\dot{q}\}' [m] \{\dot{q}\} \end{aligned} \tag{7.4-5}$$

Potential energy. In a conservative system, the forces can be derived from the potential energy U, which is a function of the generalized coordinates q_j. Expanding U in a Taylor series about the equilibrium position, we have for a system of n degrees of freedom

$$U = U_0 + \sum_{j=1}^{n} \left(\frac{\partial U}{\partial q_j} \right)_0 q_j + \frac{1}{2} \sum_{j=1}^{n} \sum_{l=1}^{n} \left(\frac{\partial^2 U}{\partial q_j \, \partial q_l} \right)_0 q_j q_l + \cdots \tag{7.4-6}$$

In this expression U_0 is an arbitrary constant which we can set equal to zero. The derivatives of U are evaluated at the equilibrium position 0 and are constants when the q_j's are small quantities equal to zero at the equilibrium position. Since U is a minimum in the equilibrium position, the first derivative $(\partial U / \partial q_j)_0$ is zero, which leaves only $(\partial^2 U / \partial q_j \, \partial q_l)_0$ and higher-order terms.

In the theory of small oscillations about the equilibrium position, terms beyond the second order are ignored and the equation for the potential energy reduces to

$$k_{jl} = \left(\frac{\partial^2 U}{\partial q_j \, \partial q_l} \right)_0 \tag{7.4-7}$$

and the potential energy is written as

$$\begin{aligned} U &= \frac{1}{2} \sum_{j=1}^{n} \sum_{l=1}^{n} k_{jl} q_j q_l \\ &= \frac{1}{2} \{q\}' [k] \{q\} \end{aligned} \tag{7.4-8}$$

Generalized force. For the development of the generalized force, we start from the virtual displacement of the coordinate $\mathbf{r}_j$,

$$\delta \mathbf{r}_j = \sum_i \frac{\delta \mathbf{r}_j}{\delta q_i} \delta q_i \tag{7.4-9}$$

and the time t is not involved.

When the system is in equilibrium, the virtual work can now be expressed in terms of the generalized coordinates q_i

$$\delta W = \sum_j \mathbf{F}_j \cdot \delta \mathbf{r}_j = \sum_j \sum_i \mathbf{F}_j \cdot \frac{\delta \mathbf{r}_j}{\delta q_i} \delta q_i \tag{7.4-10}$$

Interchanging the order of summation and letting

$$Q_i = \sum_j \mathbf{F}_j \cdot \frac{\delta \mathbf{r}_j}{\delta q_i} \tag{7.4-11}$$

be defined as the *generalized force*, the virtual work for the system, expressed in terms of the generalized coordinates, becomes

$$\delta W = \sum_i Q_i \, \delta q_i \tag{7.4-12}$$

Example 7.4-1 ***Generalized Mass***

Determine the generalized mass when the displacement at position x is represented by the equation

$$\begin{aligned} r(x,t) &= \phi_1(x) q_1(t) + \phi_2(x) q_2(t) + \cdots + \phi_N(x) q_N(t) \\ &= \sum_{i=1}^{N} \phi_i(x) q_i(t) \end{aligned} \tag{a}$$

where $\phi_i(x)$ are shape functions of only x.

Solution: The velocity is

$$v(x) = \sum_{i=1}^{N} \phi_i(x)\dot{q}_i(t) \tag{b}$$

and the kinetic energy becomes

$$T = \frac{1}{2}\sum_{i=1}^{N}\sum_{j=1}^{N} \dot{q}_i\dot{q}_j \int \phi_i(x)\phi_j(x)\, dm$$

$$= \frac{1}{2}\sum_{i=1}^{N}\sum_{j=1}^{N} m_{ij}\dot{q}_i\dot{q}_j \tag{c}$$

Thus the generalized mass is

$$m_{ij} = \int \phi_i(x)\phi_j(x)\, dm \tag{d}$$

where the integration is carried out over the entire system. In case the system consists of discrete masses, m_{ij} becomes

$$m_{ij} = \sum_{p=1}^{N} m_p\phi_i(x_p)\phi_j(x_p) \tag{e}$$

Example 7.4-2 ***Generalized Stiffness*** **(Axial Vibration)**

Determine the generalized stiffness of a rod in axial vibration. We again represent the longitudinal displacement of the rod in terms of generalized coordinates by the equation

$$u(x,t) = \sum_{i=1}^{n} \varphi_i(x)q_i(t)$$

The potential energy of the rod under axial stress is

$$U = \frac{1}{2}\int AE\left(\frac{du}{dx}\right)^2 dx$$

Substituting for $u(x,t)$ gives

$$U = \frac{1}{2}\sum_i\sum_j q_iq_j\int AE\varphi_i'\varphi_j'\, dx$$

$$= \frac{1}{2}\sum_i\sum_j k_{ij}q_iq_j$$

where the generalized stiffness is

$$k_{ij} = \int AE\varphi_i'\varphi_j'\, dx$$

Example 7.4-3 ***Generalized Stiffness*** **(Beams)**

Determine the generalized stiffness for a beam of cross-sectional property EI when the displacement $y(x, t)$ is represented by the sum

$$y(x,t) = \sum_{i=1}^{n} \varphi_i(x) q_i(t) \tag{a}$$

The potential energy of a beam in bending is

$$U = \frac{1}{2}\int EI\left(\frac{d^2y}{dx^2}\right)^2 dx \tag{b}$$

Substituting for

$$\frac{d^2y}{dx^2} = \sum_{i=1}^{n} \varphi_i''(x) q_i(t)$$

we obtain

$$U = \frac{1}{2}\sum_i \sum_j q_i q_j \int EI\varphi_i''\varphi_j''\, dx$$

$$= \frac{1}{2}\sum_i \sum_j k_{ij} q_i q_j \tag{c}$$

and the generalized stiffness is

$$k_{ij} = \int EI\varphi_i''\varphi_j''\, dx \tag{d}$$

Example 7.4-4 ***Generalized Force***

The frame of Fig. 7.1-3 with rigid members is acted upon by the forces and moments shown in Fig. 7.4-1. Determine the generalized forces.

Solution: We let δq_1 be the virtual displacement of the upper left corner and δq_2 be the translation of the right support hinge. Due to δq_1 the virtual work done is

$$Q_1\,\delta q_1 = F_1\,\delta q_1 - F_2\frac{a}{l}\,\delta q_1 + (M_1 - M_2)\frac{1}{l}\,\delta q_1$$

$$\therefore Q_1 = F_1 - \frac{a}{l}F_2 + \frac{1}{l}(M_1 - M_2)$$

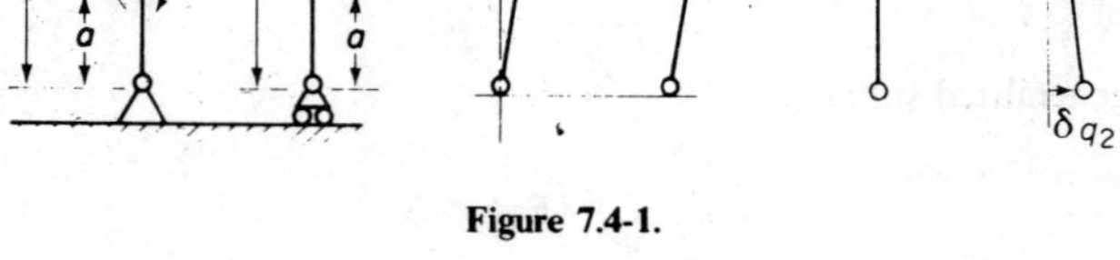

Figure 7.4-1.

The virtual work done due to δq_2 is

$$Q_2 \, \delta q_2 = -F_2(l - a)\frac{\delta q_2}{l} + M_2 \frac{\delta q_2}{l}$$

$$\therefore Q_2 = [-F_2(l - a) + M_2]\frac{1}{l}$$

It should be noted that the dimension of Q_1 and Q_2 is that of a force.

Example 7.4-5

In Fig. 7.4-2, three forces F_1, F_2, and F_3 act at discrete points x_1; x_2, and x_3 of a structure whose displacement is expressed by the equation

$$y(x,t) = \sum_{i=1}^{n} \varphi_i(x) q_i(t)$$

Determine the generalized force Q_i.

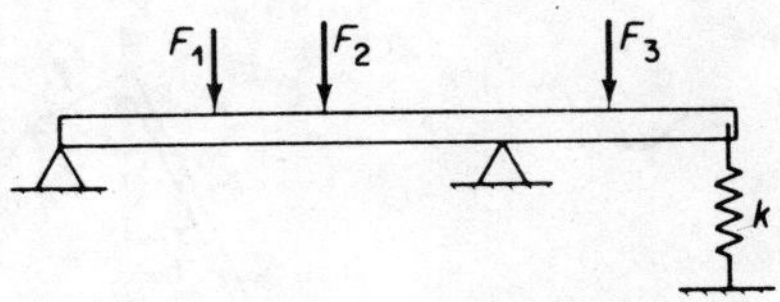

Figure 7.4-2.

Solution: The virtual displacement δy is

$$\sum_{i=1}^{n} \varphi_i(x) \, \delta q_i$$

and the virtual work due to this displacement is

$$\delta W = \sum_{j=1}^{3} F_j \cdot \left(\sum_{i=1}^{n} \varphi_i(x_j) \, \delta q_i \right)$$

$$= \sum_{i=1}^{n} \delta q_i \left(\sum_{j=1}^{3} F_j \varphi_i(x_j) \right) = \sum_{i=1}^{n} Q_i \, \delta q_i$$

The generalized force is then equal to $\delta W / \delta q_i$ or

$$Q_i = \sum_{j=1}^{3} F_j \varphi_i(x_j)$$

$$= F_1 \varphi_i(x_1) + F_2 \varphi_i(x_2) + F_3 \varphi_i(x_3)$$

PROBLEMS

7-1 List the displacement coordinates u_i for the plane frame of Fig. P7-1 and write the geometric constraint equations. State the number of degrees of freedom for the system.

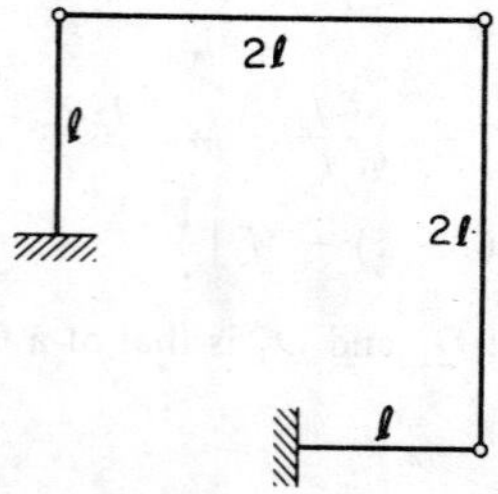

Figure P7-1.

7-2 Choose the generalized coordinates q_i for the previous problem and express the u_i coordinates in terms of q_i.

7-3 Using the method of virtual work, determine the equilibrium position of a carpenter's square hooked over a peg as shown in Fig. P7-3.

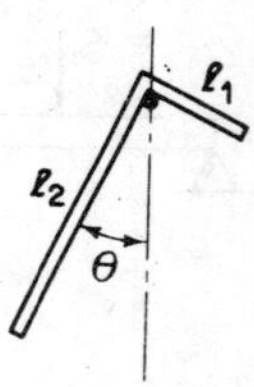

Figure P7-3.

7-4 Determine the equilibrium position of the two uniform bars shown in Fig. P7-4 when a force P is applied as shown. All surfaces are friction free.

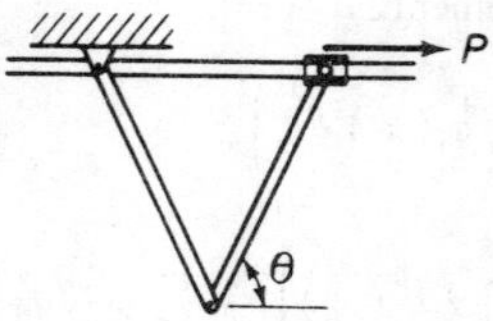

Figure P7-4.

7-5 Determine the equilibrium position of two point masses m_1 and m_2 connected by a massless rod and placed in a smooth hemispherical bowl of radius R as shown in Fig. P7-5.

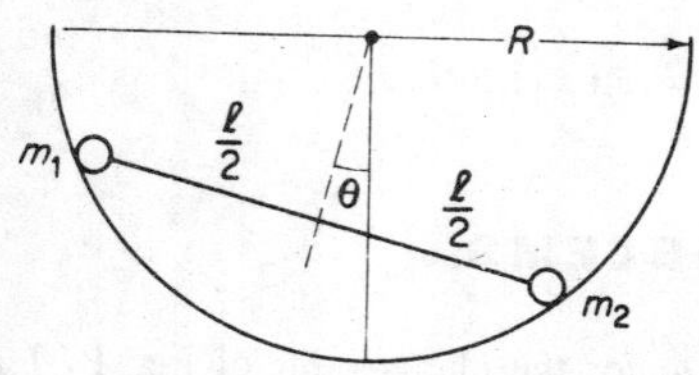

Figure P7-5.

7-6 The four masses on the string in Fig. P7-6 are displaced by a horizontal force F. Determine its equilibrium position by using virtual work.

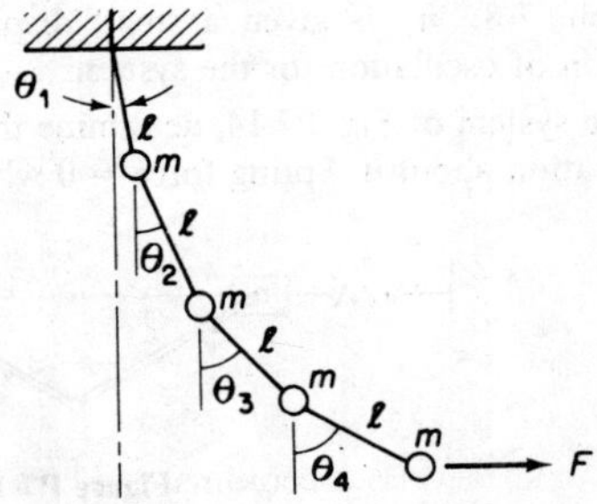

Figure P7-6.

7-7 A mass m is supported by two springs of unstretched length r_0 attached to a pin and slider as shown in Fig. P7-7. There is coulomb friction with coefficient μ between the massless slider and the rod. Determine its equilibrium position by virtual work.

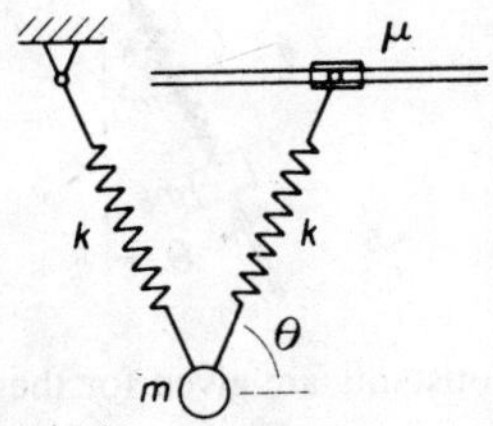

Figure P7-7.

7-8 Determine the equilibrium position of m_1 and m_2 attached of strings of equal length, as shown in Fig. P7-8.

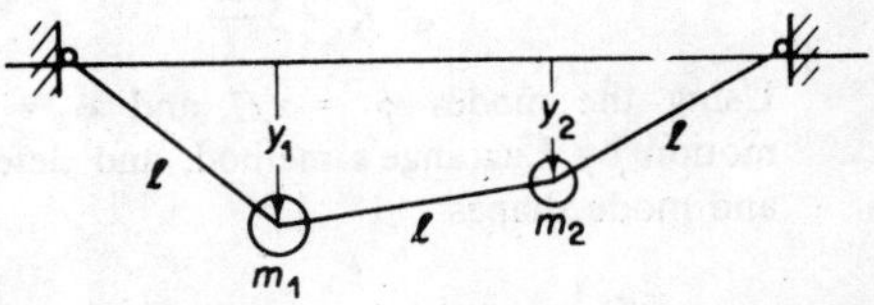

Figure P7-8.

7-9 A rigid uniform rod of length l is supported by a spring and a smooth floor as shown in Fig. P7-9. Determine its equilibrium position by virtual work. The unstretched length of the spring is $h/4$.

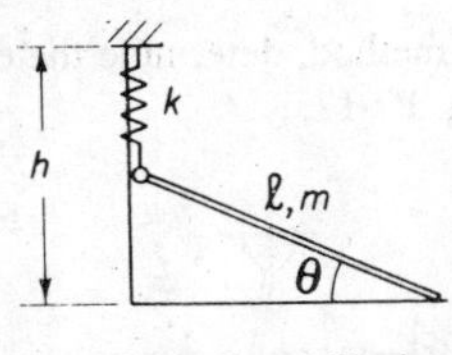

Figure P7-9.

7-10 Determine the equation of motion for small oscillation about the equilibrium position in Prob. 7-9.

7-11 The carpenter's square of Prob. 7-3 is displaced slightly from its equilibrium position and released. Determine its equation of oscillation.

7-12 Determine the equation of motion and the natural frequency of oscillation about its equilibrium position for the system in Prob. 7-5.

7-13 In Prob. 7-8, m_1 is given a small displacement and released. Determine the equation of oscillation for the system.

7-14 For the system of Fig. P7-14, determine the equilibrium position and its equation of vibration about it. Spring force = 0 when $\theta = 0$.

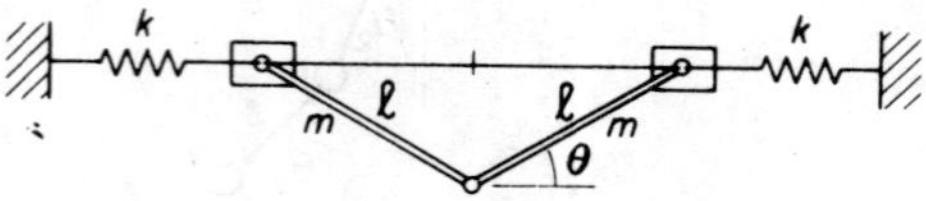

Figure P7-14.

7-15 Write Lagrange's equations of motion for the system shown in Fig. P7-15.

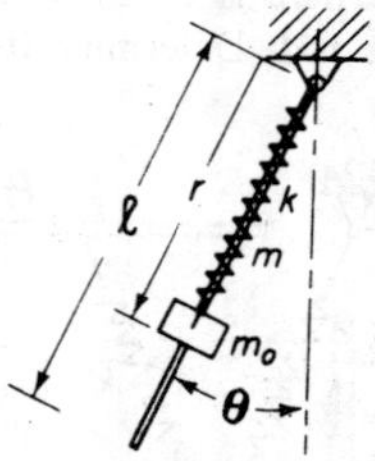

Figure P7-15.

7-16 The following constants are given for the beam of Fig. P7-16:

$$k = \frac{EI}{l^3}, \quad \frac{EI}{ml^4} = N, \quad \frac{k}{ml} = N$$

$$K = 5\frac{EI}{l}, \quad \frac{K}{ml^3} = 5N$$

Using the modes $\phi_1 = x/l$ and $\phi_2 = \sin(\pi x/l)$, determine the equation of motion by Lagrange's method, and determine the first two natural frequencies and mode shapes.

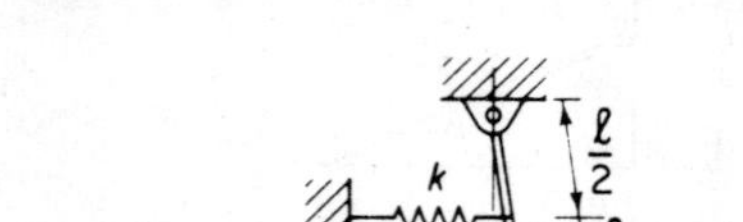

Figure P7-16.

7-17 Using Lagrange's method, determine the equations for the small oscillation of the bars shown in Fig. P7-17.

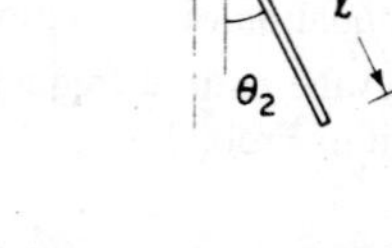

Figure P7-17.

7-18 The rigid bar linkages of Example 7.1-1 are loaded by springs and masses as shown in Fig. P7-18. Write Lagrange's equations of motion.

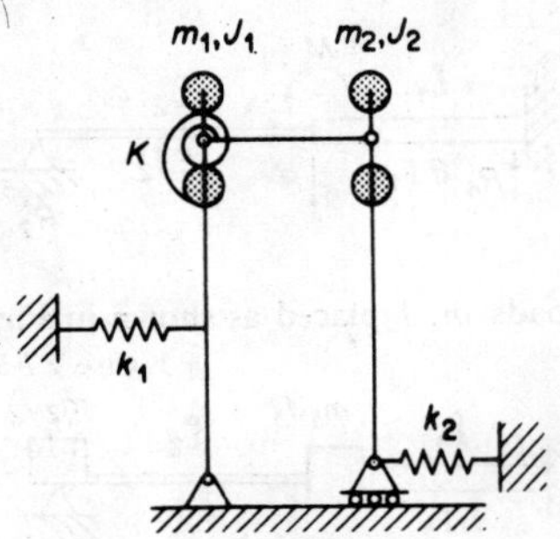

Figure P7-18.

7-19 Equal masses are placed at the nodes of the frame of Example 7.1-2 as shown in Fig. P7-19. Determine the stiffness matrix and the matrix equation of motion. (Let $l_2 = l_1$.)

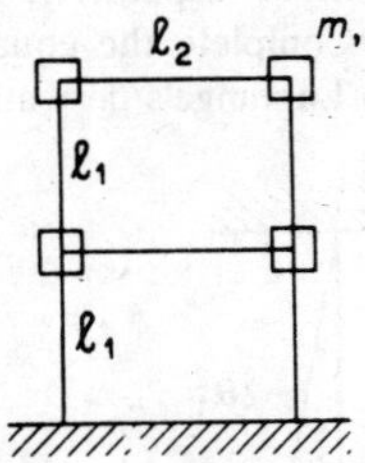

Figure P7-19.

7-20 Determine the stiffness matrix for the frame shown in Fig. P7-20.

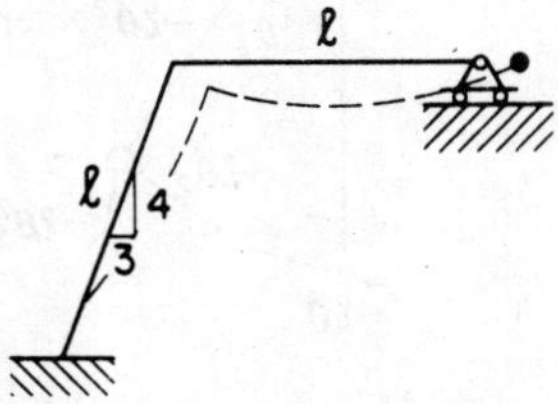

Figure P7-20.

7-21 The frame of Prob. 7-20 is loaded by springs and masses as shown in Fig. P7-21. Determine the equations of motion and the normal modes of the system.

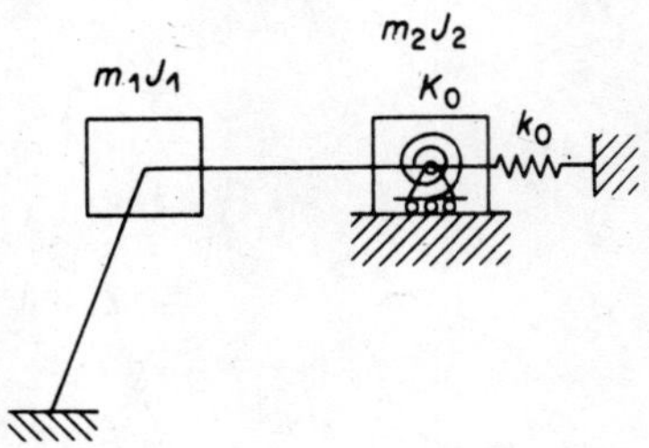

Figure P7-21.

7-22 Using area moment and superposition, determine M_1 and R_2 for the beam shown in Fig. P7-22. Let $EI_1 = 2EI_2$.

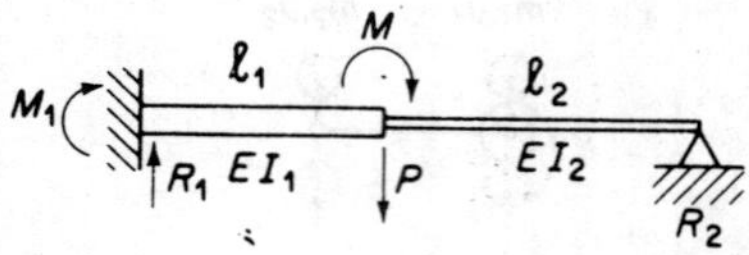

Figure P7-22.

7-23 With loads m, J placed as shown in Fig. P7-23, set up the equations of motion.

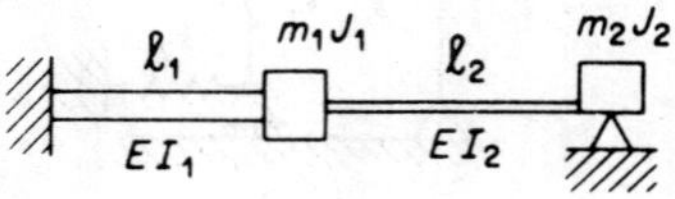

Figure P7-23.

7-24 For the extension of the double pendulum to the dynamic problem, the actual algebra can become long and tedious. Instead, draw in the components of $-\ddot{r}$ as shown. By taking each $\delta\theta$ separately, the virtual work equation can be easily determined visually. Complete the equations of motion for the system in Fig. P7-24. Compare with Lagrange's derivation.

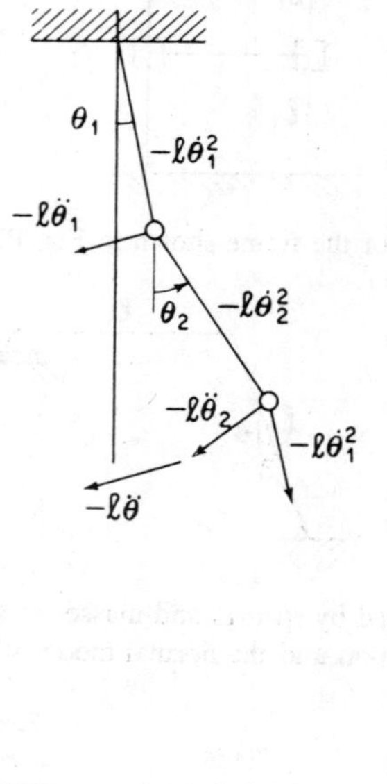

Figure P7-24.

8

NORMAL MODE VIBRATION OF CONTINUOUS SYSTEMS

The systems to be studied in this chapter have continuously distributed mass and elasticity. These bodies are assumed to be homogeneous and isotropic, obeying Hooke's law within the elastic limit. To specify the position of every particle in the elastic body, an infinite number of coordinates are necessary, and such bodies therefore possess an infinite number of degrees of freedom.

In general, the free vibration of these bodies is the sum of the principal modes as previously stated. For the principal mode of vibration every particle of the body performs simple harmonic motion at the frequency corresponding to the particular root of the frequency equation, each particle passing simultaneously through its respective equilibrium position. If the elastic curve of the body under which the motion is started coincides exactly with one of the principal modes, only that principal mode will be produced. However, the elastic curve resulting from a blow or a sudden removal of forces seldom corresponds to that of a principal mode, and thus all modes are excited. In many cases, however, a particular principal mode can be excited by proper initial conditions.

In this chapter some of the simpler problems of normal mode vibration of elastic bodies of constant sectional properties are taken up. For the normal modes of elastic bodies with varying sectional properties, the finite element approach of Chapter 10 is recommended.

8.1 VIBRATING STRING

A flexible string of mass ρ per unit length is stretched under tension T. By assuming the lateral deflection y of the string to be small, the change in tension with deflection is negligible and can be ignored.

In Fig. 8.1-1, a free-body diagram of an elementary length dx of the string is shown. Assuming small deflections and slopes, the equation of motion in the y-direction is

$$T\left(\theta + \frac{\partial \theta}{\partial x}\,dx\right) - T\theta = \rho\,dx\frac{\partial^2 y}{\partial t^2}$$

or

$$\frac{\partial \theta}{\partial x} = \frac{\rho}{T}\frac{\partial^2 y}{\partial t^2} \tag{8.1-1}$$

Since the slope of the string is $\theta = \partial y/\partial x$, the preceding equation reduces to

$$\frac{\partial^2 y}{\partial x^2} = \frac{1}{c^2}\frac{\partial^2 y}{\partial t^2} \tag{8.1-2}$$

where $c = \sqrt{T/\rho}$ can be shown to be the velocity of wave propagation along the string.

The general solution of Eq. (8.1-2) can be expressed in the form

$$y = F_1(ct - x) + F_2(ct + x) \tag{8.1-3}$$

where F_1 and F_2 are arbitrary functions. Regardless of the type of function F, the argument $(ct \pm x)$ upon differentiation leads to the equation

$$\frac{\partial^2 F}{\partial x^2} = \frac{1}{c^2}\frac{\partial^2 F}{\partial t^2} \tag{8.1-4}$$

and hence the differential equation is satisfied.

Considering the component $y = F_1(ct - x)$, its value is determined by the argument $(ct - x)$ and hence by a range of values of t and x. For example, if $c = 10$, the equation for $y = F_1(100)$ is satisfied by $t = 0$, $x = -100$; $t = 1$, $x = -90$; $t = 2$, $x = -80$, and so forth. Therefore, the wave

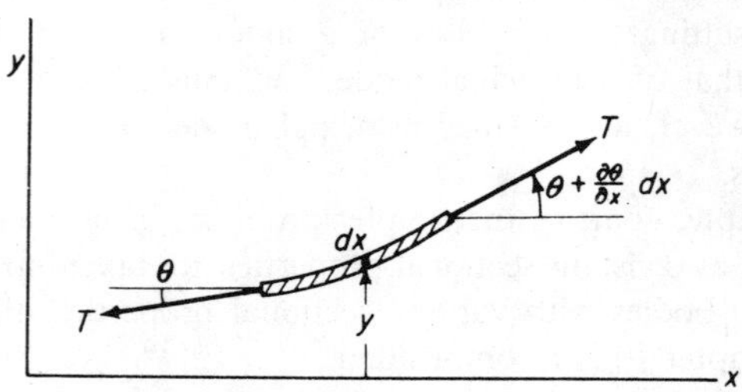

Figure 8.1-1. String element in lateral vibration.

profile moves in the positive x-direction with speed c. In a similar manner we can show that $F_2(ct + x)$ represents a wave moving toward the negative x-direction with speed c. We therefore refer to c as the velocity of wave propagation.

One method of solving partial differential equations is that of separation of variables. In this method the solution is assumed in the form

$$y(x, t) = Y(x)G(t) \tag{8.1-5}$$

By substitution into Eq. (8.1-2) we obtain

$$\frac{1}{Y}\frac{d^2Y}{dx^2} = \frac{1}{c^2}\frac{1}{G}\frac{d^2G}{dt^2} \tag{8.1-6}$$

Since the left side of this equation is independent of t, whereas the right side is independent of x, it follows that each side must be a constant. Letting this constant be $-(\omega/c)^2$ we obtain two ordinary differential equations

$$\frac{d^2Y}{dx^2} + \left(\frac{\omega}{c}\right)^2 Y = 0 \tag{8.1-7}$$

$$\frac{d^2G}{dt^2} + \omega^2 G = 0 \tag{8.1-8}$$

with the general solutions

$$Y = A\sin\frac{\omega}{c}x + B\cos\frac{\omega}{c}x \tag{8.1-9}$$

$$G = C\sin\omega t + D\cos\omega t \tag{8.1-10}$$

The arbitrary constants A, B, C, D depend on the boundary conditions and the initial conditions. For example, if the string is stretched between two fixed points with distance l between them, the boundary conditions are $y(0, t) = y(l, t) = 0$. The condition that $y(0, t) = 0$ will require that $B = 0$, so the solution will appear as

$$y = (C\sin\omega t + D\cos\omega t)\sin\frac{\omega}{c}x \tag{8.1-11}$$

The condition $y(l, t) = 0$ then leads to the equation

$$\sin\frac{\omega l}{c} = 0$$

or

$$\frac{\omega_n l}{c} = \frac{2\pi l}{\lambda} = n\pi, \qquad n = 1, 2, 3, \ldots$$

and $\lambda = c/f$ is the wavelength and f is the frequency of oscillation. Each n represents a normal mode vibration with natural frequency determined from the equation

$$f_n = \frac{n}{2l}c = \frac{n}{2l}\sqrt{\frac{T}{\rho}}, \qquad n = 1, 2, 3, \ldots \tag{8.1-12}$$

The mode shape is sinusoidal with the distribution

$$Y = \sin n\pi \frac{x}{l} \tag{8.1-13}$$

In the more general case of free vibration initiated in any manner, the solution will contain many of the normal modes and the equation for the displacement may be written as

$$y(x, t) = \sum_{n=1}^{\infty} (C_n \sin \omega_n t + D_n \cos \omega_n t) \sin \frac{n\pi x}{l} \tag{8.1-14}$$

$$\omega_n = \frac{n\pi c}{l}$$

Fitting this equation to the initial conditions of $y(x,0)$ and $\dot{y}(x,0)$, the C_n and D_n can be evaluated.

Example 8.1-1

A uniform string of length l is fixed at the ends and stretched under tension T. If the string is displaced into an arbitrary shape $y(x,0)$ and released, determine C_n and D_n of Eq. (8.1-14).

Solution: At $t = 0$, the displacement and velocity are

$$y(x,0) = \sum_{n=1}^{\infty} D_n \sin \frac{n\pi x}{l}$$

$$\dot{y}(x,0) = \sum_{n=1}^{\infty} \omega_n C_n \sin \frac{n\pi x}{l} = 0$$

If we multiply each equation by $\sin k\pi x/l$ and integrate from $x = 0$ to $x = l$, all the terms on the right side will be zero, except the term $n = k$. Thus we arrive at the result

$$D_k = \frac{2}{l} \int_0^l y(x,0) \sin \frac{k\pi x}{l}\, dx$$

$$C_k = 0, \qquad k = 1, 2, 3, \ldots$$

8.2 LONGITUDINAL VIBRATION OF RODS

The rod considered in this section is assumed to be thin and uniform along its length. Due to axial forces there will be displacements u along the rod which will be a function of both the position x and the time t. Since the rod has an infinite number of natural modes of vibration, the distribution of the displacements will differ with each mode.

Let us consider an element of this rod of length dx (Fig. 8.2-1). If u is the displacement at x, the displacement at $x + dx$ will be $u + (\partial u/\partial x)\, dx$. It is evident then that the element dx in the new position has changed in length by an amount $(\partial u/\partial x)\, dx$, and thus the unit strain is $\partial u/\partial x$. Since from Hooke's law the ratio of unit stress to unit strain is equal to the modulus of

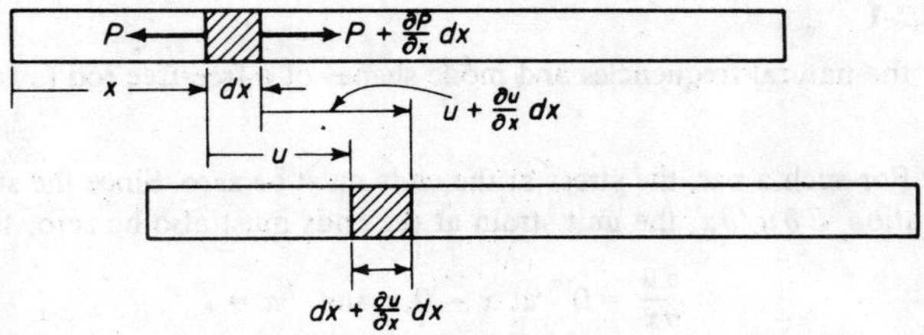

Figure 8.2-1. Displacement of rod element.

elasticity E, we can write

$$\frac{\partial u}{\partial x} = \frac{P}{AE} \tag{8.2-1}$$

where A is the cross-sectional area of the rod. Differentiating with respect to x

$$AE\frac{\partial^2 u}{\partial x^2} = \frac{\partial P}{\partial x} \tag{8.2-2}$$

We now apply Newton's law of motion for the element and equate the unbalanced force to the product of the mass and acceleration of the element

$$\frac{\partial P}{\partial x}\,dx = \rho A\,dx\frac{\partial^2 u}{\partial t^2} \tag{8.2-3}$$

where ρ is the density of the rod, mass per unit volume. Eliminating $\partial P/\partial x$ between Eqs. (8.2-2) and (8.2-3), we obtain the partial differential equation

$$\frac{\partial^2 u}{\partial t^2} = \left(\frac{E}{\rho}\right)\frac{\partial^2 u}{\partial x^2} \tag{8.2-4}$$

or

$$\frac{\partial^2 u}{\partial x^2} = \frac{1}{c^2}\frac{\partial^2 u}{\partial t^2} \tag{8.2-5}$$

which is similar to that of Eq. (8.1-2) for the string. The velocity of propagation of the displacement or stress wave in the rod is then equal to

$$c = \sqrt{\frac{E}{\rho}} \tag{8.2-6}$$

and a solution of the form

$$u(x, t) = U(x)G(t) \tag{8.2-7}$$

will result in two ordinary differential equations similar to Eqs. (8.1-7) and (8.1-8), with

$$U(x) = A\sin\frac{\omega}{c}x + B\cos\frac{\omega}{c}x \tag{8.2-8}$$

$$G(t) = C\sin\omega t + D\cos\omega t \tag{8.2-9}$$

Example 8.2-1

Determine the natural frequencies and mode shapes of a free-free rod (a rod with both ends free).

Solution: For such a bar, the stress at the ends must be zero. Since the stress is given by the equation $E\,\partial u/\partial x$, the unit strain at the ends must also be zero; that is,

$$\frac{\partial u}{\partial x} = 0 \quad \text{at } x = 0, \quad \text{and} \quad x = l$$

The two equations corresponding to these boundary conditions are therefore

$$\left(\frac{\partial u}{\partial x}\right)_{x=0} = A\frac{\omega}{c}(C \sin \omega t + D \cos \omega t) = 0$$

$$\left(\frac{\partial u}{\partial x}\right)_{x=l} = \frac{\omega}{c}\left(A \cos\frac{\omega l}{c} - B \sin\frac{\omega l}{c}\right)(C \sin \omega t + D \cos \omega t) = 0$$

Since these equations must be true for any time t, A must be equal to zero from the first equation. Since B must be finite in order to have vibration, the second equation is satisfied when

$$\sin\frac{\omega l}{c} = 0$$

or

$$\frac{\omega_n l}{c} = \omega_n l\sqrt{\frac{\rho}{E}} = \pi, 2\pi, 3\pi, \ldots, n\pi$$

The frequency of vibration is thus given by

$$\omega_n = \frac{n\pi}{l}\sqrt{\frac{E}{\rho}}, \qquad f_n = \frac{n}{2l}\sqrt{\frac{E}{\rho}}$$

where n represents the order of the mode. The solution of the free-free rod with zero initial displacement can then be written as

$$u = u_0 \cos\frac{n\pi}{l}x \sin\frac{n\pi}{l}\sqrt{\frac{E}{\rho}}\,t$$

The amplitude of the longitudinal vibration along the rod is therefore a cosine wave having n nodes.

8.3 TORSIONAL VIBRATION OF RODS

The equation of motion of a rod in torsional vibration is similar to that of longitudinal vibration of rods discussed in the preceding section.

Letting x be measured along the length of the rod, the angle of twist in any length dx of the rod due to a torque T is

$$d\theta = \frac{T\,dx}{I_pG} \tag{8.3-1}$$

where I_pG is the torsional stiffness given by the product of the polar moment

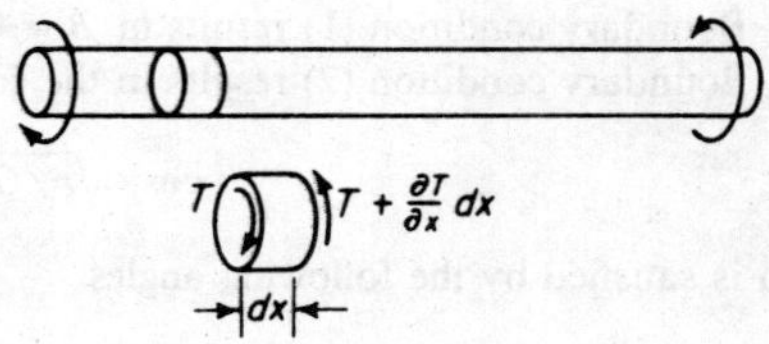

Figure 8.3-1. Torque acting on an element *dx*.

of inertia I_P of the cross-sectional area and the shear modulus of elasticity G. The torque on the two faces of the element being T and $T + (\partial T/\partial x)\, dx$, as shown in Fig. 8.3-1, the net torque from Eq. (8.3-1) becomes

$$\frac{\partial T}{\partial x}dx = I_P G \frac{\partial^2 \theta}{\partial x^2}dx \tag{8.3-2}$$

Equating this torque to the product of the mass moment of inertia $\rho I_P\, dx$ of the element and the angular acceleration $\partial^2\theta/\partial t^2$, where ρ is the density of the rod in mass per unit volume, the differential equation of motion becomes

$$\rho I_P\, dx \frac{\partial^2 \theta}{\partial t^2} = I_P G \frac{\partial^2 \theta}{\partial x^2}dx, \qquad \frac{\partial^2 \theta}{\partial t^2} = \left(\frac{G}{\rho}\right)\frac{\partial^2 \theta}{\partial x^2} \tag{8.3-3}$$

This equation is of the same form as that of longitudinal vibration of rods where θ and G/ρ replace u and E/ρ, respectively. The general solution may hence be written immediately by comparison as

$$\theta = \left(A \sin \omega\sqrt{\frac{\rho}{G}}\, x + B \cos \omega\sqrt{\frac{\rho}{G}}\, x\right)(C \sin \omega t + D \cos \omega t) \tag{8.3-4}$$

Example 8.3-1

Determine the equation for the natural frequencies of a uniform rod in torsional oscillation with one end fixed and the other end free, as in Fig. 8.3-2.

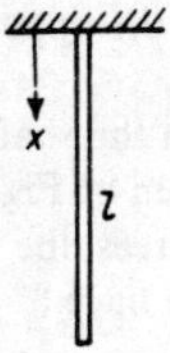

Figure 8.3-2.

Solution: Starting with equation

$$\theta = \left(A \sin \omega\sqrt{\rho/G}\, x + B \cos \omega\sqrt{\rho/G}\, x\right) \sin \omega t$$

apply the boundary conditions, which are

(1) when $x = 0$, $\theta = 0$,

(2) when $x = l$, torque = 0, or

$$\frac{\partial \theta}{\partial x} = 0$$

Boundary condition (1) results in $B = 0$.
Boundary condition (2) results in the equation

$$\cos \omega\sqrt{\rho/G}\,l = 0$$

which is satisfied by the following angles

$$\omega_n\sqrt{\frac{\rho}{G}}\,l = \frac{\pi}{2}, \frac{3\pi}{2}, \frac{5\pi}{2}, \ldots, \left(n + \frac{1}{2}\right)\pi$$

The natural frequencies of the bar are hence determined by the equation

$$\omega_n = \left(n + \frac{1}{2}\right)\frac{\pi}{l}\sqrt{\frac{G}{\rho}}$$

where $n = 0, 1, 2, 3, \ldots$.

Example 8.3-2

The drill pipe of an oil well terminates at the lower end in a rod containing a cutting bit. Derive the expression for the natural frequencies, assuming the drill pipe to be uniform and fixed at the upper end and the rod and cutter to be represented by an end mass of moment of inertia J_0, as shown in Fig. 8.3-3.

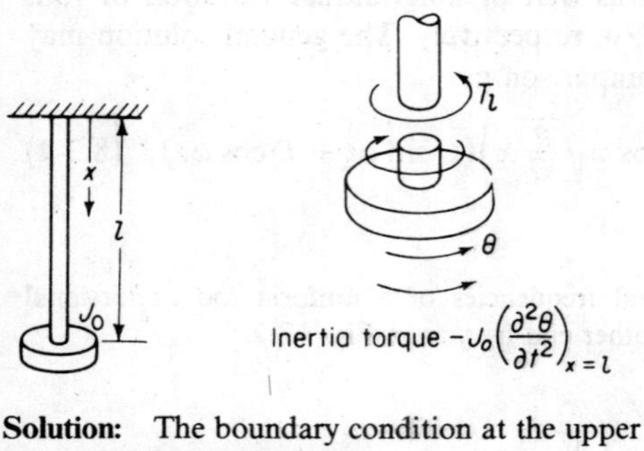

Figure 8.3-3.

Solution: The boundary condition at the upper end is $x = 0$, $\theta = 0$, which requires B to be zero in Eq. (8.3-4).

For the lower end, the torque on the shaft is due to the inertia torque of the end disk, as shown by the free-body diagram of Fig. 8.3-3. The inertia torque of the disk is $-J_0(\partial^2\theta/\partial^2 t)_{x=l} = J_0\omega^2(\theta)_{x=l}$, whereas the shaft torque from Eq. (8.3-1) is $T_l = GI_P(d\theta/dx)_{x=l}$. Equating the two, we have

$$GI_P\left(\frac{d\theta}{dx}\right)_{x=l} = J_0\omega^2\theta_{x=l}$$

Substituting from Eq. (8.3-4) with $B = 0$

$$GI_P\omega\sqrt{\frac{\rho}{G}}\cos\omega\sqrt{\frac{\rho}{G}}\,l = J_0\omega^2\sin\omega\sqrt{\frac{\rho}{G}}\,l$$

$$\tan\omega l\sqrt{\frac{\rho}{G}} = \frac{I_P}{\omega J_0}\sqrt{G\rho} = \frac{I_P\rho l}{J_0\omega l}\sqrt{\frac{G}{\rho}} = \frac{J_{\text{rod}}}{J_0\omega l}\sqrt{\frac{G}{\rho}}$$

This equation is of the form

$$\beta \tan \beta = \frac{J_{\text{rod}}}{J_0}, \qquad \beta = \omega l \sqrt{\frac{\rho}{G}}$$

which can be solved graphically or from tables.†

Example 8.3-3

Using the frequency equation developed in the previous example, determine the first two natural frequencies of an oil-well drill pipe 500 ft long, fixed at the upper end and terminating at the lower end to a drill collar 120 ft long. The average values for the drill pipe and drill collar are given as

Drill pipe: outside diameter $= 4\frac{1}{2}$ in.

inside diameter $= 3.83$ in.

$$I_P = 0.00094 \text{ ft}^4, \qquad l = 5000 \text{ ft}$$

$$J_{\text{rod}} = I_P \rho l = 0.00094 \times \frac{490}{32.2} \times 5000 = 71.4 \text{ lb} \cdot \text{ft} \cdot \text{s}^2$$

Drill collar: outside diameter $= 7\frac{5}{8}$ in.

inside diameter $= 2.0$ in.

$$J_0 = 0.244 \times 120 \text{ ft} = 29.3 \text{ lb} \cdot \text{ft} \cdot \text{s}^2$$

Solution: The equation to be solved is

$$\beta \tan \beta = \frac{J_{\text{rod}}}{J_0} = 2.44$$

From Table V, p. 32 of Jahnke and Emde, $\beta = 1.135, 3.722, \ldots$

$$\beta = \omega l \sqrt{\frac{\rho}{G}} = 5000\,\omega \sqrt{\frac{490}{12 \times 10^6 \times 12^2 \times 32.2}} = 0.470\,\omega$$

Solving for ω, the first two natural frequencies are found to be

$$\omega_1 = \frac{1.135}{0.470} = 2.41 \text{ rad/s} = 0.384 \text{ cps}$$

$$\omega_2 = \frac{3.722}{0.470} = 7.93 \text{ rad/s} = 1.26 \text{ cps}$$

8.4 EULER EQUATION FOR BEAMS

To determine the differential equation for the lateral vibration of beams, consider the forces and moments acting on an element of the beam shown in Fig. 8.4-1.

V and M are shear and bending moments, respectively, and $p(x)$ represents the loading per unit length of the beam.

By summing forces in the y-direction

$$dV - p(x)\,dx = 0 \tag{8.4-1}$$

†See Jahnke and Emde, *Tables of Functions*, 4th Ed. (Dover Publications, Inc., 1945), Table V, p. 32.

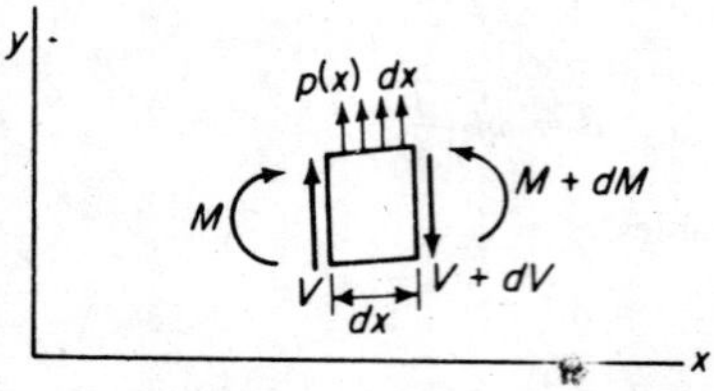

Figure 8.4-1.

By summing moments about any point on the right face of the element,

$$dM - V\,dx - \tfrac{1}{2}p(x)(dx)^2 = 0 \tag{8.4-2}$$

In the limiting process these equations result in the following important relationships

$$\frac{dV}{dx} = p(x), \qquad \frac{dM}{dx} = V \tag{8.4-3}$$

The first part of Eq. (8.4-3) states that the rate of change of shear along the length of the beam is equal to the loading per unit length, and the second states that the rate of change of the moment along the beam is equal to the shear.

From Eq. (8.4-3) we obtain the following

$$\frac{d^2M}{dx^2} = \frac{dV}{dx} = p(x) \tag{8.4-4}$$

The bending moment is related to the curvature by the flexure equation, which, for the coordinates indicated in Fig. 8.4-1, is

$$M = EI\frac{d^2y}{dx^2} \tag{8.4-5}$$

Substituting this relation into Eq. (7.4-4), we obtain

$$\frac{d^2}{dx^2}\left(EI\frac{d^2y}{dx^2}\right) = p(x) \tag{8.4-6}$$

For a beam vibrating about its static equilibrium position under its own weight, the load per unit length is equal to the inertia load due to its mass and acceleration. Since the inertia force is in the same direction as $p(x)$, as shown in Fig. 8.4-1, we have, by assuming harmonic motion

$$p(x) = \rho\omega^2 y \tag{8.4-7}$$

where ρ is the mass per unit length of the beam. Using this relation, the equation for the lateral vibration of the beam reduces to

$$\frac{d^2}{dx^2}\left(EI\frac{d^2y}{dx^2}\right) - \rho\omega^2 y = 0 \tag{8.4-8}$$

In the special case where the flexural rigidity EI is a constant, the above equation may be written as

$$EI\frac{d^4y}{dx^4} - \rho\omega^2 y = 0 \tag{8.4-9}$$

On substituting

$$\beta^4 = \rho\frac{\omega^2}{EI} \tag{8.4-10}$$

we obtain the fourth-order differential equation

$$\frac{d^4y}{dx^4} - \beta^4 y = 0 \tag{8.4-11}$$

for the vibration of a uniform beam.

The general solution of Eq. (8.4-11) can be shown to be

$$y = A\cosh\beta x + B\sinh\beta x + C\cos\beta x + D\sin\beta x \tag{8.4-12}$$

To arrive at this result, we assume a solution of the form

$$y = e^{ax}$$

which will satisfy the differential equation when

$$a = \pm\beta, \text{ and } a = \pm i\beta$$

Since

$$e^{\pm\beta x} = \cosh\beta x \pm \sinh\beta x$$

$$e^{\pm i\beta x} = \cos\beta x \pm i\sin\beta x$$

the solution in the form of Eq. (8.4-12) is readily established.

The natural frequencies of vibration are found from Eq. (8.4-10) to be

$$\omega_n = \beta_n^2\sqrt{\frac{EI}{\rho}} = (\beta_n l)^2\sqrt{\frac{EI}{\rho l^4}}$$

where the number β_n depends on the boundary conditions of the problem. The following table lists numerical values of $(\beta_n l)^2$ for typical end conditions.

Beam configuration	$(\beta_1 l)^2$ *Fundamental*	$(\beta_2 l)^2$ *Second Mode*	$(\beta_3 l)^2$ *Third Mode*
Simply supported	9.87	39.5	88.9
Cantilever	3.52	22.0	61.7
Free-free	22.4	61.7	121.0
Clamped-clamped	22.4	61.7	121.0
Clamped-hinged	15.4	50.0	104.0
Hinged-free	0	15.4	50.0

Figure 8.4-2.

Example 8.4-1

Determine the natural frequencies of vibration of a uniform beam clamped at one end and free at the other.

Solution: The boundary conditions are

$$\text{At } x = 0 \begin{cases} y = 0 \\ \dfrac{dy}{dx} = 0 \end{cases}$$

$$\text{At } x = l \begin{cases} M = 0 \quad \text{or} \quad \dfrac{d^2y}{dx^2} = 0 \\ V = 0 \quad \text{or} \quad \dfrac{d^3y}{dx^3} = 0 \end{cases}$$

Substituting these boundary conditions in the general solution, we obtain

$$(y)_{x=0} = A + C = 0, \qquad \therefore A = -C$$

$$\left(\frac{dy}{dx}\right)_{x=0} = \beta[A \sinh \beta x + B \cosh \beta x - C \sin \beta x + D \cos \beta x]_{x=0} = 0$$

$$\beta[B + D] = 0, \qquad \therefore B = -D$$

$$\left(\frac{d^2y}{dx^2}\right)_{x=l} = \beta^2[A \cosh \beta l + B \sinh \beta l - C \cos \beta l - D \sin \beta l] = 0$$

$$A(\cosh \beta l + \cos \beta l) + B(\sinh \beta l + \sin \beta l) = 0$$

$$\left(\frac{d^3y}{dx^3}\right)_{x=l} = \beta^3[A \sinh \beta l + B \cosh \beta l + C \sin \beta l - D \cos \beta l] = 0$$

$$A(\sinh \beta l - \sin \beta l) + B(\cosh \beta l + \cos \beta l) = 0$$

From the last two equations we obtain

$$\frac{\cosh \beta l + \cos \beta l}{\sinh \beta l - \sin \beta l} = \frac{\sinh \beta l + \sin \beta l}{\cosh \beta l + \cos \beta l}$$

which reduces to

$$\cosh \beta l \cos \beta l + 1 = 0$$

This last equation is satisfied by a number of values of βl, corresponding to each normal mode of oscillation, which for the first and second modes are 1.875 and 4.695, respectively. The natural frequency for the first mode is hence given by

$$\omega_1 = \frac{1.875^2}{l^2}\sqrt{\frac{EI}{\rho}} = \frac{3.515}{l^2}\sqrt{\frac{EI}{\rho}}$$

8.5 EFFECT OF ROTARY INERTIA AND SHEAR DEFORMATION

The Timoshenko theory accounts for both the rotary inertia and shear deformation of the beam. The free-body diagram and the geometry for the beam element are shown in Fig. 8.5-1. If the shear deformation is zero,

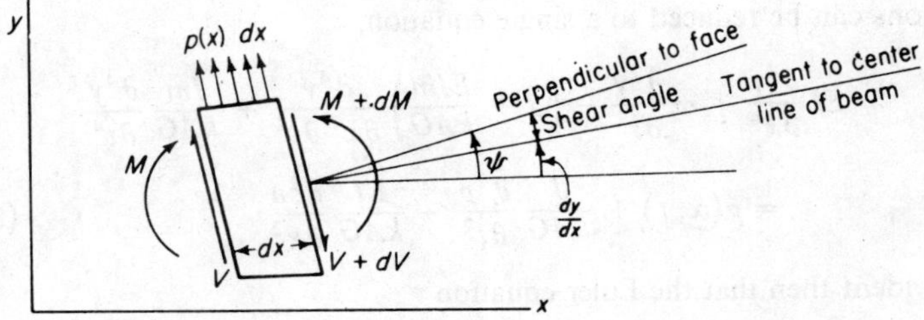

Figure 8.5-1. Effect of shear deformation.

the center line of the beam element will coincide with the perpendicular to the face of the cross section. Due to shear, the rectangular element tends to go into a diamond shape without rotation of the face, and the slope of the center line is diminished by the shear angle $(\psi - dy/dx)$. The following quantities can then be defined:

$$y = \text{deflection of the center line of the beam}$$

$$\frac{dy}{dx} = \text{slope of the center line of the beam}$$

$$\psi = \text{slope due to bending}$$

$$\psi - \frac{dy}{dx} = \text{loss of slope, equal to the shear angle}$$

There are two elastic equations for the beam, which are

$$\psi - \frac{dy}{dx} = \frac{V}{kAG} \tag{8.5-1}$$

$$\frac{d\psi}{dx} = \frac{M}{EI} \tag{8.5-2}$$

where A is the cross-sectional area, G is the shear modulus, k is a factor depending on the shape of the cross section, and EI is the bending stiffness. For rectangular and circular cross sections, the values of k are $\frac{2}{3}$ and $\frac{3}{4}$, respectively. In addition, there are two dynamical equations,

$$(\text{moment})\, J\ddot{\psi} = \frac{dM}{dx} - V \tag{8.5-3}$$

$$(\text{force})\, m\ddot{y} = -\frac{dV}{dx} + p(x, t) \tag{8.5-4}$$

where J and m are the rotary inertia and mass of the beam per unit length.

Substituting the elastic equations into the dynamical equations, we have

$$\frac{d}{dx}\left(EI\frac{d\psi}{dx}\right) + kAG\left(\frac{dy}{dx} - \psi\right) - J\ddot{\psi} = 0 \tag{8.5-5}$$

$$m\ddot{y} - \frac{d}{dx}\left[kAG\left(\frac{dy}{dx} - \psi\right)\right] - p(x, t) = 0 \tag{8.5-6}$$

which are the coupled equations of motion for the beam.

If ψ is eliminated and the cross section remains constant, these two equations can be reduced to a single equation,

$$EI\frac{\partial^4 y}{\partial x^4} + m\frac{\partial^2 y}{\partial t^2} - \left(J + \frac{EIm}{kAG}\right)\frac{\partial^4 y}{\partial x^2 \partial t^2} + \frac{Jm}{kAG}\frac{\partial^4 y}{\partial t^4}$$

$$\doteq p(x, t) + \frac{J}{kAG}\frac{\partial^2 p}{\partial t^2} - \frac{EI}{kAG}\frac{\partial^2 p}{\partial x^2} \tag{8.5-7}$$

It is evident then that the Euler equation

$$EI\frac{\partial^4 y}{\partial x^4} + m\frac{\partial^2 y}{\partial t^2} = p(x, t)$$

is a special case of the general beam equation including the rotary inertia and the shear deformation.

Runge-Kutta method. The Runge-Kutta method is a practical approach for the structural problem. It is self-starting and results in good accuracy. The error is of order h^5.

To illustrate the procedure, we consider the beam with rotary inertia and shear terms. The fourth-order equation is first written in terms of four first-order equations as follows

$$\frac{d\psi}{dx} = \frac{M}{EI} = F(x, \psi, y, M, V)$$

$$\frac{dy}{dx} = \psi - \frac{V}{kAG} = G(x, \psi, y, M, V)$$

$$\frac{dM}{dx} = V - \omega^2 J\psi = P(x, \psi, y, M, V)$$

$$\frac{dV}{dx} = \omega^2 my = K(x, \psi, y, M, V) \tag{8.5-8}$$

The Runge-Kutta procedure, discussed in Sec. 4.6 for a single coordinate, is now extended to the simultaneous solution of four variables listed next:

$$\psi = \psi_1 + \frac{h}{6}(f_1 + 2f_2 + 2f_3 + f_4)$$

$$y = y_1 + \frac{h}{6}(g_1 + 2g_2 + 2g_3 + g_4)$$

$$M = M_1 + \frac{h}{6}(p_1 + 2p_2 + 2p_3 + p_4)$$

$$V = V_1 + \frac{h}{6}(k_1 + 2k_2 + 2k_3 + k_4) \tag{8.5-9}$$

where $h = \Delta x$.

Let f_i, g_i, p_i, k_i and F, G, P, and K be represented by vectors

$$l_i = \begin{Bmatrix} f_i \\ g_i \\ p_i \\ k_i \end{Bmatrix}, \qquad L = \begin{Bmatrix} F \\ G \\ P \\ K \end{Bmatrix}$$

Then the computation proceeds as follows:

$$l_1 = L(x_1, \psi_1, y_1, M_1, V_1)$$

$$l_2 = L\left(x_1 + \frac{h}{2}, \psi_1 + f_1\frac{h}{2}, y_1 + g_1\frac{h}{2}, M_1 + p_1\frac{h}{2}, V_1 + k_1\frac{h}{2}\right)$$

$$l_3 = L\left(x_1 + \frac{h}{2}, \psi_1 + f_2\frac{h}{2}, y_1 + g_2\frac{h}{2}, M_1 + p_2\frac{h}{2}, V_1 + k_2\frac{h}{2}\right)$$

$$l_4 = L(x_1 + h, \psi_1 + f_3h, y_1 + g_3h, M_1 + p_3h, V_1 + k_3h)$$

With these quantities substituted into Eq. (8.5-9), the dependent variables at the neighboring point x_2 are found, and the procedure is repeated for the point x_3, and so on.

Returning to the beam equations, the boundary conditions at the beginning end x_1 provide a starting point. For example, in the cantilever beam with origin at the fixed end, the boundary conditions at the starting end are

$$\psi_1 = 0, \qquad M_1 = M_1$$
$$y_1 = 0, \qquad V_1 = V_1$$

These can be considered to be the linear combination of two boundary vectors as follows:

$$\begin{Bmatrix} \psi_1 \\ y_1 \\ M_1 \\ V_1 \end{Bmatrix} = \begin{Bmatrix} 0 \\ 0 \\ 1 \\ 0 \end{Bmatrix} + \alpha \begin{Bmatrix} 0 \\ 0 \\ 0 \\ 1 \end{Bmatrix} = C_1 + \alpha D_1$$

Since the system is linear, we can start with each boundary vector separately. Starting with C_1, we obtain

$$C_N = \begin{Bmatrix} \psi_N \\ y_N \\ M_N \\ V_N \end{Bmatrix}_C$$

Starting with D_1, we obtain

$$\alpha D_N = \begin{Bmatrix} \psi_N \\ y_N \\ M_N \\ V_N \end{Bmatrix}_D$$

These must now add to satisfy the actual boundary conditions at the terminal end, which for a cantilever free end are

$$\begin{Bmatrix} \psi \\ y \\ M \\ V \end{Bmatrix}_N = \begin{Bmatrix} \psi \\ y \\ 0 \\ 0 \end{Bmatrix} = C_N + \alpha D_N$$

If the frequency chosen is correct, the above boundary equations lead to

$$M_{NC} + \alpha M_{ND} = 0$$

$$V_{NC} + \alpha V_{ND} = 0$$

$$\alpha = -\frac{M_{NC}}{M_{ND}} = -\frac{V_{NC}}{V_{ND}}$$

which is satisfied by the determinant

$$\begin{vmatrix} M_{NC} & V_{NC} \\ M_{ND} & V_{ND} \end{vmatrix} = 0$$

The iteration can be started with three different frequencies, which results in three values of the determinant. A parabola is passed through these three points and the zero of the curve is chosen for a new estimate of the frequency. When the frequency is close to the correct value, the new estimate may be made by a straight line between two values of the boundary determinant.

PROBLEMS

8-1 Find the wave velocity along a rope whose mass is 0.372 kg/m when stretched to a tension of 444 N.

8-2 Derive the equation for the natural frequencies of a uniform cord of length l fixed at the two ends. The cord is stretched to a tension T and its mass per unit length is ρ.

8-3 A cord of length l and mass per unit length ρ is under tension T with the left end fixed and the right end attached to a spring-mass system as shown in Fig. P8-3. Determine the equation for the natural frequencies.

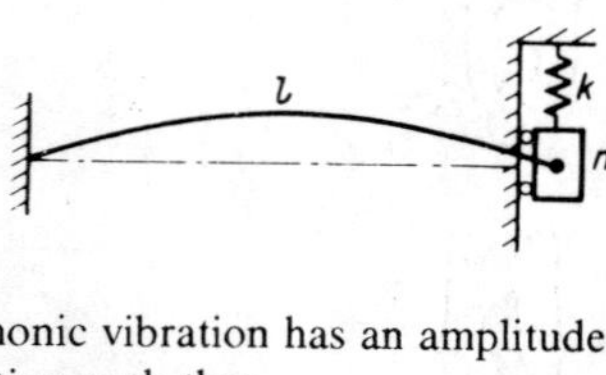

Figure P8-3.

8-4 A harmonic vibration has an amplitude that varies as a cosine function along the x-direction such that

$$y = a \cos kx \cdot \sin \omega t$$

Show that if another harmonic vibration of same frequency and equal amplitude displaced in space phase and time phase by a quarter wave length is added to the first vibration, the resultant vibration will represent a traveling wave with a propagation velocity equal to $c = \omega/k$.

8-5 Find the velocity of longitudinal waves along a thin steel bar. The modulus of elasticity and mass per unit volume of steel are 200×10^9 N/m^2 and 7810 kg/m^3.

8-6 Shown in Fig. P8-6 is a flexible cable supported at the upper end and free to oscillate under the influence of gravity. Show that the equation of lateral motion is

$$\frac{\partial^2 y}{\partial t^2} = g\left(x\frac{\partial^2 y}{\partial x^2} + \frac{\partial y}{\partial x}\right)$$

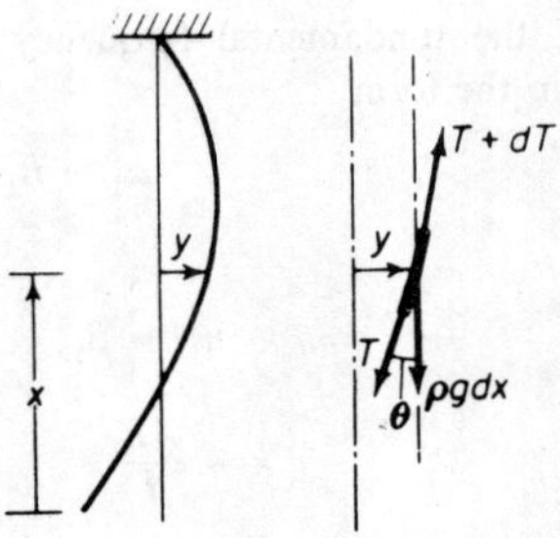

Figure P8-6.

8-7 In Prob. 8-6, assume a solution in the form $y = Y(x)\cos\omega t$ and show that $Y(x)$ can be reduced to a Bessel's differential equation

$$\frac{d^2Y(z)}{dz^2} + \frac{1}{z}\frac{dY(x)}{dz} + Y(z) = 0$$

with solution

$$Y(z) = J_0(z) \text{ or } Y(x) = J_0\left(2\omega\sqrt{\frac{x}{g}}\right)$$

by a change in variable $z^2 = 4\omega^2 x/g$.

8-8 A particular satellite consists of two equal masses m each, connected by a cable of length $2l$ and mass density ρ, as shown in Fig. P8-8. The assembly rotates in space with angular speed ω_0. Show that if the variation in the cable tension is neglected, the differential equation of lateral motion of the cable is

$$\frac{\partial^2 y}{\partial x^2} = \frac{\rho}{m\omega_0^2 l}\left(\frac{\partial^2 y}{\partial t^2} - \omega_0^2 y\right)$$

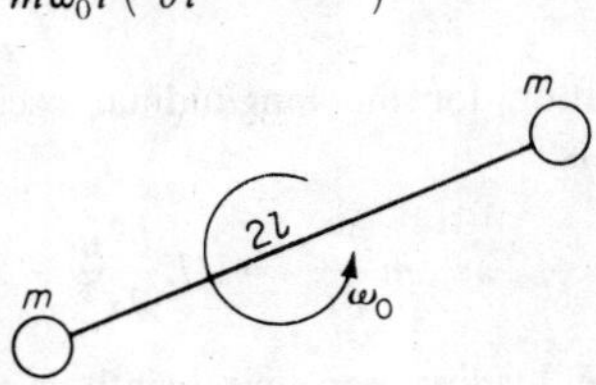

Figure P8-8.

and that its fundamental frequency of oscillation is

$$\omega^2 = \left(\frac{\pi}{2l}\right)^2\left(\frac{m\omega_0 l}{\rho}\right) - \omega_0^2.$$

8-9 A uniform bar of length l is fixed at one end and free at the other end. Show that the frequencies of normal longitudinal vibrations are $f = (n + \frac{1}{2})c/2l$, where $c = \sqrt{E/\rho}$ is the velocity of longitudinal waves in the bar, and $n = 0, 1, 2, \ldots$.

8-10 A uniform rod of length l and cross-sectional area A is fixed at the upper end and is loaded with a weight W on the other end. Show that the natural frequencies are determined from the equation

$$\omega l\sqrt{\frac{\rho}{E}} \tan \omega l\sqrt{\frac{\rho}{E}} = \frac{A\rho l g}{W}$$

8-11 Show that the fundamental frequency for the system of Prob. 8-10 can be expressed in the form

$$\omega_1 = \beta_1\sqrt{k/rM}$$

where

$$n_1 l = \beta_1, \qquad r = \frac{M_{\text{rod}}}{M}$$

$$k = \frac{AE}{l}, \qquad M = \text{end mass}$$

Reducing the above system to a spring k and an end mass equal to $M + \frac{1}{3}M_{\text{rod}}$, determine an appropriate equation for the fundamental frequency. Show that the ratio of the approximate to the exact frequency as found above is

$$(1/\beta_1)\sqrt{3r/(3 + r)}$$

8-12 The frequency of magnetostriction oscillators is determined by the length of the nickel alloy rod which generates an alternating voltage in the surrounding coils equal to the frequency of longitudinal vibration of the rod, as shown in Fig. P8-12. Determine the proper length of the rod clamped at the middle for a frequency of 20 kcps if the modulus of elasticity and density are given as $E = 30 \times 10^6$ lb/in.^{2} and $\rho = 0.31$ lb/in.^{3}

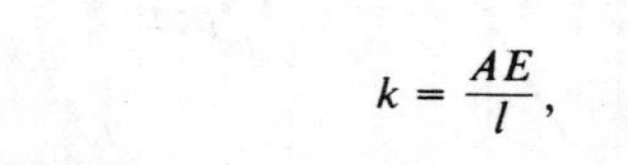

Figure P8-12.

8-13 The equation for the longitudinal oscillations of a slender rod with viscous damping is

$$m\frac{\partial^2 u}{\partial t^2} = AE\frac{\partial^2 u}{\partial x^2} - \alpha\frac{\partial u}{\partial t} + \frac{p_0}{l}p(x)f(t)$$

where the loading per unit length is assumed to be separable. Letting $u =$

$\Sigma_i \phi_i(x) q_i(t)$ and $p(x) = \Sigma_i b_i \phi_i(x)$ show that

$$u = \frac{p_0}{ml\sqrt{1-\zeta^2}} \sum_j \frac{b_j \phi_j}{\omega_j} \int_0^t f(t-\tau)\, e^{-\zeta\omega_j\tau} \sin \omega_j \sqrt{1-\zeta^2}\, \tau \, d\tau$$

$$b_j = \frac{1}{l} \int_0^l p(x) \phi_j(x)\, dx$$

Derive the equation for the stress at any point x.

8-14 Show that $c = \sqrt{G/\rho}$ is the velocity of propagation of torsional strain along the rod. What is the numerical value of c for steel?

8-15 Determine the expression for the natural frequencies of torsional oscillations of a uniform rod of length l clamped at the middle and free at the two ends.

8-16 Determine the natural frequencies of a torsional system consisting of a uniform shaft of mass moment of inertia J_s with a disk of inertia J_0 attached to each end. Check the fundamental frequency by reducing the uniform shaft to a torsional spring with end masses.

8-17 A uniform bar has these specifications: length l, mass density per unit volume ρ, and torsional stiffness $I_P G$ where I_P is the polar moment of inertia of the cross section and G the shear modulus. The end $x = 0$ is fastened to a torsional spring of stiffness K lb · in./rad, while the end l is fixed as shown in Fig. P8-17. Determine the transcendental equation from which natural frequencies can be established. Verify the correctness of this equation by considering special cases for $K = 0$ and $K = \infty$.

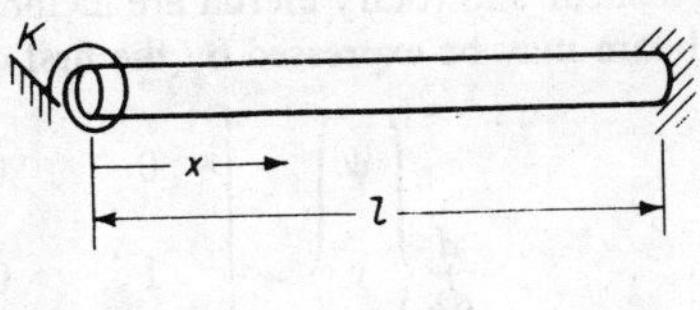

Figure P8-17.

8-18 Determine the expression for the natural frequencies of a free-free bar in lateral vibration.

8-19 Determine the node position for the fundamental mode of the free-free beam by Rayleigh's method, assuming the curve to be $y = \sin(\pi x/l) - b$. By equating the momentum to zero, determine b. Substitute this value of b to find ω_1.

8-20 A concrete test beam $2 \times 2 \times 12$ in., supported at two points $0.224l$ from the ends, was found to resonate at 1690 cps. If the density of concrete is 153 lb/ft^3, determine the modulus of elasticity, assuming the beam to be slender.

8-21 Determine the natural frequencies of a uniform beam of length l clamped at both ends.

8-22 Determine the natural frequencies of a uniform beam of length l, clamped at one end and pinned at the other end.

8-23 A uniform beam of length l and weight W_b is clamped at one end and carries a concentrated weight W_0 at the other end. State the boundary conditions and determine the frequency equation.

8-24 The pinned end of a pinned-free beam is given a harmonic motion of amplitude y_0 perpendicular to the beam. Show that the boundary conditions result in the equation

$$\frac{y_0}{y_l} = \frac{\sinh \beta l \cos \beta l - \cosh \beta l \sin \beta l}{\sinh \beta l - \sin \beta l}$$

which for $y_0 \to 0$, reduces to

$$\tanh \beta l = \tan \beta l$$

8-25 A simply supported beam has an overhang of length l_2, as shown in Fig. P8-25. If the end of the overhang is free, show that boundary conditions require the deflection equation for each span to be

$$\phi_1 = C\left(\sin \beta x - \frac{\sin \beta l_1}{\sinh \beta l_1} \sinh \beta x\right)$$

$$\phi_2 = A\left\{\cos \beta x + \cosh \beta x - \left(\frac{\cos \beta l_2 + \cosh \beta l_2}{\sin \beta l_2 + \sinh \beta l_2}\right)(\sin \beta x + \sinh \beta x)\right\}$$

where x is measured from the left and right ends.

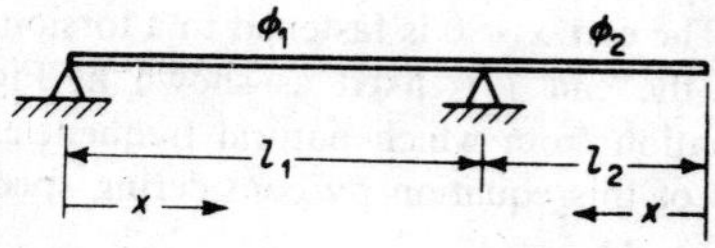

Figure P8-25.

8-31 When shear and rotary inertia are included, show that the differential equation of the beam may be expressed by the first order matrix equation

$$\frac{d}{dx}\begin{Bmatrix} \psi \\ y \\ M \\ V \end{Bmatrix} = \begin{bmatrix} 0 & 0 & \frac{1}{EI} & 0 \\ 1 & 0 & 0 & \frac{-1}{kAG} \\ -\omega^2 J & 0 & 0 & 1 \\ 0 & \omega^2 m & 0 & 0 \end{bmatrix}\begin{Bmatrix} \psi \\ y \\ M \\ V \end{Bmatrix}$$

9

MODE-SUMMATION PROCEDURES FOR CONTINUOUS SYSTEMS

Structures made up of beams are common in engineering. They constitute systems of an infinite number of degrees of freedom, and the mode summation methods make possible their analysis as systems of a finite number of degrees of freedom. The effect of rotary inertia and shear deformation is sometimes of interest in beam problems. Constraints are often found as additional supports of the structure, and they alter the normal modes of the system. In the use of the mode-summation method, convergence of the series is of importance, and the mode-acceleration method offers a varied approach. The modes used in representing the deflection of a system need not always be orthogonal. The synthesis of a system using nonorthogonal functions is illustrated.

9.1 MODE-SUMMATION METHOD

In Section 6.8 the equations of motion were decoupled by the modal matrix to obtain the solution of forced vibration in terms of the normal coordinates of the system. In this section, we apply a similar technique to continuous systems by expanding the deflection in terms of the normal modes of the system.

Consider, for example, the general motion of a beam loaded by a distributed force $p(x, t)$, whose equation of motion is

$$[EIy''(x,t)]'' + m(x)\ddot{y}(x,t) = p(x,t) \tag{9.1-1}$$

The normal modes $\phi_i(x)$ of such a beam must satisfy the equation

$$(EI\phi_i'')'' - \omega_i^2 m(x)\phi_i = 0 \tag{9.1-2}$$

and its boundary conditions. The normal modes $\phi_i(x)$ are also orthogonal functions satisfying the relation

$$\int_0^l m(x)\phi_i\phi_j\,dx = \begin{cases} 0 & \text{for } j \neq i \\ M_i & \text{for } j = i \end{cases} \tag{9.1-3}$$

By representing the solution to the general problem in terms of $\phi_i(x)$

$$y(x,t) = \sum_i \phi_i(x)q_i(t) \tag{9.1-4}$$

the generalized coordinate $q_i(t)$ can be determined from Lagrange's equation by first establishing the kinetic and potential energies.

Recognizing the orthogonality relation, Eq. (9.1-3), the kinetic energy is

$$T = \tfrac{1}{2}\int_0^l \dot{y}^2(x,t)m(x)\,dx = \tfrac{1}{2}\sum_i\sum_j \dot{q}_i\dot{q}_j\int_0^l \phi_i\phi_j m(x)\,dx$$
$$= \tfrac{1}{2}\sum_i M_i\dot{q}_i^2 \tag{9.1-5}$$

where the generalized mass M_i is defined as

$$M_i = \int_0^l \phi_i^2(x)m(x)\,dx \tag{9.1-6}$$

Similarly, the potential energy is

$$U = \tfrac{1}{2}\int_0^l EIy''^2(x,t)\,dx = \tfrac{1}{2}\sum_i\sum_j q_iq_j\int_0^l EI\phi_i''\phi_j''\,dx$$
$$= \tfrac{1}{2}\sum_i K_iq_i^2 = -\tfrac{1}{2}\sum \omega_i^2M_iq_i^2 \tag{9.1-7}$$

where the generalized stiffness is

$$K_i = \int_0^l EI[\phi_i''(x)]^2\,dx \tag{9.1-8}$$

In addition to T and U, we need the generalized force Q_i, which is determined from the work done by the applied force $p(x,t)\,dx$ in the virtual displacement δq_i.

$$\delta W = \int_0^l p(x,t)\left(\sum_i \phi_i\delta q_i\right)dx$$
$$= \sum_i \delta q_i\int_0^l p(x,t)\phi_i(x)\,dx \tag{9.1-9}$$

where the generalized force is

$$Q_i = \int_0^l p(x,t)\phi_i(x)\,dx \tag{9.1-10}$$

Substituting into Lagrange's equation,

$$\frac{d}{dt}\left(\frac{\partial T}{\partial \dot{q}_i}\right) - \frac{\partial T}{\partial q_i} + \frac{\partial U}{\partial q_i} = Q_i \tag{9.1-11}$$

we find the differential equation for $q_i(t)$ to be

$$\ddot{q}_i + \omega_i^2 q_i = \frac{1}{M_i}\int_0^l p(x, t)\phi_i(x)\,dx \tag{9.1-12}$$

It is convenient at this point to consider the case where the loading per unit length $p(x, t)$ is separable in the form

$$p(x, t) = \frac{P_0}{l}p(x)f(t) \tag{9.1-13}$$

Eq. (9.1-12) then reduces to

$$\ddot{q}_i + \omega_i^2 q_i = \frac{P_0}{M_i}\Gamma_i f(t) \tag{9.1-14}$$

where

$$\Gamma_i = \frac{1}{l}\int_0^l p(x)\phi_i(x)\,dx \tag{9.1-15}$$

is defined as the *mode participation factor* for mode i. The solution of Eq. (9.1-14) is then

$$q_i(t) = q_i(0)\cos\omega_i t + \frac{1}{\omega_i}\dot{q}_i(0)\sin\omega_i t$$

$$+\left(\frac{P_0\Gamma_i}{M_i\omega_i^2}\right)\omega_i\int_0^t f(\xi)\sin\omega_i(t-\xi)\,d\xi \tag{9.1-16}$$

Since the ith mode statical deflection [with $\ddot{q}_i(t) = 0$] expanded in terms of $\phi_i(x)$ is $P_0\Gamma_i/M_i\omega_i^2$, the quantity

$$D_i(t) = \omega_i\int_0^t f(\xi)\sin\omega_i(t-\xi)\,d\xi \tag{9.1-17}$$

can be called the *dynamic load factor* for the ith mode.

Example 9.1-1

A simply supported uniform beam of mass M_0 is suddenly loaded by the force shown in Fig. 9.1-1. Determine the equation of motion.

Solution: The normal modes of the beam are

$$\phi_n(x) = \sqrt{2}\sin\frac{n\pi x}{l}$$

$$\omega_n = (n\pi)^2\sqrt{EI/M_0 l^3}$$

and the generalized mass is

$$M_n = \frac{M_0}{l}\int_0^l 2\sin^2\frac{n\pi x}{l}\,dx = M_0$$

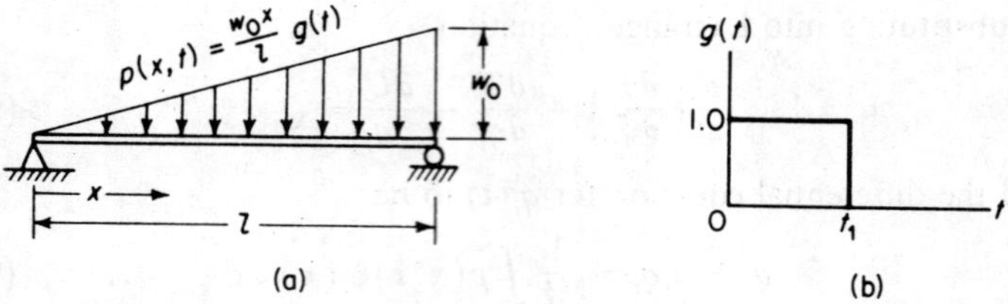

Figure 9.1-1.

The generalized force is

$$\int_0^l p(x,t)\phi_n(x)\,dx = g(t)\int_0^l \frac{w_0 x}{l}\sqrt{2}\sin\frac{n\pi x}{l}\,dx$$

$$= g(t)\frac{w_0\sqrt{2}}{l}\left[\frac{\sin(n\pi x/l)}{(n\pi/l)^2} - \frac{x\cos(n\pi x/l)}{(n\pi/l)}\right]_0^l$$

$$= -g(t)\frac{w_0\sqrt{2}\,l}{n\pi}\cos n\pi$$

$$= -\frac{\sqrt{2}\,lw_0}{n\pi}g(t)(-1)^n$$

where $g(t)$ is the time history of the load. The equation for q_n is then

$$\ddot{q}_n + \omega_n^2 q_n = -\frac{\sqrt{2}\,lw_0}{n\pi M_0}(-1)^n g(t)$$

which has the solution

$$q_n(t) = \frac{-\sqrt{2}\,lw_0}{n\pi M_0}\frac{(-1)^n}{\omega_n^2}(1-\cos\omega_n t) \qquad 0 \le t \le t_1$$

$$= \frac{-\sqrt{2}\,lw_0}{n\pi M_0}\frac{(-1)^n}{\omega_n^2}(1-\cos\omega_n t)$$

$$+\frac{2\sqrt{2}\,lw_0(-1)^n}{n\pi M_0\omega_n^2}[1-\cos\omega_n(t-t_1)] \qquad t_1 \le t \le \infty$$

Thus the deflection of the beam is expressed by the summation

$$y(x,t) = \sum_{n=1}^{\infty} q_n(t)\sqrt{2}\sin\frac{\pi nx}{l}$$

Example 9.1-2

A missile in flight is excited longitudinally by the thrust $F(t)$ of its rocket engine at the end $x = 0$. Determine the equation for the displacement $u(x,t)$ and the acceleration $\ddot{u}(x,t)$.

Solution: We assume the solution for the displacement to be

$$u(x,t) = \sum q_i(t)\varphi_i(x)$$

where $\varphi_i(x)$ are normal modes of the missile in longitudinal oscillation. The generalized coordinate q_i satisfies the differential equation

$$\ddot{q}_i + \omega_i^2 q_i = \frac{F(t)\varphi_i(0)}{M_i}$$

If, instead of $F(t)$, a unit impulse acted at $x = 0$, the preceding equation would have the solution $(\varphi_i(0)/M_i\omega_i)\sin\omega_i t$ for initial conditions $q_i(0) = \dot{q}(0) = 0$. Thus the response to the arbitrary force $F(t)$ is

$$q_i(t) = \frac{\varphi_i(0)}{M_i\omega_i}\int_0^t F(\xi)\sin\omega_i(t-\xi)\,d\xi$$

and the displacement at any point x is

$$u(x,t) = \sum_i \frac{\varphi_i(x)\varphi_i(0)}{M_i\omega_i}\int_0^t F(\xi)\sin\omega_i(t-\xi)\,d\xi$$

The acceleration $\ddot{q}_i(t)$ of mode i can be determined by rewriting the differential equation and substituting the former solution for $q_i(t)$

$$\ddot{q}_i(t) = \frac{F(t)\varphi_i(0)}{M_i} - \omega_i^2 q_i(t)$$

$$= \frac{F(t)\varphi_i(0)}{M_i} - \frac{\varphi_i(0)\omega_i}{M_i}\int_0^t F(\xi)\sin\omega_i(t-\xi)\,d\xi$$

Thus the equation for the acceleration of any point x is found as

$$\ddot{u}(x,t) = \sum_i \ddot{q}_i(t)\varphi_i(x)$$

$$= \sum_i\left\{\frac{F(t)\varphi_i(0)\varphi_i(x)}{M_i} - \frac{\varphi_i(0)\varphi_i(x)\omega_i}{M_i}\int_0^t F(\xi)\sin\omega_i(t-\xi)\,d\xi\right\}$$

Example 9.1-3

Determine the response of a cantilever beam when its base is given a motion $y_b(t)$ normal to the beam axis as shown in Fig. 9.1-2.

Solution: The differential equation for the beam with base motion is

$$[EIy''(x,t)]'' + m(x)[\ddot{y}_b(t) + \ddot{y}(x,t)] = 0$$

which can be rearranged to

$$[EIy''(x,t)]'' + m(x)\ddot{y}(x,t) = -m(x)\ddot{y}_b(t)$$

Thus instead of the force per unit length $F(x,t)$, we have the inertial force per unit

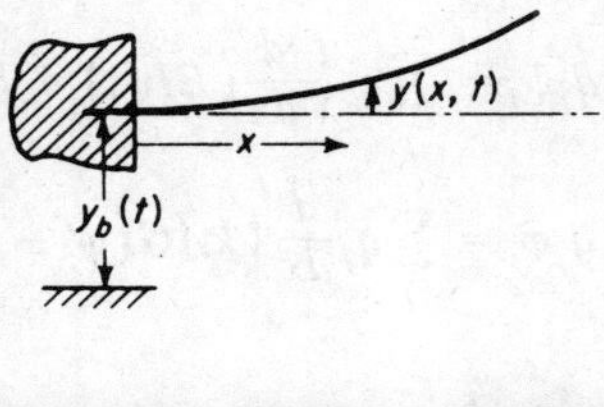

Figure 9.1-2.

length, $-m(x)\ddot{y}_b(t)$. Assuming the solution in the form

$$y(x,t) = \sum_i q_i(t)\varphi_i(x)$$

the equation for the generalized coordinate q_i becomes

$$\ddot{q}_i + \omega_i^2 q_i = -\ddot{y}_b(t)\frac{1}{M_i}\int_0^l \varphi_i(x)\,dx$$

The solution for q_i then differs from that of a simple oscillator only by the factor $-1/M_i\int_0^l \varphi_i(x)\,dx$ so that for the initial conditions $y(0) = \dot{y}(0) = 0$

$$q_i(t) = \left\{-\frac{1}{M_i}\int_0^l \varphi_i(x)\,dx\right\}\frac{1}{\omega_i}\int_0^t \ddot{y}_b(\xi)\sin\omega_i(t-\xi)\,d\xi$$

9.2 BEAM ORTHOGONALITY INCLUDING ROTARY INERTIA AND SHEAR DEFORMATION

The equations for the beam, including rotary inertia and shear deformation, were derived in Sec. 8.5. For such beams the orthogonality is no longer expressed by Eq. (9.1-3), but by the equation

$$\int\left[m(x)\varphi_j\varphi_i + J(x)\psi_j\psi_i\right]dx = \begin{cases}0 & \text{if } j \neq i\\ M_i & \text{if } j = i\end{cases} \tag{9.2-1}$$

which can be proved in the following manner.

For convenience we will rewrite Eqs. (8.5-5) and (8.5-6), including a distributed moment per unit length $\mathfrak{M}(x,t)$

$$\frac{d}{dx}\left(EI\frac{d\psi}{dx}\right) + kAG\left(\frac{dy}{dx} - \psi\right) - J\ddot{\psi} - \mathfrak{M}(x,t) = 0 \tag{8.5-5}$$

$$m\ddot{y} - \frac{d}{dx}\left[kAG\left(\frac{dy}{dx} - \psi\right)\right] - p(x,t) = 0 \tag{8.5-6}$$

For the forced oscillation with excitation $p(x,t)$ and $\mathfrak{M}(x,t)$ per unit length of beam, the deflection $y(x,t)$ and the bending slope $\psi(x,t)$ can be expressed in terms of the generalized coordinates

$$\begin{aligned} y &= \sum_j q_j(t)\varphi_j(x)\\ \psi &= \sum_j q_j(t)\psi_j(x)\end{aligned} \tag{9.2-2}$$

With these summations substituted into the two beam equations, we obtain

$$\begin{aligned} J\sum_j \ddot{q}_j\psi_j &= \sum_j q_j\left\{\frac{d}{dx}(EI\psi_j') + kAG(\varphi_j' - \psi_j)\right\} + \mathfrak{M}(x,t)\\ m\sum_j \ddot{q}_j\varphi_j &= \sum_j q_j\frac{d}{dx}\{kAG(\varphi_j' - \psi_j)\} + p(x,t)\end{aligned} \tag{9.2-3}$$

However, normal mode vibrations are of the form

$$\begin{aligned} y &= \varphi_j(x)e^{i\omega_j t} \\ \psi &= \psi_j(x)e^{i\omega_j t} \end{aligned} \tag{9.2-4}$$

which, when substituted into the beam equations with zero excitation, lead to

$$\begin{aligned} -\omega_j^2 J\psi_j &= \frac{d}{dx}(EI\psi_j') + kAG(\varphi_j' - \psi_j) \\ -\omega_j^2 m\varphi_j &= \frac{d}{dx}\{kAG(\varphi_j' - \psi_j)\} \end{aligned} \tag{9.2-5}$$

The right sides of this set of equations are the coefficients of the generalized coordinates q_j in the forced vibration equations, so that we can write Eqs. (9.2-3) as

$$\begin{aligned} J\sum_j \ddot{q}_j\psi_j &= -\sum_j q_j\omega_j^2 J\psi_i + \mathfrak{M}(x,t) \\ m\sum_j \ddot{q}_j\varphi_j &= -\sum_j q_j\omega_j^2 m\varphi_j + p(x,t) \end{aligned} \tag{9.2-6}$$

Multiplying these two equations by $\varphi_i\,dx$ and $\psi_i\,dx$, adding, and integrating, we obtain

$$\sum_j \ddot{q}_j \int_0^l (m\varphi_j\varphi_i + J\psi_j\psi_i)\,dx + \sum_j q_j\omega_j^2 \int_0^l (m\varphi_j\varphi_i + J\psi_j\psi_i)\,dx$$
$$= \int_0^l p(x,t)\varphi_i\,dx + \int_0^l \mathfrak{M}(x,t)\psi_i\,dx \tag{9.2-7}$$

If the q's in these equations are generalized coordinates, they must be independent coordinates which satisfy the equation

$$\ddot{q}_i + \omega_i^2 q_i = \frac{1}{M_i}\left\{\int_0^l p(x,t)\varphi_i\,dx + \int_0^l \mathfrak{M}(x,t)\psi_i\,dx\right\} \tag{9.2-8}$$

We see then that this requirement is satisfied only if

$$\int_0^l (m\varphi_j\varphi_i + J\psi_j\psi_i)\,dx = \begin{cases} 0 & \text{if } j \neq i \\ M_i & \text{if } j = i \end{cases} \tag{9.2-9}$$

which defines the orthogonality for the beam, including rotary inertia and shear deformation.

9.3 NORMAL MODES OF CONSTRAINED STRUCTURES

When a structure is altered by the addition of a mass or a spring, we refer to it as a *constrained structure*. For example, a spring will tend to act as a constraint on the motion of the structure at the point of its application, and possibly increase the natural frequencies of the system. An added mass, on the other hand, may decrease the natural frequencies of the system. Such problems can be formulated in terms of generalized coordinates and the mode-summation technique.

Consider the forced vibration of any one dimensional structure (i.e., the points on the structure defined by one coordinate x) excited by a force per unit length $f(x, t)$ and moment per unit length $M(x, t)$. If we know the normal modes of the structure, ω_i and $\varphi_i(x)$, its deflection at any point x can be represented by

$$y(x, t) = \sum_i q_i(t)\varphi_i(x) \tag{9.3-1}$$

where the generalized coordinate q_i must satisfy the equation

$$\ddot{q}_i(t) + \omega_i^2 q_i(t) = \frac{1}{M_i}\left[\int f(x, t)\varphi_i(x)\,dx + \int M(x, t)\varphi_i'(x)\,dx\right] \tag{9.3-2}$$

The right side of this equation is $1/M_i$ times the generalized force Q_i, which can be determined from the virtual work of the applied loads as $Q_i = \delta W/\delta q_i$.

If, instead of distributed loads, we have a concentrated force $F(a, t)$ and a concentrated moment $M(a, t)$ at some point $x = a$, the generalized force for such loads is found from

$$\begin{aligned}\delta W &= F(a, t)\,\delta y(a, t) + M(a, t)\,\delta y'(a, t)\\ &= F(a, t)\sum_i \varphi_i(a)\,\delta q_i + M(a, t)\sum_i \varphi_i'(a)\,\delta q_i\end{aligned} \tag{9.3-3}$$

$$Q_i = \frac{\delta W}{\delta q_i} = F(a, t)\varphi_i(a) + M(a, t)\varphi_i'(a)$$

Then, instead of Eq. (9.1-14), we obtain the equation

$$\ddot{q}_i(t) + \omega_i^2 q_i(t) = \frac{1}{M_i}\left[F(a, t)\varphi_i(a) + M(a, t)\varphi_i'(a)\right] \tag{9.3-4}$$

These equations form the starting point for the analysis of constrained structures, provided the constraints are expressible as external loads on the structure.

As an example, let us consider attaching a linear and torsional spring to the simply supported beam of Fig. 9.3-1. The linear spring exerts a force on the beam equal to

$$F(a, t) = -ky(a, t) = -k\sum_j q_j(t)\varphi_j(a) \tag{9.3-5}$$

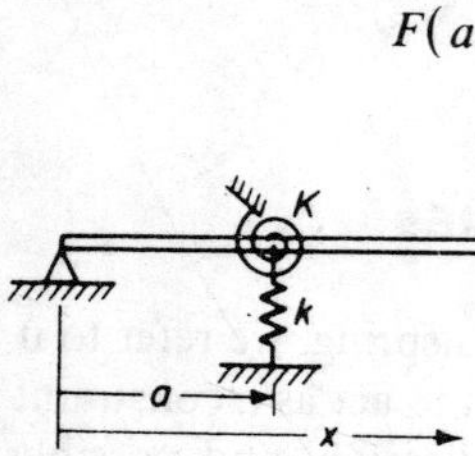

Figure 9.3-1.

whereas the torsional spring exerts a moment

$$M(a,t) = -Ky'(a,t) = -K\sum_j q_j(t)\varphi_j'(a) \tag{9.3-6}$$

Substituting these equations into Eq. (9.3-4), we obtain

$$\ddot{q}_i + \omega_i^2 q_i = \frac{1}{M_i}\left[-k\varphi_i(a)\sum_j q_j\varphi_j(a) - K\varphi_i'(a)\sum_j q_j\varphi_j'(a)\right] \tag{9.3-7}$$

The normal modes of the constrained modes are also harmonic and so we can write

$$q_i = \bar{q}_i e^{i\omega t}$$

The solution to the ith equation is then

$$\bar{q}_i = \frac{1}{M_i(\omega_i^2 - \omega^2)}\left[-k\varphi_i(a)\sum_j \bar{q}_j\varphi_j(a) - K\varphi_i'(a)\sum_j \bar{q}_j\varphi_j'(a)\right] \tag{9.3-8}$$

If we use n modes, there will be n values of $\bar{q}_j$ and n equations such as the one above. The determinant formed by the coefficients of the $\bar{q}_j$ will then lead to the natural frequencies of the constrained modes, and the mode shapes of the constrained structure are found by substituting the $\bar{q}_j$ into Eq. (9.3-1).

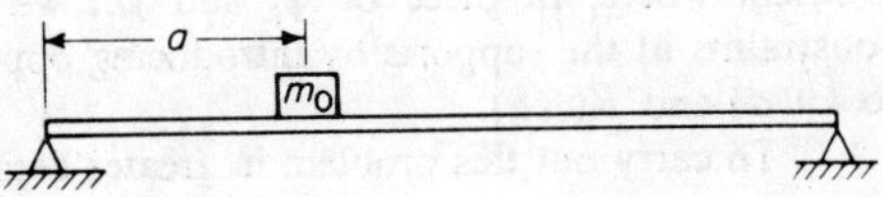

Figure 9.3-2.

If, instead of springs, a mass m_0 is placed at a point $x = a$, as shown in Fig. 9.3-2, the force exerted by m_0 on the beam is

$$F(a,t) = -m_0\ddot{y}(a,t) = -m_0\sum_j \ddot{q}_j\varphi_j(a) \tag{9.3-9}$$

Thus in place of Eq. (9.3-8), we would obtain the equation

$$\bar{q}_i = \frac{1}{M_i(\omega_i^2 - \omega^2)}\left[\omega^2 m_0\varphi_i(a)\sum_j \bar{q}_j\varphi_j(a)\right] \tag{9.3-10}$$

Example 9.3-1

Give a single mode approximation for the natural frequency of a simply supported beam when a mass m_0 is attached to it at $x = l/3$.

Solution: When only a single mode is used, Eq. (9.3-10) reduces to

$$M_1(\omega_1^2 - \omega^2) = \omega^2 m_0\varphi_1^2(a)$$

Solving for ω^2, we obtain

$$\left(\frac{\omega}{\omega_1}\right)^2 = \frac{1}{1 + \dfrac{m_0}{M_1}\varphi_1^2(a)}$$

For the first mode of the unconstrained beam, we have

$$\omega_1 = \pi^2\sqrt{\frac{EI}{Ml^3}}, \qquad \varphi_1(x) = \sqrt{2}\sin\frac{\pi x}{l}$$

$$\varphi_1\left(\frac{l}{3}\right) = \sqrt{2}\sin\frac{\pi}{3} = \sqrt{2} \times 0.866$$

$$M_1 = M = \text{mass of the beam}$$

Thus its substitution into the preceding equation gives the value for the one-mode approximation for the constrained beam of

$$\left(\frac{\omega}{\omega_1}\right)^2 = \frac{1}{1 + 1.5\dfrac{m_0}{M}}$$

Example 9.3-2

A missile is constrained in a test stand by linear and torsional springs, as shown in Fig. 9.3-3. Formulate the inverse problem of determining its free-free modes from the normal modes of the constrained missile, which are designated as Φ_i and Ω_i.

Solution: The problem is approached in a manner similar to that of the direct problem where, in place of φ_i and ω_i, we use Φ_i and Ω_i. We now relieve the constraints at the supports by introducing opposing forces $-F(a)$ and $-M(a)$ equal to $ky(a)$ and $Ky'(a)$.

To carry out this problem in greater detail, we start with the equation

$$\bar{q}_i = \frac{-F(a)\Phi_i(a) - M(a)\Phi_i'(a)}{M_i\Omega_i^2\left[1 - (\Omega/\Omega_i)^2\right]}$$

which replaces Eq. (9.3-8). Letting $D_i(\omega) = M_i\Omega_i^2[1 - (\omega/\Omega_i)^2]$, the displacement at $x = a$ is

$$y(a) = \sum_i \Phi_i(a)\bar{q}_i = \sum_i \frac{-F(a)\Phi_i^2(a) - M(a)\Phi_i'(a)\Phi_i(a)}{D_i(\omega)}$$

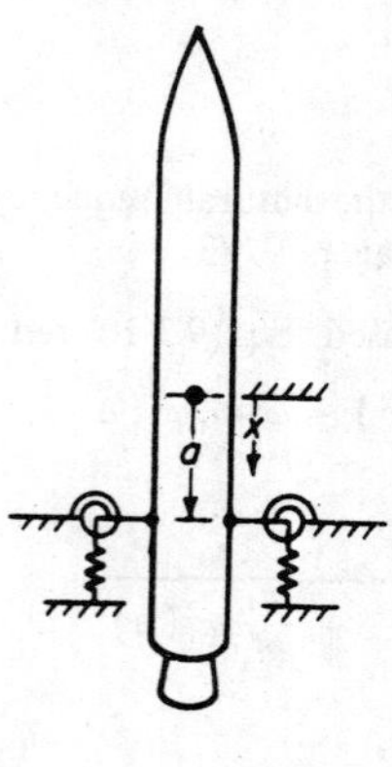

Figure 9.3-3.

We now replace $-F(a)$ and $-M(a)$ with $ky(a)$ and $Ky'(a)$ and write

$$y(a) = \sum_i \frac{ky(a)\Phi_i^2(a) + Ky'(a)\Phi_i'(a)\Phi_i(a)}{D_i(\omega)}$$

$$y'(a) = \sum_i \frac{ky(a)\Phi_i'(a)\Phi_i(a) + Ky'(a)\Phi_i'^2(a)}{D_i(\omega)}$$

These equations may now be rearranged as

$$y(a)\left[1 - k\sum_i \frac{\Phi_i^2(a)}{D_i(\omega)}\right] = y'(a)K\sum_i \frac{\Phi_i'(a)\Phi_i(a)}{D_i(\omega)}$$

$$y(a)k\sum_i \frac{\Phi_i'(a)\Phi_i(a)}{D_i(\omega)} = y'(a)\left[1 - K\sum_i \frac{\Phi_i'^2(a)}{D_i(\omega)}\right]$$

The frequency equation then becomes

$$\left[1 - k\sum_i \frac{\Phi_i^2(a)}{D_i(\omega)}\right]\left[1 - K\sum_i \frac{\Phi_i'^2(a)}{D_i(\omega)}\right] - kK\left[\sum \frac{\Phi_i'(a)\Phi_i(a)}{D_i(\omega)}\right]^2 = 0$$

The slope to deflection ratio at $x = a$ is

$$\frac{y'(a)}{y(a)} = \frac{1 - k\sum_i \dfrac{\Phi_i^2(a)}{D_i(\omega)}}{K\sum_i \dfrac{\Phi_i'(a)\Phi_i(a)}{D_i(\omega)}}$$

The free-free mode shape is then given by

$$\frac{y(x)}{y(a)} = \sum_i \frac{k\Phi_i(a)\Phi_i(x) + K\dfrac{y'(a)}{y(a)}\Phi_i'(a)\Phi_i(x)}{D_i(\omega)}$$

Example 9.3-3

Determine the constrained modes of the missile of Fig. 9.3-3, using only the first free-free mode $\varphi_1(x)$, ω_1, together with translation $\varphi_T = 1$, $\Omega_T = 0$ and rotation $\varphi_R = x$, $\Omega_R = 0$, where x is measured positively toward the tail of the missile.

Solution: The generalized mass for each of the three modes is

$$M_T = \int dm = M$$

$$M_R = \int x^2\, dm = I = M\rho^2$$

$$M_1 = \int \varphi_1^2(x)\, dm = M$$

where the $\varphi_1(x)$ mode was normalized such that $M_1 = M =$ actual mass.

The frequency dependent factors D_i are

$$D_T = -M_T\omega^2 = -M\omega^2 = -M\omega_1^2\lambda$$

$$D_R = -M\rho^2\omega^2 = -M\rho^2\omega_1^2\lambda$$

$$D_1 = M\omega_1^2\left[1 - \left(\frac{\omega}{\omega_1}\right)^2\right] = M\omega_1^2(1-\lambda)$$

$$\left(\frac{\omega}{\omega_1}\right)^2 = \lambda$$

The frequency equation for this problem is the same as that of Example 9.3-2, except that the minus k's are replaced by positive k's and $\varphi(x)$ and ω replace $\Phi(x)$ and Ω. Substituting the above quantities into the frequency equation, we have

$$\left\{1 - \frac{k}{M\omega_1^2}\left[\frac{1}{\lambda} + \frac{a^2}{\rho^2\lambda} - \frac{\varphi_1^2(a)}{(1-\lambda)}\right]\right\}\left\{1 - \frac{K}{M\omega_1^2}\left[\frac{1}{\rho^2\lambda} - \frac{\varphi_1'^2(a)}{(1-\lambda)}\right]\right\}$$

$$-\frac{kK}{M^2\omega_1^4}\left\{\frac{-a}{\rho^2\lambda} + \frac{\varphi_1'(a)\varphi_1(a)}{(1-\lambda)}\right\}^2 = 0$$

which can be simplified to

$$\lambda^2(1-\lambda) + \left(\frac{k}{M\omega_1^2}\right)\left[\varphi_1^2(a) + \frac{K}{k}\varphi_1'^2(a)\right]\lambda^2$$

$$-\left(\frac{k}{M\omega_1^2}\right)\left[1 + \frac{a^2}{\rho^2} + \frac{K}{k\rho^2}\right]\lambda(1-\lambda) + \left(\frac{k}{M\omega_1^2}\right)^2\frac{K}{k\rho^2}(1-\lambda)$$

$$-\left(\frac{k}{M\omega_1^2}\right)^2\frac{K}{k}\lambda\left\{\varphi_1'^2(a) + \frac{1}{\rho^2}[\varphi_1(a) - a\varphi_1'(a)]^2\right\} = 0$$

A number of special cases of the preceding equation are of interest, and we mention one of these. If $K = 0$, the frequency equation simplifies to

$$\lambda^2 - \left\{1 + \left(\frac{k}{M\omega_1^2}\right)\left[1 + \frac{a^2}{\rho^2} + \varphi_1^2(a)\right]\right\}\lambda + \left(\frac{k}{M\omega_1^2}\right)\left(1 + \frac{a^2}{\rho^2}\right) = 0$$

Here $x = a$ might be taken negatively so that the missile is hanging by a spring.

9.4 MODE-ACCELERATION METHOD

One of the difficulties encountered in any mode-summation method has to do with the convergence of the procedure. If this convergence is poor, a large number of modes must be used, thereby increasing the order of the frequency determinant. The mode-acceleration method tends to overcome this difficulty by improving the convergence so that a fewer number of normal modes are needed.

The mode-acceleration method starts with the same differential equation for the generalized coordinate q_i but rearranged in order. For example, we can start with Eq. (9.3-4) and write it in the order

$$q_i(t) = \frac{F(a,t)\varphi_i(a)}{M_i\omega_i^2} + \frac{M(a,t)\varphi_i'(a)}{M_i\omega_i^2} - \frac{\ddot{q}_i(t)}{\omega_i^2} \tag{9.4-1}$$

Substituting this into Eq. (9.3-1), we obtain

$$\begin{aligned} y(x,t) &= \sum_i q_i(t)\varphi_i(x) \\ &= F(a,t)\sum_i \frac{\varphi_i(a)\varphi_i(x)}{M_i\omega_i^2} + M(a,t)\sum_i \frac{\varphi_i'(a)\varphi_i(x)}{M_i\omega_i^2} \\ &\quad - \sum_i \frac{\ddot{q}_i(t)\varphi_i(x)}{\omega_i^2} \end{aligned} \tag{9.4-2}$$

We note here that, if $F(a, t)$ and $M(a, t)$ were static loads, the last term containing the acceleration would be zero. Thus the terms

$$\begin{aligned} \sum_i \frac{\varphi_i(a)\varphi_i(x)}{M_i\omega_i^2} &= \alpha(a,x) \\ \sum \frac{\varphi_i'(a)\varphi_i(x)}{M_i\omega_i^2} &= \beta(a,x) \end{aligned} \tag{9.4-3}$$

must represent influence functions, where $\alpha(a, x)$ and $\beta(a, x)$ are the deflections at x due to a unit load and unit moment at a, respectively. We can therefore rewrite Eq. (9.4-2) as

$$y(x,t) = F(\alpha,t)\alpha(a,x) + M(a,t)\beta(a,x) - \sum \frac{\ddot{q}(t)\varphi_i(x)}{\omega_i^2} \tag{9.4-4}$$

Because of ω_i^2 in the denominator of the terms summed, the convergence is improved over the mode summation method.

In the forced vibration problem where $F(a, t)$ and $M(a, t)$ are excitations, Eq. (9.3-4) is first solved for $q_i(t)$ in the conventional manner, and then substituted into Eq. (9.4-4) for the deflection. For the normal modes of constrained structures, $F(a, t)$ and $M(a, t)$ are again the forces and moments exerted by the constraints, and the problem is treated in a manner similar to those of Sec. 9.3. However, because of the improved convergence, fewer number of modes will be found to be necessary.

Example 9.4-1

Using the mode-acceleration method, solve the problem of Fig. 9.3-2 of a concentrated mass m_0 attached to the structure.

Solution: Assuming harmonic oscillations

$$F(a,t) = \bar{F}(a)e^{i\omega t}$$

$$q_i(t) = \bar{q}_i e^{i\omega t}$$

$$y(x,t) = y(x)e^{i\omega t}$$

Substituting these equations into Eq. (9.4-4) and letting $x = a$,

$$\bar{y}(a) = \bar{F}(a)\alpha(a, a) + \omega^2 \sum_j \frac{\bar{q}_j \varphi_j(a)}{\omega_j^2}$$

Since the force exerted by m_0 on the structure is

$$\bar{F}(a) = m_0 \omega^2 \bar{y}(a)$$

we can eliminate $\bar{y}(a)$ between the above two equations, obtaining

$$\frac{\bar{F}(a)}{m_0 \omega^2} = \bar{F}(a) x(a, a) + \omega^2 \sum_j \frac{\bar{q}_j \varphi_j(a)}{\omega_j^2}$$

or

$$\bar{F}(a) = \frac{\omega^2 \sum_j \dfrac{\bar{q}_j \varphi_j(a)}{\omega_j^2}}{\dfrac{1}{m_0 \omega^2} - \alpha(a, a)}$$

If we now substitute this equation into Eq. (9.3-4) and assume harmonic motion, we obtain the equation

$$(\omega_i^2 - \omega^2)\bar{q}_i = \frac{\bar{F}(a)\varphi_i(a)}{M_i} = \frac{\omega^2 \varphi_i(a) \sum_j \bar{q}_j \dfrac{\varphi_j(a)}{\omega_j^2}}{M_i \left[\dfrac{1}{m_0 \omega^2} - \alpha(a, a)\right]}$$

Rearranging, we have

$$[1 - m_0 \omega^2 \alpha(a, a)](\omega_i^2 - \omega^2)\bar{q}_i = \frac{\omega^4 m_0 \varphi_i(a)}{M_i} \sum_j \frac{\bar{q}_j \varphi_j(a)}{\omega_j^2}$$

which represents a set of linear equations in $\bar{q}_k$. The series represented by the summation will, however, converge rapidly because of ω_j^2 in the denominator. Offsetting this advantage of smaller number of modes is the disadvantage that these equations are now quartic rather than quadratic in ω.

9.5 COMPONENT MODE SYNTHESIS

We discuss here another mode-summation procedure, where the deflection of each structural subcomponent is represented by the sum of polynomials instead of normal modes. These mode functions themselves need not be orthogonal or satisfy the junction conditions of displacement and force as long as their combined sum allows these conditions to be satisfied. Lagrange's equation, and in particular the method of superfluous coordinates, form the basis for the synthesis process.

To present the basic ideas of the method of modal synthesis, we consider a simple beam with a 90° bend, an example that was used by W. Hurty.† The

†Walter C. Hurty, "Vibrations of Structural Systems by Component Synthesis," *Jour. Engr. Mech. Div., Proc. of ASCE* (August 1960), pp. 51–69.

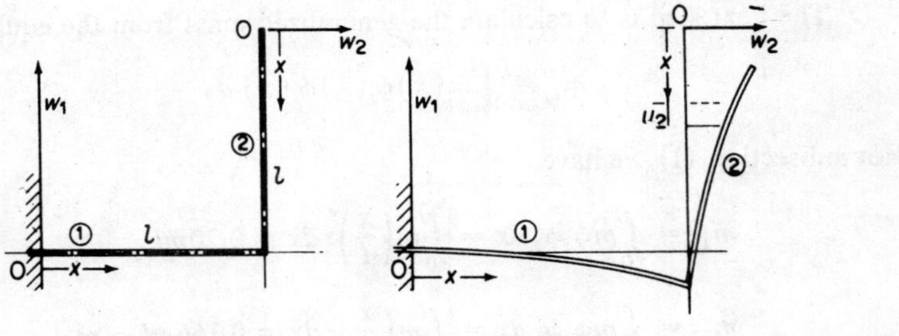

Figure 9.5-1. Beam sections 1 and 2 with their coordinates.

beam, shown in Fig. 9.5-1, is considered to vibrate only in the plane of the paper.

We separate the beam into two sections, ① and ②, whose coordinates are shown as w_1, x; w_2, x; and u_2, x. For part ① we assume the deflection to be

$$
\begin{aligned}
w_1(x,t) &= \phi_1(x)p_1(t) + \phi_2(x)p_2(t) + \cdots \\
&= \left(\frac{x}{l}\right)^2 p_1 + \left(\frac{x}{l}\right)^3 p_2 \qquad (9.5\text{-}1)
\end{aligned}
$$

Note that the two mode functions satisfy the geometric and force conditions at the boundaries of section ① as follows:

$$
\begin{aligned}
w_1(0) &= 0 & w_1(l) &= p_1 + p_2 \\
w_1'(0) &= 0 & w_1'(l) &= \frac{2}{l}p_1 + \frac{3}{l}p_3 \\
w_1''(0) &= \frac{M(0)}{EI} = \frac{2}{l^2}p_1 & w_1''(l) &= \frac{M(l)}{EI} = \frac{2}{l^2}p_1 + \frac{6}{l^2}p_2 \qquad (9.6\text{-}2) \\
w_1'''(0) &= \frac{V(0)}{EI} = \frac{6}{l^3}p_2 & w_1'''(l) &= \frac{V(l)}{EI} = \frac{6}{l^3}p_2
\end{aligned}
$$

Next consider part ② with the origin of the coordinates w_2, x at the free end. The following functions satisfy the boundary conditions of beam section ②

$$
\begin{aligned}
w_2(x,t) &= \phi_3(x)p_3(t) + \phi_4(x)p_4(t) + \phi_5(x)p_5(t) + \cdots \\
&= 1p_3 + \left(\frac{x}{l}\right)p_4 + \left(\frac{x}{l}\right)^4 p_5 \qquad (9.5\text{-}3) \\
u_2(x,t) &= \phi_6(x)p_6(t) + \cdots \\
&= 1p_6 \qquad (9.5\text{-}4)
\end{aligned}
$$

where $u_2(x,t)$ is the displacement in the x-direction.

The next step is to calculate the generalized mass from the equation

$$m_{ij} = \int_0^l m(x)\phi_i(x)\phi_j(x)\,dx$$

For subsection ① we have

$$m_{11} = \int_0^l m\phi_1\phi_1\,dx = \int_0^l m\left(\frac{x}{l}\right)^4 dx = 0.20ml$$

$$m_{12} = \int_0^l m\phi_1\phi_2\,dx = \int_0^l m\left(\frac{x}{l}\right)^5 dx = 0.166ml = m_{21}$$

$$m_{22} = \int_0^l m\phi_2\phi_2\,dx = \int_0^l m\left(\frac{x}{l}\right)^6 dx = 0.1428ml$$

The generalized mass for subsection ② is computed in a similar manner using ϕ_3 to ϕ_6:

$$\begin{aligned} m_{33} &= 1.0ml \\ m_{34} &= 0.50ml = m_{43} \\ m_{35} &= 0.20ml = m_{53} \\ m_{44} &= 0.333ml \\ m_{45} &= 0.166ml = m_{54} \\ m_{55} &= 0.111ml \\ m_{66} &= 1.0ml \end{aligned}$$

Since there is no coupling between the longitudinal displacement u_2 and the lateral displacement w_2, $m_{63} = m_{64} = m_{65} = 0$.

The generalized stiffness is found from the equation

$$k_{ij} = \int_0^l EI\phi_i''\phi_j''\,dx$$

Thus

$$k_{11} = EI\int_0^l \phi_1''\phi_1''\,dx = EI\int_0^l \left(\frac{2}{l^2}\right)^2 dx = 4\frac{EI}{l^3}$$

$$k_{12} = k_{21} = EI\int_0^l \left(\frac{2}{l^2}\right)\left(\frac{6x}{l^3}\right) dx = 6\frac{EI}{l^3}$$

$$k_{22} = 12\frac{EI}{l^3}$$

$$k_{55} = 28.8\frac{EI}{l^3}$$

All other k_{ij} are zero.

The results computed for m_{ij} and k_{ij} can now be arranged in the mass and stiffness matrices partitioned as follows:

$$[m] = ml\begin{bmatrix} 0.2000 & 0.1666 & 0 & 0 & 0 & 0 \\ 0.1666 & 0.1428 & 0 & 0 & 0 & 0 \\ 0 & 0 & 1.0000 & 0.5000 & 0.2000 & 0 \\ 0 & 0 & 0.5000 & 0.3333 & 0.1666 & 0 \\ 0 & 0 & 0.2000 & 0.1666 & 0.1111 & 0 \\ 0 & 0 & 0 & 0 & 0 & 1.0000 \end{bmatrix} \tag{9.5-5}$$

$$[k] = \frac{EI}{l^3}\begin{bmatrix} 4 & 6 & 0 & 0 & 0 & 0 \\ 6 & 12 & 0 & 0 & 0 & 0 \\ 0 & 0 & 0 & 0 & 0 & 0 \\ 0 & 0 & 0 & 0 & 0 & 0 \\ 0 & 0 & 0 & 0 & 28.8 & 0 \\ 0 & 0 & 0 & 0 & 0 & 0 \end{bmatrix} \tag{9.5-6}$$

where the upper left matrix refers to section ① and the remainder to section ②.

At the junction between sections ① and ② we have the following constraint equations:

$$\begin{aligned} &w_1(l) + u_2(l) = 0 && \text{or} \quad p_1 + p_2 + p_6 = 0 \\ &w_2(l) = 0 && \qquad p_3 + p_4 + p_5 = 0 \\ &w_1'(l) - w_2'(l) = 0 && \qquad 2p_1 + 3p_2 - p_4 - 4p_5 = 0 \\ &EI\left[w_1''(l) + w_2''(l)\right] = 0 && \qquad 2p_1 + 6p_2 + 12p_5 = 0 \end{aligned}$$

Arranged in matrix form, these are

$$\begin{bmatrix} 1 & 1 & 0 & 0 & 0 & 1 \\ 0 & 0 & 1 & 1 & 1 & 0 \\ 2 & 3 & 0 & -1 & -4 & 0 \\ 2 & 6 & 0 & 0 & 12 & 0 \end{bmatrix}\begin{Bmatrix} p_1 \\ p_2 \\ p_3 \\ p_4 \\ p_5 \\ p_6 \end{Bmatrix} = 0 \tag{9.5-7}$$

Since the total number of coordinates used are six and there are four constraint equations, the number of generalized coordinates for the system is two (i.e., there are four superfluous coordinates corresponding to the four constraint equations, see Sec. 7.1). We can thus choose any two of the coordinates to be the generalized coordinates q. Let $p_1 = q_1$ and $p_6 = q_6$ be the generalized coordinates and express $p_1, \ldots, p_6$ in terms of q_1 and q_6. This is accomplished in the following steps.

Rearrange Eq. (9.5-7) by shifting columns 1 and 6 to the right side

$$\begin{bmatrix} 1 & 0 & 0 & 0 \\ 0 & 1 & 1 & 1 \\ 3 & 0 & -1 & -4 \\ 6 & 0 & 0 & 12 \end{bmatrix}\begin{Bmatrix} p_2 \\ p_3 \\ p_4 \\ p_5 \end{Bmatrix} = \begin{bmatrix} -1 & -1 \\ 0 & 0 \\ -2 & 0 \\ -2 & 0 \end{bmatrix}\begin{Bmatrix} q_1 \\ q_6 \end{Bmatrix} \tag{9.5-8}$$

In abbreviated notation the preceding equation is

$$[s]\{p_{2-5}\} = [Q]\{q_{1,6}\}$$

Premultiply by $[s]^{-1}$ to obtain

$$\{p_{2-5}\} = [s]^{-1}[Q]\{q_{1,6}\}$$

Supply the identity $p_1 = q_1$ and $p_6 = q_6$ and write

$$\{p_{1-6}\} = [C]\{q_{1,6}\}$$

This constraint equation is now in terms of the generalized coordinates q_1 and q_6 as follows:

$$\begin{Bmatrix} p_1 \\ p_2 \\ p_3 \\ p_4 \\ p_5 \\ p_6 \end{Bmatrix} = \begin{bmatrix} 1 & 0 \\ -1 & -1 \\ 2 & 4.50 \\ -2.333 & -5.0 \\ 0.333 & 0.50 \\ 0 & 1 \end{bmatrix} \begin{Bmatrix} q_1 \\ q_6 \end{Bmatrix} = [C]\begin{Bmatrix} q_1 \\ q_6 \end{Bmatrix} \tag{9.5-9}$$

Returning to the Lagrange equation for the system, which is

$$ml[m]\{\ddot{p}\} + \frac{EI}{l^3}[k]\{p\} = 0 \tag{9.5-10}$$

substitute for $\{p\}$ in terms of $\{q\}$ from the constraint equation (9.5-9)

$$ml[m][C]\{\ddot{q}\} + \frac{EI}{l^3}[k][C]\{q\} = 0$$

Premultiply by the transpose $[C]'$

$$ml[C]'[m][C]\{\ddot{q}\} + \frac{EI}{l^3}[C]'[k][C]\{q\} = 0 \tag{9.5-11}$$

Comparing Eqs. (9.5-10) and (9.5-11), we note that in (9.5-10) the mass and stiffness matrices are 6×6 (see Eqs. 9.5-5 and 9.5-6), whereas the matrices $[C]'[m][C]$ and $[C]'[k][C]$ in Eq. (9.5-11) are 2×2. Thus we have reduced the size of the system from a 6×6 to a 2×2 problem.

Letting $\{\ddot{q}\} = -\omega^2\{q\}$, Eq. (9.5-11) is in the form

$$\left[-\omega^2 ml\begin{bmatrix} a_{11} & a_{12} \\ a_{21} & a_{22} \end{bmatrix} + \frac{EI}{l^3}\begin{bmatrix} b_{11} & b_{12} \\ b_{21} & b_{22} \end{bmatrix}\right]\begin{Bmatrix} q_1 \\ q_6 \end{Bmatrix} = 0 \tag{9.5-12}$$

The numerical values of the matrix $[a_{ij}]$ and $[b_{ij}]$ from Eqs. (9.5-5), (9.5-6), and (9.5-9) are

$$[a_{ij}] = [C]'[m][C] = \begin{bmatrix} 1.1774 & 2.6614 \\ 2.6614 & 7.3206 \end{bmatrix}$$

$$[b_{ij}] = [C]'[k][C] = \begin{bmatrix} 7.200 & 10.800 \\ 10.800 & 19.200 \end{bmatrix}$$

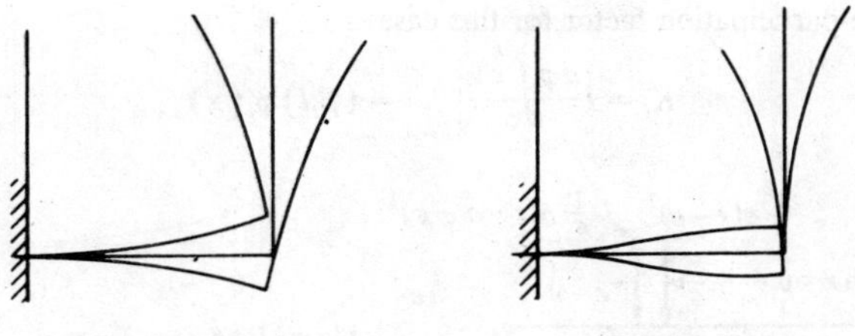

Figure 9.5-2. First and second mode shapes.

Using these numerical results, we find the two natural frequencies of the system from the characteristic equation of Eq. (9.5-12)

$$\omega_1 = 1.172\sqrt{\frac{EI}{ml^4}}$$

$$\omega_2 = 3.198\sqrt{\frac{EI}{Ml^4}}$$

Figure 9.5-2 shows the mode shapes corresponding to the above frequencies. Since Eq. (9.5-12) enables the solution of the eigenvectors only in terms of an arbitrary reference, q_6 can be solved with $q_1 = 1.0$. The coordinates p are then found from Eq. (9.5-9), and the mode shapes are obtained from Eqs. (9.5-1), (9.5-3), and (9.5-4).

PROBLEMS

9-1 Show that the dynamic load factor for a suddenly applied constant force reaches a maximum value of 2.0.

9-2 If a suddenly applied constant force is applied to a system for which the damping factor of the ith mode is $\zeta = c/c_{cr}$, show that the dynamic load factor is given approximately by the equation

$$D_i = 1 - e^{-\zeta\omega_i t}\cos\omega_i t$$

9-3 Determine the mode participation factor for a uniformly distributed force.

9-4 If a concentrated force acts at $x = a$, the loading per unit length corresponding to it can be represented by a delta function $l\,\delta(x - a)$. Show that the mode-participation factor then becomes $K_i = \varphi_i(a)$ and the deflection is expressible as

$$y(x,t) = \frac{P_0 l_3}{EI}\sum_i \frac{\varphi_i(a)\varphi_i(x)}{(\beta_i l)^4} D_i(t)$$

where $\omega_i^2 = (\beta_i l)^4 (EI/Ml^3)$ and $(\beta_i l)$ is the eigenvalue of the normal mode equation.

9-5 For a couple of moment M_0 acting at $x = a$, show that the loading $p(x)$ is the limiting case of two delta functions shown in Fig. P9-5 as $\epsilon \to 0$. Show also that the mode-participation factor for this case is

$$K_i = l\frac{d\varphi_i(x)}{dx}\bigg|_{x=a} = (\beta_i l)\,\varphi_i'(x)_{x=a}$$

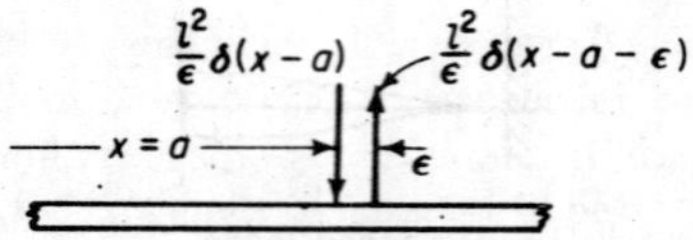

Figure P9-5.

9-6 A concentrated force $P_0 f(t)$ is applied to the center of a simply supported uniform beam, as shown in Fig. P9-6. Show that the deflection is given by

$$y(x,t) = \frac{P_0 l^3}{EI}\sum_i \frac{K_i\varphi_i(x)}{(\beta_i l)^4} D_i$$

$$= \frac{2P_0 l^3}{EI}\left\{\frac{\sin \pi \frac{x}{l}}{\pi^4} D_1(t) - \frac{\sin 3\pi\frac{x}{l}}{(3\pi)^4} D_3(t) + \frac{\sin 5\pi\frac{x}{l}}{(5\pi)^4} D_5(t) - \cdots\right\}$$

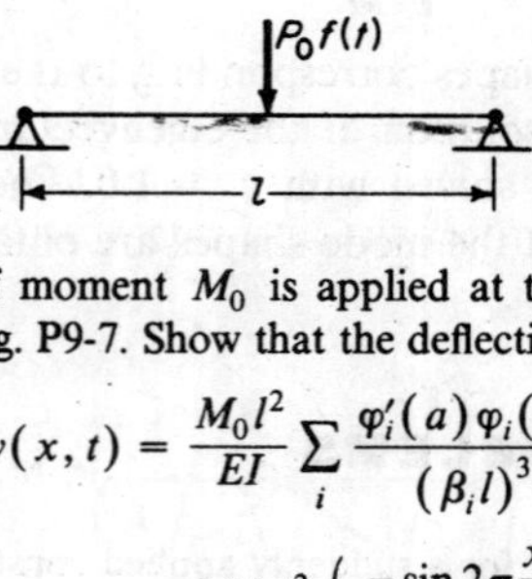

Figure P9-6.

9-7 A couple of moment M_0 is applied at the center of the beam of Prob. 9-6, as shown in Fig. P9-7. Show that the deflection at any point is given by the equation

$$y(x,t) = \frac{M_0 l^2}{EI}\sum_i \frac{\varphi_i'(a)\varphi_i(x)}{(\beta_i l)^3} D_i(t)$$

$$= \frac{2M_0 l^2}{EI}\left\{\frac{-\sin 2\pi\frac{x}{l}}{(2\pi)^3} D_2(t) + \frac{\sin 4\pi\frac{x}{l}}{(4\pi)^3} D_4(t) \right.$$

$$\left. \frac{\sin 6\pi\frac{x}{l}}{(6\pi)^3} D_6(t) + \cdots\right\}$$

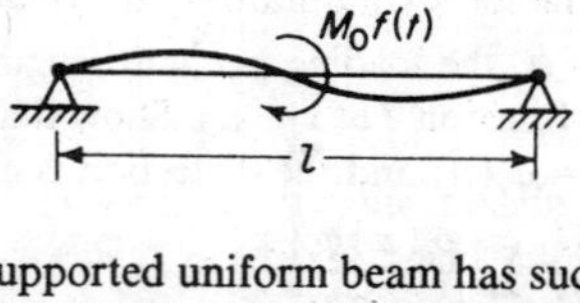

Figure P9-7.

9-8 A simply supported uniform beam has suddenly applied to it the load distribution shown in Fig. P9-8, where the time variation is a step function. Determine the

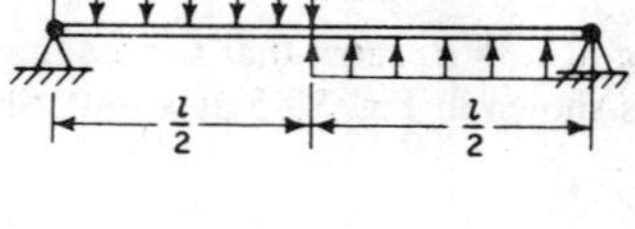

Figure P9-8.

response $y(x, t)$ in terms of the normal modes of the beam. Indicate what modes are absent and write down the first two existing modes.

9-9 A slender rod of length l, free at $x = 0$ and fixed at $x = l$, is struck longitudinally by a time-varying force concentrated at the end $x = 0$. Show that all modes are equally excited (i.e., that the mode-participation factor is independent of the mode number), the complete solution being

$$u(x, t) = \frac{2F_0 l}{AE}\left\{\frac{\cos\frac{\pi}{2}\frac{x}{l}}{\left(\frac{\pi}{2}\right)^2}D_1(t) + \frac{\cos\frac{3\pi}{2}\frac{x}{l}}{\left(\frac{3\pi}{2}\right)^2}D_3(t) + \cdots\right\}$$

9-10 If the force of Prob. 9-9 is concentrated at $x = l/3$, determine which modes will be absent in the solution.

9-11 In Prob. 9-10, determine the participation factor of the modes present and obtain a complete solution for an arbitrary time variation of the applied force.

9-12 Consider a uniform beam of mass M and length l supported on equal springs of total stiffness k, as shown in Fig. P9-12a. Assume the deflection to be

$$y(x, t) = \varphi_1(x)\,q_1(t) + \varphi_2(x)\,q_2(t)$$

and choose $\varphi_1 = \sin\frac{\pi x}{l}$ and $\varphi_2 = 1.0$.

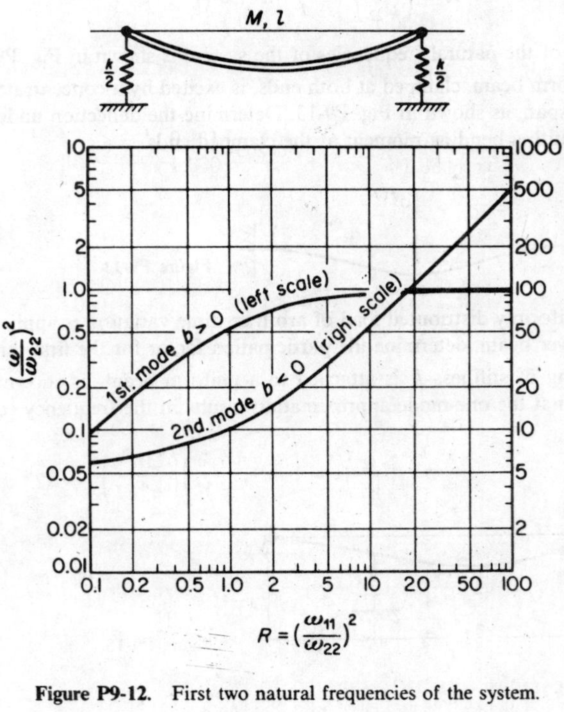

Figure P9-12. First two natural frequencies of the system.

Using Lagrange's equation, show that

$$\ddot{q}_1 + \frac{4}{\pi}\ddot{q}_2 + \omega_{11}^2 q_1 = 0$$

$$\frac{2}{\pi}\ddot{q}_1 + \ddot{q}_2 + \omega_{22}^2 q_2 = 0$$

where

$$\omega_{11}^2 = \pi^4 (EI/Ml^3) = \text{natural frequency of beam on rigid supports}$$

$$\omega_{22}^2 = k/M = \text{natural frequency of rigid beam on springs}$$

Solve these equations and show that

$$\omega^2 = \omega_{22}^2 \frac{\pi^2}{2}\left\{\frac{(R+1) \pm \sqrt{(R-1)^2 + \frac{32}{\pi^2}R}}{\pi^2 - 8}\right\}$$

Let $y(x, t) = (b + sin\frac{\pi x}{l})q$ and use Rayleigh's method to obtain

$$\frac{q_2}{q_1} = b = \frac{\pi}{8}\left\{(R-1) \mp \sqrt{(R-1)^2 + \frac{32}{\pi^2}R}\right\}$$

$$R = \left(\frac{\omega_{11}}{\omega_{22}}\right)^2$$

A plot of the natural frequencies of the system is shown in Fig. P9-12b.

9-13 A uniform beam, clamped at both ends, is excited by a concentrated force $P_0 f(t)$ at midspan, as shown in Fig. P9-13. Determine the deflection under the load and the resulting bending moment at the clamped ends.

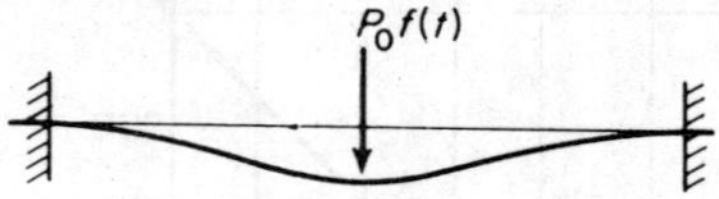

Figure P9-13.

9-14 If a uniformly distributed load of arbitrary time variation is applied to a uniform cantilever beam, determine the participation factor for the first three modes.

9-15 A spring of stiffness k is attached to a uniform beam, as shown in Fig. P9-15. Show that the one-mode approximation results in the frequency equation

$$\left(\frac{\omega}{\omega_1}\right)^2 = 1 + 1.5\left(\frac{k}{M}\right)\left(\frac{Ml^3}{\pi^4 EI}\right)$$

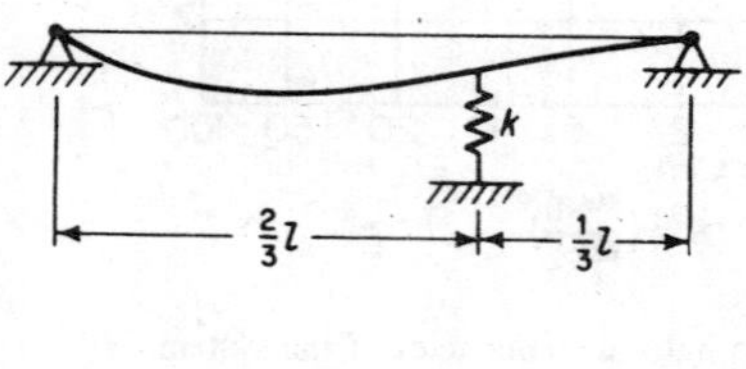

Figure P9-15.

where

$$\omega_1^2 = \frac{\pi^4 EI}{Ml^3}$$

9-16 Write the equations for the two-mode approximation of Prob. 9-15.

9-17 Repeat Prob. 9-16, using the mode acceleration method.

9-18 Show that for the problem of a spring attached to any point $x = a$ of a beam, both the constrained-mode and the mode-acceleration methods result in the same equation when only one mode is used, this equation being

$$\left(\frac{\omega}{\omega_1}\right)^2 = 1 + \frac{k}{M\omega_1^2}\varphi_1^2(a)$$

9-19 The beam shown in Fig. P9-19 has a spring of rotational stiffness K lb in./rad at the left end. Using two modes in Eq. (9.3-8), determine the fundamental frequency of the system as a function of $K/M\omega_1^2$ where ω_1 is the fundamental frequency of the simply supported beam.

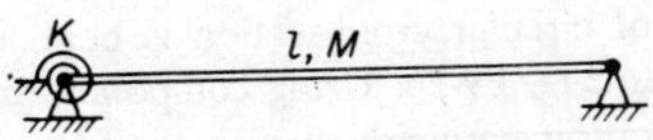

Figure P9-19.

9-20 If both ends of the beam of Fig. P9-19 are restrained by springs of stiffness K, determine the fundamental frequency. As K approaches infinity, the result should approach that of the clamped ended beam.

9-21 An airplane is idealized to a simplified model of a uniform beam of length l and mass per unit length m with a lumped mass M_0 at its center, as shown in Fig. P9-21. Using the translation of M_0 as one of the generalized coordinates, write the equations of motion and establish the natural frequency of the symmetric mode. Use first cantilever mode for the wing.

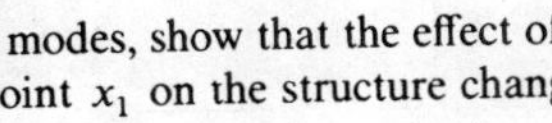

Figure P9-21.

9-22 For the system of Prob. 9-21, determine the antisymmetric mode by using the rotation of the fuselage as one of the generalized coordinates.

9-23 If wing tip tanks of mass M_1 are added to the system of Prob. 9-21, determine the new frequency.

9-24 Using the method of constrained modes, show that the effect of adding a mass m_1 with moment of inertia J_1 to a point x_1 on the structure changes the first natural frequency ω_1 to

$$\omega_1' = \frac{\omega_1}{\sqrt{1 + \dfrac{m_1}{M_1}\varphi_1^2(x_1) + \dfrac{J_1}{M_1}\varphi_1'^2(x_1)}}$$

and the generalized mass and damping to

$$M_1' = M_1\left\{1 + \frac{m_1\varphi_1^2(x_1)}{M_1} + \frac{J_1}{M_1}\varphi_1'^2(x)\right\}$$

$$\zeta_1' = \frac{\zeta_1}{\sqrt{1 + \frac{m_1}{M_1}\varphi_1^2(x_1) + \frac{J_1}{M_1}\varphi_1'^2(x_1)}}$$

where a one-mode approximation is used for the inertia forces.

9-25 Formulate the vibration problem of the bent shown in Fig. P9-25 by the component mode synthesis. Assume the corners to remain at 90°.

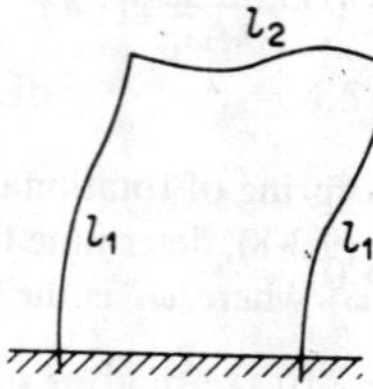

Figure P9-25.

9-26 A rod of circular cross-section is bent at right angles in a horizontal plane as shown in Fig. P9-26. Using component mode synthesis, set up the equations for the vibration perpendicular to the plane of the rod. Note that member 1 is in flexure and torsion. Assume its bending only in the vertical plane.

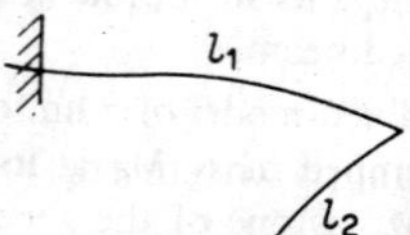

Figure P9-26.

10

INTRODUCTION TO THE FINITE ELEMENT METHOD

In Chapter 6 we were able to determine the stiffness matrix of simple frame structures by considering the structure as an assemblage of structural elements. With the forces and moments at the ends of the elements known from structural theory, the joints between the elements were matched for compatibility of displacements and the forces and moments at the joints were established by imposing the condition of equilibrium.

In the finite element method, the same procedure is followed but in a more systematic way for computer calculation. Although structures with few elements can be analyzed simply by the method outlined in Chapter 6, the "bookkeeping" for a large structure of many elements would soon overcome the patience of the analyst. In the finite element method, element coordinates and forces are transformed into global coordinates and the stiffness matrix of the entire structure is presented in a global system of common orientation.

This chapter introduces the student to the basic ideas of the finite element method and also includes the development of the corresponding mass matrix to complete the equations of motion for the dynamic problem. Only structural elements of two joints (axial and beam elements) are discussed here. For the treatment of plates and shells, the reader is referred to other texts devoted to a more extensive discussion of the subject.

10.1 ELEMENT STIFFNESS AND MASS

Axial element. An element with pinned ends can support only axial forces and hence will act like a spring. Figure 10.1-1 shows a spring and a uniform rod pinned to a stationary wall and subjected to a force F. The

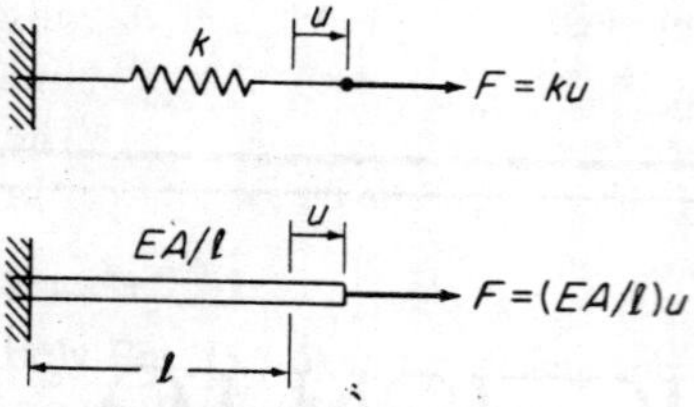

Figure 10.1-1.

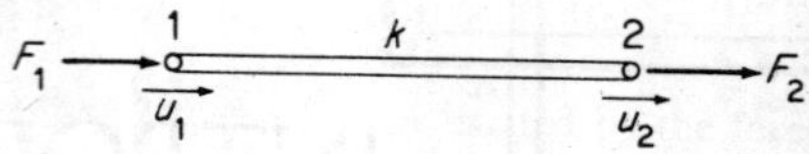

Figure 10.1-2.

force-displacement relationships for the two cases are simply

$$\text{Spring} \qquad f = ku$$

$$\text{Uniform rod} \qquad F = \left(\frac{EA}{l}\right)u \tag{10.1-1}$$

In general, these axial elements may be a part of a pin-connected structure that allows displacement of both ends. In the finite element method the displacement and force at each end of the element must be accounted for with proper sign. Figure 10.1-2 shows an axial element labeled with displacements u_1, u_2, and forces F_1, F_2, all in the positive sense. If we write the force-displacement relationship in terms of the stiffness matrix, the equation is

$$\begin{Bmatrix} F_1 \\ F_2 \end{Bmatrix} = \begin{bmatrix} k_{11} & k_{12} \\ k_{21} & k_{22} \end{bmatrix} \begin{Bmatrix} u_1 \\ u_2 \end{Bmatrix} \tag{10.1-2}$$

The elements of the first column of the stiffness matrix represent the forces at the two ends when $u_1 = 1$ and $u_2 = 0$, as shown in Fig. 10.1-3. Thus $F_1 = ku_1$ and $F_2 = -ku_1$.

Similarly, by letting $u_2 = 1$ and $u_1 = 0$, we obtain, as in Fig. 10.1-4, $F_1 = -ku_2$ and $F_2 = ku_2$. Thus Eq. 10.1-2 becomes

$$\begin{Bmatrix} F_1 \\ F_2 \end{Bmatrix} = k \begin{bmatrix} 1 & -1 \\ -1 & 1 \end{bmatrix} \begin{Bmatrix} u_1 \\ u_2 \end{Bmatrix} \tag{10.1-2'}$$

If the spring is replaced by a uniform rod, $k = AE/l$ and the equation becomes

$$\begin{Bmatrix} F_1 \\ F_2 \end{Bmatrix} = \frac{EA}{l} \begin{bmatrix} 1 & -1 \\ -1 & 1 \end{bmatrix} \begin{Bmatrix} u_1 \\ u_2 \end{Bmatrix} \tag{10.1-3}$$

These equations thus define the stiffness matrix for axial elements in terms of axial coordinates u_i and axial forces F_i, regardless of the orientation of the axial member.

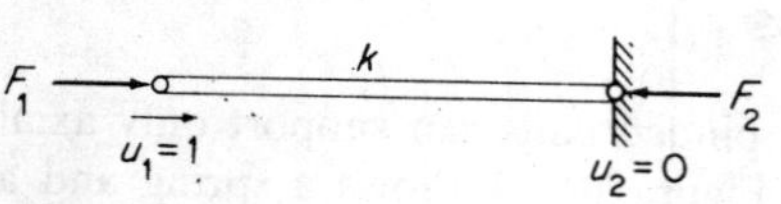

Figure 10.1-3.

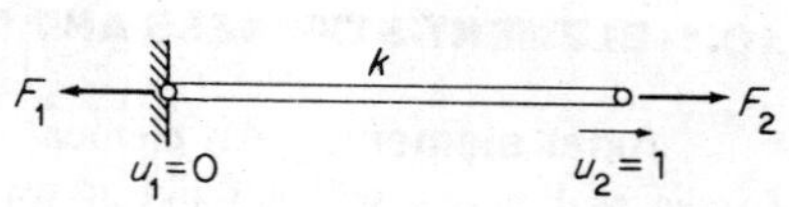

Figure 10.1-4.

Mode shape and mass matrix for axial element. With the two ends of the axial member displaced by u_1 and u_2, the displacement at any point $\xi = x/l$ is assumed to be a straight line, as shown in Fig. 10.1-5(a). The displacement is therefore the superposition of the two mode shapes shown in Fig. 10.1-5(b). The normalized mode shapes are then

$$\varphi_1 = (1 - \xi) \quad \text{and} \quad \varphi_2 = \xi \tag{10.1-4}$$

The mass matrix is found by expressing u as the sum of the two mode shapes,

$$u = (1 - \xi)u_1 + \xi u_2 \tag{10.1-5}$$

and writing the equation for the kinetic energy. We here assume uniform mass distribution m per unit length.

$$\begin{aligned} T &= \frac{1}{2}\int_0^l \dot{u}^2 m\,dx = \frac{1}{2}m\int_0^l [(1-\xi)\dot{u}_1 + \xi\dot{u}_2]^2 l\,d\xi \\ &= \frac{1}{2}ml\left(\frac{1}{3}\dot{u}_1^2 + \frac{1}{3}\dot{u}_1\dot{u}_2 + \frac{1}{3}\dot{u}_2^2\right) \end{aligned} \tag{10.1-6}$$

Since the generalized mass from Lagrange's equation is

$$\frac{d}{dt}\frac{\partial T}{\partial \dot{u}_i}$$

we find

$$\frac{d}{dt}\frac{\partial T}{\partial \dot{u}_1} = ml\left(\frac{1}{3}\ddot{u}_1 + \frac{1}{6}\ddot{u}_2\right)$$

$$\frac{d}{dt}\frac{\partial T}{\partial \dot{u}_2} = ml\left(\frac{1}{6}\ddot{u}_1 + \frac{1}{3}\ddot{u}_2\right)$$

which establishes the mass matrix for the axial element as

$$\frac{ml}{6}\begin{bmatrix} 2 & 1 \\ 1 & 2 \end{bmatrix} \tag{10.1-7}$$

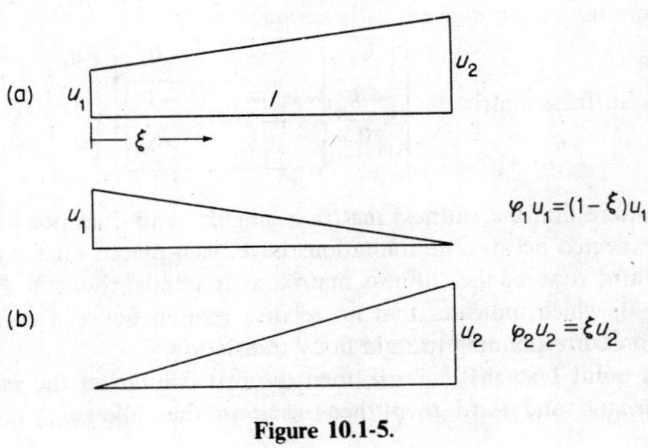

Figure 10.1-5.

The terms of the mass matrix are also available from

$$m_{ij} = \int \varphi_i \varphi_j \, dm$$

Example 10.1-1

Determine the equation of motion for the longitudinal vibration of the two-section bar shown in Fig. 10.1-6.

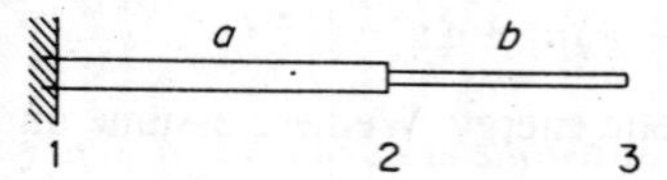

Figure 10.1-6.

Solution: Numbering the joints as 1, 2, and 3, we have two axial elements, 1–2, and 2–3, with displacements u_1, u_2, and u_3. Although u_1 is zero, we at first allow it to be unrestrained and later impose its zero value.

The element mass and stiffness terms from Eq. (10.1-7) and Eq. (10.1-3) are as follows:

$$\text{Element } a. \quad \frac{M_a}{6}\begin{bmatrix} 2 & 1 \\ 1 & 2 \end{bmatrix} \qquad k_a\begin{bmatrix} 1 & -1 \\ -1 & 1 \end{bmatrix}$$

$$\text{Element } b. \quad \frac{M_b}{6}\begin{bmatrix} 2 & 1 \\ 1 & 2 \end{bmatrix} \qquad k_b\begin{bmatrix} 1 & -1 \\ -1 & 1 \end{bmatrix}$$

where $k_a = EA_a/l_a$, $k_b = EA_b/l_b$, $M_a = m_a l_a$, and $M_b = m_b l_b$.

The element matrices have a common coordinate u_2, and by superimposing them, they can be assembled into a 3×3 matrix as follows:

$$\text{Mass Matrix} \quad \frac{1}{6}\begin{bmatrix} 2M_a & M_a & 0 \\ M_a & 2M_a + 2M_b & M_b \\ 0 & M_b & 2M_b \end{bmatrix}\begin{Bmatrix} \ddot{u}_1 \\ \ddot{u}_2 \\ \ddot{u}_3 \end{Bmatrix}$$

$$\text{Stiffness matrix} \quad \begin{bmatrix} k_a & -k_a & 0 \\ -k_a & k_a + k_b & -k_b \\ 0 & -k_b & k_b \end{bmatrix}\begin{Bmatrix} u_1 \\ u_2 \\ u_3 \end{Bmatrix}$$

We note here that the stiffness matrix is singular and does not have an inverse. This is to be expected because no limitations have been placed on the displacements. The first and third rows of the stiffness matrix, as it stands, result in $k_a(u_1 - u_2) = k_b(u_2 - u_3) = 0$, which indicates that no relative motion between coordinates takes place, a situation corresponding to rigid body translation.

If we fix point 1 so that $u_1 = 0$, then the first column of the matrices can be deleted. The second and third rows then result in the following equation for the longitudinal vibration of the two-section bar:

$$\frac{1}{6}\begin{bmatrix} 2(M_a + M_b) & M_b \\ M_b & 2M_b \end{bmatrix}\begin{Bmatrix} \ddot{u}_2 \\ \ddot{u}_3 \end{Bmatrix} + \begin{bmatrix} (k_a + k_b) & -k_b \\ -k_b & k_b \end{bmatrix}\begin{Bmatrix} u_2 \\ u_3 \end{Bmatrix} = \begin{Bmatrix} 0 \\ 0 \end{Bmatrix}$$

Special Case. If $A_a = A_b = A$, $l_a = l_b = \frac{1}{2}L$, and $M_a = M_b = \frac{1}{2}M$, the previous problem becomes that of a uniform bar of total length L and total mass M, solved as a two-DOF system with coordinates at the midpoint and the free end. The equation of the previous problem then becomes

$$\frac{M}{12}\begin{bmatrix} 4 & 1 \\ 1 & 2 \end{bmatrix}\begin{Bmatrix} \ddot{u}_2 \\ \ddot{u}_3 \end{Bmatrix} + \frac{2EA}{L}\begin{bmatrix} 2 & -1 \\ -1 & 1 \end{bmatrix}\begin{Bmatrix} u_2 \\ u_3 \end{Bmatrix} = \begin{Bmatrix} 0 \\ 0 \end{Bmatrix}$$

If we let $\lambda = \omega^2 ML/24EA$, the characteristic equation for the natural frequencies becomes

$$\begin{vmatrix} (2 - 4\lambda) & -(1 + \lambda) \\ -(1 + \lambda) & (1 - 2\lambda) \end{vmatrix} = 0$$

or

$$\lambda^2 - \tfrac{10}{7}\lambda - \tfrac{1}{7} = 0$$

Its solution results in

$$\lambda = \begin{cases} 0.1082 \\ 1.3204 \end{cases} \qquad \omega = \begin{cases} 1.6115\sqrt{\dfrac{EA}{ML}} \\ 5.6293\sqrt{\dfrac{EA}{ML}} \end{cases}$$

The natural frequencies of the uniform bar in longitudinal vibration are known and are given by the equation $\omega_i = (2n + 1)(\pi/2)\sqrt{EA/ML}$. Results computed from this equation for the first two modes are

$$\omega = \begin{cases} 1.5708\sqrt{\dfrac{EA}{ML}} \\ 4.7124\sqrt{\dfrac{EA}{ML}} \end{cases}$$

Comparison between the two indicates that the agreement between the results of the two DOF finite element model and that of the continuous model is 2.6% high for the first mode and 19.5% high for the second mode. A three element model will of course be expected to give closer agreement.

Variable properties. One simple approach to problems of variable properties is to use many elements of short length. The variation of mass or stiffness over each element is then small and can be neglected. The problem then becomes one of constant mass and stiffness for each element which simplifies the problem considerably since these terms can be placed outside of the integrals. Of course the larger numbers of elements will lead to equations of larger DOF.

10.2 STIFFNESS AND MASS FOR THE BEAM ELEMENT

Beam stiffness. If the ends of the element are rigidly connected to the adjoining structure instead of being pinned, the element will act like a beam with moments and lateral forces acting at the joints. In general, the relative axial displacement $u_2 - u_1$ will be small compared to the lateral displacement v of the beam and can be assumed to be zero. When axial forces as well as beam forces and moments must be considered, it is a simple matter to make additions to the beam stiffness matrix, as we show later.

The local coordinates for the beam element are the lateral displacements and rotations at the two ends. We consider only the planar structure in this discussion, so that each joint will have a lateral displacement v and a rotation θ, resulting in four coordinates, v_1, θ_1 and v_2, θ_2. The positive sense of these coordinates is arbitrary, but for computer bookkeeping purposes the diagram of Fig. 10.2-1 is the one accepted by most structural engineers. Positive senses of the forces and moments also follow the same diagram.

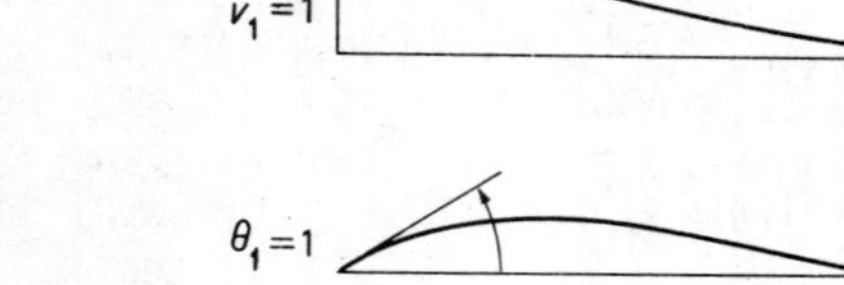

Figure 10.2-1. Positive sense of beam displacements and forces.

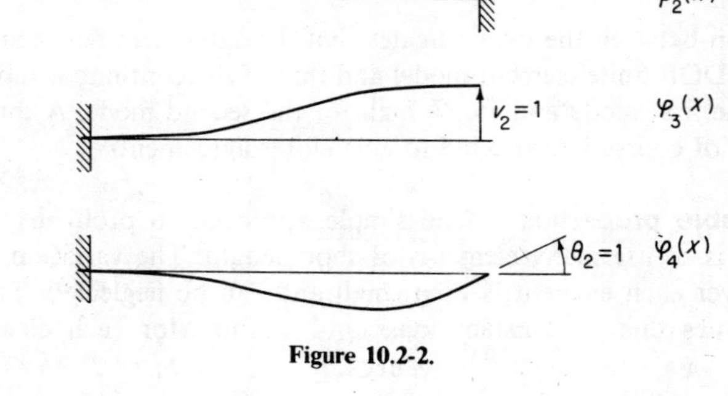

Figure 10.2-2.

The preceding displacements can be considered to be the superposition of the four shapes, labeled $\varphi_1(x)$, $\varphi_2(x)$, $\varphi_3(x)$, and $\varphi_4(x)$, shown in Fig. 10.2-2. The forces and moments required at the two ends were found in Chapter 6 and are shown in Fig. 10.2-3 with the factor EI/l^3 omitted. The diagram immediately leads to the force-stiffness equation

$$\begin{Bmatrix} F_1 \\ M_1 \\ F_2 \\ M_2 \end{Bmatrix} = \frac{EI}{l^3}\left[\begin{array}{cc|cc} 12 & 6l & -12 & 6l \\ 6l & 4l^2 & -6l & 2l^2 \\ \hline -12 & -6l & 12 & -6l \\ 6l & 2l^2 & -6l & 4l^2 \end{array}\right]\begin{Bmatrix} v_1 \\ \theta_1 \\ v_2 \\ \theta_2 \end{Bmatrix} \tag{10.2-1}$$

Equation (10.2-1) for the stiffness was obtained from the given forces and moments shown in Fig. 10.2-3. The stiffness matrix as well as the mass matrix can also be determined from the potential and kinetic energy, provided the shape functions $\varphi_i(x)$ of the beam are known.

For the development of the general equation of the beam, which is a cubic polynomial, the deflection is expressed in the form

$$v(x) = p_1 + p_2\xi + p_3\xi^2 + p_4\xi^3 \tag{10.2-2}$$

where

$$\xi = \frac{x}{l} \quad \text{and} \quad p_i = \text{constants}$$

Differentiating yields the slope equation

$$l\theta(x) = p_2 + 2p_3\xi + 3p_4\xi^2 \tag{10.2-3}$$

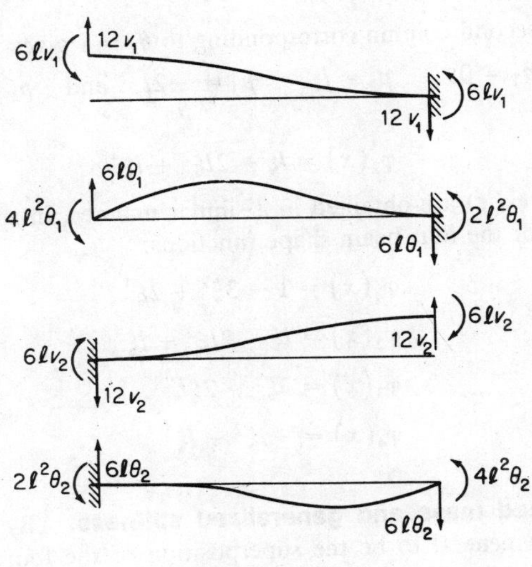

Figure 10.2-3.

If we apply the boundary conditions $\xi = 0$ and $\xi = 1$, the boundary equations can be expressed by the following matrix equation:

$$\begin{Bmatrix} v_1 \\ l\theta_1 \\ \hline v_2 \\ l\theta_2 \end{Bmatrix} = \left[\begin{array}{cc|cc} 1 & 0 & 0 & 0 \\ 0 & 1 & 0 & 0 \\ \hline 1 & 1 & 1 & 1 \\ 0 & 1 & 2 & 3 \end{array}\right] \begin{Bmatrix} p_1 \\ p_2 \\ \hline p_3 \\ p_4 \end{Bmatrix} \tag{10.2-4}$$

With the matrix partitioned as shown, it is evident that p_1 and p_2 are related to v_1 and $l\theta_1$ by a unit matrix. After substituting $p_1 = v_1$ and $p_2 = l\theta_1$, we can easily solve the last two rows of the matrix for p_3 and p_4. The desired inverse of Eq. (10.2-4) then becomes

$$\begin{Bmatrix} p_1 \\ p_2 \\ \hline p_3 \\ p_4 \end{Bmatrix} = \left[\begin{array}{cc|cc} 1 & 0 & 0 & 0 \\ 0 & 1 & 0 & 0 \\ \hline -3 & -2 & 3 & -1 \\ 2 & 1 & -2 & 1 \end{array}\right] \begin{Bmatrix} v_1 \\ l\theta_1 \\ \hline v_2 \\ l\theta_2 \end{Bmatrix} \tag{10.2-5}$$

This equation enables the determination of the p_i for each of the displacements equated to unity with all the others equal to zero. That is, for $v_1(x) = 1$ with all other displacements equal to zero, the first column of Eq. (10.2-5) gives

$$p_1 = 1, \qquad p_2 = 0, \qquad p_3 = -3, \quad \text{and} \quad p_4 = 2$$

Substituting these into Eq. (10.2-2) gives the shape function for the first configuration of Fig. 10.2-2 of

$$\varphi_1(x) = 1 - 3\xi^2 + 2\xi^3$$

Similarly, the second column corresponding to $\theta_1 = 1$ gives

$$p_1 = 0, \qquad p_2 = l, \qquad p_3 = -2l, \quad \text{and} \quad p_4 = l$$

and

$$\varphi_2(x) = l\xi - 2l\xi^2 + l\xi^3$$

The other two $\varphi_i(x)$ are obtained in a similar manner. In summary, we have the following for the four beam shape functions:

$$\begin{aligned} \varphi_1(x) &= 1 - 3\xi^2 + 2\xi^3 \\ \varphi_2(x) &= l\xi - 2l\xi^2 + l\xi^3 \\ \varphi_3(x) &= 3\xi^2 - 2\xi^3 \\ \varphi_4(x) &= -l\xi^2 + l\xi^3 \end{aligned} \tag{10.2-6}$$

Generalized mass and generalized stiffness. By considering the displacement in general to be the superposition of the four shape functions shown in Fig. (10.2-2), we have,

$$\begin{aligned} y(x) &= \varphi_1 v_1 + \varphi_2 \theta_1 + \varphi_3 v_2 + \varphi_4 \theta_2 \\ &= \varphi_1 q_1 + \varphi_2 q_2 + \varphi_3 q_3 + \varphi_4 q_4 \end{aligned} \tag{10.2-7}$$

where q_i have been substituted for the end displacements.

To determine the generalized mass, the preceding equation is substituted into the equation for the kinetic energy.

$$\begin{aligned} T &= \frac{1}{2}\int \dot{y}^2 m\,dx = \frac{1}{2}\sum_i \sum_j \dot{q}_i \dot{q}_j \int \varphi_i \varphi_j m\,dx \\ &= \frac{1}{2}\sum_i \sum_j m_{ij} \dot{q}_i \dot{q}_j \end{aligned} \tag{10.2-8}$$

Thus the generalized mass m_{ij}, which forms the elements of the mass matrix, is equal to

$$m_{ij} = \int_0^l \varphi_i \varphi_j m\,dx \tag{10.2-9}$$

Substituting the four beam functions into Eq. (10.2-9), the mass matrix for the uniform beam element is expressed in terms of the end displacements:

$$\frac{ml}{420}\left[\begin{array}{cc|cc} 156 & 22l & 54 & -13l \\ 22l & 4l^2 & 13l & -3l^2 \\ \hline 54 & 13l & 156 & -22l \\ -13l & -3l^2 & -22l & 4l^2 \end{array}\right] \tag{10.2-10}$$

The matrix is called *consistent mass*, since it is based on the same beam functions used for the stiffness matrix.†

10.3 TRANSFORMATION OF COORDINATES (GLOBAL COORDINATES)

In determining the stiffness matrix of the entire structure in terms of local elements, it is necessary first to match the displacements of the adjacent elements to ensure compatibility. In Chapter 6 this was done by examination of each joint, taking account of the orientations of the adjoining members at each joint.

In the finite element method this requirement for displacement compatibility is simplified by resolving the element displacements and forces into a common coordinate system known as the *global coordinates*.

Consider again a planar structure and examine a local element ①, ② at an angle α with the global coordinates $\bar{u}, \bar{v}$, which will be assumed in the horizontal and vertical directions, as shown in Fig. (10.3-1a).

†J. S. Archer, "Consistent Mass Matrix for Distributed Mass Systems," *Jour. Struct. Div. ASCE*, Vol. 89, No. STA4, (August 1963), pp. 161–178.

The displacement $\mathbf{r}_1$ of joint ① to ①′ must be the same in both the local and global coordinates. This requirement can be expressed by the equation

$$\mathbf{r}_1 = u_1\mathbf{i} + v_1\mathbf{j} = \bar{u}_1\bar{\mathbf{i}} + \bar{v}_1\bar{\mathbf{j}}$$

where $\mathbf{i}, \mathbf{j}$, and $\bar{\mathbf{i}}, \bar{\mathbf{j}}$ are unit vectors for the two coordinate systems. Forming the dot product of the preceding equation with $\mathbf{i}$, we obtain

$$u_1(\mathbf{i} \cdot \mathbf{i}) + v_1(\mathbf{j} \cdot \mathbf{i}) = \bar{u}_1(\bar{\mathbf{i}} \cdot \mathbf{i}) + \bar{v}_1(\bar{\mathbf{j}} \cdot \mathbf{i})$$

or

$$u_1 + 0 = \bar{u}_1 \cos\alpha + \bar{v}_1 \sin\alpha$$

Next, taking the dot product with $\mathbf{j}$, we obtain

$$0 + v_1 = -\bar{u}_1 \sin\alpha + \bar{v}_1 \cos\alpha$$

Thus we can express these results by the matrix equation

$$\begin{Bmatrix} u_1 \\ v_1 \end{Bmatrix} = \begin{bmatrix} \cos\alpha & \sin\alpha \\ -\sin\alpha & \cos\alpha \end{bmatrix} \begin{Bmatrix} \bar{u}_1 \\ \bar{v}_1 \end{Bmatrix} \tag{10.3-1}$$

The preceding equation expresses the local coordinates u_1, v_1 in terms of the global coordinates $\bar{u}_1, \bar{v}_1$. These results are readily confirmed geometrically from Fig. 10.3-1(b).

Similarly, the displacement at joint ② in local coordinates can be expressed in terms of the global coordinates by the same transformation equation. The rotation angle for the two coordinate systems must of course be the same, so that $\theta = \bar{\theta}$. We can then include θ in the transformation matrix as

$$\begin{Bmatrix} u \\ v \\ \theta \end{Bmatrix} = \begin{bmatrix} \cos\alpha & \sin\alpha & 0 \\ -\sin\alpha & \cos\alpha & 0 \\ 0 & 0 & 1 \end{bmatrix} \begin{Bmatrix} \bar{u} \\ \bar{v} \\ \bar{\theta} \end{Bmatrix}_i \qquad i = 1, 2 \tag{10.3-2}$$

Thus the transformation matrix for any element making an angle α measured

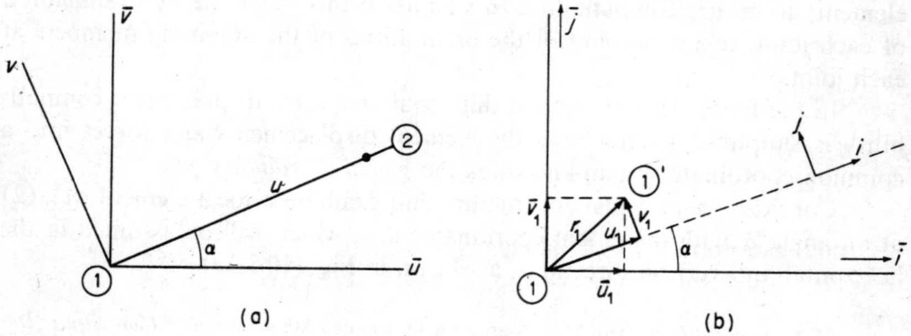

Figure 10.3-1.

counterclockwise from the horizontal is

$$\begin{Bmatrix} u_1 \\ v_1 \\ \theta_1 \\ \hline u_2 \\ v_2 \\ \theta_2 \end{Bmatrix} = \left[\begin{array}{ccc|ccc} c & s & 0 & & & \\ -s & c & 0 & & 0 & \\ 0 & 0 & 1 & & & \\ \hline & & & c & s & 0 \\ & 0 & & -s & c & 0 \\ & & & 0 & 0 & 1 \end{array}\right] \begin{Bmatrix} \bar{u}_1 \\ \bar{v}_1 \\ \theta_1 \\ \hline \bar{u}_2 \\ \bar{v}_2 \\ \theta_2 \end{Bmatrix} \tag{10.3-3}$$

where $c = \cos \alpha$ and $s = \sin \alpha$. It is easily seen that the transformation matrix developed for displacements also applies for the force vector.

In shorter notation we can rewrite the transformation equations from local to global coordinates as

$$\begin{aligned} r &= T\bar{r} \\ F &= T\bar{F} \end{aligned} \tag{10.3-4}$$

where T is the transformation matrix and r, F and $\bar{r}$, $\bar{F}$ are the displacement and force in the local and global coordinates, respectively. We add to this the relationship between r and F, which is the stiffness matrix

$$F = kr \tag{10.3-5}$$

and which we wish to write in the global system as $\bar{F} = \bar{k}\bar{r}$. From Eq. (10.3-4), we have

$$\bar{F} = T^{-1}F = T'F \tag{10.3-6}$$

Here we have taken note of the fact that transformation matrices are orthogonal matrices and $T^{-1} = T'$.† Substituting for F from the stiffness equation and replacing r in terms of $\bar{r}$, we obtain

$$\begin{aligned} \bar{F} &= T'kr \\ &= T'kT\bar{r} = \bar{k}\bar{r} \end{aligned} \tag{10.3-7}$$

Thus k for the local coordinates is transformed to $\bar{k}$ for the global coordinates by the equation

$$\bar{k} = T'kT \tag{10.3-8}$$

10.4 ELEMENT STIFFNESS AND ELEMENT MASS IN GLOBAL COORDINATES

Axial element. For axial elements the end moments are zero and the end forces and displacements are colinear with the element length. Thus for systems involving only axial elements, the 6 × 6 transformation matrix re-

†See Appendix C.

duces to the following 4 × 4 matrix.

$$T = \left[\begin{array}{cc|cc} c & s & 0 & \\ -s & c & & \\ \hline 0 & & c & s \\ & & -s & c \end{array}\right] \tag{10.4-1}$$

We note here that the stiffness and mass matrices for the axial element are of order 2 × 2 and therefore must be rewritten as a 4 × 4 matrix as follows:

$$\begin{Bmatrix} F_1 \\ F_2 \end{Bmatrix} = \frac{EA}{l}\begin{bmatrix} 1 & -1 \\ -1 & 1 \end{bmatrix}\begin{Bmatrix} u_1 \\ u_2 \end{Bmatrix} = \frac{EA}{l}\left[\begin{array}{cc|cc} 1 & 0 & -1 & 0 \\ 0 & 0 & 0 & 0 \\ \hline -1 & 0 & 1 & 0 \\ 0 & 0 & 0 & 0 \end{array}\right]\begin{Bmatrix} u_1 \\ v_1 \\ u_2 \\ v_2 \end{Bmatrix} \tag{10.4-2}$$

$$\begin{Bmatrix} F_1 \\ F_2 \end{Bmatrix} = \frac{ml}{6}\begin{bmatrix} 2 & 1 \\ 1 & 2 \end{bmatrix}\begin{Bmatrix} \ddot{u}_1 \\ \ddot{u}_2 \end{Bmatrix} = \frac{ml}{6}\left[\begin{array}{cc|cc} 2 & 0 & 1 & 0 \\ 0 & 0 & 0 & 0 \\ \hline 1 & 0 & 2 & 0 \\ 0 & 0 & 0 & 0 \end{array}\right]\begin{Bmatrix} \ddot{u}_1 \\ \ddot{v}_1 \\ \ddot{u}_2 \\ \ddot{v}_2 \end{Bmatrix}$$

These 4 × 4 matrices can then be substituted into Eq. (10.3-8) in order to convert them to the global coordinates.

$$\bar{k} = T'kT = \frac{EA}{l}\left[\begin{array}{cc|cc} c^2 & cs & -c^2 & -cs \\ cs & s^2 & -cs & -s^2 \\ \hline -c^2 & -cs & c^2 & cs \\ -cs & -s^2 & cs & s^2 \end{array}\right] \tag{10.4-3}$$

$$\bar{m} = T'mT = \frac{ml}{6}\left[\begin{array}{cc|cc} 2c^2 & 2cs & c^2 & cs \\ 2cs & 2s^2 & cs & s^2 \\ \hline c^2 & cs & 2c^2 & 2cs \\ cs & s^2 & 2cs & 2s^2 \end{array}\right] \tag{10.4-4}$$

Example 10.4-1

Determine the stiffness matrix for the 3, 4, 5 oriented pinned truss of Fig. 10.4-1

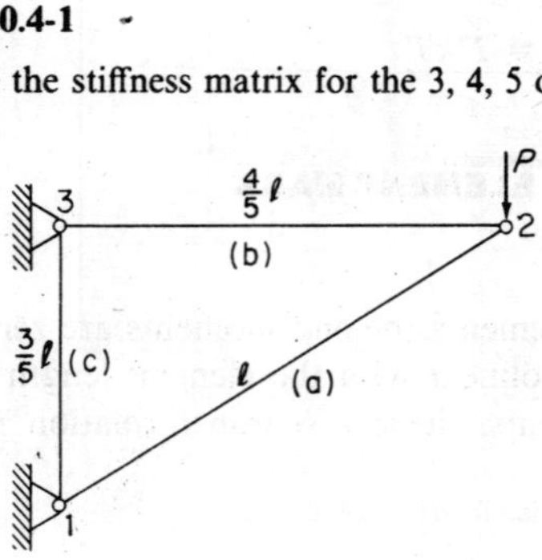

Figure 10.4-1.

Solution: The structure is composed of three pinned members a, b, c with joints 1, 2, 3. Each joint has two DOF in the global system, and the six forces and displacements are related by the following equation:

$$\begin{Bmatrix} \bar{F}_{1x} \\ \bar{F}_{1y} \\ \bar{F}_{2x} \\ \bar{F}_{2y} \\ \bar{F}_{3x} \\ \bar{F}_{3y} \end{Bmatrix} = [\bar{K}] \begin{Bmatrix} \bar{u}_1 \\ \bar{v}_1 \\ \bar{u}_2 \\ \bar{v}_2 \\ \bar{u}_3 \\ \bar{v}_3 \end{Bmatrix}$$

The global stiffness of each element is determined from Eq. (10.4-3) by substituting $\sin \alpha$ and $\cos \alpha$ for the particular member.

Element a (1 to 2).

$$c = \tfrac{4}{5}, \; s = \tfrac{3}{5}$$

$$\bar{k}_a \bar{r}_a = \frac{1}{25}\left(\frac{EA}{l}\right) \left[\begin{array}{cc|cc} 16 & 12 & -16 & -12 \\ 12 & 9 & -12 & -9 \\ \hline -16 & -12 & 16 & 12 \\ -12 & -9 & 12 & 9 \end{array}\right] \begin{Bmatrix} \bar{u}_1 \\ \bar{v}_1 \\ \bar{u}_2 \\ \bar{v}_2 \end{Bmatrix}$$

Element b (2 to 3).

$$c = -1, \; s = 0$$

$$\bar{k}_b \bar{r}_b = \left(\frac{5EA}{4l}\right) \left[\begin{array}{cc|cc} 1 & 0 & -1 & 0 \\ 0 & 0 & 0 & 0 \\ \hline -1 & 0 & 1 & 0 \\ 0 & 0 & 0 & 0 \end{array}\right] \begin{Bmatrix} \bar{u}_2 \\ \bar{v}_2 \\ \bar{u}_3 \\ \bar{v}_3 \end{Bmatrix}$$

Element c (3 to 1).

$$c = 0, \; s = -1$$

$$\bar{k}_c \bar{r}_c = \left(\frac{5EA}{3l}\right) \left[\begin{array}{cc|cc} 0 & 0 & 0 & 0 \\ 0 & 1 & 0 & -1 \\ \hline 0 & 0 & 0 & 0 \\ 0 & -1 & 0 & 1 \end{array}\right] \begin{Bmatrix} \bar{u}_3 \\ \bar{v}_3 \\ \bar{u}_1 \\ \bar{v}_1 \end{Bmatrix}$$

These must now be assembled for the 6×6 stiffness equation. The matrices for a and b have a common displacement $\begin{Bmatrix} \bar{u}_2 \\ \bar{v}_2 \end{Bmatrix}$, and it is easily seen that they fit together with an overlap of the section associated with the common displacement:

$$\left(\frac{EA}{25l}\right)\begin{bmatrix} 16 & 12 & -16 & -12 & & \\ 12 & 9 & -12 & -9 & & \\ -16 & -12 & 16+\frac{125}{4} & 12 & -\frac{125}{4} & 0 \\ -12 & -9 & 12 & 9 & 0 & 0 \\ & & -\frac{125}{4} & 0 & \frac{125}{4} & 0 \\ & & 0 & 0 & 0 & 0 \end{bmatrix}\begin{Bmatrix} \bar{u}_1 \\ \bar{v}_1 \\ \bar{u}_2 \\ \bar{v}_2 \\ \bar{u}_3 \\ \bar{v}_3 \end{Bmatrix}$$

In order to find the proper location for $\bar{k}_c$, it can be separated into four 2×2 matrices, which can be arranged as

$$\left(\frac{EA}{25l}\right)\begin{bmatrix} 0 & 0 & & & 0 & 0 \\ 0 & \frac{125}{3} & & & 0 & -\frac{125}{3} \\ & & & & & \\ & & & & & \\ 0 & 0 & & & 0 & 0 \\ 0 & -\frac{125}{3} & & & 0 & \frac{125}{3} \end{bmatrix}\begin{Bmatrix} \bar{u}_1 \\ \bar{v}_1 \\ - \\ - \\ \bar{u}_3 \\ \bar{v}_3 \end{Bmatrix}$$

Superimposing these three matrices, we see that the stiffness matrix for the truss is

$$\begin{Bmatrix} \bar{F}_{1x} \\ \bar{F}_{1y} \\ \bar{F}_{2x} \\ \bar{F}_{2y} \\ \bar{F}_{3x} \\ \bar{F}_{3y} \end{Bmatrix} = \left(\frac{EA}{25}\right)\begin{bmatrix} 16 & 12 & -16 & -12 & 0 & 0 \\ 12 & 9+\frac{125}{3} & -12 & -9 & 0 & -\frac{125}{3} \\ -16 & -12 & 16+\frac{125}{4} & 12 & -\frac{125}{4} & 0 \\ -12 & -9 & 12 & 9 & 0 & 0 \\ 0 & 0 & -\frac{125}{4} & 0 & \frac{125}{4} & 0 \\ 0 & -\frac{125}{3} & 0 & 0 & 0 & \frac{125}{3} \end{bmatrix}\begin{Bmatrix} \bar{u}_1 \\ \bar{v}_1 \\ \bar{u}_2 \\ \bar{v}_2 \\ \bar{u}_3 \\ \bar{v}_3 \end{Bmatrix}$$

Although this procedure of assembling element matrices into their final positions for the total structure is a simple matter for small frames, for large systems the assembly procedure should be automatically accomplished by the locator matrix L. For example, the locator matrix for element c of this problem is a 4×6 matrix of 1s for the diagonal with columns of 0s corresponding to the missing joints.

$$L_c = \begin{bmatrix} 1 & 0 & 0 & 0 & 0 & 0 \\ 0 & 1 & 0 & 0 & 0 & 0 \\ 0 & 0 & 0 & 0 & 1 & 0 \\ 0 & 0 & 0 & 0 & 0 & 1 \end{bmatrix}$$

The multiplication

$$L_c'\bar{k}_c L_c$$

places the coefficients of the matrix k_c into the proper space in the final matrix for the structure.

Similarly, the locator matrices for elements a and b are

$$L_a = \left[\begin{array}{cc|cc|cc} 1 & 0 & 0 & 0 & 0 & 0 \\ 0 & 1 & 0 & 0 & 0 & 0 \\ \hline 0 & 0 & 1 & 0 & 0 & 0 \\ 0 & 0 & 0 & 1 & 0 & 0 \end{array}\right] \qquad L_b = \left[\begin{array}{cc|cc|cc} 0 & 0 & 1 & 0 & 0 & 0 \\ 0 & 0 & 0 & 1 & 0 & 0 \\ \hline 0 & 0 & 0 & 0 & 1 & 0 \\ 0 & 0 & 0 & 0 & 0 & 1 \end{array}\right]$$

i.e., the unit matrices occupy the space corresponding to the desired coordinates.

We now impose the condition of zero displacement for joints 1 and 3, which wipes out columns 1, 2, 5, and 6, leaving the equation

$$\begin{Bmatrix} \bar{F}_{1x} \\ \bar{F}_{1y} \\ \hline 0 \\ -P \\ \hline \bar{F}_{3x} \\ \bar{F}_{3y} \end{Bmatrix} = \left(\frac{EA}{25l}\right) \left[\begin{array}{cc} -16 & -12 \\ -12 & -9 \\ \hline 16 + 31.25 & 12 \\ 12 & 9 \\ \hline -31.25 & 0 \\ 0 & 0 \end{array}\right] \begin{Bmatrix} \bar{u}_2 \\ \bar{v}_2 \end{Bmatrix}$$

The middle two rows are

$$\begin{Bmatrix} 0 \\ -P \end{Bmatrix} = \left(\frac{EA}{25l}\right) \begin{bmatrix} 47.25 & 12 \\ 12 & 9 \end{bmatrix} \begin{Bmatrix} \bar{u}_2 \\ \bar{v}_2 \end{Bmatrix}$$

which can be inverted to

$$\begin{Bmatrix} \bar{u}_2 \\ \bar{v}_2 \end{Bmatrix} = \left(\frac{25l}{EA}\right) \frac{1}{281.25} \begin{bmatrix} 9 & -12 \\ -12 & 47.25 \end{bmatrix} \begin{Bmatrix} 0 \\ -P \end{Bmatrix}$$

Thus the horizontal and vertical deflections of joint 2 are

$$\bar{u}_2 = \left(\frac{25l}{281.25EA}\right)(12P) = 1.066\frac{Pl}{EA}$$

$$\bar{v}_2 = \left(\frac{25l}{281.25EA}\right)(-47.25P) = -4.200\frac{Pl}{EA}$$

With these values, the reaction forces at pins 1 and 3 are found to be

$$\bar{F}_{1x} = \left(\frac{EA}{25l}\right)\left[-16 \times 1.066\frac{Pl}{EA} + 12 \times 4.200\frac{Pl}{EA}\right] = 1.333P$$

$$\bar{F}_{1y} = 1.000P$$

$$\bar{F}_{3x} = -1.333P$$

$$\bar{F}_{3y} = 0$$

Of course, these reaction forces are easily found by taking moments about the fixed pins; however, this example illustrates the general procedure to be followed for a more complex structure.

Beam element. The stiffness and mass matrices for the beam element are of order 4×4, whereas the transformation matrix is 6×6. Thus to transform these matrices for the global coordinates, we need to modify them by adding the axial components rearranged as follows:

$$\frac{EA}{l}\left[\begin{array}{ccc|ccc} 1 & 0 & 0 & -1 & 0 & 0 \\ 0 & 0 & 0 & 0 & 0 & 0 \\ 0 & 0 & 0 & 0 & 0 & 0 \\ \hline -1 & 0 & 0 & 1 & 0 & 0 \\ 0 & 0 & 0 & 0 & 0 & 0 \\ 0 & 0 & 0 & 0 & 0 & 0 \end{array}\right] \begin{matrix} u_1 \\ v_1 \\ \theta_1 \\ u_2 \\ v_2 \\ \theta_2 \end{matrix}$$

$$\frac{ml}{6}\left[\begin{array}{ccc|ccc} 2 & 0 & 0 & 1 & 0 & 0 \\ 0 & 0 & 0 & 0 & 0 & 0 \\ 0 & 0 & 0 & 0 & 0 & 0 \\ \hline 1 & 0 & 0 & 2 & 0 & 0 \\ 0 & 0 & 0 & 0 & 0 & 0 \\ 0 & 0 & 0 & 0 & 0 & 0 \end{array}\right]$$

The element matrices to be used in the transformation then become

$$k = \frac{EI}{l^3}\left[\begin{array}{ccc|ccc} R & 0 & 0 & -R & 0 & 0 \\ 0 & 12 & 6l & 0 & -12 & 6l \\ 0 & 6l & 4l^2 & 0 & -6l & 2l^2 \\ \hline -R & 0 & 0 & R & 0 & 0 \\ 0 & -12 & -6l & 0 & 12 & -6l \\ 0 & 6l & 2l^2 & 0 & -6l & 4l^2 \end{array}\right] \qquad (10.4\text{-}5)$$

Where $\quad R = \left(\dfrac{EA}{l}\right)\left(\dfrac{l^3}{EI}\right) = \dfrac{Al^2}{I}$

$$m = \frac{ml}{420}\left|\begin{array}{ccc|ccc} N & 0 & 0 & \frac{1}{2}N & 0 & 0 \\ 0 & 156 & 22l & 0 & 54 & -13l \\ 0 & 22l & 4l^2 & 0 & 13l & -3l^2 \\ \hline \frac{1}{2}N & 0 & 0 & N & 0 & 0 \\ 0 & 54 & 13l & 0 & 156 & -22l \\ 0 & -13l & -3l^2 & 0 & -22l & 4l^2 \end{array}\right| \qquad (10.4\text{-}6)$$

Where $\quad N = \left(\dfrac{ml}{3}\right)\left(\dfrac{420}{ml}\right) = 140$

These 6×6 element matrices are transformed to global coordinates (with bars over the letters) by the equations $\bar{k} = T'kT$ and $\bar{m} = T'mT$:

$$\bar{k} = \frac{EI}{l^3}\left[\begin{array}{ccc|ccc} (Rc^2+12s^2) & (R-12)cs & -6ls & (-Rc^2-12s^2) & (-R+12)cs & -6ls \\ (R-12)cs & (Rs^2+12c^2) & 6lc & (-R+12)cs & (-Rs^2-12c^2) & 6lc \\ -6ls & 6lc & 4l^2 & 6ls & -6lc & 2l^2 \\ \hline (-Rc^2-12s^2) & (-R+12)cs & 6ls & (Rc^2+12s^2) & (R-12)cs & 6ls \\ (-R+12)cs & (-Rs^2-12c^2) & -6lc & (R-12)cs & (Rs^2+12c^2) & -6lc \\ -6ls & 6lc & 2l^2 & 6ls & -6lc & 4l^2 \end{array}\right] \begin{matrix} \bar{u} \\ \bar{v} \\ \bar{\theta} \\ \\ \\ \\ \end{matrix}$$

(10.4-7)

$$\bar{m} = \frac{ml}{420}\left[\begin{array}{ccc|ccc} (Nc^2+156s^2) & (N-156)cs & -22ls & (\frac{1}{2}Nc^2+54s^2) & (\frac{1}{2}N-54)cs & 13ls \\ (N-156)cs & (Ns^2+156c^2) & 22lc & (\frac{1}{2}N-54)cs & (\frac{1}{2}Ns^2+54c^2) & -13lc \\ -22ls & 22lc & 4l^2 & -13ls & 13lc & -3l^2 \\ \hline (\frac{1}{2}Nc^2+54s^2) & (\frac{1}{2}N-54)cs & -13ls & (Nc^2+156s^2) & (N-156)cs & 22ls \\ (\frac{1}{2}N-54)cs & (\frac{1}{2}Ns^2+54c^2) & 13lc & (N-156)cs & (Ns^2+156c^2) & -22lc \\ 13ls & -13lc & -3l^2 & 22ls & -22lc & 4l^2 \end{array}\right]$$

(10.4-8)

10.5 VIBRATIONS INVOLVING BEAM ELEMENTS

To illustrate the finite element method for beams, we consider some problems solved in Chapters 6 and 7. The object here is first to show how to assemble the system equation using two elements and second to reduce the degrees of freedom of the equation by elimination of rotational coordinates.

Example 10.5-1

The beam in Fig. 10.5-1 is considered as two equal elements of length $l/2$, whose stiffness and mass matrices are given by Eqs. (10.2-1) and (10.2-10). With $l/2$ substituted for l, the element matrices are as follows:

Element *a*

$$\text{Stiffness}\left(\frac{8EI}{l^3}\right)\left[\begin{array}{cc|cc} 12 & 3l & -12 & 3l \\ 3l & l^2 & -3l & 0.5l \\ \hline -12 & -3l & 12 & -3l \\ 3l & 0.5l & -3l & l^2 \end{array}\right] \qquad \text{Displacement vector}\begin{Bmatrix} v_1 \\ \theta_1 \\ v_2 \\ \theta_2 \end{Bmatrix}$$

$$\text{Mass}\left(\frac{ml}{840}\right)\left[\begin{array}{cc|cc} 156 & 11l & 54 & -6.5l \\ 11l & l^2 & 6.5l & -0.75l \\ \hline 54 & 6.5l & 156 & -11l \\ -6.5l & -0.75l & -11l & l^2 \end{array}\right]$$

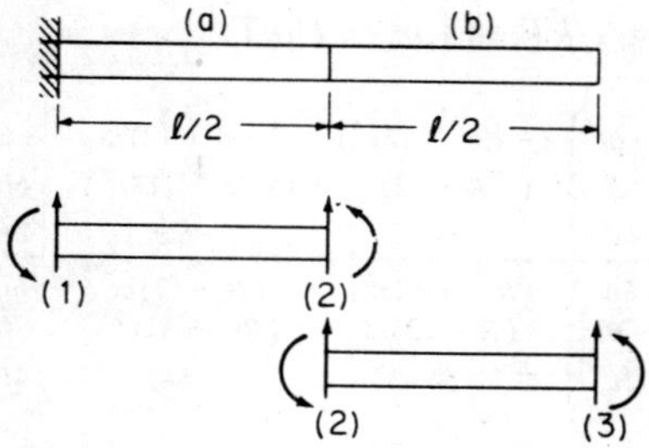

Figure 10.5-1. Uniform cantilever beam.

Element b. Element b is the same as element a except for the displacement vector, which is

$$\begin{Bmatrix} v_2 \\ \theta_2 \\ v_3 \\ \theta_3 \end{Bmatrix}$$

With the global coordinates coinciding with the beam axis, the assembly of the system matrix is simply that of superimposing the preceding matrices for elements a and b into a 6×6 matrix. That is, for the stiffness matrix we have,

$$\begin{bmatrix} \text{Element } a & & \\ & \ddots & \\ & & \text{Element } b \end{bmatrix} \begin{Bmatrix} v_1 \\ \theta_1 \\ v_2 \\ \theta_2 \\ v_3 \\ \theta_3 \end{Bmatrix}$$

Since $v_1 = \theta_1 = 0$ due to the constraint of the wall, the first two columns can be ignored. Also, we are not concerned with the force and moment F_1 and M_1 in the vibration problem. We can therefore strike out the first two rows as well as the first two columns, leaving the equation

$$\frac{ml}{840}\begin{bmatrix} 312 & 0 & 54 & -6.5l \\ 0 & 2l^2 & 6.5l & -0.75l^2 \\ 54 & 6.5l & 156 & -11l \\ -6.5l & -0.75l^2 & -11l & l^2 \end{bmatrix} \begin{Bmatrix} \ddot{v}_2 \\ \ddot{\theta}_2 \\ \ddot{v}_3 \\ \ddot{\theta}_3 \end{Bmatrix}$$

$$+\left(\frac{8EI}{l^3}\right)\begin{bmatrix} 24 & 0 & -12 & 3l \\ 0 & 2l^2 & -3l & 0.5l^2 \\ -12 & -3l & 12 & -3l \\ 3l & 0.5l & -3l & l^2 \end{bmatrix} \begin{Bmatrix} v_2 \\ \theta_2 \\ v_3 \\ \theta_3 \end{Bmatrix} = \begin{Bmatrix} F_2 \\ M_2 \\ F_3 \\ M_3 \end{Bmatrix}$$

To solve for the free vibration of the beam, the force vector is made equal to zero and the acceleration vector is replaced by ($-\omega^2$ times the displacement).

Coordinate reduction

Example 10.5-2

The solution of the preceding equation requires an eigenvalue-eigenvector computer program. However, we consider a simpler problem of replacing the uniformly distributed mass by lumped masses at joints 2 and 3. The mass matrix is then a series of 0s except for elements m_{22} and m_{33}. This suggests rearranging the preceding equation so that the displacement vector is in the rearranged order

$$\begin{Bmatrix} v_2 \\ v_3 \\ \theta_2 \\ \theta_3 \end{Bmatrix}$$

This is simply accomplished by interchanging columns 2 and 3 and rows 2 and 3, resulting in the following equation:

$$\left[\begin{array}{cc|cc} m_2 & 0 & 0 & \\ 0 & m_3 & & \\ \hline 0 & & 0 & \\ & & & \end{array}\right]\begin{Bmatrix} \ddot{v}_2 \\ \ddot{v}_3 \\ \hline \ddot{\theta}_2 \\ \ddot{\theta}_3 \end{Bmatrix} + \left(\frac{8EI}{l^3}\right)\left[\begin{array}{cc|cc} 24 & -12 & 0 & 3l \\ -12 & 12 & -3l & -3l \\ \hline 0 & -3l & 2l^2 & 0.5l^2 \\ 3l & -3l & 0.5l^2 & l^2 \end{array}\right]\begin{Bmatrix} v_2 \\ v_3 \\ \hline \theta_2 \\ \theta_3 \end{Bmatrix} = \begin{Bmatrix} 0 \\ 0 \\ \hline 0 \\ 0 \end{Bmatrix}$$

The equation is now in the form

$$\left[\begin{array}{c|c} M_{11} & 0 \\ \hline 0 & 0 \end{array}\right]\begin{Bmatrix} \ddot{V} \\ \hline \ddot{\theta} \end{Bmatrix} + \left[\begin{array}{c|c} K_{11} & K_{12} \\ \hline K_{21} & K_{22} \end{array}\right]\begin{Bmatrix} V \\ \hline \theta \end{Bmatrix} = \begin{Bmatrix} 0 \\ 0 \end{Bmatrix} \tag{10.5-1}$$

which can be written out as

$$M_{11}\ddot{V} + K_{11}V + K_{12}\theta = 0$$

$$K_{21}V + K_{22}\theta = 0$$

From the second equation θ can be expressed in terms of V.

$$\theta = -K_{22}^{-1}K_{21}V$$

Substituting into the first equation, we have

$$M_{11}\ddot{V} + \left(K_{11} - K_{12}K_{22}^{-1}K_{21}\right)V = 0 \tag{10.5-2}$$

which in terms of the original quantities becomes

$$\begin{bmatrix} m_2 & 0 \\ 0 & m_3 \end{bmatrix}\begin{Bmatrix} \ddot{v}_2 \\ \ddot{v}_3 \end{Bmatrix} + \left(\frac{8EI}{l^3}\right)\left[\begin{bmatrix} 24 & -12 \\ -12 & 12 \end{bmatrix}\right.$$

$$\left. - \begin{bmatrix} 0 & 3l \\ -3l & -3l \end{bmatrix}\begin{bmatrix} 2l^2 & 0.5l^2 \\ 0.5l^2 & l^2 \end{bmatrix}^{-1}\begin{bmatrix} 0 & -3l \\ 3l & -3l \end{bmatrix}\right]\begin{Bmatrix} v_2 \\ v_3 \end{Bmatrix} = \begin{Bmatrix} 0 \\ 0 \end{Bmatrix}$$

The term

$$K_{11} - K_{12}K_{22}^{-1}K_{21} \tag{10.5-3}$$

is the *reduced stiffness*, and its value when multiplied out is

$$\frac{8EI}{7l^3}\begin{vmatrix} 96 & -30 \\ -30 & 12 \end{vmatrix} = \frac{48EI}{7l^3}\begin{vmatrix} 16 & -5 \\ -5 & 2 \end{vmatrix}$$

Thus the original 4×4 equation has been reduced to a 2×2 equation, the final form being

$$\begin{bmatrix} m_2 & 0 \\ 0 & m_3 \end{bmatrix}\begin{Bmatrix} \ddot{v}_2 \\ \ddot{v}_3 \end{Bmatrix} + \left(\frac{48EI}{7l^3}\right)\begin{bmatrix} 16 & -5 \\ -5 & 2 \end{bmatrix}\begin{Bmatrix} v_2 \\ v_3 \end{Bmatrix} = \begin{Bmatrix} 0 \\ 0 \end{Bmatrix}$$

Example 10.5-3

Determine the equation of free vibration of the portal frame with identical elements.

Solution: Labeling the joints as shown in Fig. (10.5-2), the stiffness and mass for each element are available from Eqs. (10.4-7) and (10.4-8). Since joints 0 and 3 have zero displacements, we write only the terms corresponding to joints 1 and 2.
Element 0–1, $\alpha = 90°$, $c = 0$, $s = 1$.

$$\bar{k}_{0-1} = \frac{EI}{l^3}\left[\begin{array}{ccc|ccc} & & & -12 & 0 & -6l \\ & & & 0 & -R & 0 \\ & & & 6l & 0 & 2l^2 \\ \hline & & & 12 & 0 & 6l \\ & & & 0 & R & 0 \\ & & & 6l & 0 & 4l^2 \end{array}\right]\begin{array}{c} \\ \\ \\ \bar{u}_1 \\ \bar{v}_1 \\ \bar{\theta}_1 \end{array}$$

$$\bar{m}_{0-1} = \frac{ml}{420}\left[\begin{array}{ccc|ccc} & & & 54 & 0 & 13l \\ & & & 0 & \frac{1}{2}N & 0 \\ & & & -13l & 0 & -3l^2 \\ \hline & & & 156 & 0 & 22l \\ & & & 0 & 0 & 0 \\ & & & 22l & 0 & 4l^2 \end{array}\right]$$

Element 1–2, $\alpha = 0°$, $c = 1$, $s = 0$.

$$\bar{k}_{1-2} = \frac{EI}{l^3}\left[\begin{array}{ccc|ccc} R & 0 & 0 & -R & 0 & 0 \\ 0 & 12 & 6l & 0 & -12 & 6l \\ 0 & 6l & 4l^2 & 0 & -6l & 2l^2 \\ \hline -R & 0 & 0 & R & 0 & 0 \\ 0 & -12 & -6l & 0 & 12 & -6l \\ 0 & 6l & 2l^2 & 0 & -6l & 4l^2 \end{array}\right]\begin{array}{c} \bar{u}_1 \\ \bar{v}_1 \\ \bar{\theta}_1 \\ \bar{u}_2 \\ \bar{v}_2 \\ \bar{\theta}_2 \end{array}$$

$$\bar{m}_{1-2} = \frac{ml}{420}\left[\begin{array}{ccc|ccc} N & 0 & 0 & \frac{1}{2}N & 0 & 0 \\ 0 & 156 & 22l & 0 & 54 & -13l \\ 0 & 22l & 4l^2 & 0 & 13l & -3l^2 \\ \hline \frac{1}{2}N & 0 & 0 & N & 0 & 0 \\ 0 & 54 & 13l & 0 & 156 & -22l \\ 0 & -13l & -3l^2 & 0 & -22l & 4l^2 \end{array}\right]$$

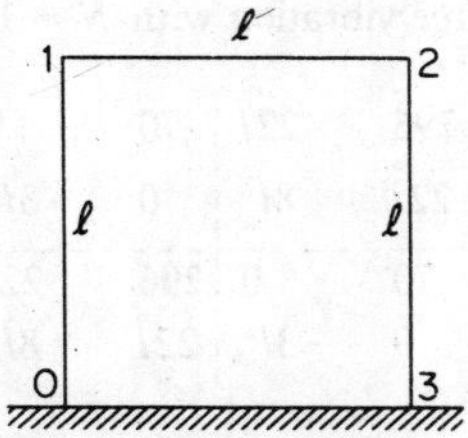

Figure 10.5-2.

Element 2–3, $\alpha = 270°$, $c = 0$, $s = -1$.

$$\bar{k}_{2-3} = \frac{EI}{l^3}\left[\begin{array}{ccc|c} 12 & 0 & 6l & \\ 0 & R & 0 & \\ 6l & 0 & 4l^2 & \\ \hline -12 & 0 & -6l & \\ 0 & -R & 0 & \\ 6l & 0 & 2l^2 & \end{array}\right] \begin{matrix} \bar{u}_2 \\ \bar{v}_2 \\ \bar{\theta}_2 \\ \bar{u}_3 \\ \bar{v}_3 \\ \bar{\theta}_3 \end{matrix}$$

$$\bar{m}_{2-3} = \frac{ml}{420}\left[\begin{array}{ccc|c} 156 & 0 & 22l & \\ 0 & N & 0 & \\ 22l & 0 & 4l^2 & \\ \hline 0 & & & \\ & & & \\ & & & \end{array}\right]$$

Assembling these matrices, we have

$$\frac{EI}{l^3}\left[\begin{array}{ccc|ccc} (12+R) & 0 & 6l & -R & 0 & 0 \\ 0 & (12+R) & 6l & 0 & -12 & 6l \\ 6l & 6l & 8l^2 & 0 & -6l & 2l^2 \\ \hline -R & 0 & 0 & (12+R) & 0 & 6l \\ 0 & -12 & -6l & 0 & (12+R) & -6l \\ 0 & 6l & 2l^2 & 6l & -6l & 8l^2 \end{array}\right] \begin{matrix} \bar{u}_1 \\ \bar{v}_1 \\ \bar{\theta}_1 \\ \bar{u}_2 \\ \bar{v}_2 \\ \bar{\theta}_2 \end{matrix} \quad \text{(a)}$$

$$\frac{ml}{420}\left[\begin{array}{ccc|ccc} (156+N) & 0 & 22l & \frac{1}{2}N & 0 & 0 \\ 0 & 156 & 22l & 0 & 54 & -13l \\ 0 & 22l & 8l^2 & 0 & 13l & -3l^2 \\ \hline \frac{1}{2}N & 0 & 0 & (156+N) & 0 & 22l \\ 0 & 54 & 13l & 0 & (156+N) & -22l \\ 0 & -13l & -3l^2 & 22l & -22l & 8l^2 \end{array}\right] \quad \text{(b)}$$

We next note that $v_1 = v_2 = 0$, which eliminates columns 2 and 5 as well as rows 2 and

5. The equation for free vibration with $N = 140$ substituted then becomes

$$
-\frac{\omega^2 ml}{420}\left[\begin{array}{cc|cc} 296 & 22l & 70 & 0 \\ 22l & 8l^2 & 0 & -3l^2 \\ \hline 70 & 0 & 296 & 22l \\ 0 & -3l^2 & 22l & 8l^2 \end{array}\right]\begin{Bmatrix} \bar{u}_1 \\ \bar{\theta}_1 \\ \bar{u}_2 \\ \bar{\theta}_2 \end{Bmatrix}
$$

$$
+\frac{EI}{l^3}\left[\begin{array}{cc|cc} (12+R) & 6l & -R & 0 \\ 6l & 8l^2 & 0 & 2l^2 \\ \hline -R & 0 & (12+R) & 6l \\ 0 & 2l^2 & 6l & 8l^2 \end{array}\right]\begin{Bmatrix} \bar{u}_1 \\ \bar{\theta}_1 \\ \bar{u}_2 \\ \bar{\theta}_2 \end{Bmatrix} = \begin{Bmatrix} 0 \\ 0 \\ 0 \\ 0 \end{Bmatrix} \tag{c}
$$

Example 10.5-4

Figure 10.5-3 shows the lowest antisymmetric and the lowest symmetric modes of free vibration for the portal frame.
Determine the natural frequencies for the given modes.

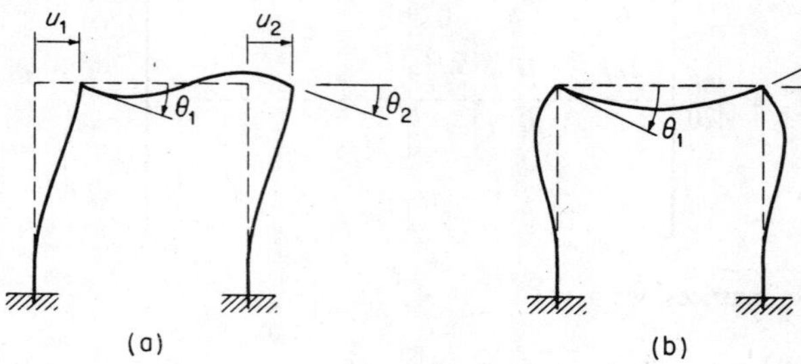

Figure 10.5-3.

Solution: *Antisymmetric Mode*. The deflection and slope of stations 1 and 2 are identical; i.e., $\bar{u}_1 = \bar{u}_2$ and $\bar{\theta}_1 = \bar{\theta}_2$. These conditions are imposed on the preceding equation by adding column 3 to column 1 and column 4 to column 2. This results in identical equations for $\begin{Bmatrix} \bar{u}_1 \\ \bar{\theta}_1 \end{Bmatrix}$ and $\begin{Bmatrix} \bar{u}_2 \\ \bar{\theta}_2 \end{Bmatrix}$:

$$
\left[-\frac{\omega^2 ml}{420}\begin{bmatrix} 366 & 22l \\ 22l & 5l^2 \end{bmatrix} + \frac{EI}{l^3}\begin{bmatrix} 12 & 6l \\ 6l & 10l^2 \end{bmatrix}\right]\begin{Bmatrix} \bar{u}_1 \\ \bar{\theta}_1 \end{Bmatrix} = \begin{Bmatrix} 0 \\ 0 \end{Bmatrix}
$$

Letting $\lambda = \omega^2 ml^4/420EI$, the determinant of this equation,

$$
\begin{vmatrix} (12-366\lambda) & (6-22\lambda)l \\ (6-22\lambda)l & (10-5\lambda)l^2 \end{vmatrix} = 0
$$

results in two roots:

$$\lambda_1 = 0.0245 \qquad \omega_1 = 3.21\sqrt{\frac{EI}{ml^4}}$$

$$\lambda_2 = 2.543 \qquad \omega_2 = 32.68\sqrt{\frac{EI}{ml^4}}$$

The lowest natural frequency corresponds to the simple shape displayed in Fig. 10.5-3(a) and is of acceptable accuracy. However, the second antisymmetric mode would be of more complex shape and ω_2 computed with the few stations used in this example would not be accurate. Several more stations would be necessary to adequately represent the higher modes.

Symmetric Mode: For the symmetric mode, we have $u_1 = u_2 = 0$ and $\theta_2 = -\theta_1$. Deleting columns 1 and 3 and subtracting column 4 from column 2, we obtain just one equation for θ_1:

$$\left[-\frac{\omega^2 ml}{420}(11l^2) + \frac{EI}{l^3}(6l^2)\right]\theta_1 = 0$$

λ and ω are then found as

$$\lambda = \frac{6}{11}, \qquad \omega = 15.14\sqrt{\frac{EI}{ml^4}}$$

Example 10.5-5

Figure 10.5-4 shows external loads acting on the portal frame. Examine the boundary condition and determine the stiffness matrix in terms of the coordinates given.

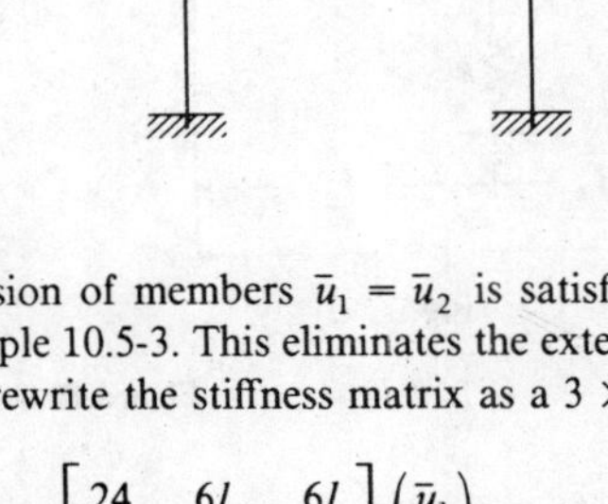

Figure 10.5-4.

Solution: The condition of no extension of members $\bar{u}_1 = \bar{u}_2$ is satisfied by adding column 3 to column 1 in Eq. (c), Example 10.5-3. This eliminates the extension term R. We can also add row 3 to row 1 and rewrite the stiffness matrix as a 3×3 matrix:

$$\begin{Bmatrix} \bar{F}_{1x} + \bar{F}_{2x} \\ \bar{M}_1 \\ \bar{M}_2 \end{Bmatrix} = \frac{EI}{l^3}\begin{bmatrix} 24 & 6l & 6l \\ 6l & 8l^2 & 2l^2 \\ 6l & 2l^2 & 8l^2 \end{bmatrix}\begin{Bmatrix} \bar{u}_1 \\ \bar{\theta}_1 \\ \bar{\theta}_2 \end{Bmatrix}$$

Comparing the external loads of Fig. 10.5-4 with those of the global system, we have

$$\begin{Bmatrix} \bar{F}_{1x} + \bar{F}_{2x} \\ \bar{M}_1 \\ \bar{M}_3 \end{Bmatrix} = \begin{Bmatrix} \mathscr{F} \\ -\mathscr{M}_1 \\ -\mathscr{M}_2 \end{Bmatrix}, \qquad \begin{Bmatrix} \bar{u}_1 \\ \bar{\theta}_1 \\ \bar{\theta}_2 \end{Bmatrix} = \begin{Bmatrix} u \\ -\theta_1 \\ -\theta_2 \end{Bmatrix}$$

With $\bar{F}_{2x} = 0$ and $\bar{F}_{1x} = \mathscr{F}$, the stiffness matrix in terms of the given coordinates and given loads is

$$\begin{Bmatrix} \mathscr{F} \\ \mathscr{M}_1 \\ \mathscr{M}_2 \end{Bmatrix} = \frac{EI}{l^3}\begin{bmatrix} 24 & -6l & -6l \\ -6l & 8l^2 & 2l^2 \\ -6l & 2l^2 & 8l^2 \end{bmatrix}\begin{Bmatrix} u \\ \theta_1 \\ \theta_2 \end{Bmatrix}$$

10.6 SPRING CONSTRAINTS ON STRUCTURE

In Chapter 9 spring constraints were treated by virtual work as generalized forces. The same concept applies in the finite element approach. The point of application of the spring must here be chosen as a joint station. Thus the load on the original structure in global coordinates is supplemented by the spring force.

Since the spring force is always opposite to the displacement, the force or moment at the joint is decreased by $-kv_i$ or $-K\theta_i$. Thus these terms, when shifted to the other side of the equation, become additions to the corresponding stiffness term.

Example 10.6-1

Determine the stiffness matrix for the uniform beam with a linear and rotational spring, shown in Fig. 10.6-1(a)

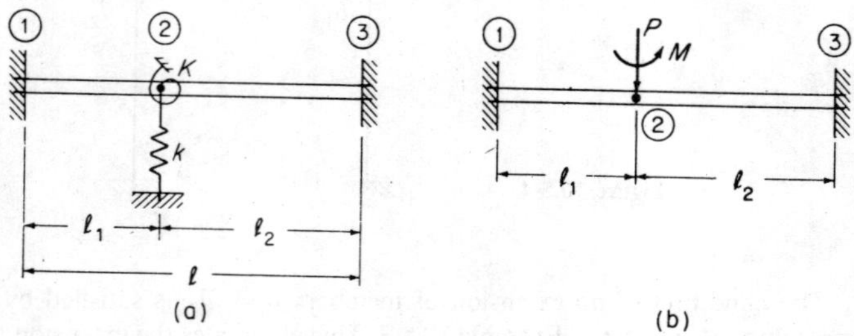

Figure 10.6-1.

Solution: We first establish the stiffness matrix of the beam without the springs [Fig. 10.6-1(b)] but with loads P and M acting at station 2. The stiffness matrix for each section 1–2 and 2–3 can be assembled from the beam element matrix, Eq. (10.2-1).

Noting that $v_1 = \theta_1 = v_3 = \theta_3 = 0$, we need only evaluate the portion of the matrix associated with the coordinates v_2 and θ_2, which becomes

$$\begin{Bmatrix} \bar{F}_2 \\ \bar{M}_2 \end{Bmatrix} = EI \begin{bmatrix} 12\left(\frac{1}{l_1^3} + \frac{1}{l_2^3}\right) & -6\left(\frac{1}{l_1^2} - \frac{1}{l_2^2}\right) \\ -6\left(\frac{1}{l_1^2} - \frac{1}{l_2^2}\right) & 4\left(\frac{1}{l_1} + \frac{1}{l_2}\right) \end{bmatrix} \begin{Bmatrix} \bar{v}_2 \\ \bar{\theta}_2 \end{Bmatrix}$$

With the springs acting at station 2, the force vector is replaced by

$$\begin{Bmatrix} \bar{F}_2 - k\bar{v}_2 \\ \bar{M}_2 - K\bar{\theta}_2 \end{Bmatrix}$$

Shifting the spring forces to the right side of the equation, we obtain

$$\begin{Bmatrix} \bar{F}_2 \\ \bar{M}_2 \end{Bmatrix} = EI \begin{bmatrix} 12\left(\frac{1}{l_1^3} + \frac{1}{l_2^3}\right) + \frac{k}{EI} & -6\left(\frac{1}{l_1^2} - \frac{1}{l_2^2}\right) \\ -6\left(\frac{1}{l_1^2} - \frac{1}{l_2^2}\right) & 4\left(\frac{1}{l_1} + \frac{1}{l_2}\right) + \frac{K}{EI} \end{bmatrix} \begin{Bmatrix} \bar{v}_2 \\ \bar{\theta}_2 \end{Bmatrix}$$

Since the force $\bar{F}_2$ in the global system is positive in the upward direction, whereas $\bar{M}_2$ is positive counterclockwise, the above equation can be rearranged to

$$\begin{Bmatrix} -P \\ M \end{Bmatrix} = \frac{EI}{l^3} \begin{bmatrix} 12\left(\frac{l^3}{l_1^3} + \frac{l^3}{l_2^3}\right) + \frac{kl^3}{EI} & -6l\left(\frac{l^2}{l_1^2} - \frac{l^2}{l_2^2}\right) \\ -6l\left(\frac{l^2}{l_1^2} - \frac{l^2}{l_2^2}\right) & 4l^2\left(\frac{l}{l_1} + \frac{l}{l_2}\right) + \frac{Kl^3}{EI} \end{bmatrix} \begin{Bmatrix} \bar{v}_2 \\ \bar{\theta}_2 \end{Bmatrix}$$

which defines the stiffness matrix for the beam with the spring constraints.

The equation indicates that the system is decoupled for $l_1 = l_2 = l/2$, in which case the equation reduces to

$$\begin{Bmatrix} -P \\ M \end{Bmatrix} = \frac{EI}{l^3} \begin{bmatrix} \left(192 + \frac{kl^3}{EI}\right) & \\ & \left(16l^2 + \frac{Kl^3}{EI}\right) \end{bmatrix} \begin{Bmatrix} \bar{v}_2 \\ \bar{\theta}_2 \end{Bmatrix}$$

The deflection at the center is then found as

$$\bar{v}_2 = \frac{-(Pl^3/EI)}{192 + kl^3/EI}, \qquad \theta_2 = \frac{(Ml^3/EI)}{16l^2 + Kl^3/EI}$$

Example 10.6-2

Determine the natural frequencies of the constrained beam of Example 10.6-1 when $l_1 = l_2 = l/2$.

Solution: For this determination we need the mass matrix, which can be assembled from Eq. (10.2-10) as

$$\left(\frac{m}{420}\right)\begin{bmatrix} 156(l_1 + l_2) & -22(l_1^2 - l_2^2) \\ -22(l_1^2 - l_2^2) & 4(l_1^3 + l_2^3) \end{bmatrix} = \left(\frac{ml}{420}\right)\begin{bmatrix} 156 & 0 \\ 0 & l^2 \end{bmatrix}$$

The equation of motion then becomes

$$\left[-\frac{\omega^2 ml}{420}\begin{bmatrix} 156 & 0 \\ 0 & l^2 \end{bmatrix} + \frac{EI}{l^3}\begin{bmatrix} \left(192 + \dfrac{kl^3}{EI}\right) & \\ & \left(16l^2 + \dfrac{Kl^3}{EI}\right) \end{bmatrix}\right]\begin{Bmatrix} \bar{v}_2 \\ \bar{\theta}_2 \end{Bmatrix} = \begin{Bmatrix} 0 \\ 0 \end{Bmatrix}$$

Again, the coordinates $\bar{v}_2$ and $\bar{\theta}_2$ are decoupled. Letting $\lambda = \omega^2 ml^4/420EI$, the equation for $\bar{v}_2$ gives

$$\lambda = \frac{1}{156}\left(192 + \frac{kl^3}{EI}\right) = 1.231 + .00641\frac{kl^3}{EI}$$

and the natural frequency for this mode is

$$\omega_1 = 22.73\sqrt{\frac{EI}{ml^4}}\sqrt{1 + 0.00521\left(\frac{kl^3}{EI}\right)}$$

Similarly, the equation for θ_2 results in

$$\lambda = 16 + \frac{Kl}{EI}$$

and

$$\omega_2 = 81.98\sqrt{\frac{EI}{ml^4}}\sqrt{1 + 0.0625\left(\frac{kl}{EI}\right)}$$

Thus both natural frequencies are increased by the constraining springs. If $k = K = 0$, the exact natural frequencies for the beam with fixed ends are

$$\omega_1 = 22.37\sqrt{\frac{EI}{ml^4}}, \qquad \omega_2 = 61.67\sqrt{\frac{EI}{ml^4}}$$

so that the error in the finite element approach is 1.61% for the first mode and 33.9% for the second mode. Dividing the beam into shorter elements reduces these errors.

10.7 GENERALIZED FORCE FOR DISTRIBUTED LOAD

As discussed in Chapters 7 and 8, the generalized force is found from the virtual work of the applied forces. With the displacement expressed as

$$y(x) = \phi_1(x)v_1 + \phi_2(x)\theta_1 + \phi_3(x)v_2 + \phi_4(x)\theta_2 \qquad (10.7\text{-}1)$$

the virtual work of the applied distributed force $p(x)$ is

$$\delta W = \int_0^l p(x)\,\delta y(x)\,dx$$
$$= \delta v_1 \int_0^l p(x)\phi_1(x)\,dx + \delta\theta_1 \int_0^l p(x)\phi_2(x)\,dx \qquad (10.7\text{-}2)$$
$$+\delta v_2 \int_0^l p(x)\phi_3(x)\,dx + \delta\theta_2 \int_0^l p(x)\phi_4(x)\,dx$$

The integrals in Eq. (10.7-2) are the generalized forces.

If the same procedure is applied to the end forces, F_1, M_1, F_2, and M_2, the virtual work is represented by

$$\delta W = F_1\,\delta v_1 + M_1\,\delta\theta_1 + F_2\,\delta v_2 + M_2\,\delta\theta_2 \qquad (10.7\text{-}3)$$

Equating the virtual work in the above two cases, we obtain the following relationships.

$$F_1 = \int_0^l p(x)\phi_1(x)\,dx \qquad F_2 = \int_0^l p(x)\phi_3(x)\,dx$$
$$M_1 = \int_0^l p(x)\phi_2(x)\,dx \qquad M_2 = \int_0^l p(x)\phi_4(x)\,dx \qquad (10.7\text{-}4)$$

Thus for the distributed load, the equivalent finite element loads are the generalized forces just given.

Example 10.7-1

Figure 10.7.1 shows a cantilever beam of length l_1 with a uniform load $p(x) = p$ lb/in. over the outer half of the beam. Determine the deflection and slope at the free end using the method of this section.

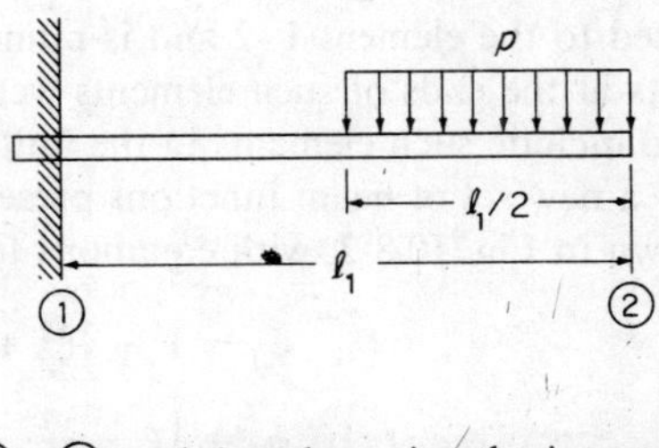

Figure 10.7-1.

Solution: We use a single element ①–② and first determine the inverse of the stiffness matrix. Since $v_1 = \theta_1 = 0$, the stiffness equation from Eq. (10.2-1) is

$$\begin{Bmatrix} F_2 \\ M_2 \end{Bmatrix} = \frac{EI}{l_1^3}\begin{bmatrix} 12 & -6l_1 \\ -6l_1 & 4l_1^2 \end{bmatrix}\begin{Bmatrix} v_2 \\ \theta_2 \end{Bmatrix}$$

Its inverse, using the adjoint method, is

$$\begin{Bmatrix} v_2 \\ \theta_2 \end{Bmatrix} = \frac{l_1^3}{EI}\frac{1}{12l_1^2}\begin{bmatrix} 4l_1^2 & 6l_1 \\ 6l_1 & 12 \end{bmatrix}\begin{Bmatrix} F_2 \\ M_2 \end{Bmatrix}$$

The equivalent finite element forces, from Eq. (10.7-4), are

$$F_2 = \int_{1/2\,l_1}^{l_1} -p\phi_3(x)\,dx = -p\int_{1/2}^{1}\phi_3(\xi)l_1\,d\xi = -pl_1\int_{1/2}^{1}(3\xi^2 - 2\xi^3)\,d\xi = -\frac{13}{32}pl_1$$

$$M_2 = \int_{1/2}^{1} -p\phi_4(\xi)l_1\,d\xi = -pl_1^2\int_{1/2}^{1}(-\xi^2+\xi^3)\,d\xi = \frac{88}{1536}pl_1^2$$

Substituting these values into the inverted equation, we have

$$\begin{Bmatrix} v_2 \\ \theta_2 \end{Bmatrix} = \frac{l_1}{12EI}\begin{bmatrix} 4l_1^2 & 6l_1 \\ 6l_1 & 12 \end{bmatrix}\begin{Bmatrix} -\frac{13}{32}pl_1 \\ \frac{88}{1536}pl_1^2 \end{Bmatrix}$$

$$= \frac{pl_1^4}{12EI}\begin{Bmatrix} -\frac{52}{32} + \frac{528}{1536} \\ -\frac{78}{32l_1} + \frac{1056}{1536l_1} \end{Bmatrix} = \frac{-pl_1^4}{48EI}\begin{Bmatrix} 5.125 \\ \frac{7.000}{l_1} \end{Bmatrix}$$

These results agree with those calculated from the area-moment method.

10.8 BEAM ELEMENT WITH ONE PINNED END

Structures frequently have beam elements with pin-ended termination at one end. The portal frame of Chapter 6 with one leg pinned to the ground is such an example. As shown in Fig. 10.8-1, the third element, 2–3, is rigidly connected to the element 1–2 and is pinned at the other end. The forces and moments at the ends of such elements were shown in Table 6.3-1.

To include such elements in the finite element method, it is necessary to develop a new set of beam functions presented here. The four shapes required are shown in Fig. 10.8-2, with equations for the beam functions as follows:

$$\begin{aligned} \psi_1 &= 1 - \tfrac{3}{2}\xi^2 + \tfrac{1}{2}\xi^3 \\ \psi_2 &= [\xi - \tfrac{3}{2}\xi^2 + \tfrac{1}{2}\xi^3]l \\ \psi_3 &= \tfrac{3}{2}\xi - \tfrac{1}{2}\xi^3 \\ \psi_4 &= [-\xi + \tfrac{3}{2}\xi - \tfrac{1}{2}\xi^3]l \end{aligned} \tag{10.8-1}$$

The element stiffness equation assembled from these components can be

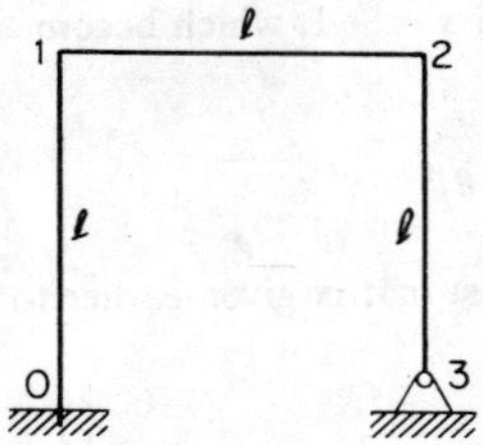

Figure 10.8-1.

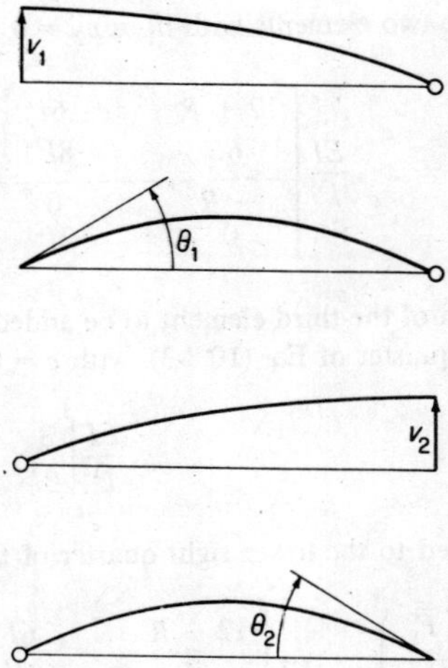

Figure 10.8-2.

shown to be

$$\begin{Bmatrix} F_1 \\ M_1 \\ F_2 \\ M_2 \end{Bmatrix} = \frac{EI}{l^3}\left[\begin{array}{cc|cc} 3 & 3l & -3 & 3l \\ 3l & 3l^2 & 0 & 0 \\ \hline -3 & -3l & 3 & -3l \\ 0 & 0 & -3l & 3l^2 \end{array}\right]\begin{Bmatrix} v_1 \\ \theta_1 \\ v_2 \\ \theta_2 \end{Bmatrix} \tag{10.8-2}$$

Adding the axial components in order to expand the k matrix to a 6×6 for global transformation, the stiffness $\bar{k} = T'kT$ in the global coordinates becomes

$$\bar{k} = \frac{EI}{l^3}\left[\begin{array}{ccc|ccc} (Rc^2+3s^2) & (R-3)sc & -3ls & (-Rc^2-3s^2) & (-R+3)sc & -3ls \\ (R-3)sc & (Rs^2+3c^2) & 3lc & (-R+3)sc & (-Rs^2-3c^2) & 3lc \\ -3ls & 3lc & 3l^2 & 0 & 0 & 0 \\ \hline (-Rc^2-3s^2) & (-R+3)sc & 3ls & (Rc^2+3s^2) & (R-3)sc & 3ls \\ (-R+3)sc & (-Rs^2-3c^2) & -3lc & (R-3)sc & (Rs^2+3c^2) & -3lc \\ 0 & 0 & 0 & 3ls & -3lc & 3l^2 \end{array}\right] \tag{10.8-3}$$

Example 10.8-1

Verify the stiffness matrix of the portal frame with one pinned end, located as in Chapter 6.

Solution: We have already solved part of this problem in Example 10.5-3, and we need only to add the new matrix for the third element. The assembled stiffness matrix for the first two elements with $v_1 = v_2 = 0$ is

$$\frac{EI}{l^3}\left[\begin{array}{cc|cc} 12+R & 6l & -R & 0 \\ 6l & 8l^2 & 0 & 2l^2 \\ \hline -R & 0 & R & 0 \\ 0 & 2l^2 & 0 & 4l^2 \end{array}\right] \begin{matrix} \bar{u}_1 \\ \bar{\theta}_1 \\ \bar{u}_2 \\ \bar{\theta}_2 \end{matrix}$$

The portion of the third element to be added to the preceding matrix is found from the upper left quarter of Eq. (10.8-3), with $c = 0$ and $s = -1$, which becomes

$$\frac{EI}{l^3}\begin{bmatrix} 3 & 3l \\ 3l & 3l^2 \end{bmatrix} \begin{matrix} \bar{u}_2 \\ \bar{\theta}_2 \end{matrix}$$

This is added to the lower right quarter of the first matrix given earlier to obtain

$$\begin{Bmatrix} \bar{F}_{1x} \\ \bar{M}_1 \\ \bar{F}_{2x} \\ \bar{M}_2 \end{Bmatrix} = \frac{EI}{l^3}\left[\begin{array}{cc|cc} 12+R & 6l & -R & 0 \\ 6l & 8l^2 & 0 & 2l^2 \\ \hline -R & 0 & R+3 & 3l \\ 0 & 2l^2 & 3l & 7l^2 \end{array}\right] \begin{Bmatrix} \bar{u}_1 \\ \bar{\theta}_1 \\ \bar{u}_2 \\ \bar{\theta}_2 \end{Bmatrix}$$

Again noting that $\bar{u}_1 = \bar{u}_2$, the matrix is reduced to a 3×3 matrix (see Example 10.5-5):

$$\begin{Bmatrix} \bar{F}_{1x} + \bar{F}_{2x} \\ \bar{M}_1 \\ \bar{M}_2 \end{Bmatrix} = \frac{EI}{l^3}\begin{bmatrix} 15 & 6l & 3l \\ 6l & 8l^2 & 2l^2 \\ 3l & 2l^2 & 7l^2 \end{bmatrix} \begin{Bmatrix} \bar{u}_1 \\ \bar{\theta}_1 \\ \bar{\theta}_2 \end{Bmatrix}$$

Replacing the global units by the actual loads and displacements specified for the problem, the stiffness equation for the portal frame becomes

$$\begin{Bmatrix} \mathscr{F} \\ \mathscr{M}_1 \\ \mathscr{M}_2 \end{Bmatrix} = \frac{EI}{l^3}\begin{bmatrix} 15 & -6l & -3l \\ -6l & 8l^2 & 2l^2 \\ -3l & 2l^2 & 7l^2 \end{bmatrix} \begin{Bmatrix} u \\ \theta_1 \\ \theta_2 \end{Bmatrix}$$

PROBLEMS

10-1 Determine the two natural frequencies in axial vibration for the uniform rod, fixed at one end and free at the other end, using two elements with the intermediate station at $l/3$ from the fixed end. Compare results with those when the station is chosen at mid-length. What conclusion do you come to regarding choice of station location?

10-2 A tapered rod is modelled as two uniform sections, as shown in Fig. P10-2, where $EA_1 = 2EA_2$ and $m_1 = 2m_2$. Determine the two natural frequencies in longitudinal vibration.

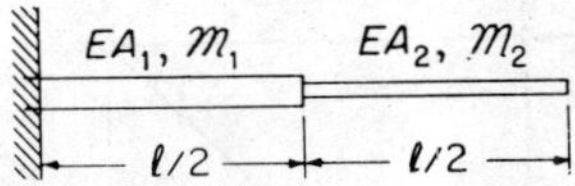

Figure P10-2.

10-3 Set up the equation for the free-free vibration of a uniform rod of length l, using three axial elements of length $l/3$ each.

10-4 Assuming linear variation for the twist of a uniform shaft, determine the finite element stiffness and mass matrices for the torsional problem. The problem is identical to that of the axial vibration.

10-5 Using two equal elements, determine the first two natural frequencies of a fixed-free shaft in torsional oscillation.

10-6 Using two uniform sections in torsional vibration, describe its finite element relationship to the 2 DOF lumped mass torsional system.

10-7 Fig. P10-7 shows a conical tube of constant thickness fixed at the large end and free at the other end. Using one element, determine the equation for its longitudinal vibration.

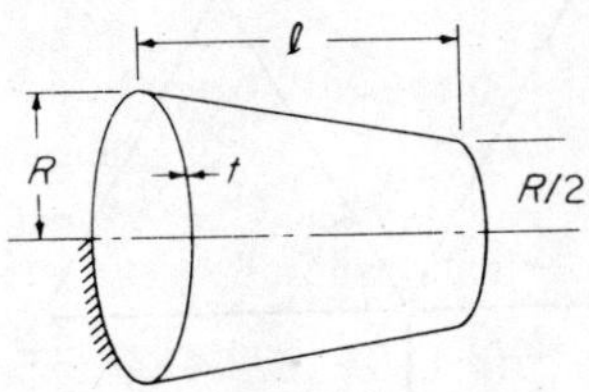

Figure P10-7.

10-8 Treat the tube of Fig. P10-7 as a two element problem of equal length in longitudinal vibration.

10-9 Determine the equation for the tube of Fig. P10-7 in torsional vibration using (a) two elements, (b) N-stepped uniform elements.

10-10 The simple frame of Fig. P10-10 has pinned joints. Determine its stiffness matrix.

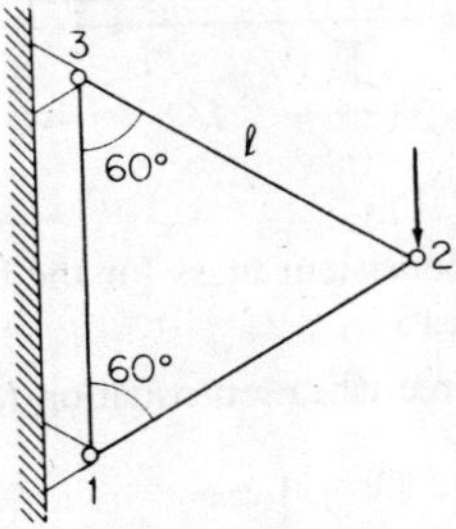

Figure P10-10.

10-11 For the pinned square truss of Fig. P10-11, determine the element stiffness matrices in global coordinates and indicate how they are assembled for the entire structure.

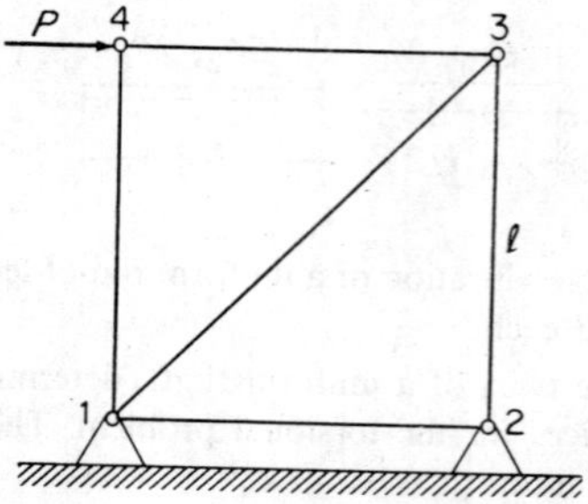

Figure P10-11.

10-12 For the pinned truss of Fig. P10-12, there are just three orientations of the elements. Determine the stiffness matrix for each orientation and indicate how each element matrix is assembled in the global system.

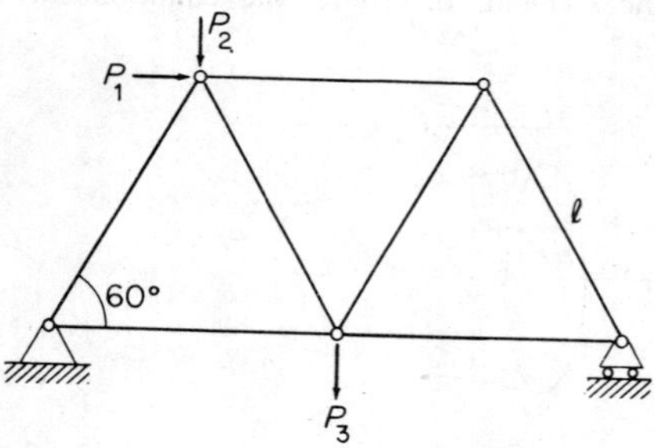

Figure P10-12.

10-13 Using two elements determine the deflection and slope at mid-span of a uniform beam, fixed at both ends, when loaded as shown in Fig. P10-13.

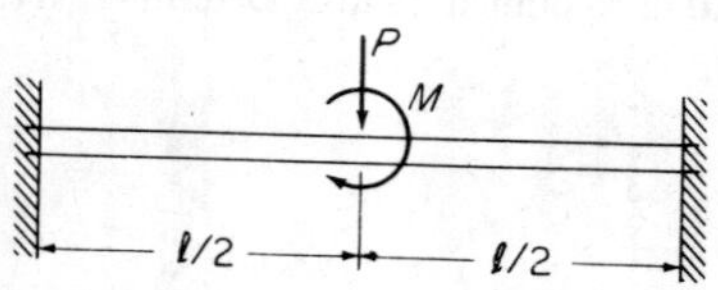

Figure P10-13.

10-14 Determine the consistent mass for the beam of Problem 10-13 and calculate its natural frequencies.

10-15 Determine the free vibration equation for the beam of Fig. P10-15.

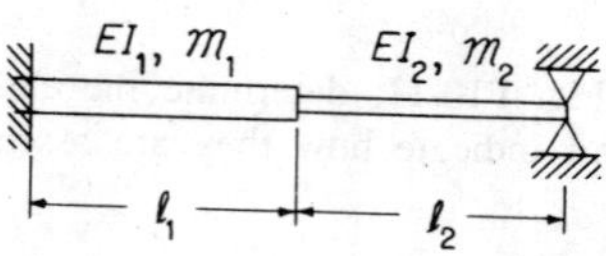

Figure P10-15.

10-16 Determine the stiffness matrix for the frame of Fig. P10-16. The upper right end is restricted from rotating but is free to slide in and out.

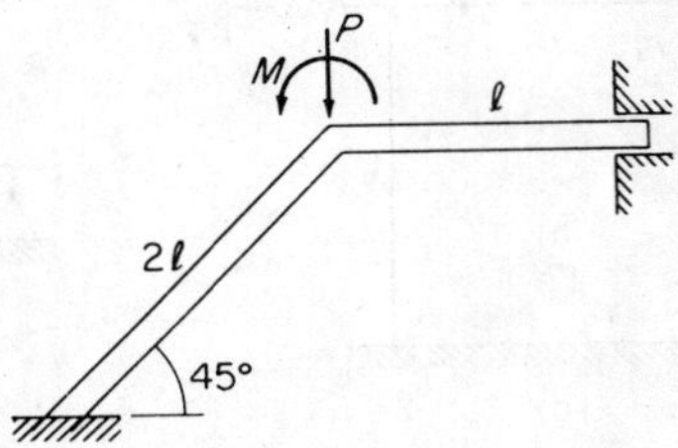

Figure P10-16.

10-17 The frame of Fig. P10-17 is free to rotate and translate at the upper right end. Determine its stiffness matrix.

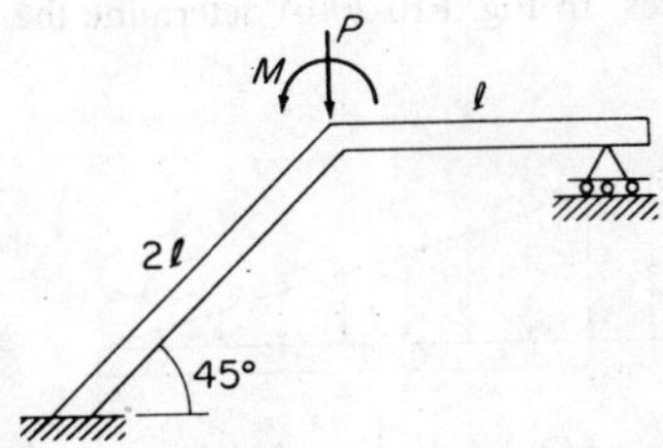

Figure P10-17.

10-18 Determine the deflection and slope at the load for the following frames. Consider the corners to be rigid.

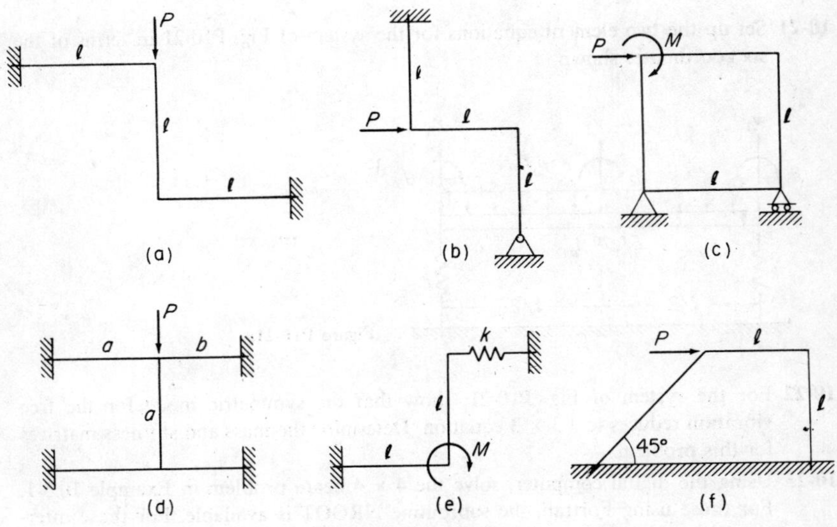

Figure P10-18.

10-19 Determine the free vibration equation for the following.

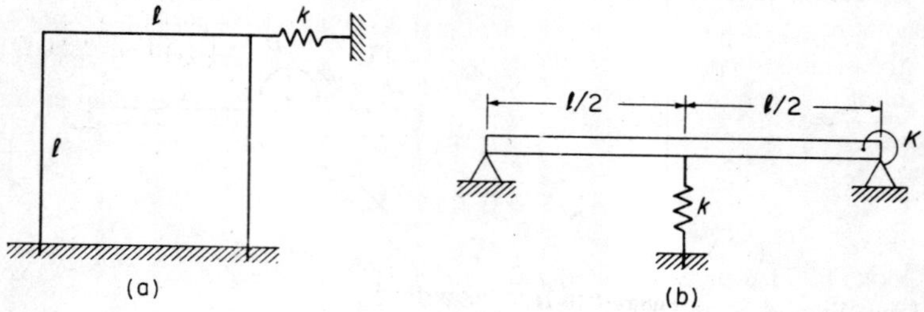

Figure P10-19.

10-20 Using two elements, determine the equivalent junction loads for the distributed forces. In Fig. P10-20(b) determine the deflection and slope at mid-span.

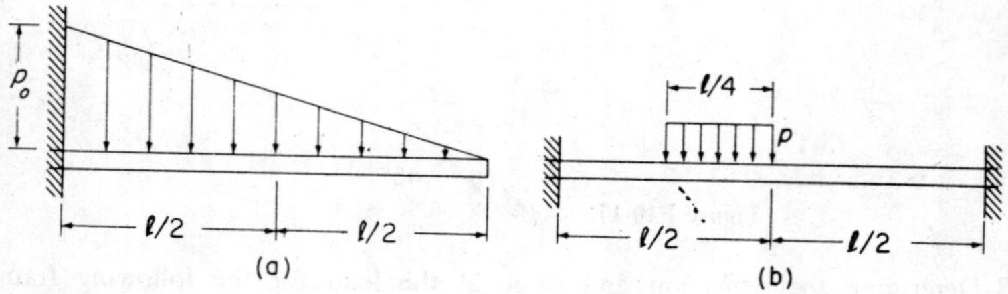

Figure P10-20.

10-21 Set up the two element equations for the system of Fig. P10-21 in terms of the six coordinates shown.

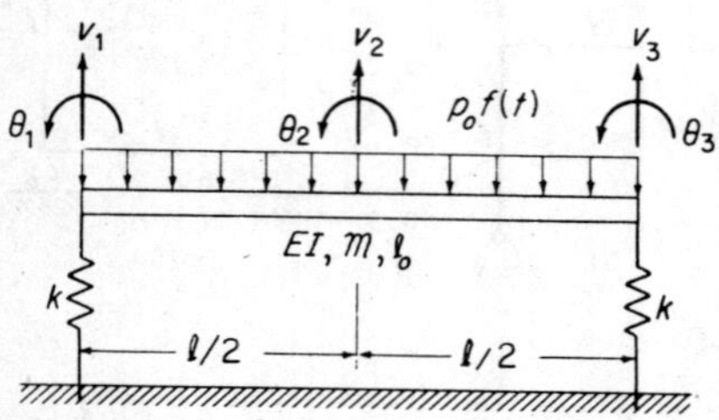

Figure P10-21.

10-22 For the system of Fig. P10-21, show that the symmetric mode for the free vibration reduces to a 3×3 equation. Determine the mass and stiffness matrices for this problem.

10-23 Using the digital computer, solve the 4×4 beam problem in Example 10.5-1. For those using Fortran, the subroutine NROOT is available. For those interested in using the standard form of the eigen-problem $[I\lambda - A]\ X = 0$, the method presented in Example 6.4-1 can be used. However, with M a full matrix, a separate determination of $Q = M^{\frac{1}{2}}$ must first be made from the relationship $Q = \phi\Lambda^{\frac{1}{2}}\phi^T$, where Λ is the diagonal matrix of the eigenvalues of M and ϕ is the orthonormal matrix of the eigenvectors of M, i.e. $\phi^T\phi = I$. The matrix M must here be positive definite.

REFERENCES

[1] COOK, R. D. *Concepts and Applications of Finite Element Analysis*. New York: John Wiley & Sons, 1974.

[2] GALLAGHER, R. H. *Finite Element Analysis Fundamentals*. Englewood Cliffs, N. J.: Prentice-Hall, Inc., 1975.

[3] ROCKEY, K. C., EVANS, H. R., GRIFFITH, D. W. AND NETHEROOT, D. A. *The Finite Element Method*. New York: Halsted Press Book, John Wiley & Sons, 1975.

[4] YANG, T. Y., *Finite Element Structural Analysis*. Englewood Cliffs, N.J.: Prentice-Hall, Inc., 1986.

11

APPROXIMATE NUMERICAL METHODS

The exact analysis for the vibration of systems of many degrees of freedom is generally difficult, and its associated calculations are laborious. In many cases, all the normal modes of the system are not required, and an estimate of the fundamental and a few of the lower modes is sufficient. In this chapter some of the approximate methods for determining the natural frequencies and mode shapes of the first few vibration modes will be presented.

11.1 RAYLEIGH METHOD

The fundamental frequency of multi-DOF systems is often of greater interest than its higher natural frequencies because its forced response in many cases is the largest. In Chapter 2, under the energy method, Rayleigh's method was introduced to obtain a better estimate of the fundamental frequency of systems which contained flexible elements such as springs and beams. In this section we wish to examine the Rayleigh method in light of the matrix techniques presented in previous chapters and show that the Rayleigh frequency approaches the fundamental frequency from the high side.

Let M and K be the mass and stiffness matrices and X the assumed displacement vector for the amplitude of vibration. Then for harmonic motion, the maximum kinetic and potential energies can be written as

$$T_{\max} = \frac{1}{2}\omega^2 X'MX \qquad (11.1\text{-}1)$$

and

$$U_{max} = \frac{1}{2} X'KX \tag{11.1-2}$$

Equating the two and solving for ω^2, we obtain the Rayleigh quotient

$$\omega^2 = \frac{X'KX}{X'MX} \tag{11.1-3}$$

This quotient approaches the lowest natural frequency (or fundamental frequency) from the high side, and its value is somewhat insensitive to the choice of the assumed amplitudes. To show these qualities, we will express the assumed displacement curve in terms of the normal modes X_i as follows

$$X = X_1 + C_2 X_2 + C_3 X_3 + \cdots \tag{11.1-4}$$

Then

$$X'KX = X_1'KX_1 + C_2^2 X_2'KX_2 + C_3^2 X_3'KX_3 + \cdots$$

and

$$X'MX = X_1'MX_1 + C_2^2 X_2'MX_2 + C_3^2 X_3'MX_3 + \cdots$$

where cross terms of the form $X_i'KX_j$ and $X_i'MX_j$ have been eliminated by the orthogonality conditions.

Noting that

$$X_i'KX_i = \omega_i^2 X_i'MX_i \tag{11.1-5}$$

the Rayleigh quotient becomes

$$\omega^2 = \omega_1^2 \left[1 + C_2^2 \left(\frac{\omega_2^2}{\omega_1^2} - 1 \right) \frac{X_2'MX_2}{X_1'MX_1} + \cdots \right] \tag{11.1-6}$$

If $X_i'MX_i$ is normalized to the same number, this equation reduces to

$$\omega^2 = \omega_1^2 \left[1 + C_2^2 \left(\frac{\omega_2^2}{\omega_1^2} - 1 \right) + \cdots \right] \tag{11.1-7}$$

It is evident, then, that ω^2 is greater than ω_1^2 because $\omega_2^2/\omega_1^2 > 1$. Since C_2 represents the deviation of the assumed amplitudes from the exact amplitudes X_1, the error in the computed frequency is only proportional to the square of the deviation of the assumed amplitudes from their exact values.

This analysis shows that if the exact fundamental deflection (or mode) X_1 is assumed, the fundamental frequency found by this method will be the correct frequency, since C_2, C_3, and so on, will then be zero. For any other curve, the frequency determined will be higher than the fundamental. This fact can be explained on the basis that any deviation from the natural curve requires additional constraints, a condition that implies greater stiffness and higher frequency. In general, the use of the static deflection curve of the elastic body results in a fairly accurate value of the fundamental frequency. If greater accuracy is desired, the approximate curve can be repeatedly improved.

In our previous discussion of the Rayleigh method the potential energy was determined by the work done by the static weights in the assumed deformation. This work is, of course, stored in the flexible member as strain energy. For beams, the elastic strain energy can be calculated in terms of its flexural rigidity EI.

Letting M be the bending moment and θ the slope of the elastic curve, the strain energy stored in an infinitesimal beam element is

$$dU = \frac{1}{2} M \, d\theta \tag{11.1-8}$$

Since the deflection in beams is generally small, the following geometric relations are assumed to hold (see Fig. 11.1-1).

$$\theta = \frac{dy}{dx} \qquad \frac{1}{R} = \frac{d\theta}{dx} = \frac{d^2y}{dx^2}$$

In addition, we have, from the theory of beams, the flexure equation

$$\frac{1}{R} = \frac{M}{EI} \tag{11.1-9}$$

where R is the radius of curvature. Substituting for $d\theta$ and $1/R$, U may be written as

$$U_{\max} = \frac{1}{2} \int \frac{M^2}{EI} \, dx = \frac{1}{2} \int EI \left(\frac{d^2y}{dx^2} \right)^2 dx \tag{11.1-10}$$

where the integration is carried out over the entire beam.

The kinetic energy is simply

$$T_{\max} = \frac{1}{2} \int \dot{y}^2 \, dm = \frac{1}{2} \omega^2 \int y^2 \, dm \tag{11.1-11}$$

where y is the assumed deflection curve. Thus, by equating the kinetic and potential energies, an alternative equation for the fundamental frequency of

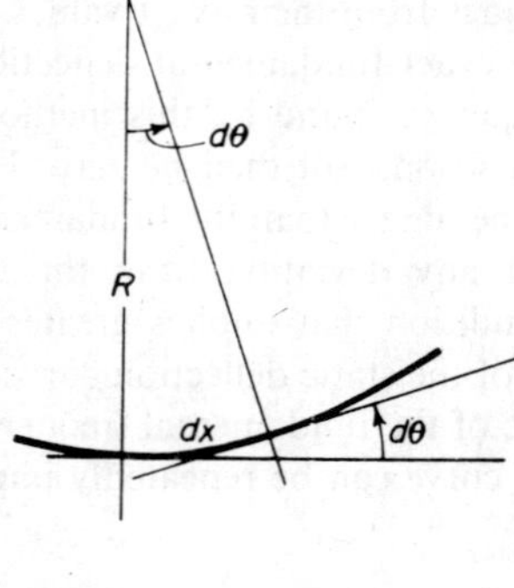

Figure 11.1-1.

the beam is

$$\omega^2 = \frac{\int EI\left(\frac{d^2y}{dx^2}\right)^2 dx}{\int y^2\, dm} \tag{11.1-12}$$

Example 11.1-1

In applying this procedure to a simply supported beam of uniform cross section, shown in Fig. 11.1-2, we assume the deflection to be represented by a sine curve as follows:

$$y = \left(y_0 \sin\frac{\pi x}{l}\right)\sin\omega t$$

where y_0 is the maximum deflection at midspan. The second derivative then becomes

$$\frac{d^2y}{dx^2} = -\left(\frac{\pi}{l}\right)^2 y_0 \sin\frac{\pi x}{l}\sin\omega t$$

Substituting into Eq. (11.1-12) we obtain

$$\omega^2 = \frac{EI\left(\frac{\pi}{l}\right)^4 \int_0^l \sin^2\frac{\pi x}{l}\,dx}{m\int_0^l \sin^2\frac{\pi x}{l}\,dx} = \pi^4\frac{EI}{ml^4}$$

The fundamental frequency is therefore found to be

$$\omega_1 = \pi^2\sqrt{\frac{EI}{ml^4}}$$

In this case, the assumed curve happended to be the natural vibration curve, and the exact frequency is obtained by Rayleigh's method. Any other curve assumed for the case can be considered to be the result of additional constraints, or stiffness, which will result in a constant greater than π^2 in the frequency equation.

Example 11.1-2

If the distance between the ends of the beam of Fig. 11.1-2 is rigidly fixed, a tensile stress σ will be developed by the lateral deflection. Account for this additional strain energy in the frequency equation.

Solution: Due to the lateral deflection, the length dx of the beam is increased by an amount

$$\left[\sqrt{1 + (dy/dx)^2} - 1\right] dx \cong \frac{1}{2}\left(\frac{dy}{dx}\right)^2 dx$$

The additional strain energy in the element dx is

$$dU = \tfrac{1}{2}\sigma A\epsilon\, dx = \tfrac{1}{2}EA\epsilon^2\, dx$$

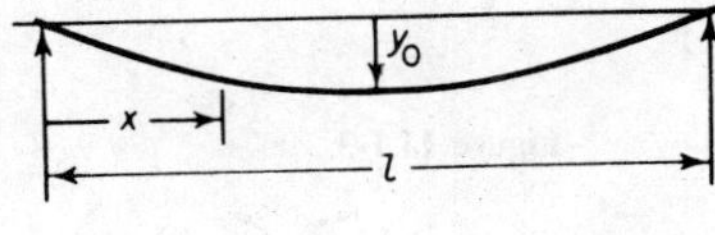

Figure 11.1-2.

where A is the cross-sectional area, σ is the stress due to tension, and $\epsilon = \frac{1}{2}(dy/dx)^2$ is the unit strain.

Equating the kinetic energy to the total strain energy of bending and tension, we obtain

$$\frac{1}{2}\omega^2 \int y^2\, dm = \frac{1}{2}\int EI\left(\frac{d^2y}{dx^2}\right)^2 dx + \frac{1}{2}\int \frac{EA}{4}\left(\frac{dy}{dx}\right)^4 dx$$

The preceding equation then leads to the frequency equation

$$\omega_1^2 = \frac{\int EI\left(\frac{d^2y}{dx^2}\right)^2 dx + \int \frac{EA}{4}\left(\frac{dy}{dx}\right)^4 dx}{\int y^2\, dm}$$

which contains an additional term due to the tension.

Accuracy of the integral method over differentiation. In using Rayleigh's method of determining the fundamental frequency, we must choose an assumed curve. Although the deviation of this assumed deflection curve compared to the exact curve may be slight, its derivative could be in error by a large amount and hence the strain energy computed from the equation

$$U = \frac{1}{2}\int EI\left(\frac{d^2y}{dx^2}\right)^2 dx$$

may be inaccurate. To avoid this difficulty, the following integral method for the evaluation of U is recommended for some beam problems.

We first recognize that the shear V is the integral of the inertia loading $m\omega^2 y$ from the free end of the beam, as indicated by both Figs. 11.1-3 and 11.1-4.

$$V(\xi) = \omega^2 \int_\xi^l m(\xi)\, y(\xi)\, d\xi \tag{11.1-13}$$

Since the bending moment is related to the shear by the equation

$$\frac{dM}{dx} = V \tag{11.1-14}$$

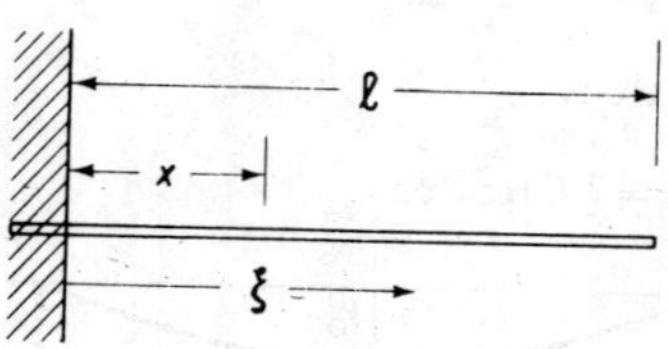

Figure 11.1-3.

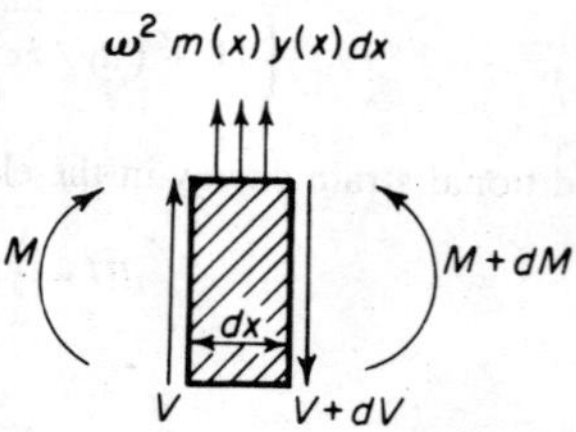

Figure 11.1-4. Free-body diagram of beam element.

the moment at x is found from the integral

$$M(x) = \int_x^l V(\xi)\, d\xi \tag{11.1-15}$$

The strain energy of the beam is then found from

$$U = \frac{1}{2}\int_0^l \frac{M(x)^2}{EI}\, dx \tag{11.1-16}$$

which avoids any differentiation of the assumed deflection curve.

Example 11.1-3

Determine the fundamental frequency of the uniform cantilever beam shown in Fig. 11.1-5, using the simple curve $y = cx^2$.

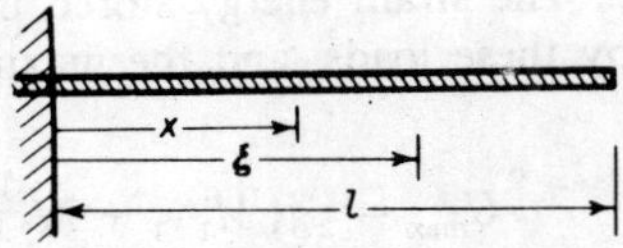

Figure 11.1-5.

Solution: If we use Eq. (11.1-12), we will find the result to be very much in error since the above curve does not satisfy the boundary conditions at the free end. By using Eq. (11.1-12) we obtain

$$\omega = 4.47\sqrt{\frac{EI}{ml^4}}$$

whereas the exact value is

$$\omega_1 = 3.52\sqrt{\frac{EI}{ml^4}}$$

Acceptable results using the given curve can be found by the procedure outlined in the previous section.

$$V(\xi) = \omega^2 \int_\xi^l mc\xi^2\, d\xi = \frac{\omega^2 mc}{3}(l^3 - \xi^3)$$

and the bending moment becomes

$$M(x) = \int_x^l V(\xi)\, d\xi = \frac{\omega^2 mc}{3}\int_x^l (l^3 - \xi^3)\, d\xi$$

$$= \frac{\omega^2 mc}{12}(3l^4 - 4l^3x + x^4)$$

The maximum strain energy is found by substituting $M(x)$ into U_{max}:

$$U_{max} = \frac{1}{2EI}\left(\frac{\omega^2 mc}{12}\right)^2 \int_0^l (3l^4 - 4l^3x + x^4)^2\, dx$$

$$= \frac{\omega^4}{2EI}\frac{m^2c^2}{144}\frac{312}{135}l^9$$

The maximum kinetic energy is

$$T_{max} = \tfrac{1}{2}\int_0^l \dot{y}^2 m\,dx = \tfrac{1}{2}c^2\omega^2 m\int_0^l x^4\,dx = \tfrac{1}{2}c^2\omega^2 m\frac{l^5}{5}$$

By equating these results, we obtain

$$\omega_1 = \sqrt{\frac{12.47EI}{ml^4}} = 3.53\sqrt{\frac{EI}{ml^4}}$$

which is very close to the exact result.

Lumped masses. The Rayleigh method can be used to determine the fundamental frequency of a beam or shaft represented by a series of lumped masses. As a first approximation we will assume a static deflection curve due to loads M_1g, M_2g, M_3g, and so on, with corresponding deflections $y_1, y_2, y_3, \ldots$. The strain energy stored in the beam is determined from the work done by these loads, and the maximum potential and kinetic energies become

$$U_{max} = \tfrac{1}{2}g(M_1y_1 + M_2y_2 + M_3y_3 + \cdots) \tag{11.1-17}$$

$$T_{max} = \tfrac{1}{2}\omega^2(M_1y_1^2 + M_2y_2^2 + M_3y_3^2 + \cdots) \tag{11.1-18}$$

Equating the two, the frequency equation is established as

$$\omega_1^2 = \frac{g\Sigma_i M_i y_i}{\Sigma_i M_i y_i^2} \tag{11.1-19}$$

Example 11.1-4

Calculate the first approximation to the fundamental frequency of lateral vibration for the system shown in Fig. 11.1-6.

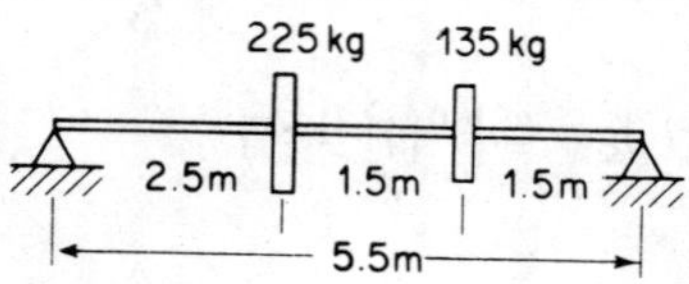

Figure 11.1-6.

Solution: Referring to the table at the end of Chapter 2, we see that the deflection of the beam at any point x from the left end due to a single load W at a distance b from the right end is

$$y(x) = \frac{Wbx}{6EIl}(l^2 - x^2 - b^2) \qquad x \le (l - b)$$

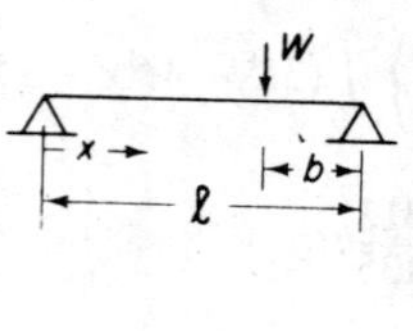

The deflections at the loads can be obtained from the superposition of the deflections due to each load acting separately.

Due to the 135-kg mass, we have

$$y_1' = \frac{(9.81 \times 135) \times 1.5 \times 2.5}{6 \times 5.5EI}(5.5^2 - 2.5^2 - 1.5^2) = 3.273 \times \frac{10^3}{EI} m$$

$$y_2' = \frac{(9.81 \times 135) \times 1.5 \times 4}{6 \times 5.5EI}(5.5^2 - 4.0^2 - 1.5^2) = 2.889 \times \frac{10^3}{EI} m$$

Due to the 225-kg mass, the deflections at the corresponding points are

$$y_1'' = \frac{(9.81 \times 225) \times 2.5 \times 3.0}{6 \times 5.5EI}(5.5^2 - 3.0^2 - 2.5^2) = 7.524 \times \frac{10^3}{EI} m$$

$$y_2'' = \frac{(9.81 \times 225) \times 2.5 \times 1.5}{6 \times 5.5EI}(5.5^2 - 1.5^2 - 2.5^2) = 5.455 \times \frac{10^3}{EI} m$$

Adding y' and y'' the deflections at 1 and 2 become

$$y_1 = 10.797 \times \frac{10^3}{EI} m, \qquad y_2 = 8.344 \times \frac{10^3}{EI} m$$

Substituting into Eq. (11.1-19), the first approximation to the fundamental frequency is

$$\omega_1 = \sqrt{\frac{9.81(225 \times 10.797 + 135 \times 8.344)\,EI}{(225 \times 10.797^2 + 135 \times 8.344^2)10^3}}$$

$$= 0.03129\sqrt{EI}\ \text{rad/s}$$

If further accuracy is desired, a better approximation to the dynamic curve can be made by using the dynamic loads $m\omega^2 y$. Since the dynamic loads are proportional to the deflection y, we can recalculate the deflection with the modified loads gm_1 and $gm_2(y_2/y_1)$.

11.2 DUNKERLEY'S EQUATION

The Rayleigh method, which gives the upper bound to the fundamental frequency, can now be complemented by Dunkerley's† equation, which results in a lower bound to the fundamental frequency. For the basis of the Dunkerley equation, we examine the characteristic equation (6.4-12) formulated from the flexibility coefficients, which is

$$\begin{vmatrix} \left(a_{11}m_1 - \frac{1}{\omega^2}\right) & a_{12}m_2 & a_{13}m_3 \\ a_{21}m_1 & \left(a_{22}m_2 - \frac{1}{\omega^2}\right) & a_{23}m_3 \\ a_{31}m_1 & a_{32}m_2 & \left(a_{33}m_3 - \frac{1}{\omega^2}\right) \end{vmatrix} = 0$$

†S. Dunkerley, "On the Whirling and Vibration of Shafts," *Phil. Trans. Roy. Soc.*, 185 (1895), pp. 269–360.

Expanding this determinant, we obtain the third-degree equation in $1/\omega^2$.

$$\left(\frac{1}{\omega^2}\right)^3 - (a_{11}m_1 + a_{22}m_2 + a_{33}m_3)\left(\frac{1}{\omega^2}\right)^2 + \cdots = 0 \qquad (11.2\text{-}1)$$

If the roots of this equation are $1/\omega_1^2$, $1/\omega_2^2$, and $1/\omega_3^2$, the above equation can be factored into the following form:

$$\left(\frac{1}{\omega^2} - \frac{1}{\omega_1^2}\right)\left(\frac{1}{\omega^2} - \frac{1}{\omega_2^2}\right)\left(\frac{1}{\omega^2} - \frac{1}{\omega_3^2}\right) = 0$$

or

$$\left(\frac{1}{\omega^2}\right)^3 - \left(\frac{1}{\omega_1^2} + \frac{1}{\omega_2^2} + \frac{1}{\omega_3^2}\right)\left(\frac{1}{\omega^2}\right)^2 + \cdots = 0 \qquad (11.2\text{-}2)$$

As is well known in algebra, the coefficient of the second highest power is equal to the sum of the roots of the characteristic equation. It is also equal to the sum of the diagonal terms of the matrix A^{-1}, which is called the *trace* of the matrix (see Appendix C):

$$\text{trace}\, A^{-1} = \sum_{i=1}^{3}\left(\frac{1}{\omega_i^2}\right)$$

These relationships are true for n greater than 3, and we can write for an n-DOF system the following equation:

$$\frac{1}{\omega_1^2} + \frac{1}{\omega_2^2} + \cdots + \frac{1}{\omega_n^2} = a_{11}m_1 + a_{22}m_2 + \cdots + a_{nn}m_n \qquad (11.2\text{-}3)$$

The estimate to the fundamental frequency is made by recognizing that $\omega_2, \omega_3, \ldots,$ are natural frequencies of higher modes and hence $1/\omega_2^2, 1/\omega_3^2, \ldots,$ can be neglected in the left side of Eq. (11.2-3). The term $1/\omega_1^2$ is consequently larger than the true value, and therefore ω_1 is smaller than the exact value of the fundamental frequency. Dunkerley's estimate of the fundamental frequency is then made from the equation

$$\frac{1}{\omega_1^2} < a_{11}m_1 + a_{22}m_2 + \cdots + a_{nn}m_n \qquad (11.2\text{-}4)$$

Since the left side of the equation has the dimension of the reciprocal of the frequency squared, each term on the right side must also be of the same dimension. Each term on the right side must then be considered to be the contribution to $1/\omega_1^2$ in the absence of all other masses, and for convenience we will let $a_{ii}m_i = 1/\omega_{ii}^2$, or

$$\frac{1}{\omega_1^2} < \frac{1}{\omega_{11}^2} + \frac{1}{\omega_{22}^2} + \cdots + \frac{1}{\omega_{nn}^2} \qquad (11.2\text{-}5)$$

Thus, the right side becomes the sum of the effect of each mass acting in the absence of all other masses.

Example 11.2-1

Dunkerley's equation is useful for estimating the fundamental frequency of a structure undergoing vibration testing. Natural frequencies of structures are often determined by attaching to the structure an eccentric mass exciter and noting the frequencies corresponding to the maximum amplitude. The frequencies so measured represent those of the structure plus exciter and may deviate considerably from the natural frequencies of the structure itself when the mass of the exciter is a substantial percentage of the total mass. In such cases the fundamental frequency of the structure by itself may be determined by the following equation:

$$\frac{1}{\omega_1^2} = \frac{1}{\omega_{11}^2} + \frac{1}{\omega_{22}^2} \tag{a}$$

where ω_1 = fundamental frequency of structure plus exciter

ω_{11} = fundamental frequency of the structure by itself

ω_{22} = natural frequency of exciter mounted on the structure in the absence of other masses

It is sometimes convenient to express the equation in another form, for instance

$$\frac{1}{\omega_1^2} = \frac{1}{\omega_{11}^2} + a_{22} m_2 \tag{b}$$

where m_2 is the mass of the concentrated weight or exciter and a_{22} the influence coefficient of the structure at the point of attachment of the exciter.

Example 11.2-2

An airplane rudder tab showed a resonant frequency of 30 cps when vibrated by an eccentric mass shaker weighing 1.5 lb. By attaching an additional weight of 1.5 lb to the shaker, the resonant frequency was lowered to 24 cps. Determine the true natural frequency of the tab.

Solution: The measured resonant frequencies are those due to the total mass of the tab and shaker. Letting f_{11} be the true natural frequency of the tab and substituting into Eq. (b) of Example 11.2-1, we obtain

$$\frac{1}{(2\pi \times 30)^2} = \frac{1}{(2\pi f_{11})^2} + \frac{1.5}{386} a_{22}$$

$$\frac{1}{(2\pi \times 24)^2} = \frac{1}{(2\pi f_{11})^2} + \frac{3.0}{386} a_{22}$$

Eliminating a_{22}, the true natural frequency is

$$f_{11} = 45.3 \text{ cps}$$

The rigidity of stiffness of the tab at the point of attachment of the shaker may be determined from $1/a_{22}$, which from the same equations is found to be

$$k_2 = \frac{1}{a_{22}} = \frac{1}{0.00407} = 246 \text{ lb/in.}$$

Example 11.2-3

Determine the fundamental frequency of a uniformly loaded cantilever beam with a concentrated mass M at the end, equal to the mass of the uniform beam (see Fig. 11.2-1).

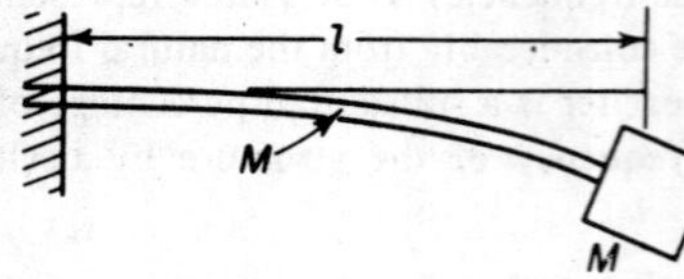

Figure 11.2-1.

Solution: The frequency equation for the uniformly loaded beam by itself is

$$\omega_{11}^2 = 3.515^2\left(\frac{EI}{Ml^3}\right)$$

For the concentrated mass by itself attached to a weightless cantilever beam, we have

$$\omega_{22}^2 = 3.00\left(\frac{EI}{Ml^3}\right)$$

Substituting into Dunkerley's formula rearranged in the following form, the natural frequency of the system is determined as

$$\omega_1^2 = \frac{\omega_{11}^2\omega_{22}^2}{\omega_{11}^2 + \omega_{22}^2} = \frac{3.515^2 \times 3.0}{3.515^2 + 3.0}\left(\frac{EI}{Ml^3}\right) = 2.41\left(\frac{EI}{Ml^3}\right)$$

This result may be compared to the frequency equation obtained by Rayleigh's method, which is

$$\omega_1^2 = \frac{3EI}{\left(1 + \frac{33}{140}\right)Ml^3} = 2.43\left(\frac{EI}{Ml^3}\right)$$

Example 11.2-4

The natural frequency of a given airplane wing in torsion is 1600 cpm. What will be the new torsional frequency if a 1000-lb fuel tank is hung at a position one-sixth of the semispan from the center line of the airplane such that its moment of inertia about the torsional axis is 1800 lb in s^2? The torsional stiffness of the wing at this point is 60×10^6 lb in/rad.

Solution: The frequency of the tank attached to the weightless wing is

$$f_{22} = \frac{1}{2\pi}\sqrt{\frac{60 \times 10^6}{1800}} = 29.1 \text{ cps} = 1745 \text{ cpm}$$

The new torsional frequency with the tank, from Eq. (a) of Example 11.2-1 then becomes

$$\frac{1}{f_1^2} = \frac{1}{1600^2} + \frac{1}{1745^2}, \qquad f_1 = 1180 \text{ cpm}$$

Example 11.2-5

The fundamental frequency of a uniform beam of mass M, simply supported as in Fig. 11.2-2, is equal to $\pi^2\sqrt{EI/Ml^3}$. If a lumped mass m_0 is attached to the beam at $x = l/3$, determine the new fundamental frequency.

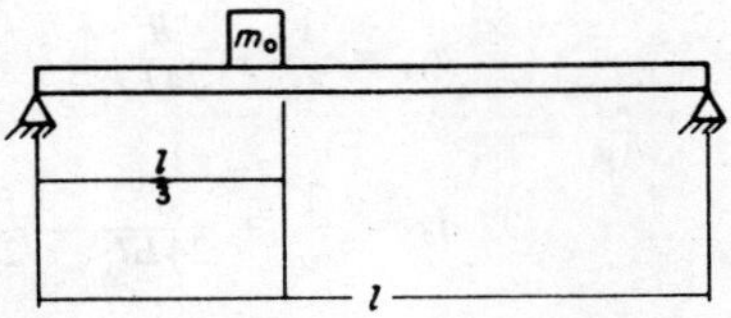

Figure 11.2-2.

Solution: Starting with Eq. (b) of Example 11.2-1, we let ω_{11} be the fundamental frequency of the uniform beam and ω_1 the new fundamental frequency with m_0 attached to the beam. Multiplying through Eq. (b) by ω_1^2, we have

$$1 = \left(\frac{\omega_1}{\omega_{11}}\right)^2 + a_{22}m_0\omega_{11}^2\left(\frac{\omega_1}{\omega_{11}}\right)^2$$

or

$$\left(\frac{\omega_1}{\omega_{11}}\right)^2 = \frac{1}{1 + a_{22}m_0\omega_{11}^2}$$

The quantity a_{22} is the influence coefficient at $x = l/3$ due to a unit load applied at the same point. It can be found from the beam formula in Example 11.1-4 to be

$$a_{22} = \frac{8}{6 \times 81}\frac{l^3}{EI}$$

Substituting $\omega_{11}^2 = \pi^4 EI/Ml^3$ together with a_{22}, we obtain the convenient equation

$$\left(\frac{\omega_1}{\omega_{11}}\right)^2 = \frac{1}{1 + \dfrac{8\pi^4}{6 \times 81}\dfrac{m_0}{M}} = \frac{1}{1 + 1.6\dfrac{m_0}{M}}$$

Example 11.2-6

Determine the fundamental frequency of the three-story building shown in Fig. 11.2-3, where the foundation is capable of translation.

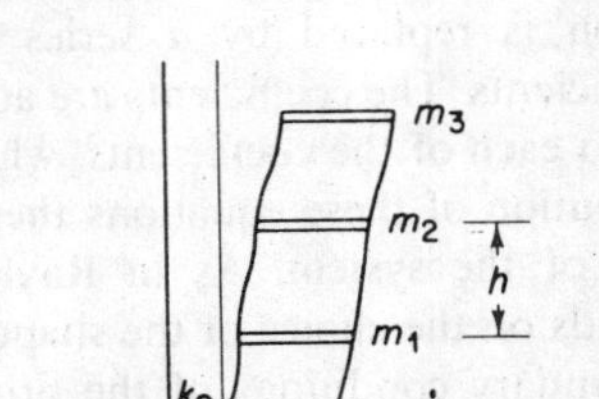

Figure 11.2-3.

Solution: If a unit force is placed at each floor, the influence coefficients are found to be

$$a_{00} = \frac{1}{k_0}$$

$$a_{11} = \frac{1}{k_0} + \frac{h^3}{24EI_1}$$

$$a_{22} = \frac{1}{k_0} + \frac{h^3}{24EI_1} + \frac{h^3}{24EI_2}$$

$$a_{33} = \frac{1}{k_0} + \frac{h^3}{24EI_1} + \frac{h^3}{24EI_2} + \frac{h^3}{24EI_3}$$

The Dunkerley equation then becomes

$$\frac{1}{\omega_1^2} = \frac{m_0}{k_0} + \left(\frac{1}{k_0} + \frac{h^3}{24EI_1}\right)m_1 = \left(\frac{1}{k_0} + \frac{h^3}{24EI_1} + \frac{h^3}{24EI_2}\right)m_2$$

$$+\left(\frac{1}{k_0} + \frac{h^3}{24EI_1} + \frac{h^3}{24EI_2} + \frac{h^3}{24EI_3}\right)m_3$$

If the columns are of equal stiffness, the above equation reduces to

$$\frac{1}{\omega_1^2} = \frac{1}{k_0}(m_0 + m_1 + m_2 + m_3) + m_1\frac{h^3}{24EI} + m_2\frac{2h^3}{24EI} + m_3\frac{3h^3}{24EI}$$

11.3 RAYLEIGH-RITZ METHOD

W. Ritz developed a method that is an extension of Rayleigh's method. It not only provides a means of obtaining a more accurate value for the fundamental frequency, but it also gives approximations to the higher frequencies and mode shapes.

The Ritz method is essentially the Rayleigh method in which the single shape function is replaced by a series of shape functions multiplied by constant coefficients. The coefficients are adjusted by minimizing the frequency with respect to each of the coefficients, which results in n algebraic equations in ω^2. The solution of these equations then gives the natural frequencies and mode shapes of the system. As in Rayleigh's method, the success of the method depends on the choice of the shape functions which should satisfy the geometric boundary conditions of the problem. The method should also be differentiable, at least to the order of the derivatives appearing in the energy equations. The functions, however, can disregard discontinuities such as those of shear due to concentrated masses which involve third derivatives in beams.

We now outline in a general manner the procedure of the Rayleigh-Ritz method, starting with Rayleigh's equation

$$\omega^2 = \frac{U_{max}}{T^*_{max}} \tag{11.3-1}$$

where the kinetic energy is expressed as $\omega^2 T^*_{max}$. In the Rayleigh method a single function is chosen for the deflection; Ritz, however, assumed the deflection to be a sum of several functions multiplied by constants, as follows:

$$v(x) = C_1\phi_1(x) + C_2\phi_2(x) + \cdots + C_n\phi_n(x) \tag{11.3-2}$$

where $\phi_i(x)$ are any admissible functions satisfying the boundary conditions. U_{max} and T_{max} are expressible in the form of Eqs. (7.4-5) and (7.4-8).

$$\begin{aligned} U &= \tfrac{1}{2}\sum_i \sum_j k_{ij} C_i C_j \\ T^* &= \tfrac{1}{2}\sum_i \sum_j m_{ij} C_i C_j \end{aligned} \tag{11.3-3}$$

where k_{ij} and m_{ij} depend on the type of problem. For example, for the beam we have

$$k_{ij} = \int EI\phi_i''\phi_j''\,dx \quad \text{and} \quad m_{ij} = \int m\phi_i\phi_j\,dx$$

whereas for the longitudinal oscillation of slender rods

$$k_{ij} = \int EA\phi_i'\phi_j'\,dx \quad \text{and} \quad m_{ij} = \int m\phi_i\phi_j\,dx$$

We now minimize ω^2 by differentiating it with respect to each of the constants. For example, the derivative of ω^2 with respect to C_i is

$$\frac{\partial \omega^2}{\partial C_i} = \frac{\partial}{\partial C_i}\left(\frac{U_{max}}{T^*_{max}}\right) = \frac{T^*_{max}\dfrac{\partial U_{max}}{\partial C_i} - U_{max}\dfrac{\partial T^*_{max}}{\partial C_i}}{T^{*2}_{max}} = 0 \tag{11.3-4}$$

which is satisfied by

$$\frac{\partial U_{max}}{\partial C_i} - \frac{U_{max}}{T^*_{max}}\frac{\partial T^*_{max}}{\partial C_i} = 0$$

or since $U_{max}/T^*_{max} = \omega^2$,

$$\frac{\partial U_{max}}{\partial C_i} - \omega^2 \frac{\partial T^*_{max}}{\partial C_i} = 0 \tag{11.3-5}$$

The two terms in this equation are then

$$\frac{\partial U_{max}}{\partial C_i} = \sum_j^n k_{ij} C_j \quad \text{and} \quad \frac{\partial T^*_{max}}{\partial C_i} = \sum_j^n m_{ij} C_j$$

and so Eq. (11.3-5) becomes

$$C_1(k_{i1} - \omega^2 m_{i1}) + C_2(k_{i2} - \omega^2 m_{i2}) + \cdots + C_n(k_{in} - \omega^2 m_{in}) = 0 \tag{11.3-6}$$

With i varying from 1 to n there will be n such equations, which can be arranged in matrix form as

$$\begin{bmatrix} (k_{11} - \omega^2 m_{11}) & (k_{12} - \omega^2 m_{12}) & \cdots & (k_{1n} - \omega^2 m_{1n}) \\ (k_{21} - \omega^2 m_{21}) & \cdots & & \vdots \\ \vdots & & & \\ (k_{n1} - \omega^2 m_{n1}) & \cdots & & (k_{nn} - \omega^2 m_{nn}) \end{bmatrix} \begin{Bmatrix} C_1 \\ C_2 \\ \vdots \\ C_n \end{Bmatrix} = 0 \tag{11.3-7}$$

The determinant of this equation is an n-degree algebraic equation in ω^2, and its solution results in the n natural frequencies. The mode shape is also obtained by solving for the C_s for each natural frequency and substituting into Eq. (11.3-2) for the deflection.

Example 11.3-1

Figure 11.3-1 shows a wedge-shaped plate of constant thickness fixed into a rigid wall. Determine the first two natural frequencies and mode shapes in longitudinal oscillation by using the Rayleigh-Ritz method.

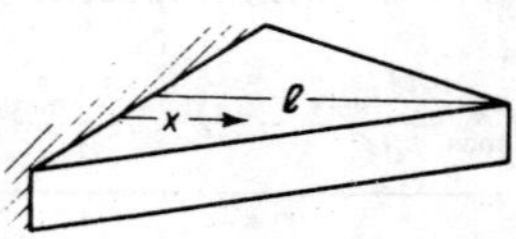

Figure 11.3-1.

Solution: For the displacement function we choose the first two longitudinal modes of a uniform rod clamped at one end.

$$\begin{aligned} u(x) &= C_1 \sin\frac{\pi x}{2l} + C_2 \sin\frac{3\pi x}{2l} \\ &= C_1\phi_1(x) + C_2\phi_2(x) \end{aligned} \tag{a}$$

The mass per unit length and the stiffness at x are

$$m(x) = m_0\left(1 - \frac{x}{l}\right) \quad \text{and} \quad EA(x) = EA_0\left(1 - \frac{x}{l}\right)$$

The k_{ij} and the m_{ij} for the longitudinal modes are calculated from the equations

$$\begin{aligned} k_{ij} &= \int_0^l EA(x)\phi_i'\phi_j'\,dx \\ m_{ij} &= \int_0^l m(x)\phi_i\phi_j\,dx \end{aligned} \tag{b}$$

$$k_{11} = \frac{\pi^2}{4l^2}EA_0\int_0^l\left(1 - \frac{x}{l}\right)\cos^2\frac{\pi x}{2l}\,dx = \frac{EA_0}{2l}\left(\frac{\pi^2}{8} + \frac{1}{2}\right)$$

$$= 0.86685\frac{EA_0}{l}$$

$$k_{12} = \frac{3\pi^2}{4l^2}EA_0\int_0^l\left(1 - \frac{x}{l}\right)\cos\frac{\pi x}{2l}\cos\frac{3\pi x}{2l}\,dx = 0.750\frac{EA_0}{l}$$

$$k_{22} = \frac{9\pi^2}{4l^2}EA_0\int_0^l\left(1 - \frac{x}{l}\right)\cos^2\frac{3\pi x}{2l}\,dx = \frac{EA_0}{2l}\left(\frac{9\pi^2}{8} + \frac{1}{2}\right)$$

$$= 5.80165\frac{EA_0}{l}$$

$$m_{11} = m_0\int_0^l\left(1 - \frac{x}{l}\right)\sin^2\frac{\pi x}{2l}\,dx = m_0 l\left(\frac{1}{4} - \frac{1}{\pi^2}\right) = 0.148679 m_0 l$$

$$m_{12} = m_0\int_0^l\left(1 - \frac{x}{l}\right)\sin\frac{\pi x}{2l}\sin\frac{3\pi x}{2l}\,dx = m_0 l\left(\frac{1}{\pi^2}\right) = 0.101321 m_0 l$$

$$m_{22} = m_0\int_0^l\left(1 - \frac{x}{l}\right)\sin^2\frac{3\pi x}{l}\,dx = m_0 l\left(\frac{1}{4} - \frac{1}{9\pi^2}\right) = 0.238742 m_0 l \tag{b}$$

Substituting into Eq. (9.3-7), we obtain

$$\begin{bmatrix} \left(0.86685\frac{EA_0}{l} - 0.14868 m_0 l\omega^2\right) & \left(0.750\frac{EA_0}{l} - 0.10132 m_0 l\omega^2\right) \\ \left(0.750\frac{EA_0}{l} - 0.10132 m_0 l\omega^2\right) & \left(5.80165\frac{EA_0}{l} - 0.23874 m_0 l\omega^2\right) \end{bmatrix} \begin{Bmatrix} C_1 \\ C_2 \end{Bmatrix} = 0 \tag{c}$$

Setting the determinant of the preceding equation to zero, we obtain the frequency equation

$$\omega^4 - 36.3676\alpha\omega^2 + 177.0377\alpha^2 = 0 \tag{d}$$

where

$$\alpha = \frac{EA_0}{m_0 l^2} \tag{e}$$

The two roots of this equation are

$$\omega_1^2 = 5.7898\alpha \quad \text{and} \quad \omega_2^2 = 30.5778\alpha$$

Using these results in Eq. (c), we obtain

$$C_2 = 0.03689C_1 \qquad \text{for mode 1}$$

$$C_1 = -0.63819C_2 \qquad \text{for mode 2}$$

The two natural frequencies and mode shapes are then

$$\omega_1 = 2.4062\sqrt{\frac{EA_0}{m_0 l^2}} \qquad u_1(x) = 1.0\sin\frac{\pi x}{2l} + 0.03689\sin\frac{3\pi x}{2l}$$

$$\omega_2 = 5.5297\sqrt{\frac{EA_0}{m_0 l^2}} \qquad u_2(x) = -0.63819\sin\frac{\pi x}{2l} + 1.0\sin\frac{3\pi x}{2l}$$

11.4 METHOD OF MATRIX ITERATION

The equations of motion, formulated either on the basis of the stiffness equation or the flexibility equation, are similar in form and appear as

$$\begin{Bmatrix} x_1 \\ x_2 \\ \vdots \\ x_n \end{Bmatrix} = \lambda \begin{bmatrix} a_{11} & a_{12} & \cdots & a_{1n} \\ a_{21} & a_{22} & \cdots & a_{2n} \\ \vdots & \vdots & & \vdots \\ a_{n1} & a_{n2} & \cdots & a_{nn} \end{bmatrix} \begin{Bmatrix} x_1 \\ x_2 \\ \vdots \\ x_n \end{Bmatrix} \tag{11.4-1}$$

where λ is equal to $1/\omega^2$ for the stiffness formulation and to ω^2 for the flexibility formulation.

The iteration procedure is started by assuming a set of deflections for the right column of Eq. (11.4-1) and performing the indicated operation, which results in a column of numbers. This is then normalized by making one of the amplitudes equal to unity and dividing each term of the column by the particular amplitude which was normalized. The procedure is then repeated with the normalized column until the amplitudes stabilize to a definite pattern.

It will be shown in Sec. 11.5 that the iteration process converges to the lowest value of λ, so that for the equation formulated on the flexibility influence coefficients, the fundamental or the lowest mode of vibration is found. Likewise, for the equation formulated on the basis of the stiffness influence coefficients, the convergence is to the highest mode that corresponds to the lowest value of $\lambda = 1/\omega^2$.

Example 11.4-1

The uniform beam of Fig. 11.4-1, free to vibrate in the plane shown, has two concentrated masses, $m_1 = 500$ kg and $m_2 = 100$ kg. Determine the fundamental frequency of the system.

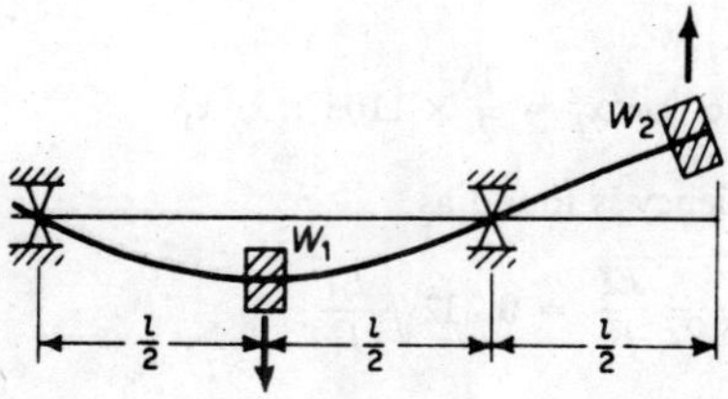

Figure 11.4-1.

Solution: The influence of coefficients for this problem, determined from deflection equations of beams by placing a unit load at positions 1 and 2, are

$$a_{11} = \frac{l^3}{48EI} = \tfrac{1}{6}a_{22}, \qquad a_{12} = a_{21} = \frac{l^3}{32EI} = \tfrac{1}{4}a_{22}, \qquad a_{22} = \frac{l^3}{8EI}.$$

Substituting into Eq. 11.4-1, we have

$$\begin{bmatrix} x_1 \\ x_2 \end{bmatrix} = \frac{\omega^2 l^3}{8EI}\begin{bmatrix} \frac{500}{6} & \frac{100}{4} \\ \frac{500}{4} & 100 \end{bmatrix}\begin{bmatrix} x_1 \\ x_2 \end{bmatrix}$$

Starting with $x_1 = x_2 = 1.0$ for the right column, we obtain

$$\begin{bmatrix} x_1 \\ x_2 \end{bmatrix} = \frac{\omega^2 l^3}{8EI}\begin{bmatrix} 108.3 \\ 225.0 \end{bmatrix} = \frac{108.3\omega^2 l^3}{8EI}\begin{bmatrix} 1.00 \\ 2.08 \end{bmatrix}$$

If the procedure is repeated with $x_1 = 1.0$ and $x_2 = 2.08$, the second result is

$$\begin{bmatrix} x_1 \\ x_2 \end{bmatrix} = \frac{\omega^2 l^3}{8EI}\begin{bmatrix} 135.3 \\ 333.0 \end{bmatrix} = \frac{135.3\omega^2 l^3}{8EI}\begin{bmatrix} 1.00 \\ 2.46 \end{bmatrix}$$

By repeating the procedure a few more times the deflections will converge to

$$\begin{bmatrix} 1.00 \\ 2.60 \end{bmatrix} = \frac{148.3\omega^2 l^3}{8EI}\begin{bmatrix} 1.00 \\ 2.60 \end{bmatrix}$$

The fundamental frequency from this equation is

$$\omega = \sqrt{\frac{8EI}{148.3l^3}} = 0.232\sqrt{\frac{EI}{l^3}}$$

and the amplitude ratio is found to be

$$\frac{x_1}{x_2} = \frac{1.0}{2.60}$$

If only the fundamental frequency is of interest, sufficient accuracy can be obtained from the results of the first and second iterations. From the first iteration the inertia forces are $500\omega^2$ and $208\omega^2$. These forces produce deflections obtained in the second iteration, which are $x_1 = 135.3\omega^2 l^3/8EI = 16.92\omega^2 l^3/EI$ and $x_2 = 2.46x_1$. The work done by these forces is then

$$U = \frac{1}{2}(500 + 208 \times 2.46)\,\omega^2 x_1 = \frac{1}{2} \times 1012 \times \omega^2 x_1$$

and the corresponding kinetic energy is

$$T = \frac{1}{2}(500 + 100 \times 2.46^2)\,\omega^2 x_1^2 = \frac{1}{2} \times 1105 \times \omega^2 x_1^2$$

Equating the two, the fundamental frequency is found as

$$\omega = \sqrt{\frac{1012}{1105 \times 16.92}\frac{EI}{l^3}} = 0.232\sqrt{\frac{EI}{l^3}}$$

11.5 CALCULATION OF HIGHER MODES

When the equations of motion are formulated in terms of the flexibility influence coefficients, the iteration procedure converges to the lowest mode present in the assumed deflection. It is evident that if the lowest mode is absent in the assumed deflection, the iteration technique will converge to the next lowest, or the second, mode.

Letting the assumed curve X be expressed by the sum of the normal modes X_i

$$X = C_1 X_1 + C_2 X_2 + C_3 X_3 + \cdots \tag{11.5-1}$$

To distinguish between the assumed curve X and the normal modes X_i in the above equation, we designate the normal modes as

$$X_i = \begin{Bmatrix} x_1 \\ x_2 \\ x_3 \end{Bmatrix}_i$$

and the assumed curve as

$$X = \begin{Bmatrix} \bar{x}_1 \\ \bar{x}_2 \\ \bar{x}_3 \end{Bmatrix}$$

We now impose the condition $C_1 = 0$ to remove the first mode from the assumed deflection X. For this, we introduce the orthogonality relationship by premultiplying Eq. (11.5-1) by $X_1'M$, which eliminates all terms on the right side except the first term.

$$X_1'MX = C_1 X_1'MX_1 \tag{11.5-2}$$

Equating the left side of Eq. (11.5-2) to zero, yields C_1 of zero, and the first mode is eliminated from Eq. (11.5-1):

$$X_1'MX = (x_1 \quad x_2 \quad x_3)\begin{bmatrix} m_1 & 0 & 0 \\ 0 & m_2 & 0 \\ 0 & 0 & m_3 \end{bmatrix}\begin{Bmatrix} \bar{x}_1 \\ \bar{x}_2 \\ \bar{x}_3 \end{Bmatrix} = 0$$

$$= m_1 x_1 \bar{x}_1 + m_2 x_2 \bar{x}_2 + m_3 x_3 \bar{x}_3 = 0 \tag{11.5-3}$$

From this equation, we obtain

$$\begin{aligned}\bar{x}_1 &= -\frac{m_2}{m_1}\left(\frac{x_2}{x_1}\right)\bar{x}_2 - \frac{m_3}{m_1}\left(\frac{x_3}{x_1}\right)\bar{x}_3 \\ \bar{x}_2 &= \bar{x}_2 \\ \bar{x}_3 &= \bar{x}_3\end{aligned} \tag{11.5-4}$$

where the last two equations are introduced merely as identities. Rewriting Eq. (11.5-4) in matrix form gives

$$\{X\} = \begin{bmatrix} 0 & -\frac{m_2}{m_1}\left(\frac{x_2}{x_1}\right) & -\frac{m_3}{m_1}\left(\frac{x_3}{x_1}\right) \\ 0 & 1 & 0 \\ 0 & 0 & 1 \end{bmatrix}\{X\} = SX \tag{11.5-5}$$

Since this equation is the result of $C_1 = 0$, the first mode has been swept out of the assumed deflection by the *sweeping matrix S*. When this equation is substituted into the original matrix equation

$$X = \omega^2 aMX \tag{11.5-6}$$

the result is

$$X = \omega^2 aMSX \tag{11.5-7}$$

The iteration procedure applied to Eq. (11.5-7) will converge to the second mode.

For the third and higher modes, the sweeping procedure is repeated, making $C_1 = C_2 = 0$, and so on. This reduces the order of the matrix equation by one each time; however, the convergence for higher modes becomes more critical if impurities are introduced through the sweeping matrices. It is well to check the highest mode by the inversion of the original matrix equation, which should be equal to the equation formulated in terms of the stiffness influence coefficients.

Example 11.5-1

Write the matrix equation, based on flexibility influence coefficients, for the system shown in Fig. 11.5-1 and determine all the natural modes.

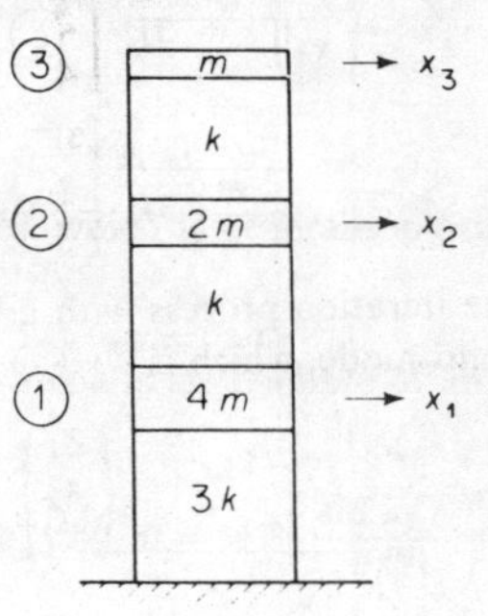

Figure 11.5-1.

Solution: The flexibility coefficients are found by applying a unit load, one at a time, to points 1, 2, and 3.

$$a_{11} = a_{21} = a_{12} = a_{31} = a_{13} = \frac{1}{3k}$$

$$a_{22} = a_{32} = a_{23} = \left(\frac{1}{3k} + \frac{1}{k}\right) = \frac{4}{3k}$$

$$a_{33} = \frac{1}{3k} + \frac{1}{k} + \frac{1}{k} = \frac{7}{3k}$$

The equations of motion in matrix form are then

$$\begin{Bmatrix} x_1 \\ x_2 \\ x_3 \end{Bmatrix} = \frac{\omega^2}{3k}\begin{bmatrix} 1 & 1 & 1 \\ 1 & 4 & 4 \\ 1 & 4 & 7 \end{bmatrix}\begin{bmatrix} 4m & 0 & 0 \\ 0 & 2m & 0 \\ 0 & 0 & m \end{bmatrix}\begin{Bmatrix} x_1 \\ x_2 \\ x_3 \end{Bmatrix}$$

$$\begin{Bmatrix} x_1 \\ x_2 \\ x_3 \end{Bmatrix} = \frac{\omega^2 m}{3k}\begin{bmatrix} 4 & 2 & 1 \\ 4 & 8 & 4 \\ 4 & 8 & 7 \end{bmatrix}\begin{Bmatrix} x_1 \\ x_2 \\ x_3 \end{Bmatrix}$$

Starting with arbitrary values of $x_1 x_2 x_3$, the above equation converges to the first mode, which is

$$\begin{Bmatrix} x_1 \\ x_2 \\ x_3 \end{Bmatrix} = \frac{\omega^2 m}{3k} 14.32 \begin{Bmatrix} 0.25 \\ 0.79 \\ 1.00 \end{Bmatrix}$$

The fundamental frequency is then found to be

$$\omega_1 = \sqrt{\frac{3k}{14.32m}} = 0.457\sqrt{\frac{k}{m}}$$

To determine the second mode, we form the sweeping matrix given by Eq. (11.5-5)

$$S = \begin{bmatrix} 0 & -\frac{1}{2}\left(\frac{0.79}{0.25}\right) & -\frac{1}{4}\left(\frac{1.00}{0.25}\right) \\ 0 & 1 & 0 \\ 0 & 0 & 1 \end{bmatrix} = \begin{bmatrix} 0 & -1.58 & -1 \\ 0 & 1 & 0 \\ 0 & 0 & 1 \end{bmatrix}$$

The new equation for the second mode iteration is from Eq. (11.5-6).

$$\begin{Bmatrix} x_1 \\ x_2 \\ x_3 \end{Bmatrix} = \frac{\omega^2 m}{3k}\begin{bmatrix} 4 & 2 & 1 \\ 4 & 8 & 4 \\ 4 & 8 & 7 \end{bmatrix}\begin{bmatrix} 0 & -1.58 & -1 \\ 0 & 1 & 0 \\ 0 & 0 & 1 \end{bmatrix}\begin{Bmatrix} x_1 \\ x_2 \\ x_3 \end{Bmatrix}$$

$$= \frac{\omega^2 m}{3k}\begin{bmatrix} 0 & -4.32 & -3.0 \\ 0 & 1.67 & 0 \\ 0 & 1.67 & 3.0 \end{bmatrix}\begin{Bmatrix} x_1 \\ x_2 \\ x_3 \end{Bmatrix}$$

Starting the iteration process with arbitrary amplitudes, the above equation converges to the second mode, which is

$$\begin{Bmatrix} x_1 \\ x_2 \\ x_3 \end{Bmatrix} = \frac{\omega^2 m}{3k} 3 \begin{Bmatrix} -1.0 \\ 0 \\ 1.0 \end{Bmatrix}$$

The natural frequency of the second mode is therefore found to be

$$\omega_2 = \sqrt{\frac{k}{m}}$$

For the determination of third mode, we impose the conditions $C_1 = C_2 = 0$ from the orthogonality equation (11.5-3)

$$C_1 = \sum_{i=1}^{3} m_i(x_i)_1 \bar{x}_i = 4(0.25)\bar{x}_1 + 2(0.79)\bar{x}_2 + 1(1.0)\bar{x}_3 = 0$$

$$C_2 = \sum_{i=1}^{3} m_i(x_i)_2 \bar{x}_i = 4(-1.0)\bar{x}_1 + 2(0)\bar{x}_2 + 1(1.0)\bar{x}_3 = 0$$

From these two equations we obtain

$$\bar{x}_1 = 0.25\bar{x}_3 \qquad \bar{x}_2 = -0.79\bar{x}_3$$

which can be expressed by the matrix equation

$$\begin{Bmatrix} x_1 \\ x_2 \\ x_3 \end{Bmatrix} = \begin{bmatrix} 0 & 0 & 0.25 \\ 0 & 0 & -0.79 \\ 0 & 0 & 1.00 \end{bmatrix} \begin{Bmatrix} \bar{x}_1 \\ \bar{x}_2 \\ \bar{x}_3 \end{Bmatrix}$$

This matrix is devoid of the first two modes and can be used as a sweeping matrix for the third mode. Applying this to the original equation, we obtain

$$\begin{Bmatrix} x_1 \\ x_2 \\ x_3 \end{Bmatrix} = \frac{\omega^2 m}{3k} \begin{bmatrix} 4 & 2 & 1 \\ 4 & 8 & 4 \\ 4 & 8 & 7 \end{bmatrix} \begin{bmatrix} 0 & 0 & 0.25 \\ 0 & 0 & -0.79 \\ 0 & 0 & 1.00 \end{bmatrix} \begin{Bmatrix} \bar{x}_1 \\ \bar{x}_2 \\ \bar{x}_3 \end{Bmatrix}$$

The above equation results immediately in the third mode, which is

$$\begin{Bmatrix} x_1 \\ x_2 \\ x_3 \end{Bmatrix} = \frac{\omega^2 m}{3k} 1.68 \begin{Bmatrix} 0.25 \\ -0.79 \\ 1.00 \end{Bmatrix}$$

The natural frequency of the third mode is then found to be

$$\omega_3 = \sqrt{\frac{3k}{1.68m}} = 1.34\sqrt{\frac{k}{m}}$$

These natural frequencies were checked by solving the stiffness equation, which is

$$m\begin{bmatrix} 4 & 0 & 0 \\ 0 & 2 & 0 \\ 0 & 0 & 1 \end{bmatrix} \begin{Bmatrix} \ddot{x}_1 \\ \ddot{x}_2 \\ \ddot{x}_3 \end{Bmatrix} + k\begin{bmatrix} 4 & -1 & 0 \\ -1 & 2 & -1 \\ 0 & -1 & 1 \end{bmatrix} \begin{Bmatrix} x_1 \\ x_2 \\ x_3 \end{Bmatrix} = 0$$

With $\lambda = m\omega^2/k$, the determinant of this equation, set equal to zero gives

$$8(1-\lambda)^3 - 5(1-\lambda) = 0$$

Its solution is found to be

$$\lambda_1 = 0.2094 \qquad \omega_1 = 0.4576\sqrt{\frac{k}{m}}$$

$$\lambda_2 = 1.0000 \qquad \omega_2 = 1.0000\sqrt{\frac{k}{m}}$$

$$\lambda_3 = 1.7906 \qquad \omega_3 = 1.3381\sqrt{\frac{k}{m}}$$

PROBLEMS

11-1 Write the kinetic and potential energy expressions for the system of Fig. P11-1 and determine the equation for ω^2 by equating the two energies. Letting $x_2/x_1 = n$, plot ω^2 versus n. Pick off the maximum and minimum values of ω^2 and the corresponding values of n, and show that they represent the two natural modes of the system.

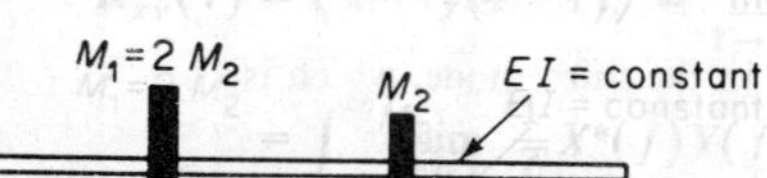

Figure P11-1.

11-2 Using Rayleigh's method, estimate the fundamental frequency of the lumped mass system shown in Fig. P11-2.

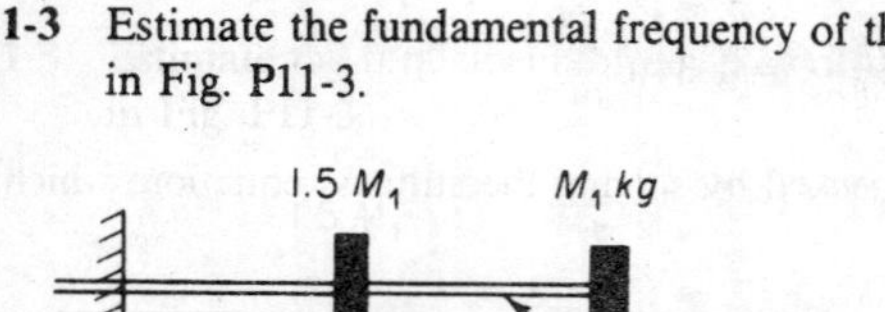

Figure P11-2.

11-3 Estimate the fundamental frequency of the lumped mass cantilever beam shown in Fig. P11-3.

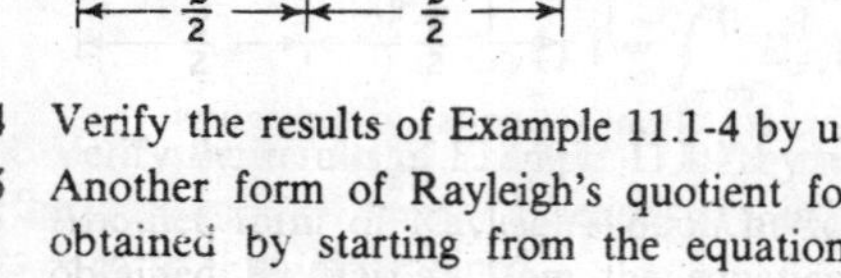

Figure P11-3.

11-4 Verify the results of Example 11.1-4 by using Eq. (11.1-3).

11-5 Another form of Rayleigh's quotient for the fundamental frequency can be obtained by starting from the equation of motion based on the flexibility influence coefficient

$$X = aM\ddot{X}$$
$$= \omega^2 aMX$$

Premultiplying by $X'M$ we obtain

$$X'MX = \omega^2 X'MaMX$$

and the Rayleigh quotient becomes

$$\omega^2 = \frac{X'MX}{X'MaMX}$$

Solve for ω_1 in Example 11.1-4 by using the above equation and compare results with Prob. 11-4.

11-6 Using the curve

$$y(x) = \frac{l^3}{3EI}\left(\frac{x}{l}\right)^2$$

solve Prob. 11-3 by using the method of integration. *Hint*: Draw shear and moment diagrams based on inertia loads.

11-7 Using the deflection $y(x) = y_{max}\sin(\pi x/l)$, determine the fundamental frequency of the beam shown in Fig. P11-7 (a) if $EI_2 = EI_1$ and (b) if $EI_2 = 4EI_1$.

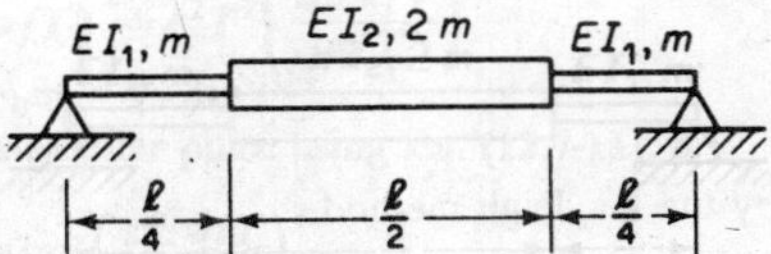

Figure P11-7.

11-8 Repeat Prob. 11-7 but use the curve

$$y(x) = y_{max}\frac{4x}{l}\left(1 - \frac{x}{l}\right)$$

11-9 A uniform cantilever beam of mass m per unit length has its free end pinned to two springs of stiffness k and mass m_0 each as shown in Fig. P11-9. Using Rayleigh's method, find its natural frequency ω_1.

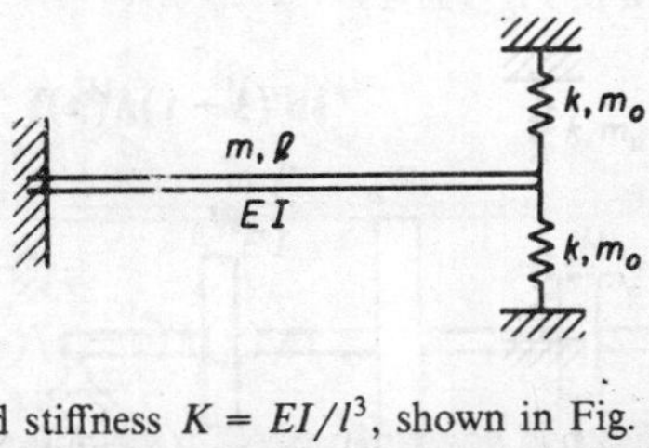

Figure P11-9.

11-10 A uniform beam of mass M and stiffness $K = EI/l^3$, shown in Fig. P11-10, is supported on equal springs with total vertical stiffness of k lb/in. Using Rayleigh's method with the deflection $y_{max} = \sin(\pi x/l) + b$, show that the

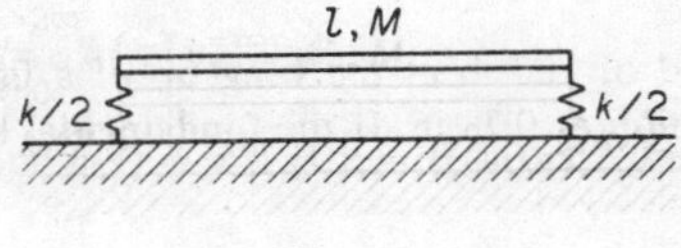

Figure P11-10.

frequency equation becomes

$$\omega^2 = \frac{2k}{M}\left[\frac{\frac{K}{k}\frac{\pi^4}{4} + \frac{b^2}{2}}{\frac{1}{2} + \frac{4b}{\pi} + b^2}\right]$$

By $\partial\omega^2/\partial b = 0$, show that the lowest frequency results when

$$b = -\frac{\pi}{4}\left(\frac{1}{2} - \frac{K\pi^4}{2k}\right) \pm \sqrt{\left[\frac{\pi}{2}\left(\frac{1}{2} - \frac{K\pi^4}{2k}\right)\right]^2 + \frac{\pi^4 K}{2k}}$$

11-11 Assuming a static deflection curve

$$y(x) = y_{max}\left[3\left(\frac{x}{l}\right) - 4\left(\frac{x}{l}\right)^3\right], \qquad 0 \le x \le \frac{1}{2}$$

determine the lowest natural frequency of a simply supported beam of constant EI and a mass distribution of

$$m(x) = m_0 \frac{x}{l}\left(1 - \frac{x}{l}\right)$$

by the Rayleigh method.

11-12 Using Dunkerley's equation, determine the fundamental frequency of the three-mass cantilever beam shown in Fig. P11-12.

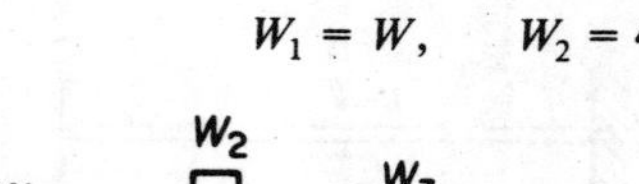

Figure P11-12.

11-13 Using Dunkerley's equation, determine the fundamental frequency of the beam shown in Fig. P11-13.

$$W_1 = W, \qquad W_2 = 4W, \qquad W_3 = 2W$$

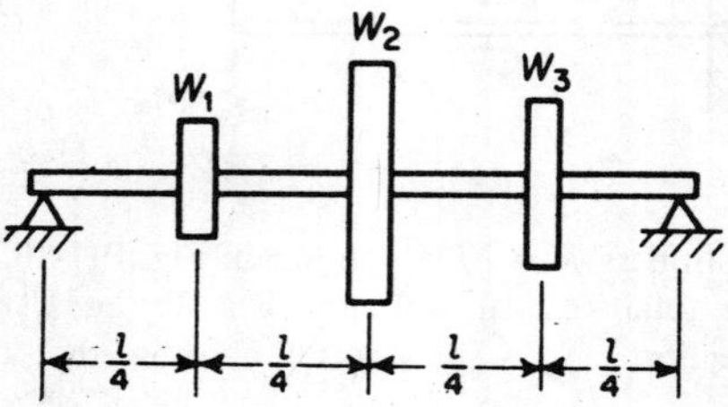

Figure P11-13.

11-14 A load of 100 lb at the wing tip of a fighter plane produced a corresponding deflection of 0.78 in. If the fundamental bending frequency of the same wing is

622 cpm, approximate the new bending frequency when a 320-lb fuel tank (including fuel) is attached to the wing tip.

11-15 A given beam was vibrated by an eccentric mass shaker of mass 5.44 kg at the midspan, and resonance was found at 435 cps. With an additional mass of 4.52 kg, the resonant frequency was lowered to 398 cps. Determine the natural frequency of the beam.

11-16 Using the Rayleigh-Ritz method and assuming modes x/l and $\sin(\pi x/l)$; determine the two natural frequencies and modes of a uniform beam pinned at the right end and attached to a spring of stiffness k at the left end.

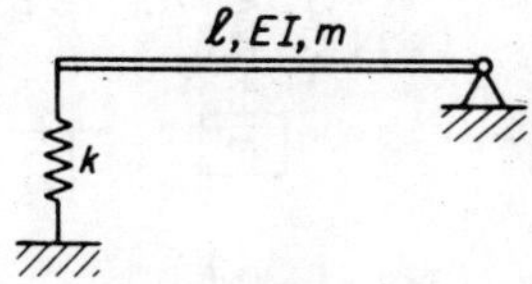

Figure P11-16.

11-17 For the wedge-shaped plate of Example 11.3-1, determine the first two natural frequencies and mode shapes for bending vibration by using the Ritz deflection function $y = C_1 x^2 + C_2 x^3$.

11-18 Using the Rayleigh-Ritz method, determine the first two natural frequencies and mode shapes for the longitudinal vibration of a uniform rod with a spring of stiffness k_0 attached to the free end as shown in Fig. P11-18. Use the first two normal modes of the fixed-free rod in longitudinal motion.

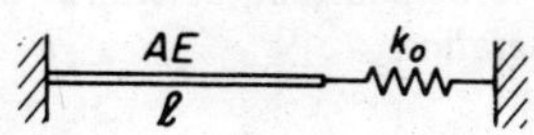

Figure P11-18.

11-19 Repeat Prob. 11-18 but this time the spring is replaced by a mass m_0, as shown in Fig. P11-19.

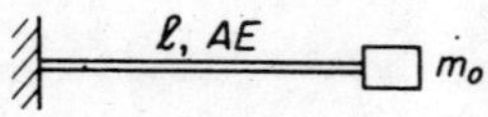

Figure P11-19.

11-20 For the simply supported variable mass beam of Prob. 11-11, assume the deflection to be made up of the first two modes of the uniform beam and solve for the two natural frequencies and mode shapes by the Rayleigh-Ritz method.

11-21 A uniform rod hangs freely from a hinge at the top. Using the three modes $\phi_1 = x/l$, $\phi_2 = \sin(\pi x/l)$, and $\phi_3 = \sin(2\pi x/l)$ determine the characteristic equation by using the Rayleigh-Ritz method.

11-22 Using matrix iteration, determine the three natural frequencies and modes for the cantilever beam of Prob. 11-12.

11-23 Determine the influence coefficients for the three-mass system of Fig. P11-23, and calculate the principal modes by matrix iteration.

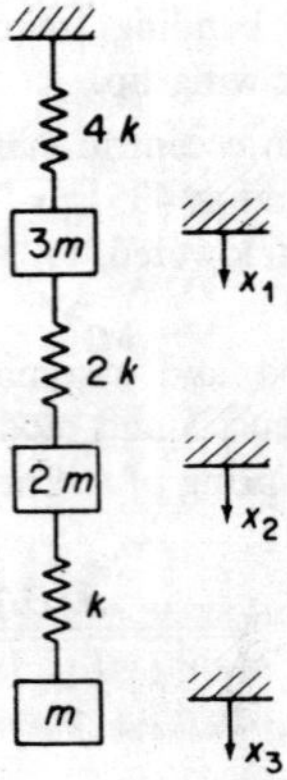

Figure P11-23.

11-24 Using matrix iteration, determine the natural frequencies and mode shapes of the torsional system of Fig. P11-24.

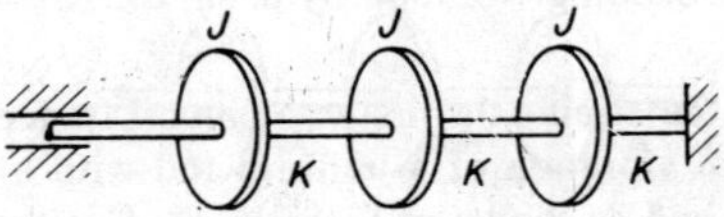

Figure P11-24.

11-25 In Fig. P11-25 four masses are strung along strings of equal lengths. Assuming the tension to be constant, determine the natural frequencies and mode shapes by matrix iteration.

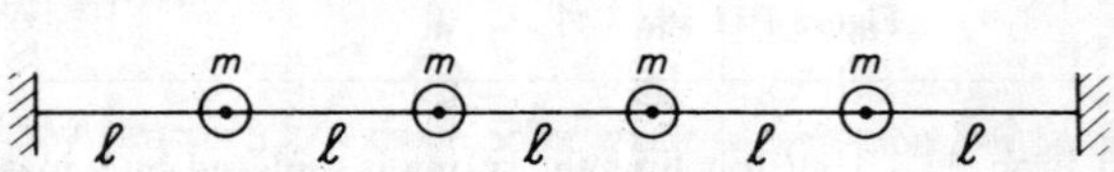

Figure P11-25.

12 NUMERICAL PROCEDURES FOR LUMPED MASS SYSTEMS

There are many vibrational systems where we can consider the mass to be lumped and located along straight lines. A shaft transmitting torque with several wheels along its length is such a system. Although the finite element method of Chapter 10 is capable of solving such problems, in this chapter we describe another procedure known as the *Holzer method*, which has enjoyed prior popularity as a simple and practical method. The Holzer method was extended to beam vibrations by Myklestad, and both methods have been matricized by the transfer matrix.

The procedure for solution in the Holzer method is quite different from that of the finite element method. In the finite element method, the structural elements are assembled into one large system, and its equations are solved for all the natural frequencies and mode shapes. In the Holzer method, an assumed frequency is chosen and the torque and displacement of the stations are progressively calculated from one end to the other. The torque or displacement at the boundary is then plotted against the chosen frequency. The natural frequencies and mode shapes are established only when these boundary values correspond to the actual boundary conditions of the problem.

12.1 HOLZER METHOD

When an undamped system is vibrating freely at any one of its natural frequencies, no external force, torque, or moment is necessary to maintain the vibration. Also, the amplitude of the mode shape is immaterial to the vibration. Recognizing these facts, Holzer† proposed a method of calculation for the natural frequencies and mode shapes of torsional systems by assuming a frequency and starting with a unit amplitude at one end of the system and progressively calculating the torque and angular displacement to the other end. The frequencies that result in zero external torque or compatible boundary conditions at the other end are the natural frequencies of the system. The method can be applied to any lumped mass system, linear spring-mass systems, beams modeled by discrete masses and beam springs, and so on.

Holzer's procedure for torsional systems. Figure 12.1-1 shows a torsional system represented by a series of disks connected by shafts. Assuming a frequency ω and amplitude $\theta_1 = 1$, the inertia torque of the first disk is

$$-J_1\ddot{\theta}_1 = J_1\omega^2\theta_1 = J_1\omega^2 1$$

where harmonic motion is implied. This torque acts through shaft 1 and twists it by

$$\frac{J_1\omega^2}{K_1} = \theta_1 - \theta_2 = 1 - \theta_2$$

or

$$\theta_2 = 1 - \frac{J_1\omega^2}{K_1}$$

With θ_2 known, the inertia torque of the second disk is calculated as $J_2\omega^2\theta_2$. The sum of the first two inertia torques acts through the shaft K_2, causing it to twist by

$$\frac{J_1\omega^2 + J_2\omega^2\theta_2}{K_2} = \theta_2 - \theta_3$$

In this manner, the amplitude and torque at every disk can be calculated. The resulting torque at the far end,

$$T_{\text{ext}} = \sum_{i=1}^{4} J_i\omega^2\theta_i$$

can then be plotted for the chosen ω. By repeating the calculation with other values of ω, the natural frequencies are found when $T_{\text{ext}} = 0$. The angular displacements θ_i corresponding to the natural frequencies are the mode shapes.

†H. Holzer, *Die Berechnung der Drehschwingungen* (Berlin: Springer-Verlag, 1921).

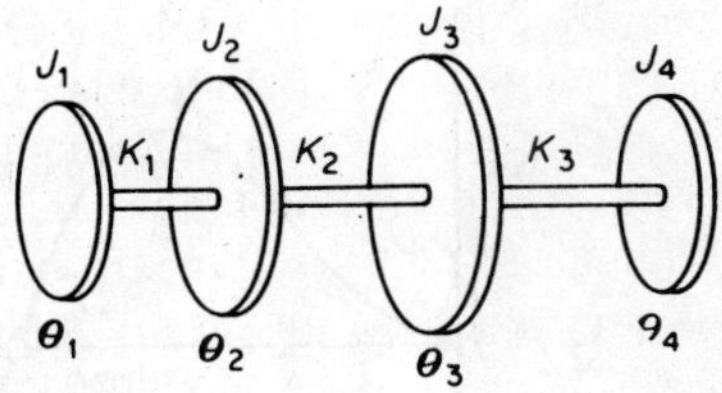

Figure 12.1-1.

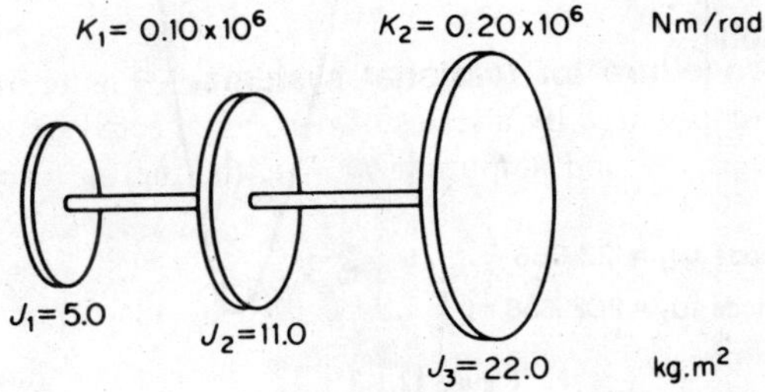

Figure 12.1-2.

Example 12.1-1

Determine the natural frequencies and mode shapes of the system shown in Fig. 12.1-2.

Solution: The following table defines the parameters of the system and the sequence of calculations, which can be easily carried out on any programmable calculator.

	Parameters of the System		
	Station 1	*Station 2*	*Station3*
	$J_1 = 5$ $K_1 = 0.10 \times 10^6$	$J_2 = 11$ $K_2 = 0.20 \times 10^6$	$J_3 = 22$ $K_3 = 0$
	Calculation Program		
ω ω^2	$\theta_1 = 1.0$ $T_1 = \omega^2\theta_1 J_1$	$\theta_2 = 1 - T_1/k_1$ $T_2 = T_1 + \omega^2\theta_2 J_2$	$\theta_3 = \theta_2 - T_2/k_2$ $T_3 = T_2 + \omega^2\theta_3 J_3$
20 400	1.0 2.0×10^3	0.980 6.312×10^3	0.9484 14.66×10^3
40 1600	1.0 8.0×10^3	0.920 24.19×10^3	0.799 52.32×10^3

Presented are calculations for $\omega = 20$ and 40. The quantity T_3 is the torque to the right of disk 3, which must be zero at the natural frequencies. Figure 12.1-3 shows a plot of T_3 versus ω. Several frequencies in the vicinity of $T_3 = 0$ were inputted to obtain accurate values of the first and second mode shapes displayed in Fig. 12.1-4.

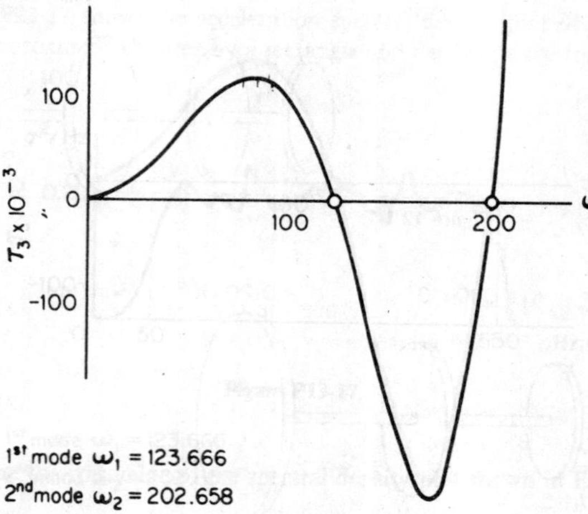

Figure 12.1-3.

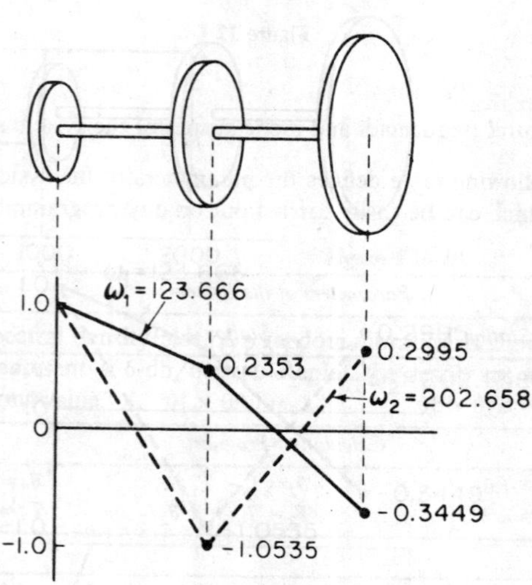

Figure 12.1-4.

12.2 DIGITAL COMPUTER PROGRAM FOR THE TORSIONAL SYSTEM

The calculations for the Holzer problem can be greatly speeded up by using a high-speed digital computer. The problem treated is the general torsional system of Fig. 12.2-1. The program is written in such a manner that by changing the data it is applicable to any other torsional system.

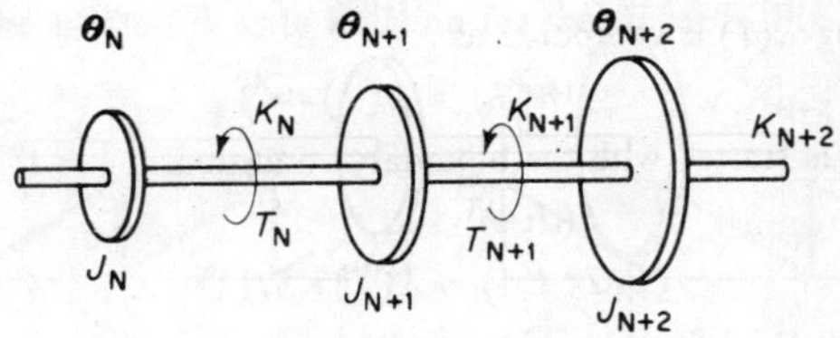

Figure 12.2-1.

The quantities of concern here are the torsional displacement θ of each disk and the torque T carried by each shaft. We adopt two indexes: N to define the position along the structure and I for the frequency to be used. For the computer program, some notation changes are required to conform to the Fortran language. For example, the stiffness K and the moment of inertia J of the disk are designated as SK and SJ.

The equations relating the displacement and torque at the Nth and $(N+1)$st stations are

$$\theta(I, N+1) = \theta(I, N) - T(I, N)/SK(N) \tag{12.2-1}$$

$$T(I, N+1) = T(I, N) + \lambda(I) * SJ(N+1) * \theta(I, N+1) \tag{12.2-2}$$

where $\lambda = \omega^2$, $\theta(I, 1) = 1$, $T(I, 1) = \lambda(I) * SJ(1)$.

Starting at $N = 1$, these two equations are to be solved for θ and T at each point N of the structure and for various values of λ. At the natural frequencies, θ must be zero at the fixed end or T must be zero at the free end.

Example 12.2-1

Determine the natural frequencies and mode shapes for the torsional system of Fig. 12.2-2.

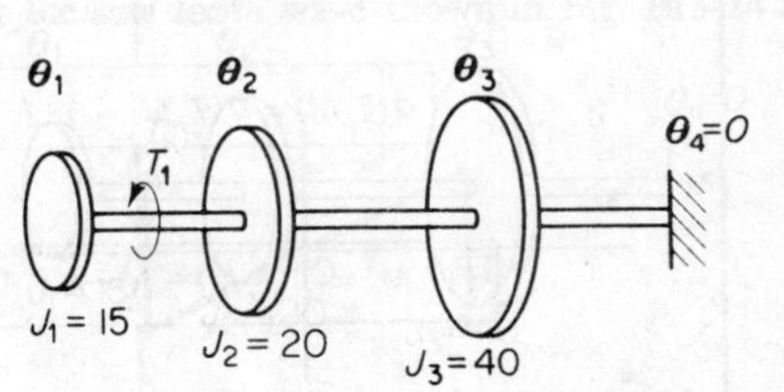

Figure 12.2-2. $K_1 = 2 \times 10^6$ $K_2 = 2 \times 10^6$ $K_3 = 3 \times 10^6$

Solution: The frequency range can be scanned by choosing an initial ω and an increment $\Delta\omega$. We choose for this problem the frequencies

$$\omega = 40, 60, 80, \ldots, 620$$

which can be programmed as

$$\omega(I) = 40 + (I-1) * 20, \qquad I = 1 \text{ to } 30$$

The corresponding $\lambda(I)$ is computed as

$$\lambda(I) = \omega(I) * * 2$$

The computation is started with the boundary conditions, $N = 1$

$$\theta(I,1) = 1$$
$$T(I,1) = \lambda(I) * SJ(1)$$

Eqs. (12.2-1) and (12.2-2) then give the values of θ and T at the next station $M = N + 1 = 2$. This loop is repeated until $M = 4$, at which time I is advanced an

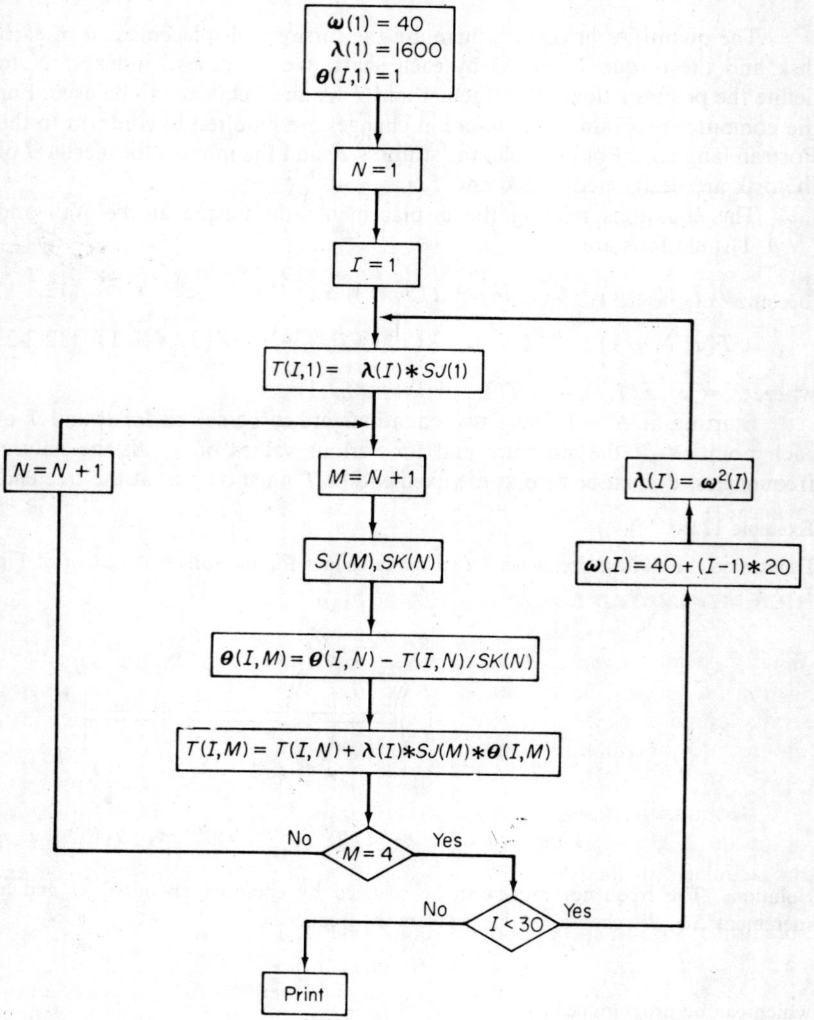

Figure 12.2-3.

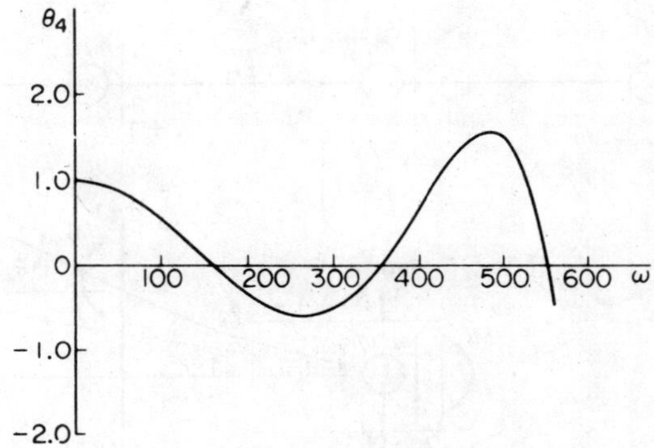

Figure 12.2-4.

integer to the next frequency. The process is then repeated. These operations are clearly seen in the flow diagram of Fig. 12.2-3.

Figure 12.2-4 shows the results of the computer study where θ_4 is plotted against ω. The natural frequencies of the system correspond to frequencies for which θ_4 becomes zero, which are approximately

$$\omega_1 = 160$$

$$\omega_2 = 356$$

$$\omega_3 = 552$$

The mode shapes can be found by printing out θ_N for each of the preceding frequencies.

12.3 MYKLESTAD'S METHOD FOR BEAMS

When a beam is replaced by lumped masses connected by massless beam sections, a method developed by N. O. Myklestad† can be used to progressively compute the deflection, slope, moment, and shear from one section to the next, in a manner similar to the Holzer method.

Uncoupled flexural vibration. Figure 12.3-1 shows a typical section of an idealized beam with lumped masses. By taking the free-body section in the manner indicated, it will be possible to write equations for the shear and moment at $i + 1$ entirely in terms of quantities at i. These can then be substituted into the geometric equations for θ and y.

†N. O. Myklestad, "A New Method of Calculating Natural Modes of Uncoupled Bending Vibration of Airplane Wings and Other Types of Beams," *Jour. Aero. Sci.* (April 1944), pp. 153–62.

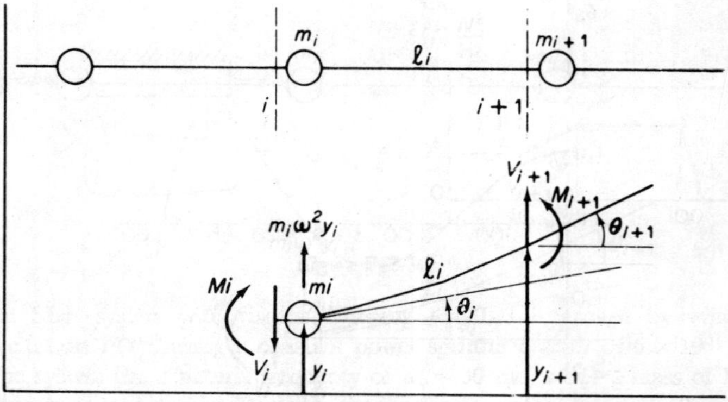

Figure 12.3-1.

From equilibrium considerations, we have

$$V_{i+1} = V_i - m_i\omega^2 y_i \tag{12.3-1}$$

$$M_{i+1} = M_i - V_{i+1} l_i \tag{12.3-2}$$

From geometric considerations, using influence coefficients of uniform beam sections, we have

$$\theta_{i+1} = \theta_i + M_{i+1}\left(\frac{l}{EI}\right)_i + V_{i+1}\left(\frac{l^2}{2EI}\right)_i \tag{12.3-3}$$

$$y_{i+1} = y_i + \theta_i l_i + M_{i+1}\left(\frac{l^2}{2EI}\right)_i + V_{i+1}\left(\frac{l^3}{3EI}\right)_i \tag{12.3-4}$$

where $(l/EI)_i$ = slope at $i+1$ measured from a tangent at i due to a unit moment at $i+1$;

$(l^2/2EI)_i$ = slope at $i+1$ measured from a tangent at i due to a unit shear at $i+1$ = deflection at $i+1$ measured from a tangent at i due to a unit moment at $i+1$;

$(l^3/3EI)_i$ = deflection at $i+1$ measured from a tangent at i due to a unit shear at $i+1$.

Thus Eqs. (12.3-1) through (12.3-4) in the sequence given enables the calculations to proceed from i to $i+1$.

Boundary conditions. Of the four boundary conditions at each end, two are generally known. For example, a cantilever beam with $i = 1$ at the free end would have $V_1 = M_1 = 0$. Since the amplitude is arbitrary, we can choose $y_1 = 1.0$. Having done so, the slope θ_1 is fixed to a value which is yet to be determined. Because of the linear character of the problem, the four quantities at the far end will be in the form

$$V_n = a_1 + b_1\theta_1$$
$$M_n = a_2 + b_2\theta_1$$
$$\theta_n = a_3 + b_3\theta_1$$
$$y_n = a_4 + b_4\theta_1$$

where a_i, b_i are constants and θ_1 is unknown. Thus the frequencies that satisfy the boundary condition $\theta_n = y_n = 0$ for the cantilever beam will establish θ_1 and the natural frequencies of the beam, i.e., $\theta_1 = -a_3/b_3$ and $y_n = a_4 - (a_3/b_3)b_4 = 0$. Hence, by plotting y_n versus ω, the natural frequencies of the beam can be found.

Example 12.3-1

To illustrate the computational procedure we will determine the natural frequencies of the cantilever beam shown in Fig. 12.3-2. The massless beam sections are assumed to be identical so that the influence coefficients for each section are equal. The numerical constants for the problem are given as

$$m_1 = 100 \text{ kg} \qquad \frac{l}{EI} = 5 \times 10^{-6}\,\frac{1}{Nm}$$

$$l = 0.5 \text{ m} \qquad \frac{l^2}{2EI} = 1.25 \times 10^{-6}\,\frac{1}{N}$$

$$EI = 0.10 \times 10^{-6} \text{ Nm}^2 \qquad \frac{l^3}{3EI} = 0.41666 \times 10^{-6}\,\frac{m}{N}$$

Th· computation is started at 1. Since each of the quantities V, M, θ, and y will be in the form $a + b$, they are arranged into two columns, each of which can be

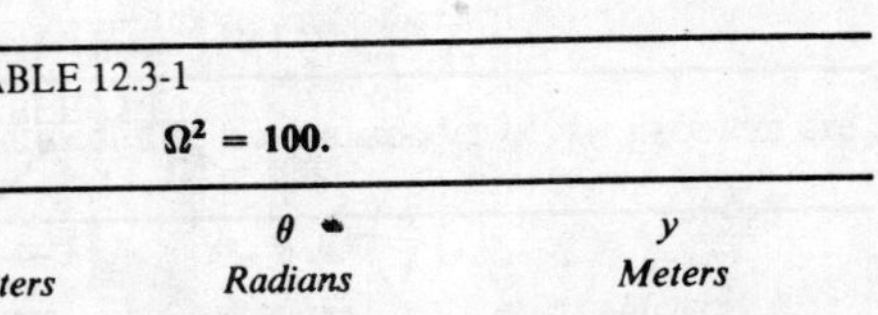

Figure 12.3-2.

TABLE 12.3-1

Ω = 10. **Ω² = 100.**

i	V Newtons		M Newton meters		θ Radians		y Meters	
1	0	0	0	0	0	θ	1.0	0
2	−10,000.	0	5000.	0	0.0125	1.0θ	1.002084	0.5θ
3	−25031.	-7500θ	17515.	3750θ	0.06879	1.00937θ	1.0198	1.001563θ
4	−45427.	-27532θ	40228.	17516θ	0.21315	1.0625θ	1.08555	1.5167θ

$\theta_4 = 0.21315 + 1.0625\theta = 0 \qquad \theta_1 = -0.2006117$

$y_4 = 1.08555 + 1.5167(-0.2006117) = 0.78128$ plot vs. $\omega = 10$

m	ℓ/EI	$\ell^2/2EI$	$\ell^3/3EI$
100.0000	0.500000E-05	0.125000E-05	0.416666E-06
150.0000	0.500000E-05	0.125000E-05	0.416666E-06
200.0000	0.500000E-05	0.125000E-05	0.416666E-06

ω	y_4	
10.	0.781295E 00	
20.	0.265886E 00	← ω_1
30.	-0.292211E 00	
40.	-0.737204E 00	
50.	-0.102850E 01	
60.	-0.118001E 01	
70.	-0.121920E 01	
80.	-0.117191E 01	
90.	-0.105892E 01	
100.	-0.896440E 00	
110.	-0.697179E 00	
120.	-0.471451E 00	
130.	-0.227783E 00	
140.	0.263367E-01	← ω_2
150.	0.284378E 00	
160.	0.540451E 00	
170.	0.789139E 00	
180.	0.102559E 01	
190.	0.124516E 01	
200.	0.144368E 01	
210.	0.161714E 01	
220.	0.176195E 01	
230.	0.187518E 01	
240.	0.195322E 01	
250.	0.199371E 01	
260.	0.199416E 01	
270.	0.195226E 01	
280.	0.186609E 01	
290.	0.173399E 01	
300.	0.155418E 01	
310.	0.132565E 01	
320.	0.104712E 01	
330.	0.718018E 00	
340.	0.336182E 00	
350.	-0.974121E-01	← ω_3
360.	-0.584229E 00	
370.	-0.112500E 01	
380.	-0.171924E 01	
390.	-0.236646E 01	
400.	-0.306812E 01	

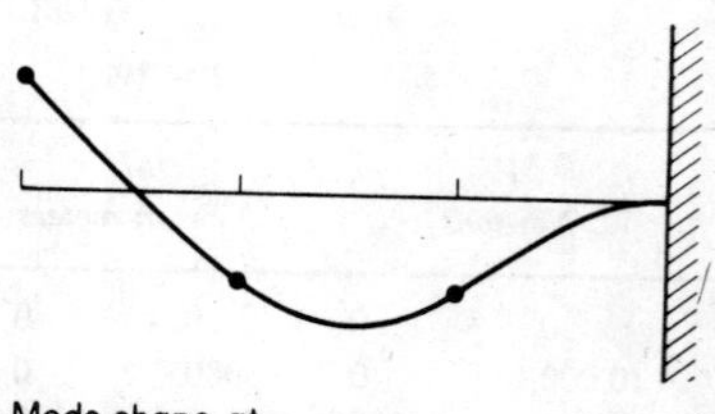

Mode shape at
$\omega_2 = 138.98$

Figure 12.3-3.

computed separately. The calculation for the left column is started with $V_I = 0$, $M_1 = 0$, $\theta_1 = 0$, and $y_1 = 1.0$. The right columns, which are proportional to θ, are started with the initial values of $V_1 = 0$, $M_1 = 0$, $\theta_1 = 1\theta$, and $y_1 = 0$.

Table 12.3-1 shows how the computation for Eqs. (10.4-1) through (12.3-4) can be carried out with any programmable calculator. The frequency chosen for this table is $\omega = 10$.

To start the computation we note that the moment and shear at station 1 are zero. We can choose the deflection at station 1 to be 1.0, in which case the slope at this point becomes an unknown θ. We therefore carry out two columns of calculations for each quantity starting with $y_1 = 1.0$, $\theta_1 = 0$, and $y_1 = 0$, $\theta_1 = \theta$. The unknown slope $\theta_1 = \theta$ is found by forcing θ_4 at the fixed end to be zero, after which the deflection y_4 can be calculated and plotted against ω.. The natural frequencies of the system are those for which $y_4 = 0$.

To search for the natural frequencies, computer calculations were made between $\omega = 10$ to $\omega = 400$ at frequency steps of 10 rad/s. Tabulation of y_4 versus ω indicates natural frequencies in the frequency regions $20 \leq \omega_1 \leq 30$, $130 \leq \omega_2 \leq 140$, and $340 \leq \omega_3 \leq 350$. Further calculations were carried out in each of these regions with a much smaller frequency step. Since the lumped mass model of only three masses could hardly give reliable results for the third mode, only the first two modes were recomputed; these were found to be $\omega_1 = 25.03$ and $\omega_2 = 138.98$. The mode shape at ω_2 is plotted in Fig. 12.3-3.

12.4 COUPLED FLEXURE-TORSION VIBRATION

Natural modes of vibration of airplane wings and other beam structures are often coupled flexure-torsion vibration which for higher modes differ considerably from those of uncoupled modes. To treat such problems we must model the beam as shown in Fig. 12.4-1. The elastic axis of the beam about which the torsional rotation takes place is assumed to be initially straight. It is able to twist, but its bending displacement is restricted to the vertical plane. The principal axes of bending for all cross sections are parallel in the undeformed

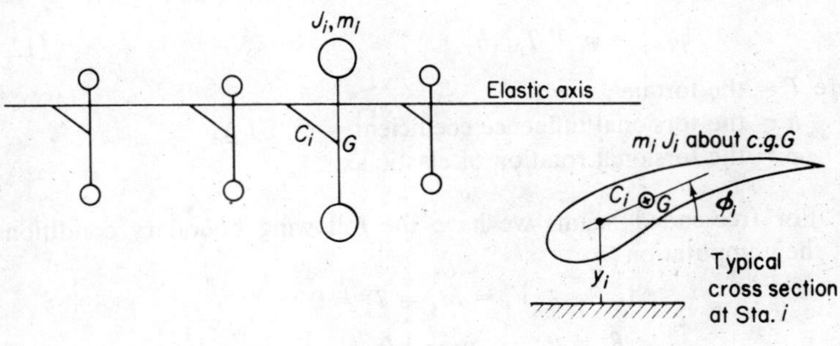

Figure 12.4-1.

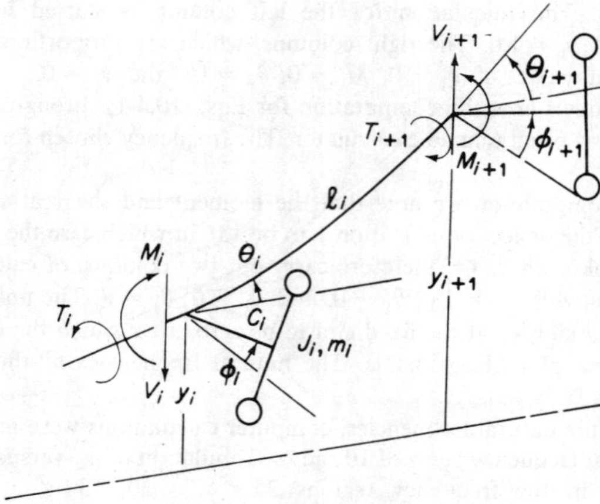

Figure 12.4-2.

state. Masses are lumped at each station with its center of gravity at distance c_i from the elastic axis and J_i is the mass moment of inertia of the section about the elastic axis, i.e., $J_i = J_{cg} + m_i c_i^2$.

Figure 12.4-2 shows the ith section, from which the following equations can be written:

$$V_{i+1} = V_i - m_i\omega^2(y_i + c_i\varphi_i) \tag{12.4-1}$$

$$M_{i+1} = M_i - V_{i+1}l_i \tag{12.4-2}$$

$$T_{i+1} = T_i + J_i\omega^2\varphi_i + m_i c_i \omega^2 y_i \tag{12.4-3}$$

$$\theta_{i+1} = \theta_i + V_{i+1}\left(\frac{l^2}{2EI}\right)_i + M_{i+1}\left(\frac{l}{EI}\right)_i \tag{12.4-4}$$

$$y_{i+1} = y_i + \theta_i l_i + V_{i+1}\left(\frac{l^3}{3EI}\right)_i + M_{i+1}\left(\frac{l^2}{2EI}\right)_i \tag{12.4-5}$$

$$\varphi_{i+1} = \varphi_i + T_{i+1}h_i \tag{12.4-6}$$

where T = the torque
h = the torsional influence coefficient = (l/GI_p)
φ = the torsional rotation of elastic axis

For free-ended beams we have the following boundary conditions to start the computation:

$$V_1 = M_1 = T_1 = 0$$

$$\theta_1 = \theta, \qquad y_1 = 1.0, \qquad \varphi_1 = \varphi_1$$

Here again, the quantities of interest at any station are linearly related to θ_1 and φ_1 and can be be expressed in the form

$$a + b\theta_1 + c\varphi_1 \tag{12.4-7}$$

Natural frequencies are established by the satisfaction of the boundary conditions at the other end. Often for symmetric beams, such as the airplane wing, only one-half the beam need be considered. The satisfaction of the boundary conditions for the symmetric and antisymmetric modes enables sufficient equations for the solution.

12.5 TRANSFER MATRICES

The Holzer and the Myklestad methods can be recast in terms of transfer matrices. The transfer matrix defines the geometric and dynamic relationships of the element between the two stations and allows the state vector for the force and displacement to be transferred from one station to the next station.

Torsional system. Signs are often a source of confusion in rotating systems, and it is necessary to clearly define the sense of positive quantities. The coordinate along the rotational axis is considered positive towards the right. If a cut is made across the shaft, the face with the outward normal towards the positive coordinate direction is called the positive face. Positive torques and positive angular displacements are indicated on the positive face by arrows pointing positively according to the right-hand screw rule as shown in Fig. 12.5-1.

With the stations numbered from left to right, the nth element is represented by the massless shaft of torsional stiffness K_n and the mass of polar moment of inertia J_n, as shown in Fig. 12.5-2.

Separating the shaft from the rotating mass, we can write the following equations and express them in matrix form. The superscripts L and R represent the left and right sides of the members.

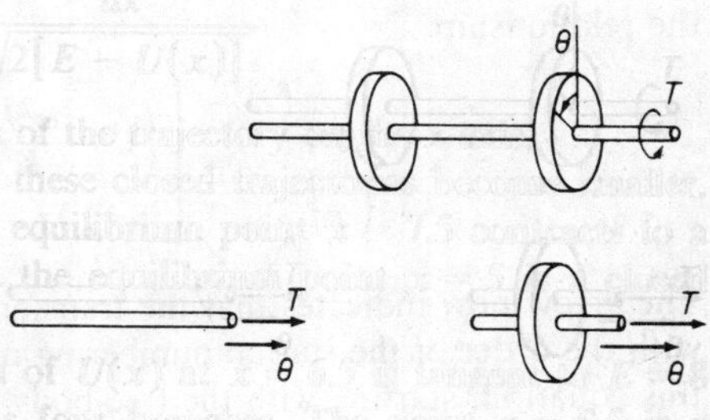

Figure 12.5-1.

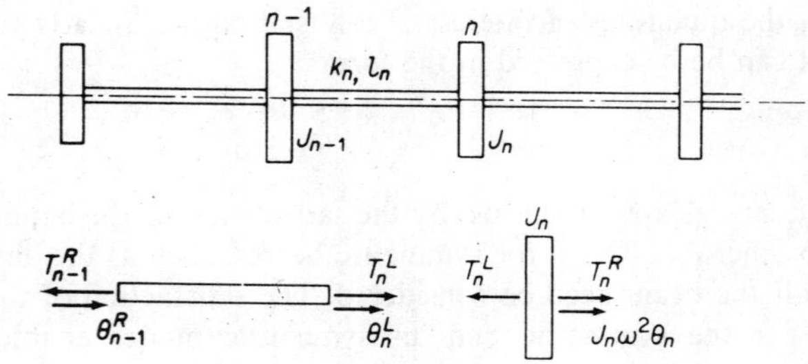

Figure 12.5-2.

For the mass:

$$\left.\begin{aligned} \theta_n^R &= \theta_n^L \\ T_n^R - T_n^L &= -\omega^2 J_n \theta_n \end{aligned}\right\} \qquad \begin{Bmatrix} \theta \\ T \end{Bmatrix}_n^R = \begin{bmatrix} 1 & 0 \\ -\omega^2 J & 1 \end{bmatrix}_n \begin{Bmatrix} \theta \\ T \end{Bmatrix}_n^L \tag{12.5-1}$$

For the shaft:

$$\left.\begin{aligned} K_n\left(\theta_n^L - \theta_{n-1}^R\right) &= T_{n-1}^R \\ T_n^L &= T_{n-1}^R \end{aligned}\right\} \qquad \begin{Bmatrix} \theta \\ T \end{Bmatrix}_n^L = \begin{bmatrix} 1 & \dfrac{1}{K} \\ 0 & 1 \end{bmatrix}_n \begin{Bmatrix} \theta \\ T \end{Bmatrix}_{n-1}^R \tag{12.5-2}$$

The matrix pertaining to the mass is called the *point matrix* and the matrix associated with the shaft, the *field matrix*. The two can be combined to establish the transfer matrix for the nth element, which is

$$\begin{Bmatrix} \theta \\ T \end{Bmatrix}_n^R \underset{\rightarrow}{=} \begin{bmatrix} 1 & \dfrac{1}{K} \\ -\omega^2 J & \left(1 - \dfrac{\omega^2 J}{K}\right) \end{bmatrix}_n \begin{Bmatrix} \theta \\ T \end{Bmatrix}_{n-1}^R \tag{12.5-3}$$

In the development so far, the stations were numbered in increasing order from left to right with the transfer matrix also progressing to the right. The arrow under the equal sign indicates this direction of progression. In some problems it is convenient to proceed with the transfer matrix in the opposite direction, in which case we need only to invert Eq. (12.5-3). We then obtain the relationship

$$\begin{Bmatrix} \theta \\ T \end{Bmatrix}_{n-1}^R \underset{\leftarrow}{=} \begin{bmatrix} \left(1 - \dfrac{\omega^2 J}{K}\right) & -\dfrac{1}{K} \\ \omega^2 J & 1 \end{bmatrix} \begin{Bmatrix} \theta \\ T \end{Bmatrix}_n^R \tag{12.5-4}$$

The arrow now indicates that the transfer matrix progresses from right to left with the order of the station numbering unchanged. The student should verify this equation, starting with the free-body development.

12.6 SYSTEMS WITH DAMPING

When damping is included, the form of the transfer matrix is not altered, but the mass and stiffness elements become complex quantities. This can be easily shown by writing the equations for the nth subsystem shown in Fig. 12.6-1. The torque equation for disk n is

$$-\omega^2 J_n \theta_n = T_n^R - T_n^L - i\omega c_n \theta_n$$

or

$$\left(i\omega c_n - \omega^2 J_n\right)\theta_n = T_n^R - T_n^L \tag{12.6-1}$$

The elastic equation for the nth shaft is

$$\begin{aligned} T_n^L &= K_n(\theta_n - \theta_{n-1}) + i\omega g_n(\theta_n - \theta_{n-1}) \\ &= (K_n + i\omega g_n)(\theta_n - \theta_{n-1}) \end{aligned} \tag{12.6-2}$$

Thus the point matrix and the field matrix for the damped system become

$$\begin{Bmatrix} \theta \\ T \end{Bmatrix}_n^R = \begin{bmatrix} 1 & 0 \\ (i\omega c - \omega^2 J) & 1 \end{bmatrix}_n \begin{Bmatrix} \theta \\ T \end{Bmatrix}_n^L \tag{12.6-3}$$

$$\begin{Bmatrix} \theta \\ T \end{Bmatrix}_n^L = \begin{bmatrix} 1 & \dfrac{1}{(K + i\omega g)} \\ 0 & 1 \end{bmatrix}_n \begin{Bmatrix} \theta \\ T \end{Bmatrix}_{n-1}^R \tag{12.6-4}$$

which are identical to the undamped case except for the mass and stiffness elements; these elements are now complex.

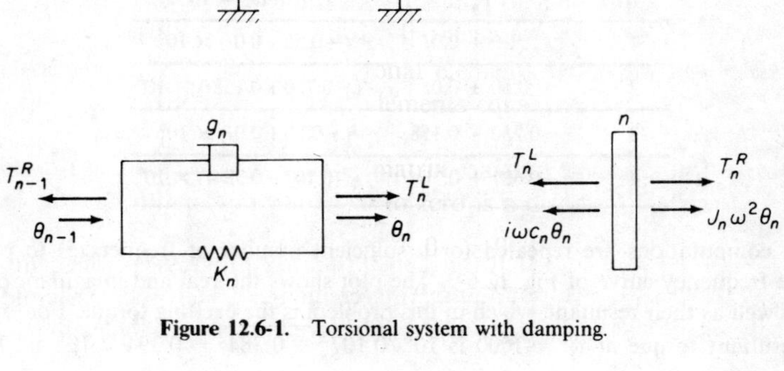

Figure 12.6-1. Torsional system with damping.

Example 12.6-1

The torsional system of Fig. 12.6-2 is excited by a harmonic torque at a point to the right of disk 4. Determine the torque-frequency curve and establish the first natural frequency of the system.

$$J_1 = J_1 = 500 \text{ lb in. s}^2$$
$$J_3 = J_4 = 1000 \text{ lb in. s}^2$$
$$K_2 = K_3 = K_4 = 10^6 \text{ lb in./rad}$$
$$c_2 = 10^4 \text{ lb in. s/rad}$$
$$g_4 = 2 \times 10^4 \text{ lb in. s/rad}$$

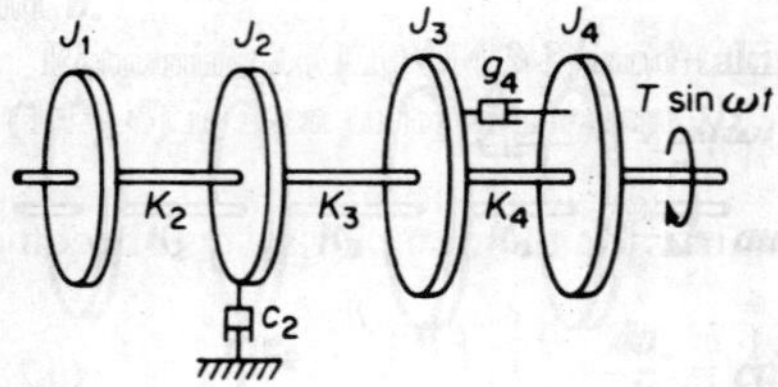

Figure 12.6-2.

Solution: The numerical computations for $\omega^2 = 1000$ are shown in the first accompanying table. The complex mass and stiffness terms are first tabulated for each station n. Substituting into the point and field matrices, i.e., Eqs. (12.6-3) and (12.6-4), the complex amplitude and torque for each station are found, as in the second table.

n	$(\omega^2 J_n - i\omega c_n)10^{-6}$	$(K_n + i\omega g_n)10^{-6}$
1	$0.50+0.0i$	
2	$0.50-0.316i$	$1.0 + 0.0i$
3	$1.0+0.0i$	$1.0 + 0.0i$
4	$1.0+0.0i$	$1.0 + 0.635i$

n	θ_n	T_n^R(for $\omega^2 = 1000$)
1	$1.0 + 0.0i$	$(-0.50+0.0i) \times 10^6$
2	$0.50 + 0.0i$	$(-0.750+0.158i) \times 10^6$
3	$-0.250 + 0.158i$	$(-0.50+0.0i) \times 10^6$
4	$-0.607 + 0.384i$	$(0.107-0.384i) \times 10^4$

These computations are repeated for a sufficient number of frequencies to plot the torque-frequency curve of Fig. 12.6-3. The plot shows the real and imaginary parts of T_4^R as well as their resultant, which in this problem is the exciting torque. For example, the resultant torque at $\omega^2 = 1000$ is $10^6\sqrt{0.107^2 + 0.384^2} = 0.394 \times 10^6$ in. lb. The

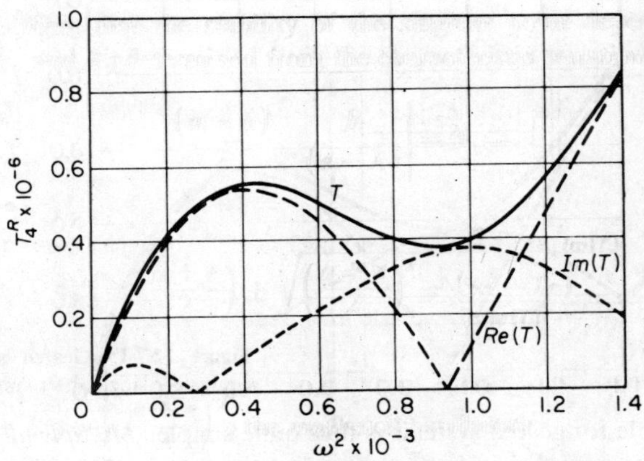

Figure 12.6-3. Torsional-frequency curve for damped torsional system of Figure 12.6-2.

first natural frequency of the system from this diagram is found to be approximately $\omega = \sqrt{930} = 30.5$ rad/s, where the natural frequency is defined as that frequency of the undamped system which requires no torque to sustain the motion.

Example 12.6-2

In Fig. 12-6-2 if $T = 2000$ in. lb and $\omega = 31.6$ rad/s, determine the amplitude of the second disk.

Solution: The table above indicates that a torque of 394,000 in. lb will produce an amplitude of $\theta_2 = 0.50$ rad. Since amplitude is proportional to torque, the amplitude of the second disk for the specified torque is $0.50 \times \frac{2}{394} = 0.00254$ rad.

12.7 GEARED SYSTEM

Consider the geared torsional system of Fig. 12.7-1, where the speed ratio of shaft 2 to shaft 1 is n. The system can be reduced to an equivalent single shaft system as follows.

With the speed of shaft 2 equal to $\dot{\theta}_2 = n\dot{\theta}_1$, the kinetic energy of the system is

$$T = \tfrac{1}{2}J_1\dot{\theta}_1^2 + \tfrac{1}{2}J_2 n^2\dot{\theta}_1^2 \tag{12.7-1}$$

Thus the equivalent inertia of disk 2 referred to shaft 1 is $n^2 J_2$.

To determine the equivalent stiffness of shaft 2 referred to shaft 1, clamp disks 1 and 2 and apply a torque to gear 1, rotating it through an angle θ_1. Gear 2 will then rotate through the angle $\theta_2 = n\theta_1$, which will also be the twist in shaft 2. The potential energy of the system is then

$$U = \tfrac{1}{2}K_1\theta_1^2 + \tfrac{1}{2}K_2 n^2\theta_1^2 \tag{12.7-2}$$

and the equivalent stiffness of shaft 2 referred to shaft 1 is $n^2 K_2$.

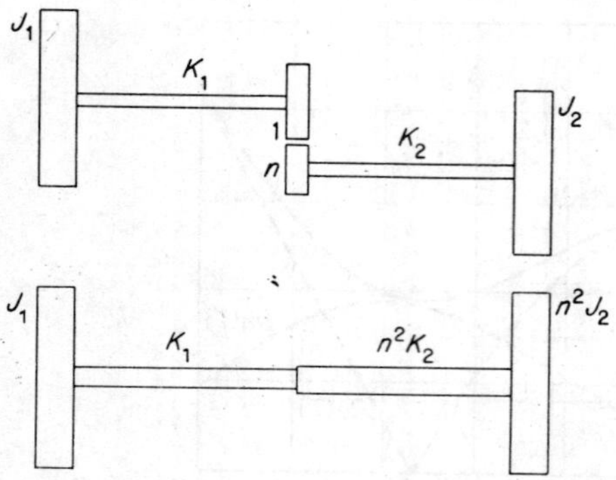

Figure 12.7-1. Geared system and its equivalent single-shaft system.

The rule for geared systems is thus quite simple: *Multiply all stiffness and inertias of the geared shaft by n^2, where n is the speed ratio of the geared shaft to the reference shaft.*

12.8 BRANCHED SYSTEMS

Branched systems are frequently encountered; some common examples are the dual propeller system of a marine installation and the drive shaft and differential of an automobile, which are shown in Fig. 12.8-1.

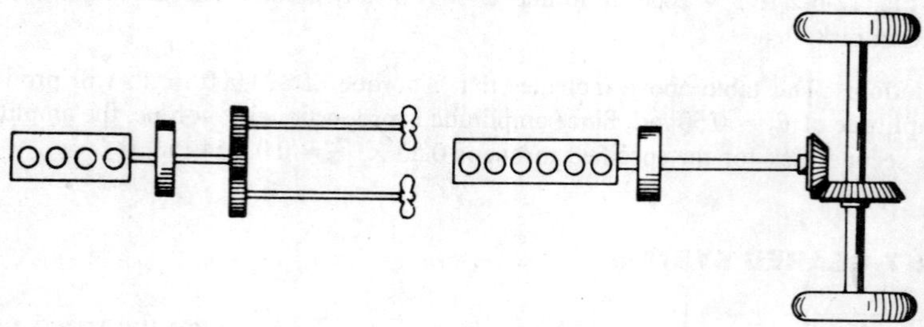

Figure 12.8-1. Examples of branched torsional systems.

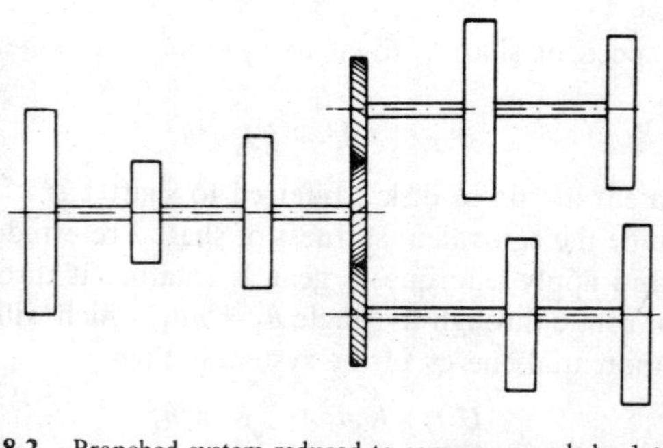

Figure 12.8-2. Branched system reduced to common speeds by 1 to 1 gears.

Such systems can be reduced to the form with one-to-one gears shown in Fig. 12.8-2 by multiplying all the inertias and stiffnesses of the branches by the squares of their speed ratios.

Example 12.8-1

Outline the matrix procedure for solving the torsional branched system of Fig. 12.8-3.

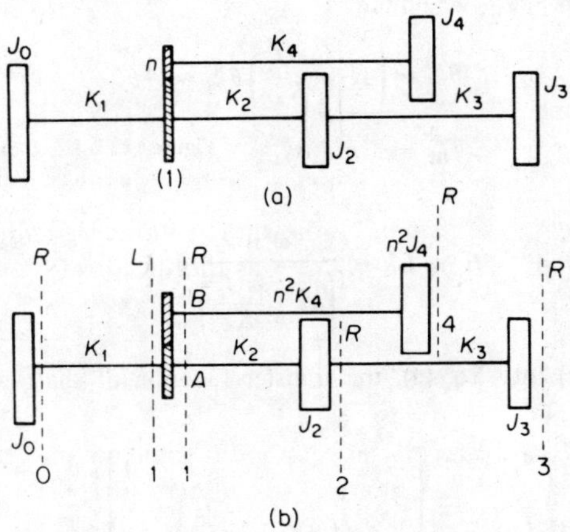

Figure 12.8-3. Branched system and reduced system.

Solution: We first convert to a system having one-to-one gears by multiplying the stiffness and inertia of branch B by n^2, as shown in Fig. 12.8-3(b). We can then proceed from station 0 through to station 3, taking note of the fact that gear B introduces a torque T_{B1}^R on gear A.

Figure 12.8-4 shows the free-body diagram of the two gears. With T_{B1}^R shown as positive torque, the torque exerted on gear A by gear B is negative as shown. The torque balance on gear A is then

$$T_{A1}^R = T_{A1}^L + T_{B1}^R \tag{a}$$

and we need now to express T_{B1}^R in terms of the angular displacement θ_1 of shaft A.

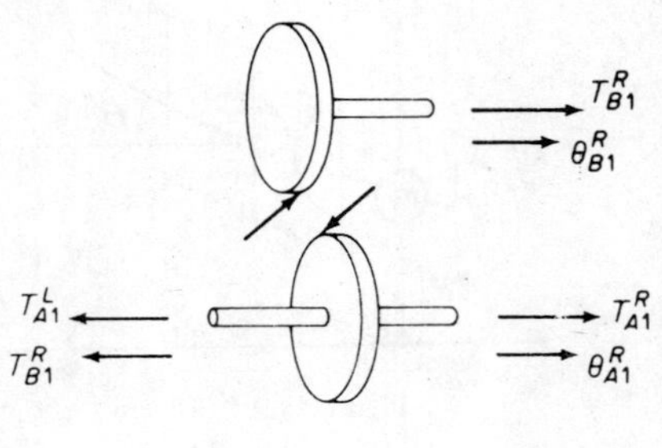

Figure 12.8-4.

Using Eq. (12.5-4) and noting that $T_{B4}^R = 0$, we have for shaft B

$$\begin{Bmatrix} \theta_B \\ T_B \end{Bmatrix}_1^R = \begin{bmatrix} \left(1 - \dfrac{\omega^2 n^2 J_4}{n^2 K_4}\right) & -\dfrac{1}{n^2 K_4} \\ \omega^2 n^3 J_4 & 1 \end{bmatrix} \begin{Bmatrix} \theta_B \\ 0 \end{Bmatrix}_4^R \tag{b}$$

Since $\theta_{B1}^R = -\theta_{A1}^L = -\theta_{A1}^R$, we obtain

$$\theta_{B1}^R = \left(1 - \frac{\omega^2 J_4}{K_4}\right) \theta_{B4}^R = -\theta_{A1}^L \tag{c}$$

$$T_{B1}^R = \omega^2 n^2 J_4 \theta_{B4}^R \tag{d}$$

Eliminating θ_{B4}^R

$$T_{B1}^R = \frac{-\omega^2 n^2 J_4}{\left(1 - \dfrac{\omega^2 J_4}{K_4}\right)} \theta_{A1}^L \tag{e}$$

Substituting Eq. (e) into Eq. (a), the transfer function of shaft A across the gears becomes

$$\begin{Bmatrix} \theta_A \\ T_A \end{Bmatrix}_1^R \Rightarrow \begin{bmatrix} 1 & 0 \\ -\omega^2 J_4 \Big/ \left(1 - \dfrac{\omega^2 J_4}{K_4}\right) & 1 \end{bmatrix} \begin{Bmatrix} \theta_A \\ T_A \end{Bmatrix}_1^L \tag{f}$$

It is now possible to proceed along shaft A from $1R$ to $3R$ in the usual manner.

12.9 TRANSFER MATRICES FOR BEAMS

The algebraic equations of Sec. 10.3 can be rearranged so that the four quantities at station $i + 1$ are expressed in terms of the same four quantities at station i. When such equations are presented in matrix form they are known

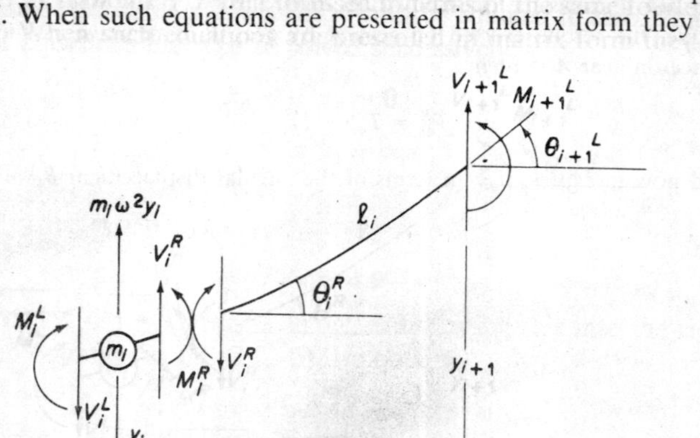

Figure 12.9-1. Beam sections for transfer matrices.

as *transfer matrices*. In this section we present a procedure for the formulation and assembly of the matrix equation in terms of its boundary conditions.

Figure 12.9-1 shows the same ith section of the beam of Fig. 12.3-1 broken down further into a point mass and a massless beam by cutting the beam just right of the mass. We designate the quantities to the left and right of the mass by superscripts L and R, respectively.

Considering first the massless beam section, the following equations can be written:

$$\begin{aligned} V_{i+1}^L &= V_i^R \\ M_{i+1}^L &= M_i^R + V_i^R l_i \\ \theta_{i+1}^L &= \theta_i^R + M_{i+1}^L\left(\frac{l}{EI}\right)_i + V_{i+1}^L\left(\frac{l^2}{2EI}\right)_i \\ y_{i+1}^L &= y_i^R + \theta_i^R l_i + M_{i+1}^L\left(\frac{l^2}{2EI}\right)_i + V_{i+1}^L\left(\frac{l^3}{3EI}\right)_i \end{aligned} \tag{12.9-1}$$

Substituting for V_{i+1}^L and M_{i+1}^L from the first two equations into the last two and arranging the results in matrix form, we obtain what is referred to as the *field matrix*:

$$\begin{Bmatrix} -V \\ M \\ \theta \\ y \end{Bmatrix}_{i+1}^{L} = \begin{bmatrix} 1 & 0 & 0 & 0 \\ l & 1 & 0 & 0 \\ \dfrac{l^2}{2EI} & \dfrac{l}{EI} & 1 & 0 \\ \dfrac{l^3}{6EI} & \dfrac{l^2}{2EI} & l & 1 \end{bmatrix} \begin{Bmatrix} -V \\ M \\ \theta \\ y \end{Bmatrix}_{i}^{R} \tag{12.9-2}$$

In the above equation a minus sign has been inserted for V in order to make the elements of the field matrix all positive.

Next consider the point mass for which the following equations can be written:

$$\begin{aligned} V_i^R &= V_i^L - m_i\omega^2 y_i \\ M_i^R &= M_i^L \\ \theta_i^R &= \theta_i^L \\ y_i^R &= y_i^L \end{aligned} \tag{12.9-3}$$

In matrix form these equations become

$$\begin{Bmatrix} -V \\ M \\ \theta \\ y \end{Bmatrix}_{i}^{R} = \begin{bmatrix} 1 & 0 & 0 & m\omega^2 \\ 0 & 1 & 0 & 0 \\ 0 & 0 & 1 & 0 \\ 0 & 0 & 0 & 1 \end{bmatrix} \begin{Bmatrix} -V \\ M \\ \theta \\ y \end{Bmatrix}_{i}^{L} \tag{12.9-4}$$

which is known as the *point matrix*.

Substituting Eq. (12.9-4) into Eq. (12.9-2) and multiplying, we obtain the assembled equation for the ith section

$$\begin{Bmatrix} -V \\ M \\ \theta \\ y \end{Bmatrix}_{i+1}^{R} = \begin{bmatrix} 1 & 0 & 0 & m\omega^2 \\ l & 1 & 0 & m\omega^2 l \\ \dfrac{l^2}{2EI} & \dfrac{l}{EI} & 1 & m\omega^2 \dfrac{l^2}{2EI} \\ \dfrac{l^3}{6EI} & \dfrac{l^2}{2EI} & l & \left(1 + \dfrac{m\omega^2 l^3}{6EI}\right) \end{bmatrix} \begin{Bmatrix} -V \\ M \\ \theta \\ y \end{Bmatrix}_{i}^{L} \tag{12.9-5}$$

The square matrix in this equation is called the *transfer matrix*, since the state vector at i is transferred to the state vector at $i + 1$ through this matrix. It is evident then that it is possible to progress through the structure so that the state vector at the far end is related to the state vector at the starting end by an equation of the form

$$\begin{Bmatrix} -V \\ M \\ \theta \\ y \end{Bmatrix}_{n} = \begin{bmatrix} u_{11} & u_{12} & \cdots & u_{14} \\ \vdots & \vdots & & \vdots \\ u_{41} & u_{42} & \cdots & u_{44} \end{bmatrix} \begin{Bmatrix} -V \\ M \\ \theta \\ y \end{Bmatrix}_{1} \tag{12.9-6}$$

where the matrix $[u]$ is the product of all the transfer matrices of the structure.

The advantage of the transfer matrix lies in the fact that the unknown quantity at 1, i.e., θ_1, for the cantilever beam, need not be carried through each station as in the algebraic set of equations. The multiplication of the 4×4 matrices by the digital computer is a routine problem. Also, the boundary equations are clearly evident in the matrix equation. For example, the assembled equation for the cantilever beam is

$$\begin{Bmatrix} -V \\ M \\ 0 \\ 0 \end{Bmatrix}_{n} = \begin{bmatrix} - & - & - & - \\ - & - & - & - \\ - & - & u_{33} & u_{34} \\ - & - & u_{43} & u_{44} \end{bmatrix} \begin{Bmatrix} 0 \\ 0 \\ \theta \\ 1 \end{Bmatrix}_{1} \tag{12.9-7}$$

and the natural frequencies must satisfy the equations

$$0 = u_{33}\theta + u_{34}$$

$$0 = u_{43}\theta + u_{44}$$

or

$$y_n = -\frac{u_{34}}{u_{33}} \cdot u_{43} + u_{44} = 0 \tag{12.9-8}$$

In a plot of y_n versus ω, the natural frequencies correspond to the zeros of the curve.

12.10 DIFFERENCE EQUATION

For systems with identical repeated sections, the method of difference equations leads to simple analytic equations for the natural frequencies and mode shapes. As an example of repeating sections, consider the N-story building shown in Fig. 12.10-1, where the mass of each floor is m and the lateral or shear stiffness of each section between floors is k lb/in. The equation of motion for the nth mass is then

$$m\ddot{x}_n = k(x_{n+1} - x_n) - k(x_n - x_{n-1}) \tag{12.10-1}$$

which for harmonic motion can be represented in terms of the amplitudes as

$$X_{n+1} - 2\left(1 - \frac{\omega^2 m}{2k}\right)X_n + X_{n-1} = 0 \tag{12.10-2}$$

The solution to this equation is found by substituting

$$X_n = e^{i\beta n} \tag{12.10-3}$$

which leads to the relationship

$$\left(1 - \frac{\omega^2 m}{2k}\right) = \frac{e^{i\beta} + e^{-i\beta}}{2} = \cos\beta$$

$$\frac{\omega^2 m}{k} = 2(1 - \cos\beta) = 4\sin^2\frac{\beta}{2} \tag{12.10-4}$$

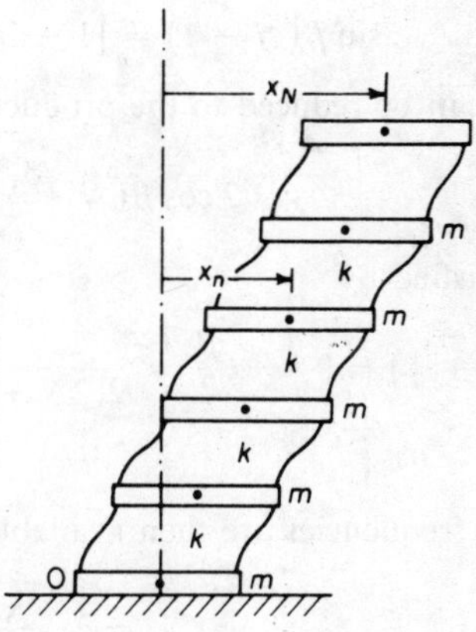

Figure 12.10-1. Repeated structure for difference equation analysis.

The general solution for X_n is

$$X_n = A\cos\beta n + B\sin\beta n \tag{12.10-5}$$

where A and B are evaluated from the boundary conditions.

The difference equation (12.10-2) is restricted to $1 \leqslant n \leqslant (N-1)$ and must be extended to $n = 0$ and $n = N$ by the boundary conditions.

At the ground the amplitude of motion is zero. With $n = 1$, Eq. (12.10-2) with $X_0 = 0$ becomes

$$X_2 - 2\left(1 - \frac{m\omega^2}{2k}\right)X_1 = 0$$

Substituting Eq. (12.10-4) and Eq. (12.10-5) into the above equation, we obtain

$$(A\cos 2\beta + B\sin 2\beta) - 2\cos\beta(A\cos\beta + B\sin\beta) = 0$$

$$A(\cos 2\beta - 2\cos^2\beta) - B(\sin 2\beta - 2\sin\beta\cos\beta) = 0$$

$$A(1) - \beta(0) = 0$$

$$\therefore A = 0$$

Thus the general solution is reduced to $X_n = B\sin\beta n$.

At the top the boundary equation is

$$m\ddot{x}_N = -k(x_N - x_{N-1})$$

which, in terms of the amplitude, becomes

$$X_{N-1} = \left(1 - \frac{\omega^2 m}{k}\right)X_N \tag{12.10-6}$$

Substituting from the general solution, we obtain the following relationship for the evaluation of β:

$$\sin\beta(N-1) = [1 - 2(1-\cos\beta)]\sin\beta N$$

This result can be reduced to the product form

$$2\cos\beta(N + \tfrac{1}{2})\sin\frac{\beta}{2} = 0 \tag{12.10-7}$$

which is satisfied by

$$\cos\beta(N + \tfrac{1}{2}) = 0, \qquad \frac{\beta}{2} = \frac{\pi}{2(2N+1)}, \frac{3\pi}{2(2N+1)}, \frac{5\pi}{2(2N+1)}, \ldots \tag{12.10-8}$$

The natural frequencies are then available from Eq. 12.10-4 as

$$\omega = 2\sqrt{\frac{k}{m}}\sin\frac{\beta}{2} \tag{12.10-9}$$

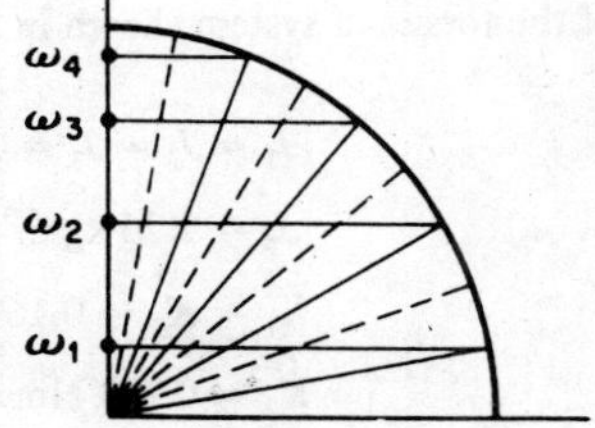

Figure 12.10-2. Natural frequencies of a repeated structure with $N = 4$.

which lead to

$$\omega_1 = 2\sqrt{\frac{k}{m}}\sin\frac{\pi}{2(2N+1)}$$

$$\omega_2 = 2\sqrt{\frac{k}{m}}\sin\frac{3\pi}{2(2N+1)}$$

$$\vdots$$

$$\omega_N = 2\sqrt{\frac{k}{m}}\sin\frac{(2N-1)\pi}{2(2N+1)}$$

Figure 12.10-2 shows a graphical representation of these natural frequencies when $N = 4$.

The method of difference equation presented here is applicable to many other dynamical systems where repeating sections are present. The natural frequencies are always given by Eq. (12.10-9); however, the quantity β must be established for each problem from its boundary conditions.

PROBLEMS

12-1 Write a computer program for your programmable calculator for the torsional system given in Sec. 12.1. Fill in the actual algebraic operations performed in the program steps.

12-2 Using Holzer's method, determine the natural frequencies and mode shapes of the torsional system of Fig. P12-2 when $J = 1.0$ kg m^2 and $K = 0.20 \times 10^6$ Nm/rad.

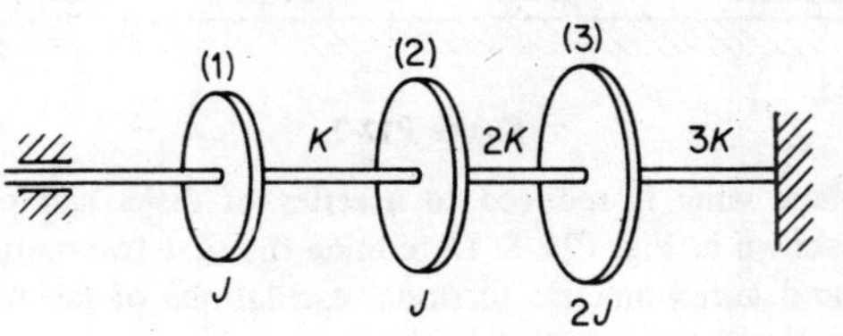

Figure P12-2.

12-3 Using Holzer's method, determine the first two natural frequencies and mode shapes of the torsional system shown in Fig. P12-3 with the following values of J and K.

$$J_1 = J_2 = J_3 = 1.13 \text{ kg m}^2$$

$$J_4 = 2.26 \text{ kg m}^2$$

$$K_1 = K_2 = 0.169 \text{ Nm/rad} \times 10^6$$

$$K_3 = 0.226 \text{ Nm/rad} \times 10^6$$

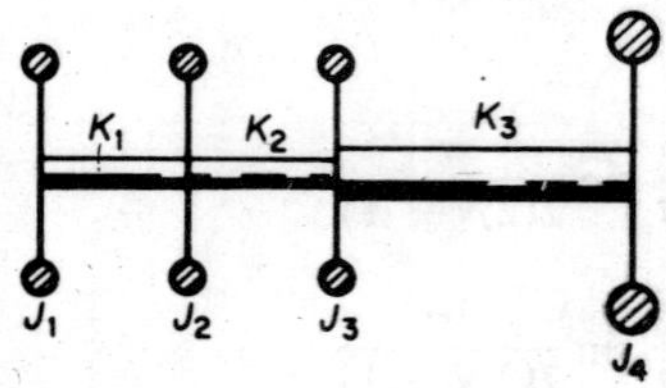

Figure P12-3.

12-4 Determine the natural frequencies and mode shapes of the three-story building of Fig. P12-4 by using Holzer's method for all $m_s = m$ and all $k_s = k$.

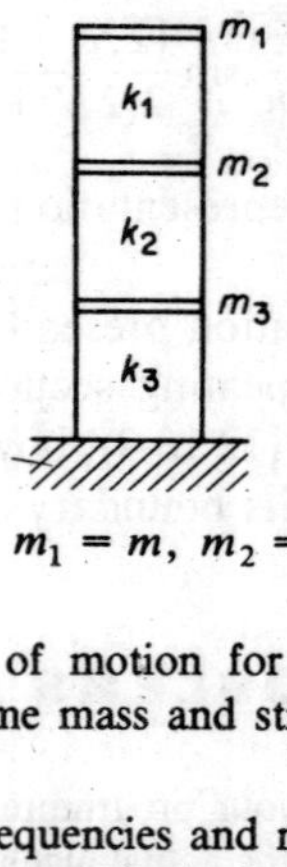

Figure P12-4.

12-5 Repeat Prob. 12-4 when $m_1 = m$, $m_2 = 2m$, $m_3 = 3m$, $k_1 = k$, $k_2 = k$, and $k_3 = 2k$.

12-6 Compare the equations of motion for the linear spring-mass system versus torsional system with same mass and stiffness distribution. Show that they are similar.

12-7 Determine the natural frequencies and mode shapes of the spring-mass system of Fig. P12-7 by the Holzer method when all masses are equal and all stiffnesses are equal.

Figure P12-7.

12-8 A fighter-plane wing is reduced to a series of disks and shafts for Holzer's analysis as shown in Fig. P12-8. Determine the first two natural frequencies for symmetric and antisymmetric torsional oscillations of the wings, and plot the torsional mode corresponding to each.

n	J lb. in. sec.^{2}	K lb. in./rad.
1	50	15×10^6
2	138	30
3	145	22
4	181	36
5	260	120
6	$\frac{1}{2} \times 140{,}000$	

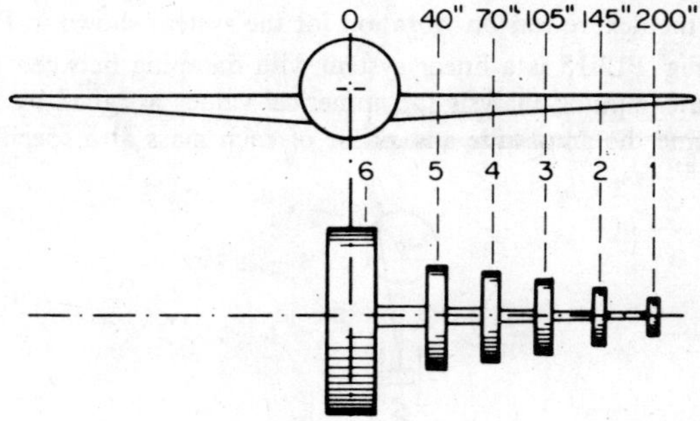

Figure P12-8.

2-9 Determine the natural modes of the simplified model of an airplane shown in Fig. P12-9 where $M/m = n$ and the beam of length l is uniform.

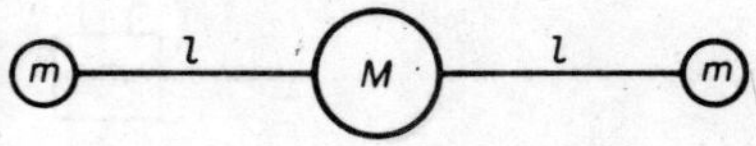

Figure P12-9.

12-10 Using Myklestad's method, determine the natural frequencies and mode shapes of the two-lumped-mass cantilever beam of Fig. P12-10. Compare with previous results by using influence coefficients.

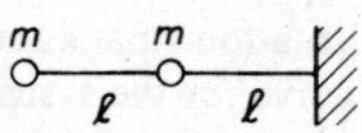

Figure P12-10.

12-11 Determine the first two natural frequencies and mode shapes of the three-mass cantilever of Fig. P12-11

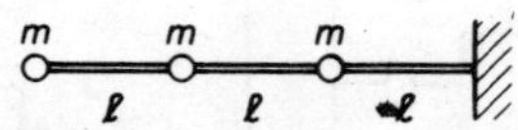

Figure P12-11.

12-12 Using Myklestad's method, determine the boundary equations for the simply supported beam of Fig. P12-12.

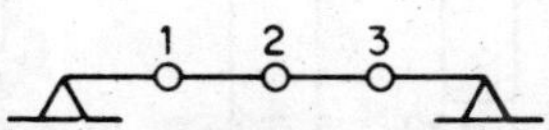

Figure P12-12.

12-13 The beam of Fig. P12-13 has been previously solved by the method of matrix iteration. Check that the boundary condition of zero deflection at the left end is satisfied for these natural frequencies when Myklestad's method is used. That is, check the deflection for change in sign when frequencies above and below the natural frequency are used.

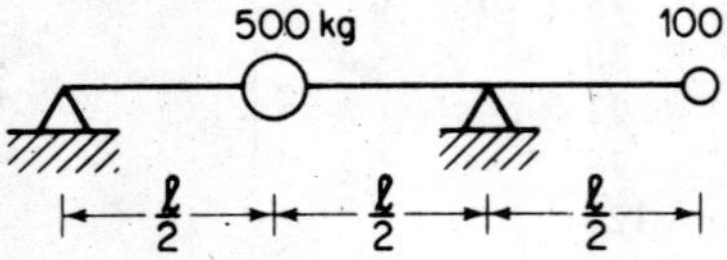

Figure 12-13.

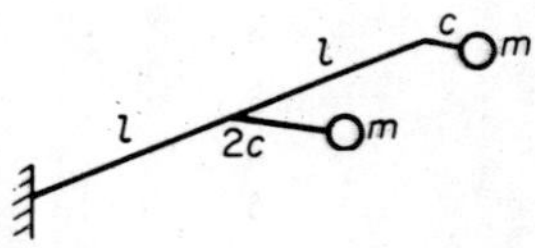

Figure 12-14.

12-14 Determine the flexure-torsion vibration for the system shown in Fig. P12-14.

12-15 Shown in Fig. P12-15 is a linear system with damping between mass 1 and 2. Carry out a computer analysis for numerical values assigned by the instructor, and determine the amplitude and phase of each mass at a specified frequency.

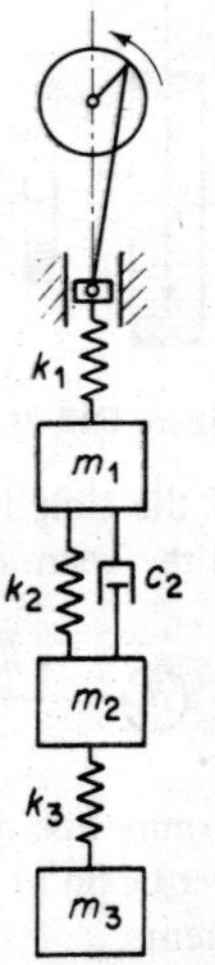

Figure P12-15.

12-16 A torsional system with a torsional damper is shown in Fig. P12-16. Determine the torque-frequency curve for the system.

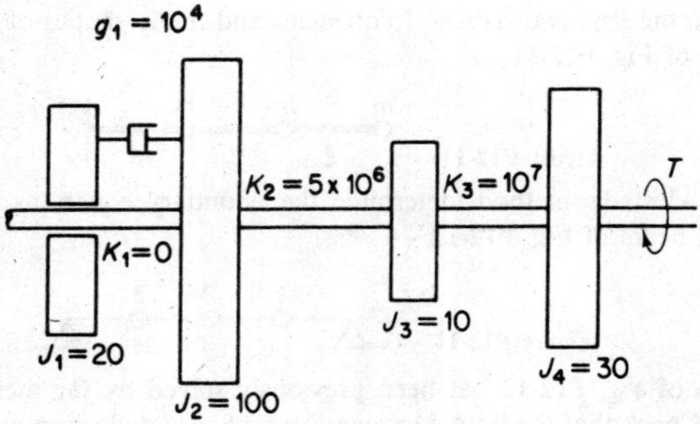

Figure P12-16.

12-17 Determine the equivalent torsional system for the geared system shown in Fig. P12-17 and find its natural frequency.

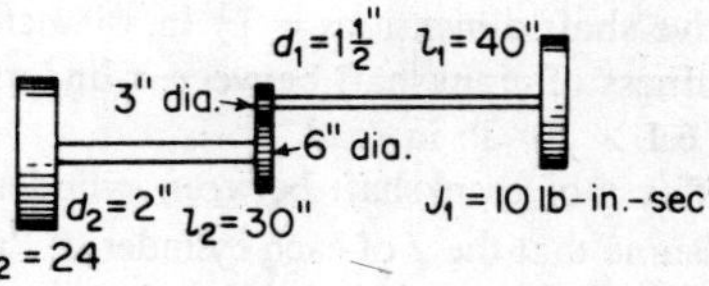

Figure P12-17.

12-18 If the small and large gears of Prob. 12-17 have the inertias $J' = 2$, $J'' = 6$, determine the equivalent single shaft system and establish the natural frequencies.

12-19 Determine the two lowest natural frequencies of the torsional system shown in Fig. P12-19 for the following values of J, K, and n

$J_1 = 15$ lb in. s^2 $\quad K_1 = 2 \times 10^6$ lb in./rad

$J_2 = 10$ lb in. s^2 $\quad K_2 = 1.6 \times 10^6$ lb in./rad

$J_3 = 18$ lb in. s^2 $\quad K_3 = 1 \times 10^6$ lb in./rad

$J_4 = 6$ lb in. s^2 $\quad K_4 = 4 \times 10^6$ lb in./rad

Speed ratio of drive shaft to axle = 4 to 1.

What are the amplitude ratios of J_2 to J_1 at the natural frequencies?

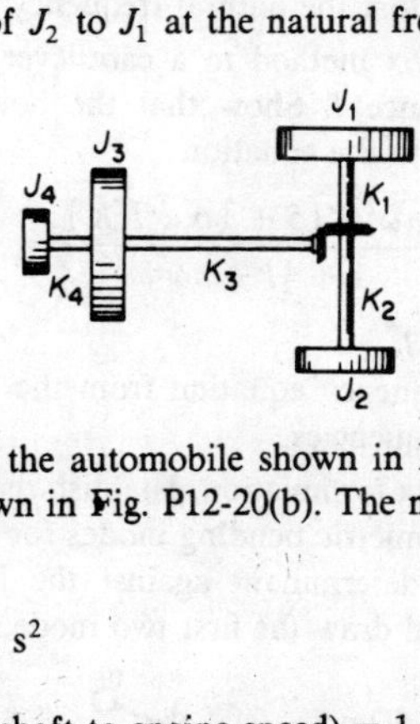

Figure P12-19.

12-20 Reduce the torsional system of the automobile shown in Fig. P12-20(a) to the equivalent torsional system shown in Fig. P12-20(b). The necessary information is given as follows:

J of each rear wheel = 9.2 lb in. s^2
J of flywheel = 12.3 lb in. s^2
Transmission speed ratio (drive shaft to engine speed) = 1.0 to 3.0

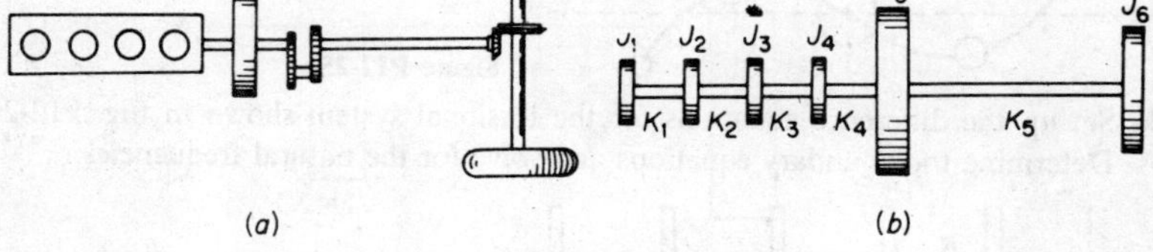

Figure P12-20.

Differential speed ratio (axle to drive shaft) = 1.0 to 3.5
Axle dimensions = $1\frac{1}{4}$ in. diameter, 25 in. long (each)
Drive shaft dimensions = $1\frac{1}{2}$ in. diameter, 74 in. long
Stiffness of crankshaft between cylinders, measured experimentally = 6.1×10^6 lb in./rad
Stiffness of crankshaft between cylinder 4 and flywheel = 4.5×10^6 lb in./rad

12-21 Assume that the J of each cylinder of Prob. 12-20 = 0.20 lb in. s^2 and determine the natural frequencies of the system.

12-22 Determine the equations of motion for the torsional system shown in Fig. P12-22, and arrange them into the matrix iteration form. Solve for the principal modes of oscillation.

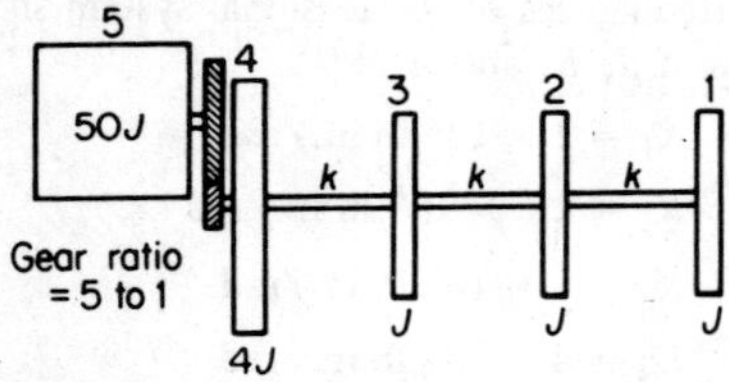

Figure P12-22.

12-23 Apply the matrix method to a cantilever beam of length l and mass m at the end, and show that the natural frequency equation is directly obtained.

12-24 Apply the matrix method to a cantilever beam with two equal masses spaced equally a distance l. Show that the boundary conditions of zero slope and deflection lead to the equation

$$\theta_1 = \frac{\frac{1}{2}m\omega^2 lK\left(5 + \frac{1}{6}m\omega^2 l^2 K\right)}{1 + \frac{1}{2}l^2 K m\omega^2} = \frac{1 + \frac{3}{2}m\omega^2 l^2 K + \left(\frac{1}{6}m\omega^2 l^2 K\right)^2}{2l + \frac{1}{6}m\omega^2 l^3 K}$$

where $K = l/EI$.
Obtain the frequency equation from the above relationship and determine the two natural frequencies.

12-25 Using the matrix formulation, establish the boundary conditions for the symmetric and antisymmetric bending modes for the system shown in Fig. P12-25. Plot the boundary determinant against the frequency ω to establish the natural frequencies, and draw the first two mode shapes.

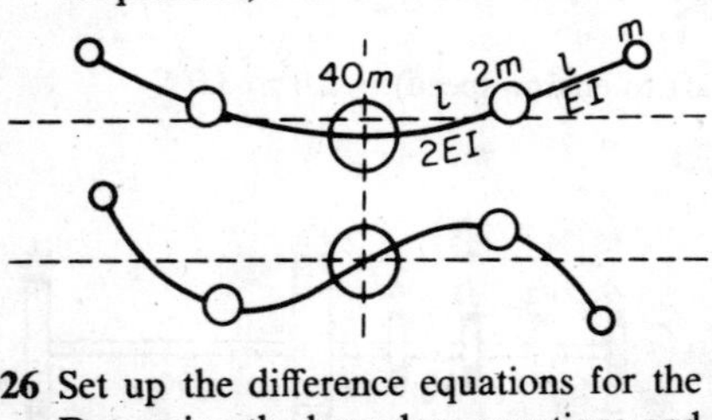

Figure P12-25.

12-26 Set up the difference equations for the torsional system shown in Fig. P12-26. Determine the boundary equations and solve for the natural frequencies.

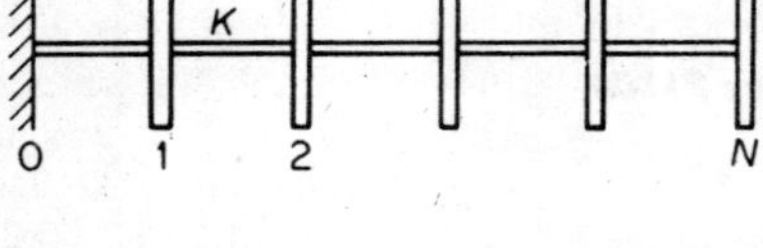

Figure P12-26.

12-27 Set up the difference equations for N equal masses on a string with tension T, as shown in Fig. P12-27. Determine the boundary equations and the natural frequencies.

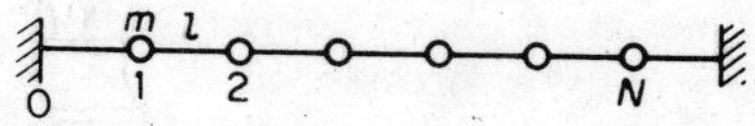

Figure P12-27.

12-28 Write the difference equations for the spring-mass system shown in Fig. P12-28 and find the natural frequencies of the system.

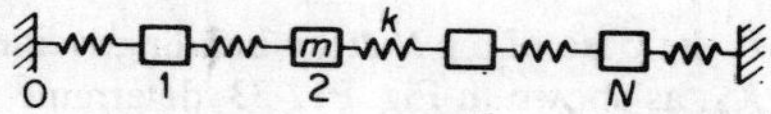

Figure P12-28.

12-29 An N-mass pendulum is shown in Fig. P12-29. Determine the difference equations, boundary conditions, and the natural frequencies.

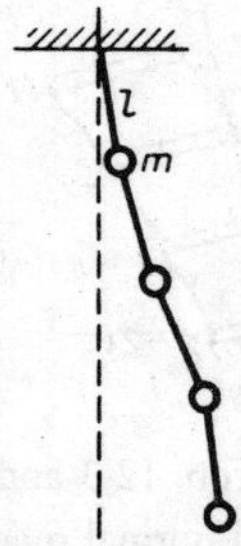

Figure P12-29.

12-30 If the left end of the system of Prob. 12-26 is connected to a heavy flywheel, as shown in Fig. P12-30, show that the boundary conditions lead to the equation

$$(-\sin N\beta \cos \beta + \sin N\beta)\left(1 + 4\frac{K}{K_a}\frac{J_a}{J}\sin^2\frac{\beta}{2}\right)$$

$$= -2\frac{J_a}{J}\sin^2\frac{\beta}{2}\sin \beta \cos N\beta$$

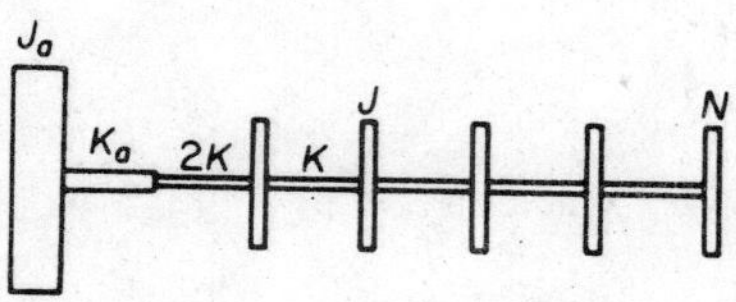

Figure P12-30.

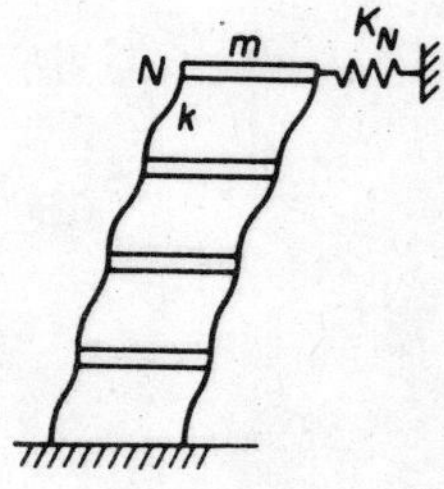

Figure P12-31.

12-31 If the top story of a building is restrained by a spring of stiffness K_N, as shown in Fig. P12-31, determine the natural frequencies of the N-story building

12-32 A ladder-type structure is fixed at both ends, as shown in Fig. P12-32. Determine the natural frequencies

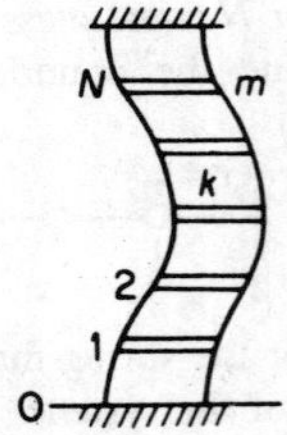

Figure P12-32.

12-33 If the base of an N-story building is allowed to rotate against a resisting spring K_θ, as shown in Fig. P12-33, determine the boundary equations and the natural frequencies.

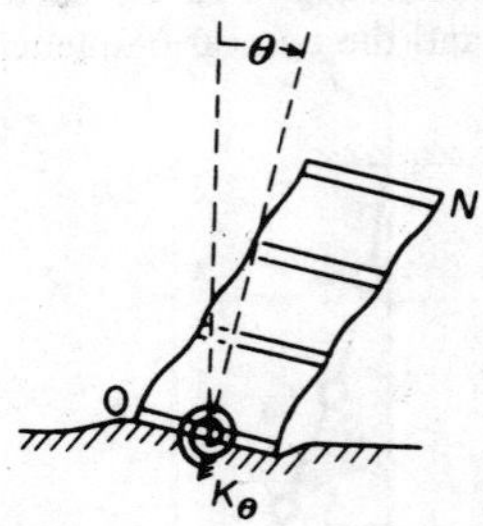

Figure P12-33.

12-34 Draw a flow diagram for Prob. 12-3 and write the Fortran program.

12-35 The natural frequencies and normal modes presented for the 10-DOF system of Example 6.9-1 were obtained from the eigenvalue-eigenvector computer program. Verify these numbers by the use of Eq. (12.10-9) for ω_n and $X_n = B \sin \beta n$ for the amplitude.

13 RANDOM VIBRATIONS

The types of functions we have considered up to now can be classified as deterministic, i.e., mathematical expressions can be written which will determine their instantaneous values at any time t. There are, however, a number of physical phenomena that result in nondeterministic data where future instantaneous values cannot be predicted in a deterministic sense. As examples, we can mention the noise of a jet engine, the heights of waves in a choppy sea, ground motion during an earthquake, and pressure gusts encountered by an airplane in flight. These phenomena all have one thing in common: the unpredictability of their instantaneous value at any future time. Nondeterministic data of this type are referred to as *random time functions*.

13.1 RANDOM PHENOMENA

A sample of a typical random time function is shown in Fig. 13.1-1. In spite of the irregular character of the function, many random phenomena exhibit some degree of statistical regularity, and certain averaging procedures can be applied to establish gross characteristics useful in engineering design.

In any statistical method, a large amount of data is necessary to establish reliability. For example, to establish the statistics of the pressure fluctuation due to air turbulence over a certain air route, an airplane may collect hundreds of records of the type shown in Fig. 13.1-2.

Each record is called a *sample*, and the total collection of samples is called the *ensemble*. We can compute the ensemble average of the instanta-

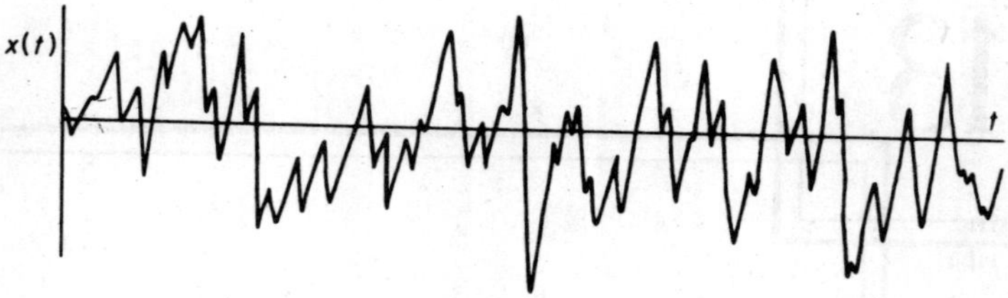

Figure 13.1-1. A record of random time function.

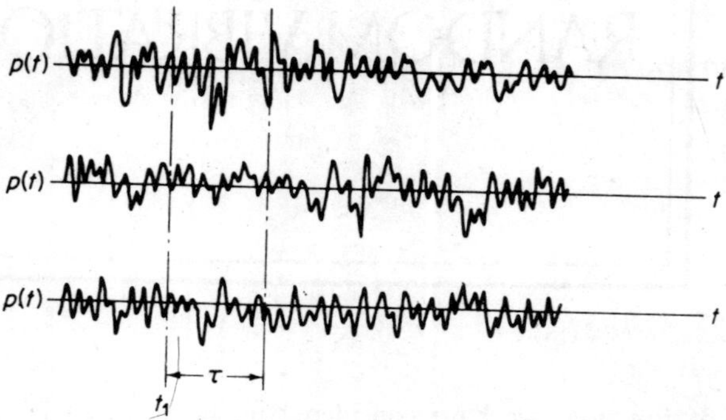

Figure 13.1-2. An ensemble of random time functions.

neous pressure at time t_1. We can also multiply the instantaneous pressures in each sample at times t_1 and $t_1 + \tau$, and average these results for the ensemble. If such averages do not differ as we choose different values of t_1, then the random process described by the above ensemble is said to be *stationary*.

If the ensemble averages are replaced next by time averages, and if the results computed from each sample are the same as those of any other sample and equal to the ensemble average, then the random process is said to be *ergodic*.

Thus for a stationary ergodic random phenomenon its statistical properties are available from a single time function of sufficiently long time period. Although such random phenomena may exist only theoretically, its assumption greatly simplifies the task of dealing with random variables. This chapter treats only this class of stationary ergodic random functions.

13.2 TIME AVERAGING AND EXPECTED VALUE

Expected value. In random vibrations we will repeatedly encounter the concept of time averaging over a long period of time. The most common notation for this operation is defined by the following equation in which $x(t)$ is the variable.

$$\overline{x(t)} = \langle x(t) \rangle = \lim_{T \to \infty} \frac{1}{T} \int_0^T x(t)\, dt \qquad (13.2\text{-}1)$$

The above number is also equal to the *expected value* of $x(t)$, which is written as

$$E[x(t)] = \lim_{T\to\infty} \frac{1}{T}\int_0^T x(t)\,dt \tag{13.2-2}$$

It is the average or mean value of a quantity sampled over a long time. In the case of discrete variables x_i, the expected value is given by the equation

$$E[x] = \lim_{n\to\infty} \frac{1}{n}\sum_{i=1}^{n} x_i \tag{13.2-3}$$

Mean square value. These average operations can be applied to any variable such as $x^2(t)$ or $x(t)\cdot y(t)$. The *mean square value*, designated by the notation $\overline{x^2}$ or $E[x^2(t)]$, is found by integrating $x^2(t)$ over a time interval T and taking its average value according to the equation

$$E\left[x^2(t)\right] = \overline{x^2} = \lim_{T\to\infty} \frac{1}{T}\int_0^T x^2\,dt \tag{13.2-4}$$

Variance and standard deviation. It is often desirable to consider the time series in terms of the mean and its fluctuation from the mean. A property of importance describing the fluctuation is the *variance* σ^2, which is the mean square value about the mean, given by the equation

$$\sigma^2 = \lim_{T\to\infty} \frac{1}{T}\int_0^T (x-\bar{x})^2\,dt \tag{13.2-5}$$

By expanding the above equation, it is easily seen that

$$\sigma^2 = \overline{x^2} - (\bar{x})^2 \tag{13.2-6}$$

so that the variance is equal to the mean square value minus the square of the mean. The positive square root of the variance is the *standard deviation* σ.

Fourier series. Generally random time functions contain oscillations of many frequencies, which approach a continuous spectrum. Although random time functions are generally not periodic, their representations by Fourier series, where the periods are extended to a large value approaching infinity, offers a logical approach.

In Chapter 1, the exponential form of the Fourier series was shown to be

$$x(t) = \sum_{-\infty}^{\infty} c_n e^{in\omega_1 t} = c_0 + \sum_{n=1}^{\infty}\left(c_n e^{in\omega_1 t} + c_n^* e^{-in\omega_1 t}\right) \tag{13.2-7}$$

This series, which is a real function, involves a summation over negative and positive frequencies, and it also contains a constant term c_0. The constant term c_0 is the average value of $x(t)$ and since it can be dealt with separately, we will exclude it in future considerations. Moreover, actual measurements are made in terms of positive frequencies, and it would be more desirable to work with the equation

$$x(t) = \text{Re} \sum_{n=1}^{\infty} C_n e^{in\omega_1 t} \tag{13.2-8}$$

The one-sided summation in the above equation is complex and hence the real part of the series must be stipulated for $x(t)$ real. Since the real part of a vector is one-half the sum of the vector and its conjugate [see Eq. (1.1-9)],

$$x(t) = \text{Re} \sum_{n=1}^{\infty} C_n e^{in\omega_1 t} = \frac{1}{2} \sum_{n=1}^{\infty} \left(C_n e^{in\omega_1 t} + C_n^* e^{-in\omega_1 t} \right)$$

By comparison with Eq. (1.2-6), we find

$$C_n = 2c_n = \frac{2}{T} \int_{-T/2}^{T/2} x(t) e^{-in\omega_1 t}\, dt$$

$$= a_n - ib_n \tag{13.2-9}$$

Example 13.2-1

Determine the mean square value of a record of random vibration $x(t)$ containing many discrete frequencies.

Solution: Since the record is periodic, we can represent it by the real part of the Fourier series

$$x(t) = \text{Re} \sum_{1}^{\infty} C_n e^{in\omega_0 t}$$

$$= \frac{1}{2} \sum_{1}^{\infty} \left(C_n e^{in\omega_0 t} + C_n^* e^{-in\omega_0 t} \right)$$

where C_n is a complex number and C_n^* is its complex conjugate. [See Eq. (13.2-9).] Its mean square value is

$$\overline{x^2} = \lim_{T \to \infty} \frac{1}{T} \int_0^T \frac{1}{4} \sum_{n=1}^{\infty} \left(C_n e^{in\omega_0 t} + C_n^* e^{-in\omega_0 t} \right)^2 dt$$

$$= \lim_{T \to \infty} \sum_{n=1}^{\infty} \frac{1}{4} \left[\frac{C_n^2 e^{i2n\omega_0 t}}{i2n\omega_0 T} + 2C_n C_n^* + \frac{C_n^{*2} e^{-i2n\omega_0 t}}{-i2n\omega_0 T} \right]_0^T$$

$$= \sum_{n=1}^{\infty} \frac{1}{2} C_n C_n^* = \sum_{n=1}^{\infty} \frac{1}{2} |C_n|^2 = \sum_{n=1}^{\infty} \overline{C_n^2}$$

In this equation, $e^{\pm i2n\omega_0 t}$ for any t, is bounded between ± 1, and due to $T \to \infty$ in the denominator, the first and last terms become zero. The middle term, however, is independent of T. Thus the mean square value of the periodic function is simply the sum of the mean square value of each harmonic component present.

13.3 FREQUENCY RESPONSE FUNCTION

In any linear system there is a direct linear relationship between the input and the output. This relationship, which also holds for random functions, is represented by the block diagram of Fig. 13.3-1.

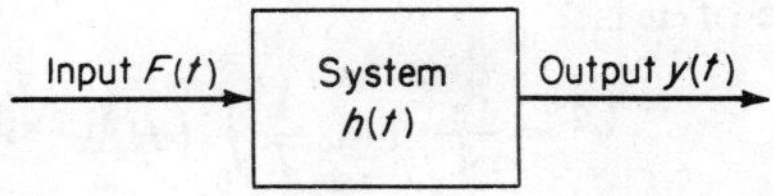

Figure 13.3-1. Input-output relationship of a linear system.

In the time domain the system behavior may be determined in terms of the system impulse response $h(t)$ used in the convolution integral of Eq. (4.2-1).

$$y(t) = \int_0^t x(\xi)h(t-\xi)\,d\xi \tag{13.3-1}$$

A much simpler relationship is available for the frequency domain in terms of the frequency response function $H(\omega)$, which we can define as the ratio of the output to the input under steady-state conditions, with the input equal to a harmonic time function of unit amplitude. The transient solution is thus excluded in this consideration. In random vibrations, the initial conditions and the phase have little meaning and are therefore ignored. We are mainly concerned with the average energy, which we can associate with the mean square value.

Applying this definition to a single-DOF system,

$$m\ddot{y} + c\dot{y} + ky = x(t) \tag{13.3-2}$$

let the input be $x(t) = e^{i\omega t}$. The steady-state output will then be $y = H(\omega)e^{i\omega t}$, where $H(\omega)$ is a complex function. Substituting these into the differential equation and canceling $e^{i\omega t}$ from each side, we obtain

$$(-m\omega^2 + ic\omega + k)H(\omega) = 1$$

The frequency response function is then

$$\begin{aligned} H(\omega) &= \frac{1}{k - m\omega^2 + ic\omega} \\ &= \frac{1}{k}\,\frac{1}{1 - \left(\dfrac{\omega}{\omega_n}\right)^2 + i2\zeta\left(\dfrac{\omega}{\omega_n}\right)} \end{aligned} \tag{13.3-3}$$

As mentioned in Chapter 3, we will absorb the factor $1/k$ in with the force. $H(\omega)$ is then a nondimensional function of ω/ω_n and the damping factor ζ.

The input-output relationship in terms of the frequency response function can be written as

$$y(t) = H(\omega)F_0 e^{i\omega t} \tag{13.3-4}$$

where $F_0 e^{i\omega t}$ is a harmonic function.

For the mean square response, we follow the procedure of Example 13.2-1 and write

$$y = \tfrac{1}{2}F_0(He^{i\omega t} + H^* e^{-i\omega t}) \tag{13.3-5}$$

Thus by squaring and substituting into Eq. (13.2-4), we find the mean square value of y is

$$\overline{y^2} = \frac{F_0^2}{4} \lim_{T\to\infty} \frac{1}{T}\int_0^T (H^2 e^{i2\omega t} + 2HH^* + H^* e^{-i2\omega t})\, dt$$

$$= \frac{F_0^2}{2} H(\omega)H^*(\omega) = \overline{F_0^2}|H(\omega)|^2 \qquad (13.3\text{-}6)$$

In the preceding equation, the first and last terms become zero because of $T \to \infty$ in the denominator, whereas the middle term is independent of T. Equation (13.3-6) indicates that the mean square value of the response is equal to the mean square excitation multiplied by the square of the absolute values of the frequency response function. For excitations expressed in terms of Fourier series with many frequencies, the response is the sum of terms similar to Eq. (13.3-6).

Example 13.3-1

A single-DOF system with natural frequency $\omega_n = \sqrt{k/m}$ and damping $\zeta = 0.20$ is excited by the force

$$F(t) = F\cos\frac{1}{2}\omega_n t + F\cos\omega_n t + F\cos\frac{3}{2}\omega_n t$$

$$= \sum_{m=1/2,1,3/2} F\cos m\omega_n t$$

Determine the mean square response and compare the output spectrum with that of the input.

Solution: The response of the system is simply the sum of the response of the single-DOF system to each of the harmonic components of the exciting force.

$$x(t) = \sum_{m=1/2,1,3/2} |H(m\omega)| F\cos(m\omega_n t - \phi_m)$$

where

$$|H(\tfrac{1}{2}\omega_n)| = \frac{\frac{1}{k}}{\sqrt{\frac{9}{16} + (0.20)^2}} = \frac{1.29}{k}$$

$$|H(\omega_n)| = \frac{\frac{1}{k}}{\sqrt{4(0.20)^2}} = \frac{2.50}{k}$$

$$|H(\tfrac{3}{2}\omega_n)| = \frac{\frac{1}{k}}{\sqrt{\frac{25}{16} + 9(0.20)^2}} = \frac{0.72}{k}$$

$$\phi_{1/2} = \tan^{-1}\frac{4\zeta}{3} = 0.083\pi$$

$$\phi_1 = \tan^{-1}\infty = 0.50\pi$$

$$\phi_{3/2} = \tan^{-1}\frac{-12\zeta}{5} = -0.142\pi$$

Substituting these values into $x(t)$, we obtain the equation

$$x(t) = \frac{F}{k}[1.29\cos(0.5\omega_n - 0.083\pi) + 2.50\cos(\omega_n t - 0.50\pi) + 0.72\cos(1.5\omega_n t + 0.142\pi)]$$

The mean square response is then

$$\overline{x^2} = \frac{F^2}{2k^2}[(1.29)^2 + (2.50)^2 + (0.72)^2]$$

Figure 13.3-2 shows the input and output spectra for the problem. The components of the mean square input are the same for each frequency and equal to $F^2/2$. The output spectrum is modified by the system frequency response function.

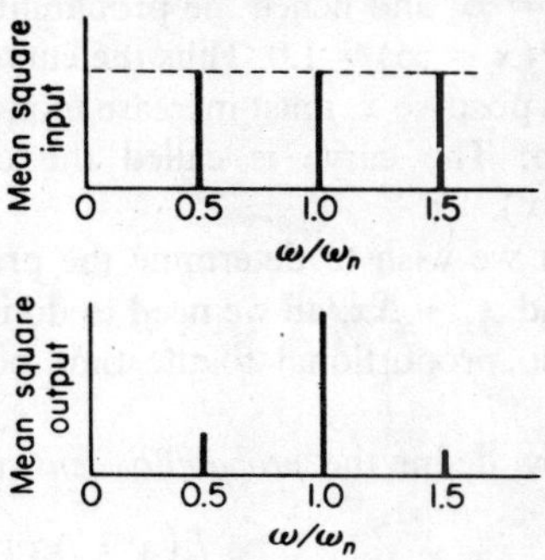

Figure 13.3-2. Input and output spectra with discrete frequencies.

13.4 PROBABILITY DISTRIBUTION

Referring to the random time function of Fig. 13.4-1, what is the probability of its instantaneous value being less than (more negative than) some specified value x_1? To answer this question, we draw a horizontal line at the specified value x_1 and sum the time intervals Δt_i during which $x(t)$ is less than x_1. This

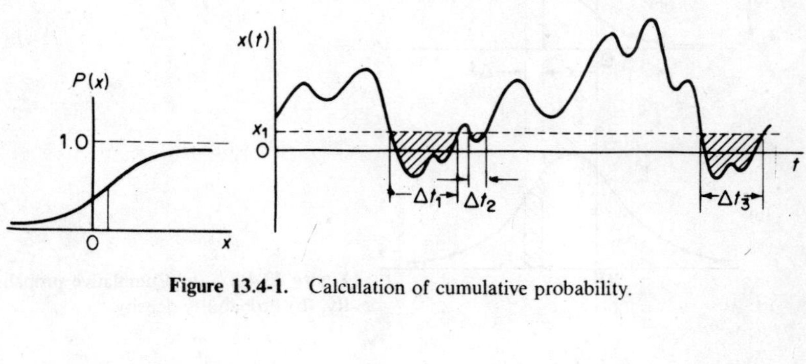

Figure 13.4-1. Calculation of cumulative probability.

sum divided by the total time then represents the fraction of the total time that $x(t)$ is less than x_1, which is the probability that $x(t)$ will be found less than x_1.

$$\begin{aligned} P(x_1) &= \text{Prob}[x(t) < x_1] \\ &= \lim_{t \to \infty} \frac{1}{t} \sum \Delta t_i \end{aligned} \tag{13.4-1}$$

If a large negative number is chosen for x_1, none of the curve will extend negatively beyond x_1, and hence $P(x_1 \to -\infty) = 0$. As the horizontal line corresponding to x_1 is moved up, more of $x(t)$ will extend negatively beyond x_1, and the fraction of the total time in which $x(t)$ extends below x_1 must increase as shown in Fig. 13.4-2(a). As $x \to \infty$, all $x(t)$ will lie in the region less than $x = \infty$, and hence the probability of $x(t)$ being less than $x = \infty$ is certain, or $P(x = \infty) = 1.0$. Thus the curve of Fig. 13.4-2(a), which is cumulative towards positive x, must increase monotonically from 0 at $x = -\infty$ to 1.0 at $x = +\infty$. The curve is called the cumulative probability distribution function $P(x)$.

If next we wish to determine the probability of $x(t)$ lying between the values x_1 and $x_1 + \Delta x$, all we need to do is subtract $P(x_1)$ from $P(x_1 + \Delta x)$, which is also proportional to the time occupied by $x(t)$ in the zone x_1 to $x_1 + \Delta x$.

We now define the *probability density function* $p(x)$ as

$$p(x) = \lim_{\Delta x \to 0} \frac{P(x + \Delta x) - P(x)}{\Delta x} = \frac{dP(x)}{dx} \tag{13.4-2}$$

and it is evident from Fig. 13.4-2(b) that $p(x)$ is the slope of the cumulative probability distribution $P(x)$. From the preceding equation we can also write

$$P(x_1) = \int_{-\infty}^{x_1} p(x)\, dx \tag{13.4-3}$$

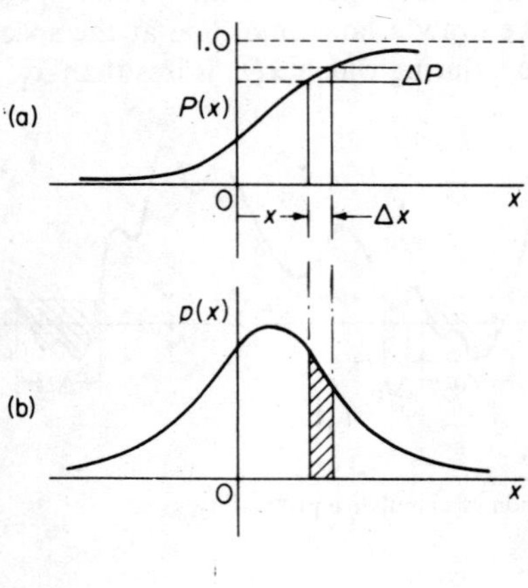

Figure 13.4-2. (a) Cumulative probability, (b) Probability density.

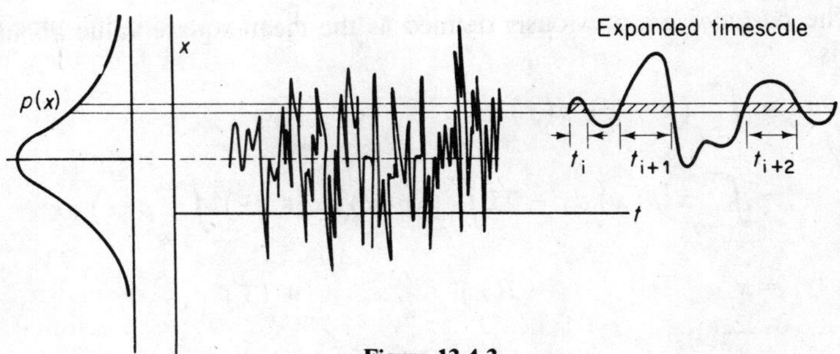

Figure 13.4-3.

The area under the probability density curve of Fig. 13.4-2(b) between two values of x represents the probability of the variable being in this interval. Since the probability of $x(t)$ being between $x = \pm\infty$ is certain

$$P(\infty) = \int_{-\infty}^{+\infty} p(x)\,dx = 1.0 \tag{13.4-4}$$

and the total area under the $p(x)$ curve must be unity. Figure 13.4-3 again illustrates the probability density $p(x)$, which is the fraction of the time occupied by $x(t)$ in the interval x to $x + dx$.

The mean and the mean square value, previously defined in terms of the time average, are related to the probability density function in the following manner. The mean value $\bar{x}$ coincides with the centroid of the area under the probability density curve $p(x)$, as shown in Fig. 13.4-4. It can therefore be determined by the first moment

$$\bar{x} = \int_{-\infty}^{\infty} xp(x)\,dx \tag{13.4-5}$$

Likewise, the mean square value is determined from the second moment

$$\overline{x^2} = \int_{-\infty}^{\infty} x^2p(x)\,dx \tag{13.4-6}$$

which is analogous to the moment of inertia of the area under the probability density curve about $x = 0$.

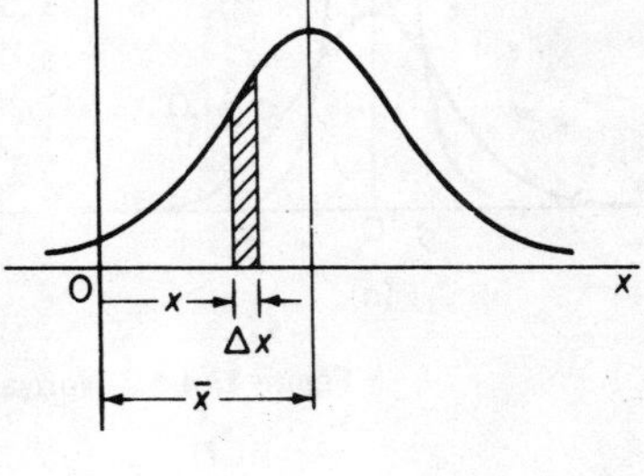

Figure 13.4-4. First and second moments of $p(x)$.

The *variance* σ^2, previously defined as the mean square value about the mean, is

$$
\begin{aligned}
\sigma^2 &= \int_{-\infty}^{\infty} (x-\bar{x})^2 p(x)\,dx \\
&= \int_{-\infty}^{\infty} x^2 p(x)\,dx - 2\bar{x}\int_{-\infty}^{\infty} xp(x)\,dx + (\bar{x})^2\int_{-\infty}^{\infty} p(x)\,dx \\
&= \overline{x^2} \qquad - 2(\bar{x})^2 \qquad + (\bar{x})^2 \\
&= \overline{x^2} \qquad - (\bar{x})^2
\end{aligned}
\tag{13.4-7}
$$

The *standard deviation* σ is the positive square root of the variance. When the mean value is zero, $\sigma = \sqrt{\overline{x^2}}$ and the standard deviation is equal to the root mean square (rms) value.

Gaussian and Rayleigh distributions. Certain distributions that occur frequently in nature are the *Gaussian* (or normal) distribution and the *Rayleigh* distribution, both of which can be expressed mathematically. The Gaussian distribution is a bell-shaped curve, symmetric about the mean value (which will be assumed to be zero) with the following equation:

$$p(x) = \frac{1}{\sigma\sqrt{2\pi}} e^{-x^2/2\sigma^2} \tag{13.4-8}$$

The standard deviation σ is a measure of the spread about the mean value; the smaller the value of σ, the narrower the $p(x)$ curve (remember that the total area = 1.0), as shown in Fig. 13.4-5(a).

In Fig. 13.4-5(b) the Gaussian distribution is plotted nondimensionally in terms of x/σ. The probability of $x(t)$ being between $\pm\lambda\sigma$, where λ is any

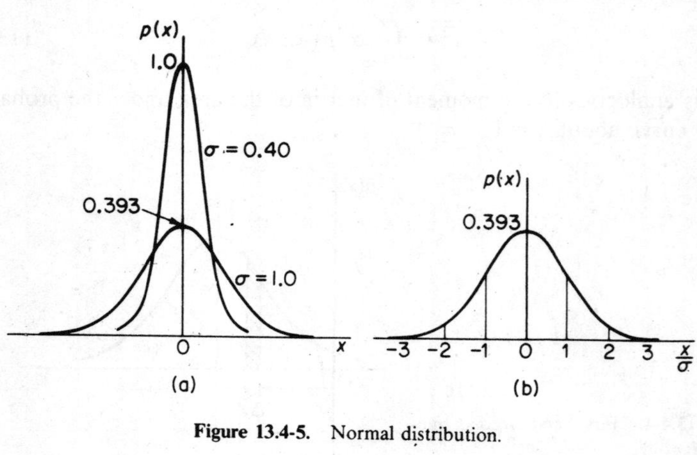

Figure 13.4-5. Normal distribution.

positive number, is found from the equation

$$\text{Prob}[-\lambda\sigma \le x(t) \le \lambda\sigma] = \frac{1}{\sigma\sqrt{2\pi}} \int_{-\lambda\sigma}^{\lambda\sigma} e^{-x^2/2\sigma^2}\, dx \qquad (13.4\text{-}9)$$

The following table presents numerical values associated with $\lambda = 1$, 2, and 3.

λ	$Prob[-\lambda\sigma \le x(t) \le \lambda\sigma]$	$Prob[\lvert x\rvert > \lambda\sigma]$
1	68.3%	31.7%
2	95.4%	4.6%
3	99.7%	0.3%

The probability of $x(t)$ lying outside $\pm\lambda\sigma$ is the probability of $|x|$ exceeding $\lambda\sigma$, which is 1.0 minus the preceding values, or the equation

$$\text{Prob}[|x| > \lambda\sigma] = \frac{2}{\sigma\sqrt{2\pi}} \int_{\lambda\sigma}^{\infty} e^{-x^2/2\sigma^2}\, dx = erfc\left(\frac{\lambda}{\sqrt{2}}\right) \qquad (13.4\text{-}10)$$

Random variables restricted to positive values, such as the absolute value A of the amplitude, often tend to follow the Rayleigh distribution, which is defined by the equation

$$p(A) = \frac{A}{\sigma^2} e^{-A^2/2\sigma^2} \qquad A > 0 \qquad (13.4\text{-}11)$$

The probability density $p(A)$ is zero here for $A < 0$ and has the shape shown in Fig. 13.4-6.

The mean and mean square values for the Rayleigh distribution can be found from the first and second moments to be

$$\begin{aligned} \bar{A} &= \int_0^\infty A p(A)\, dA = \int_0^\infty \frac{A^2}{\sigma^2} e^{-A^2/2\sigma^2}\, dA = \sqrt{\frac{\pi}{2}}\,\sigma \\ \overline{A^2} &= \int_0^\infty A^2 p(A)\, dA = \int_0^\infty \frac{A^3}{\sigma^2} e^{-A^2/2\sigma^2}\, dA = 2\sigma^2 \end{aligned} \qquad (13.4\text{-}12)$$

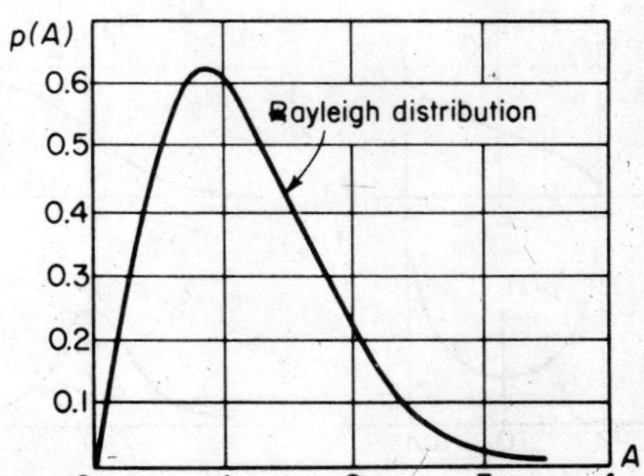

Figure 13.4-6. Rayleigh distribution.

The variance associated with the Rayleigh distribution is

$$\sigma_A^2 = \overline{A^2} - (\overline{A})^2 = \left(\frac{4-\pi}{2}\right)\sigma^2$$
$$\therefore \sigma_A \cong \frac{2}{3}\sigma \tag{13.4-13}$$

Also, the probability of A exceeding a specified value $\lambda\sigma$ is

$$\text{Prob}[A > \lambda\sigma] = \int_{\lambda\sigma}^{\infty} \frac{A}{\sigma^2} e^{-A^2/2\sigma^2}\, dA \tag{13.4-14}$$

which has the following numerical values:

λ	$P[A > \lambda\sigma]$
0	100%
1	60.7%
2	13.5%
3	1.2%

Three important examples of time records frequently encountered in practice is shown in Fig. 13.4-7, where the mean value is arbitrarily chosen to be zero. The cumulative probability distribution for the sine wave is easily shown to be

$$P(x) = \frac{1}{2} + \frac{1}{\pi}\sin^{-1}\frac{x}{A}$$

and its probability density, by differentiation, is

$$p(x) = \frac{1}{\pi\sqrt{A^2 - x^2}} \qquad |x| < A$$
$$= 0 \qquad |x| > A$$

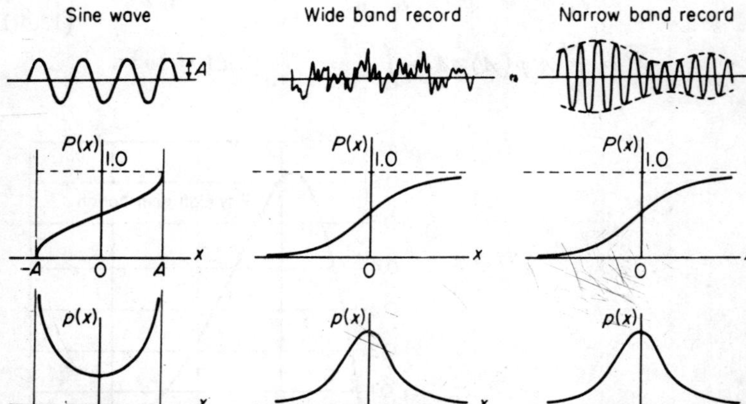

Figure 13.4-7. Probability functions for three types of records.

For the wide-band record, the amplitude, phase, and frequency all vary randomly and an analytical expression is not possible for its instantaneous value. Such functions are encountered in radio noise, jet engine pressure fluctuation, atmospheric turbulence, and so on, and a most likely probability distribution for such records is the *Gaussian distribution*.

When a wide-band record is put through a narrow-band filter, or a resonance system where the filter bandwidth is small compared to its central frequency f_0, we obtain the third type of wave, which is essentially a constant frequency oscillation with slowly varying amplitude and phase. The probability distribution for its instantaneous values is the same as that for the wide-band random function. However, the absolute values of its peaks, corresponding to the envelope, will have a Rayleigh distribution.

Another quantity of great interest is the distribution of the peak values. Rice† shows that the distribution of the peak values depends on a quantity $N_0/2M$, where N_0 is the number of zero crossings and $2M$ is the number of positive and negative peaks. For a sine wave or a narrow band, N_0 is equal to $2M$ so that the ratio $N_0/2M = 1$. For a wide-band random record, the number of peaks will greatly exceed the number of zero crossings so that $N_0/2M$ tends to approach zero. When $N_0/2M = 0$, the probability density distribution of peak values turns out to be Gaussian, whereas when $N_0/2M = 1$, as in the narrow-band case, the probability density distribution of the peak values tends to a *Rayleigh distribution*.

13.5 CORRELATION

Correlation is a measure of the similarity between two quantities. As it applies to vibration waveforms, correlation is a time-domain analysis useful for detecting hidden periodic signals buried in measurement noise, propagation time through the structure, and for determining other information related to the structure's spectral characteristics, which are better discussed under Fourier transforms.

Suppose we have two records, $x_1(t)$ and $x_2(t)$, as shown in Fig. 13.5-1. The *correlation* between them is computed by multiplying the ordinates of the two records at each time t and determining the average value $\langle x_1(t)x_2(t)\rangle$ by dividing the sum of the products by the number of products. It is evident that the correlation so found will be largest when the two records are similar or identical. For dissimilar records, some of the products will be positive and others will be negative, so their sum will be smaller.

Next consider the case where $x_2(t)$ is identical to $x_1(t)$ but shifted to the left by a time τ, as shown in Fig. 13.5-2. Then at time t, when x_1 is $x(t)$, the value of x_2 is $x(t+\tau)$, and the correlation will be given by $\langle x(t)x(t+\tau)\rangle$.

†See Ref. [8], at the end of the chapter.

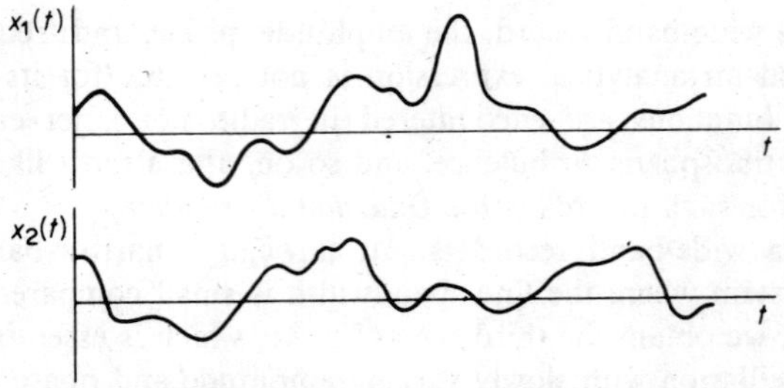

Figure 13.5-1. Correlation between $x_1(t)$ and $x_2(t)$.

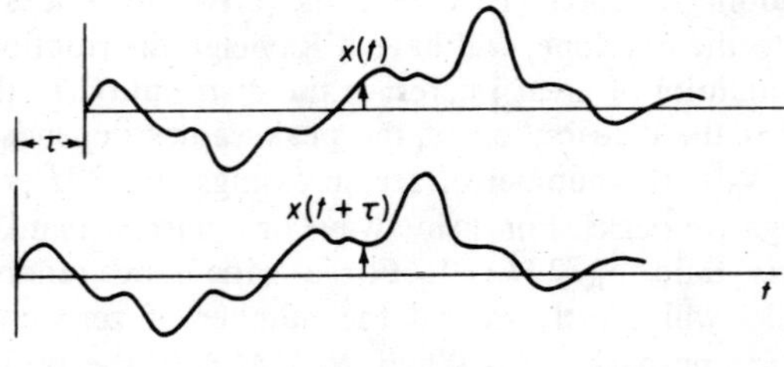

Figure 13.5-2. Function $x(t)$ shifted by τ.

Here, if $\tau = 0$, we have complete correlation. As τ increases the correlation decreases.

It is evident that the above result can be computed from a single record by multiplying the ordinates at time t and $t + \tau$ and determining the average. We then call this result the *autocorrelation* and designate it by $R(\tau)$. It is also the expected value of the product $x(t)x(t + \tau)$, or

$$\begin{aligned} R(\tau) &= E[x(t)x(t+\tau)] = \langle x(t)x(t+\tau)\rangle \\ &= \lim_{T\to\infty} \frac{1}{T}\int_{-T/2}^{T/2} x(t)x(t+\tau)\,dt \end{aligned} \tag{13.5-1}$$

When $\tau = 0$, the above definition reduces to the mean square value

$$R(0) = \overline{x^2} = \sigma^2 \tag{13.5-2}$$

Since the second record of Fig. 13.5-2 can be considered to be delayed with respect to the first record, or the first record advanced with respect to the second record, it is evident that $R(\tau) = R(-\tau)$ is symmetric about the origin $\tau = 0$ and is always less than $R(0)$.

Highly random functions, such as the wide-band noise shown in Fig. 13.5-3, soon lose their similarity within a short time shift. Its autocorrelation, therefore, is a sharp spike at $\tau = 0$ that drops off rapidly with $\pm\tau$ as shown. It implies that wide-band random records have little or no correlation except near $\tau = 0$.

For the special case of a periodic wave, the autocorrelation must be periodic of the same period, since shifting the wave one period brings the wave

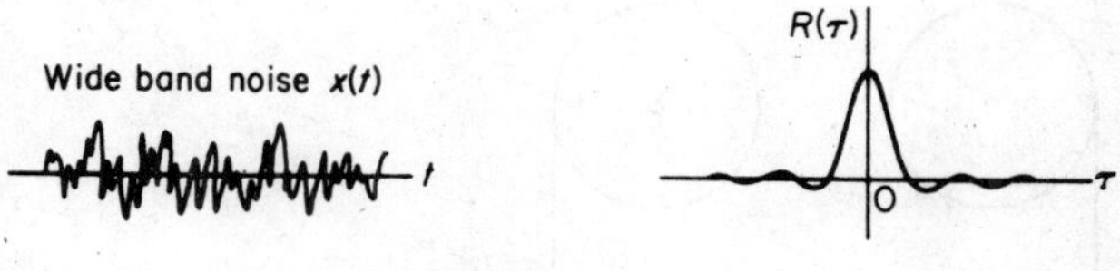

Figure 13.5-3. Highly random function and its autocorrelation.

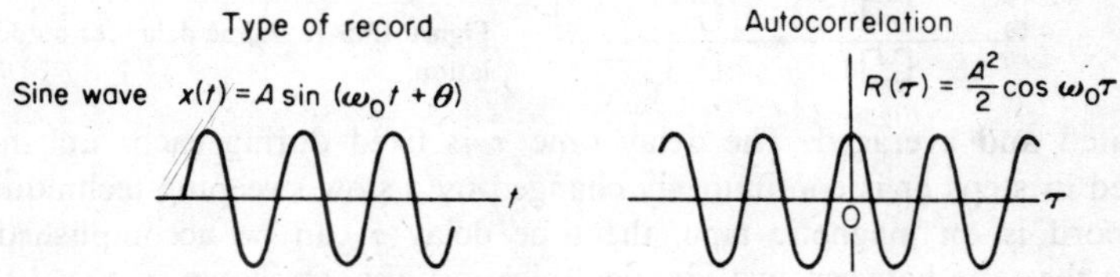

Figure 13.5-4. Sine wave and its autocorrelation.

into coincidence again. Figure 13.5-4 shows a sine wave and its autocorrelation.

For the narrow-band record shown in Fig. 13.5-5, the autocorrelation has some of the characteristics found for the sine wave in that it is again an even function with a maximum at $\tau = 0$ and frequency ω_0 corresponding to the dominant or central frequency. The difference appears in the fact that $R(\tau)$ approaches zero for large τ for the narrow-band record. It is evident from this discussion that hidden periodicities in a noisy random record can be detected by correlating the record with a sinusoid. There will be almost no correlation between the sinusoid and the noise that will be suppressed. By exploring with sinusoids of differing frequencies the hidden periodic signal can be detected. Figure 13.5-6 shows a block diagram for the determination of the autocorrelation. The signal $x(t)$ is delayed by τ and multiplied, after which it is

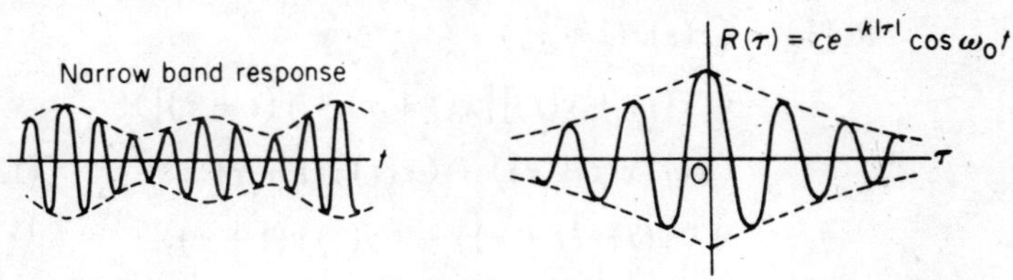

Figure 13.5-5. Autocorrelation for the narrow-band record.

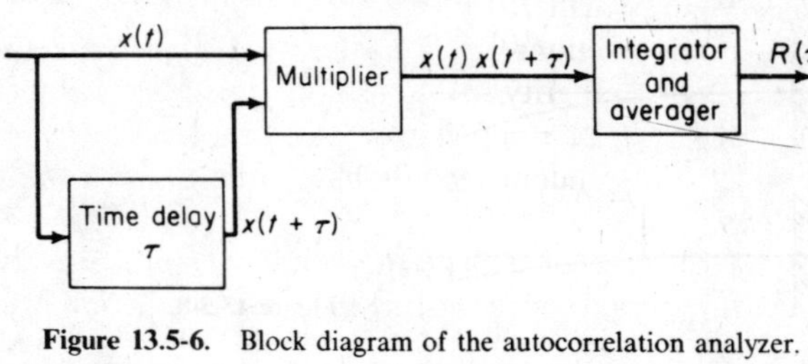

Figure 13.5-6. Block diagram of the autocorrelation analyzer.

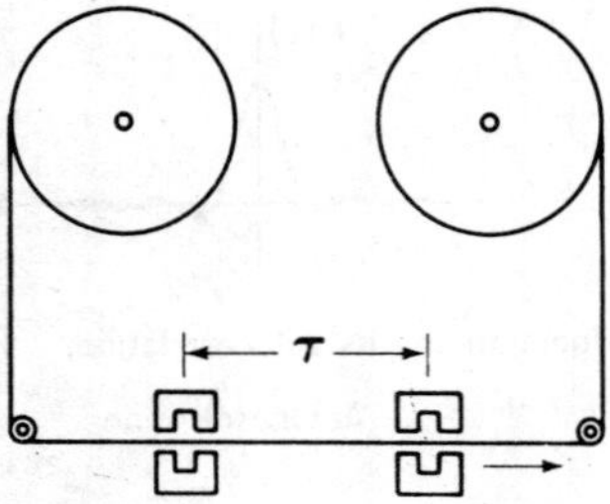

Figure 13.5-7. Time delay for autocorrelation.

integrated and averaged. The delay time τ is fixed during each run and is changed in steps or is continuously changed by a slow sweeping technique. If the record is on magnetic tape, the time delay τ can be accomplished by passing the tape between two identical pickup units, as shown in Fig. 13.5-7.

Cross correlation. Consider two random quantities $x(t)$ and $y(t)$. The correlation between these two quantities is defined by the equation

$$\begin{aligned} R_{xy}(\tau) &= E[x(t)y(t+\tau)] = \langle x(t)y(t+\tau)\rangle \\ &= \lim_{T\to\infty}\frac{1}{T}\int_{-T/2}^{T/2} x(t)y(t+\tau)\,dt \end{aligned} \tag{13.5-3}$$

that can also be called the *cross correlation* between the quantities x and y.

Such quantities often arise in dynamical problems. For example, let $x(t)$ be the deflection at the end of a beam due to a load $F_1(t)$ at some specified point. $y(t)$ is the deflection at the same point, due to a second load $F_2(t)$ at a different point than the first, as illustrated in Fig. 13.5-8. The deflection due to both loads is then $z(t) = x(t) + y(t)$, and the autocorrelation of $z(t)$ as a result of the two loads is

$$\begin{aligned} R_z(\tau) &= \langle z(t)z(t+\tau)\rangle \\ &= \langle [x(t)+y(t)][x(t+\tau)+y(t+\tau)]\rangle \\ &= \langle x(t)x(t+\tau)\rangle + \langle x(t)y(t+\tau)\rangle \\ &\quad + \langle y(t)x(t+\tau)\rangle + \langle y(t)y(t+\tau)\rangle \\ &= R_x(\tau) + R_{xy}(\tau) + R_{yx}(\tau) + R_y(\tau) \end{aligned} \tag{13.5-4}$$

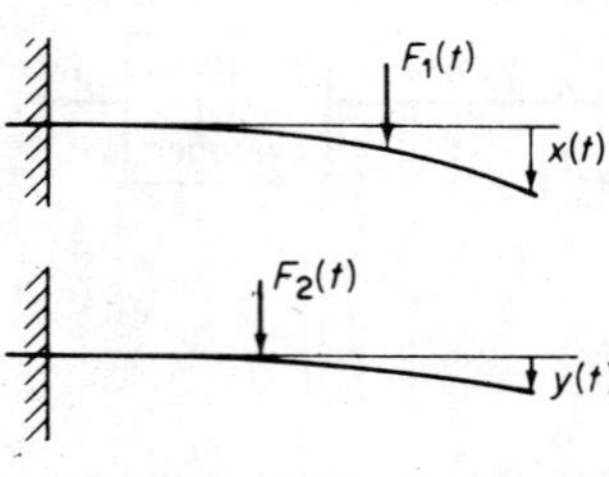

Figure 13.5-8.

Thus the autocorrelation of a deflection at a given point due to separate loads $F_1(t)$ and $F_2(t)$ cannot be determined simply by adding the autocorrelations $R_x(\tau)$ and $R_y(\tau)$ resulting from each load acting separately. $R_{xy}(\tau)$ and $R_{yx}(\tau)$ are here referred to as *cross correlation*, and, in general, they are not equal.

Example 13.5-1

Show that the autocorrelation of the rectangular gating function shown in Fig. 13.5-9 is a triangle.

Solution: If the rectangular pulse is shifted in either direction by τ, its product with the original pulse is $A^2(T - \tau)$. It is easily seen then that starting with $\tau = 0$, the autocorrelation curve is a straight line that forms a triangle with height A^2 and base equal to $2T$.

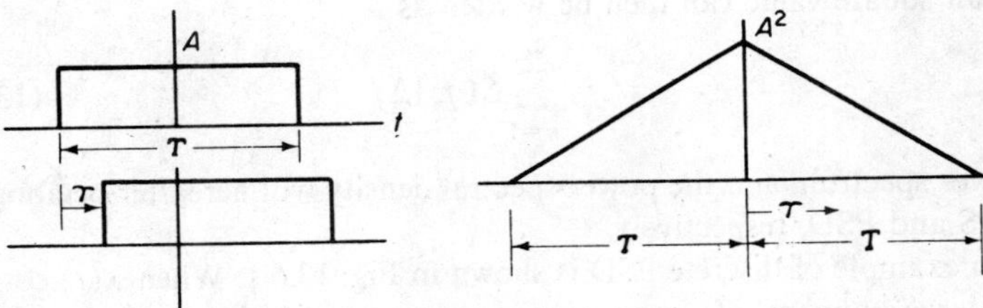

Figure 13.5-9. Autocorrelation of a rectangle is a triangle.

13.6 POWER SPECTRUM AND POWER SPECTRAL DENSITY

The frequency composition of a random function can be described in terms of the spectral density of the mean square value. We found in Example 13.5-1 that the mean square value of a periodic time function is the sum of the mean square value of the individual harmonic component present.

$$\overline{x^2} = \sum_{n=1}^{\infty} \tfrac{1}{2} C_n C_n^*$$

Thus $\overline{x^2}$ is made up of discrete contributions in each frequency interval Δf.

We will first define the contribution to the mean square in the frequency interval Δf as the *power spectrum* $G(f_n)$.

$$G(f_n) = \tfrac{1}{2} C_n C_n^* \tag{13.6-1}$$

The mean square value is then

$$\overline{x^2} = \sum_{n=1}^{\infty} G(f_n) \tag{13.6-2}$$

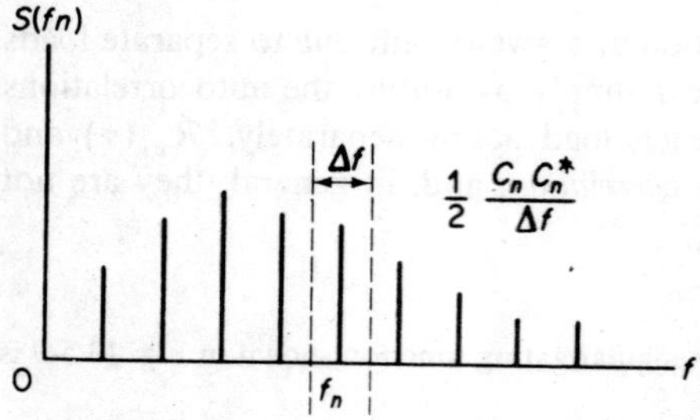

Figure 13.6-1. Discrete spectrum.

We now define the discrete *power spectral density* $S(f_n)$ as power spectrum divided by the frequency interval Δf:

$$S(f_n) = \frac{G(f_n)}{\Delta f} = \frac{C_n C_n^*}{2\Delta f} \tag{13.6-3}$$

The mean square value can then be written as

$$\overline{x^2} = \sum_{n=1}^{\infty} S(f_n)\Delta f \tag{13.6-4}$$

The power spectrum and the power spectral density will hereafter be abbreviated as PS and PSD, respectively.

An example of discrete PSD is shown in Fig. 13.6.1. When $x(t)$ contains a very large number of frequency components, the lines of the discrete spectrum become closer together and they more nearly resemble a continuous spectrum, as shown in Fig. 13.6-2. We now define the PSD, $S(f)$, for a continuous spectrum as the limiting case of $S(f_n)$ as $\Delta f \to 0$.

$$\lim_{\Delta f \to 0} S(f_n) = S(f) \tag{13.6-5}$$

The mean square value is then

$$\overline{x^2} = \int_0^{\infty} S(f)\, df \tag{13.6-6}$$

To illustrate the meaning of PS and PSD, the following experiment is described. An Xtal accelerometer is attached to a shaker, and its output is

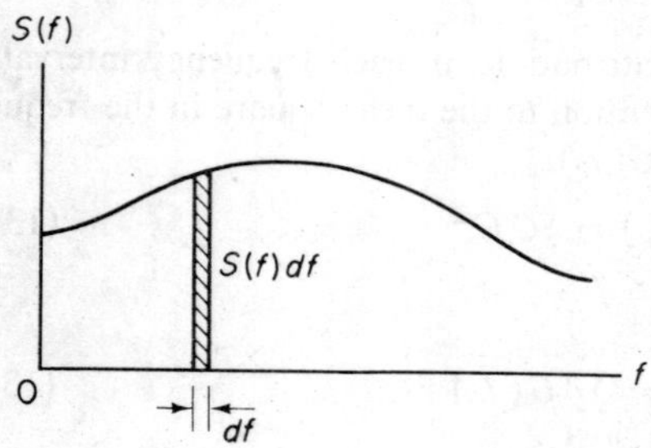

Figure 13.6-2. Continuous spectrum.

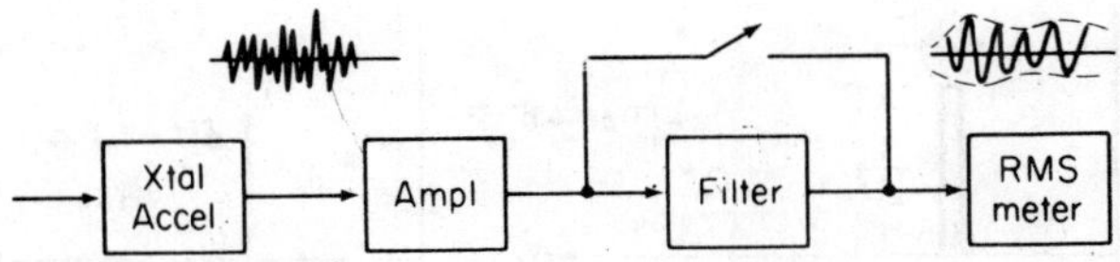

Figure 13.6-3. Measurement of random data.

amplified, filtered, and read by a rms voltmeter as shown by the block diagram of Fig. 13.6-3. The rms voltmeter should have a long time constant, which corresponds to a long averaging time.

We will excite the shaker by a wide-band random input that is constant over the frequency range 0 Hz to 2000 Hz. If the filter is bypassed, the rms voltmeter will read the rms vibration in the entire frequency spectrum. Assuming an ideal filter that will pass all vibrations of frequencies within the passband, the output of the filter represents a narrow-band vibration.

We consider a central frequency of 500 Hz and first set the upper and lower cutoff frequencies at 580 Hz and 420 Hz. The rms meter will now read only the vibration within this 160-Hz band. Let us say that the reading is 8 g. The mean square value is then $G(f_n) = 64\ g^2$, and its spectral density is $S(f_n) = 64\ g^2/160 = 0.40\ g^2/\text{Hz}$.

We next reduce the passband to 40 Hz by setting tne upper and lower filter frequencies to 520 Hz and 480 Hz. The mean square value passed by the filter is now one-quarter of the previous value, or $16\ g^2$, and the rms meter will read 4 g.

Reducing the passband further to 10 Hz, betwcen 505 Hz and 495 Hz, the rms meter reading becomes 2 g, as shown in the following tabulation.

Frequencies f	*Band width* Δf	rms *Meter Reading* $\sqrt{\Delta(\overline{x^2})}$	*Filtered Mean Square* $G(f_n) = \Delta(\overline{x^2})$	*Spectral Density* $S(f_n) = \dfrac{\Delta(\overline{x^2})}{\Delta f}$
580–420	160	8 g	64 g^2	0.40 g^2/Hz
520–480	40	4 g	16 g^2	0.40 g^2/Hz
505–495	10	2 g	4 g^2	0.40 g^2/Hz

Note that as the bandwidth is reduced, the mean square value passed by the filter, or $G(f_n)$, is reduced proportionally. However, by dividing by the bandwidth, the density of the mean square value, $S(f_n)$ remains constant. The example clearly points out the advantage of plotting $S(f_n)$ instead of $G(f_n)$.

The PSD can also be expressed in terms of the delta function. As seen from Fig. 13.6-4, the area of a rectangular pulse of height $1/\Delta f$ and width Δf is always unity, and in the limiting case when $\Delta f \to 0$, it becomes a delta

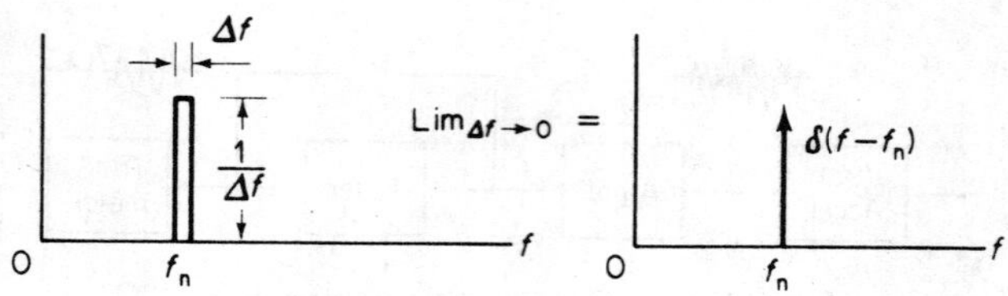

Figure 13.6-4. $\text{Lim}_{\Delta f \to 0} \dfrac{1}{\Delta f} = \delta(f - f_n)$

function. Thus $S(f)$ becomes

$$S(f) = \lim_{\Delta f \to 0} S(f_{\hat{n}}) = \lim_{\Delta f \to 0} \frac{G(f_{\hat{n}})}{\Delta f} = G(f)\delta(f - f_{\hat{n}})$$

Typical spectral density functions for two common types of random records are shown in Figs. 13.6-5 and 13.6-6. The first is a wide-band noise-type of record which has a broad spectral density function. The second is a narrow-band random record, which is typical of a response of a sharply resonant system to a wide-band input. Its spectral density function is concentrated around the frequency of the instantaneous variation within the envelope.

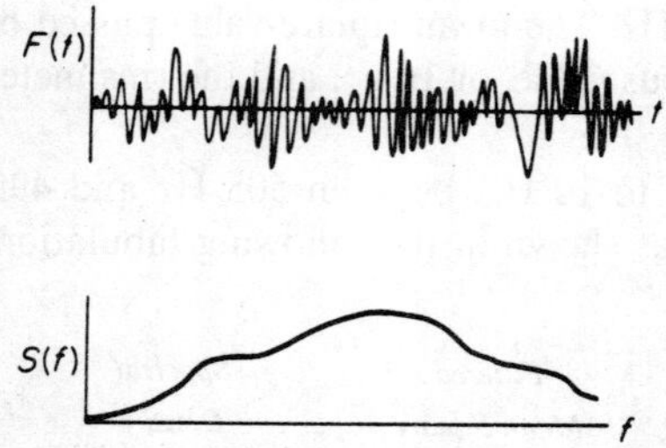

Figure 13.6-5. Wide-band record and its spectral density.

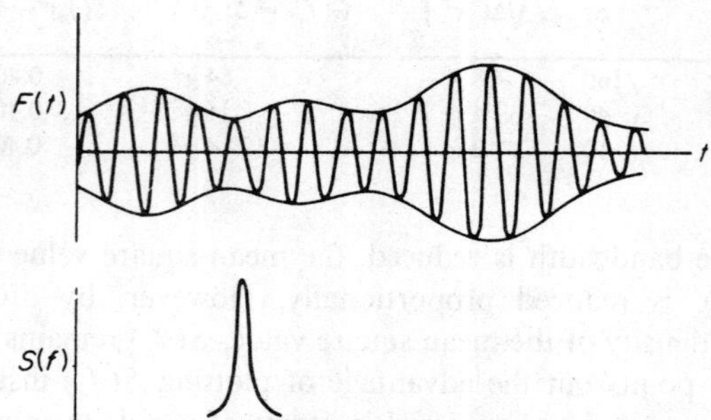

Figure 13.6-6. Narrow-band record and its spectral density.

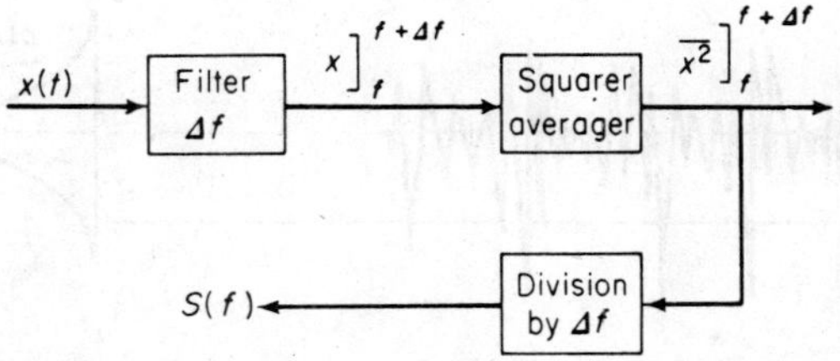

Figure 13.6-7. Power spectral density analyzer.

The spectral density of a given record can be measured electronically by the circuit of Fig. 13.6-7. Here the spectral density is noted as the contribution of the mean square value in the frequency interval Δf divided by Δf.

$$S(f) = \lim_{\Delta f \to 0} \frac{\Delta(\overline{x^2})}{\Delta f} \tag{13.6-7}$$

The band-pass filter of passband $B = \Delta f$ passes $x(t)$ in the frequency interval f to $f + \Delta f$, and the output is squared, averaged, and divided by Δf.

For high resolution, Δf should be made as narrow as possible; however, the passband of the filter cannot be reduced indefinitely without losing the reliability of the measurement. Also, a long record is required for the true estimate of the mean square value, but actual records are always of finite length. It is evident now that a parameter of importance is the product of the record length and the bandwidth, $2BT$, which must be sufficiently large.†

Example 13.6-1

A random signal has a spectral density that is a constant

$$S(f) = 0.004 \text{ cm}^2/\text{cps}$$

between 20 cps and 1200 cps and that is zero outside this frequency range. Its mean value is 2.0 cm. Determine its rms value and its standard deviation.

Solution: The mean square value is found from

$$\overline{x^2} = \int_0^\infty S(f)\, df = \int_{20}^{1200} 0.004\, df = 4.72$$

and the rms value is

$$\text{rms} = \sqrt{\overline{x^2}} = \sqrt{4.72} = 2.17 \text{ cm}$$

The variance σ^2 is defined by Eq. (13.2-6)

$$\sigma^2 = \overline{x^2} - (\bar{x})^2$$
$$= 4.72 - (2)^2 = 0.72$$

†See J. S. Bendat, and A. G. Piersol, *Random Data* (New York: Wiley Interscience, 1971), p. 96.

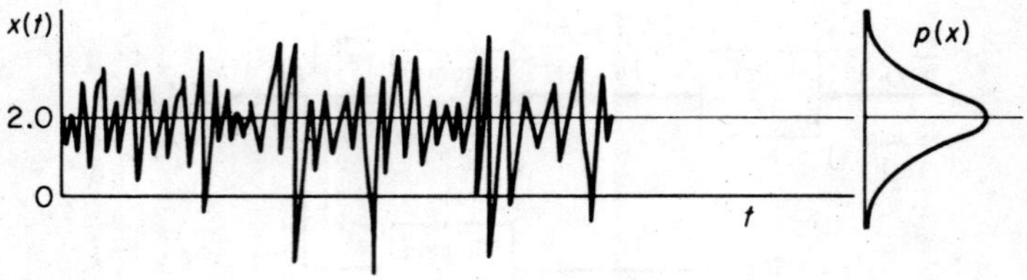

Figure 13.6-8.

and the standard deviation becomes

$$\sigma = \sqrt{0.72} = 0.85 \text{ cm}$$

The problem is graphically displayed by Fig. 13.6-8, which shows the time variation of the signal and its probability distribution.

Example 13.6-2

Determine the Fourier coefficients C_n and the power spectral density of the periodic function shown in Fig. 13.6-9.

Solution: The period is $2T$ and C_n are

$$C_0 = \frac{2}{2T}\int_{-T/2}^{T/2} F_0\, d\xi = F_0$$

$$C_n = \frac{2}{2T}\int_{-T/2}^{T/2} F_0 e^{-in\omega_0\xi}\, d\xi = F_0\left[\frac{\sin(n\pi/2)}{n\pi/2}\right]$$

Numerical values of C_n are computed as follows and plotted in Fig. 13.6-10.

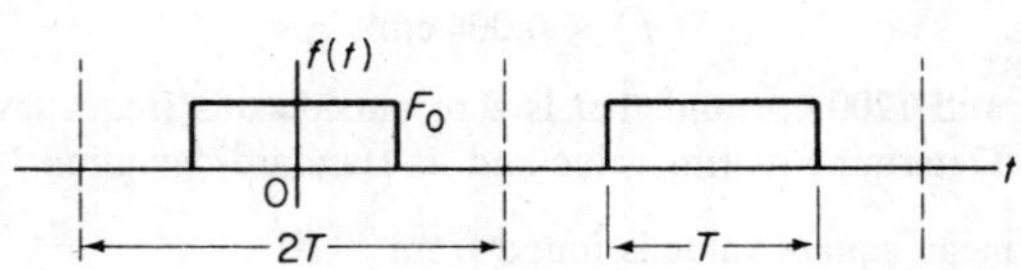

Figure 13.6-9.

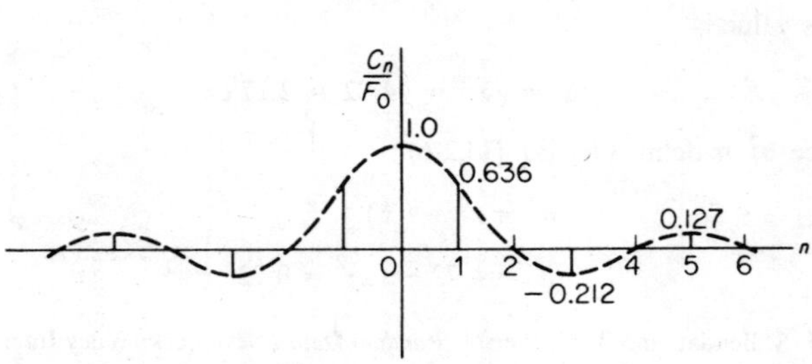

Figure 13.6-10. Fourier coefficients versus n.

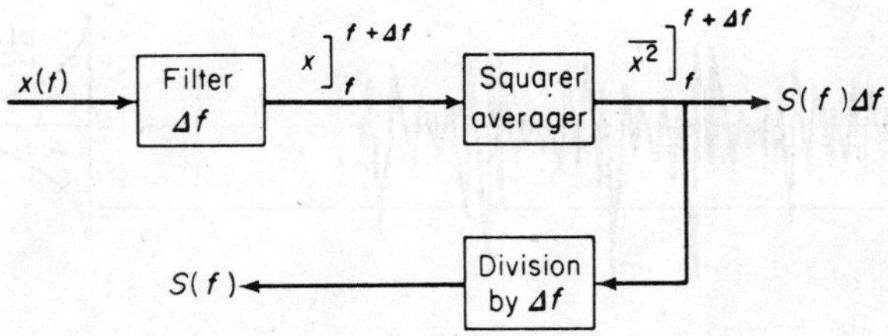

Figure 13.6-7. Power spectral density analyzer.

The spectral density of a given record can be measured electronically by the circuit of Fig. 13.6-7. Here the spectral density is noted as the contribution of the mean square value in the frequency interval Δf divided by Δf.

$$S(f) = \lim_{\Delta f \to 0} \frac{\Delta(\overline{x}^2)}{\Delta f} \tag{13.6-7}$$

The band-pass filter of passband $B = \Delta f$ passes $x(t)$ in the frequency interval f to $f + \Delta f$, and the output is squared, averaged, and divided by Δf.

For high resolution, Δf should be made as narrow as possible; however, the passband of the filter cannot be reduced indefinitely without losing the reliability of the measurement. Also, a long record is required for the true estimate of the mean square value, but actual records are always of finite length. It is evident now that a parameter of importance is the product of the record length and the bandwidth, $2BT$, which must be sufficiently large.†

Example 13.6-1

A random signal has a spectral density that is a constant

$$S(f) = 0.004 \text{ cm}^2/\text{cps}$$

between 20 cps and 1200 cps and that is zero outside this frequency range. Its mean value is 2.0 cm. Determine its rms value and its standard deviation.

Solution: The mean square value is found from

$$\overline{x^2} = \int_0^\infty S(f)\, df = \int_{20}^{1200} 0.004\, df = 4.72$$

and the rms value is

$$\text{rms} = \sqrt{\overline{x^2}} = \sqrt{4.72} = 2.17 \text{ cm}$$

The variance σ^2 is defined by Eq. (13.2-6)

$$\sigma^2 = \overline{x^2} - (\overline{x})^2$$

$$= 4.72 - (2)^2 = 0.72$$

†See J. S. Bendat, and A. G. Piersol, *Random Data* (New York: Wiley Interscience, 1971), p. 96.

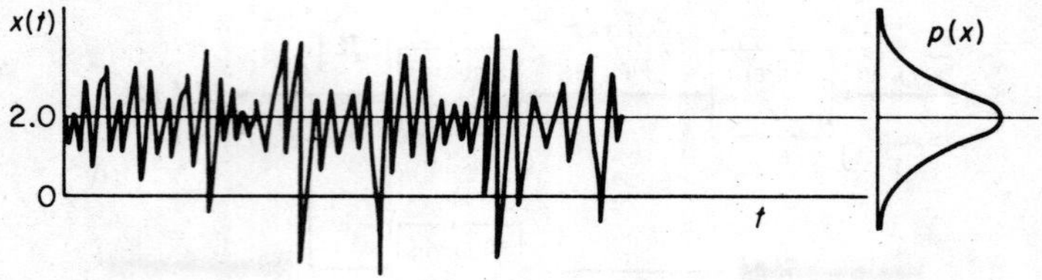

Figure 13.6-8.

and the standard deviation becomes

$$\sigma = \sqrt{0.72} = 0.85 \text{ cm}$$

The problem is graphically displayed by Fig. 13.6-8, which shows the time variation of the signal and its probability distribution.

Example 13.6-2

Determine the Fourier coefficients C_n and the power spectral density of the periodic function shown in Fig. 13.6-9.

Solution: The period is $2T$ and C_n are

$$C_0 = \frac{2}{2T}\int_{-T/2}^{T/2} F_0 \, d\xi = F_0$$

$$C_n = \frac{2}{2T}\int_{-T/2}^{T/2} F_0 e^{-in\omega_0\xi} \, d\xi = F_0 \left[\frac{\sin(n\pi/2)}{n\pi/2}\right]$$

Numerical values of C_n are computed as follows and plotted in Fig. 13.6-10.

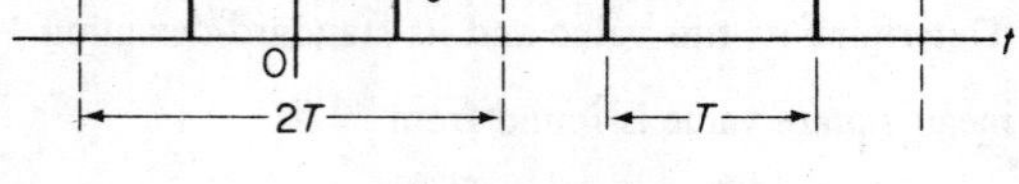

Figure 13.6-9.

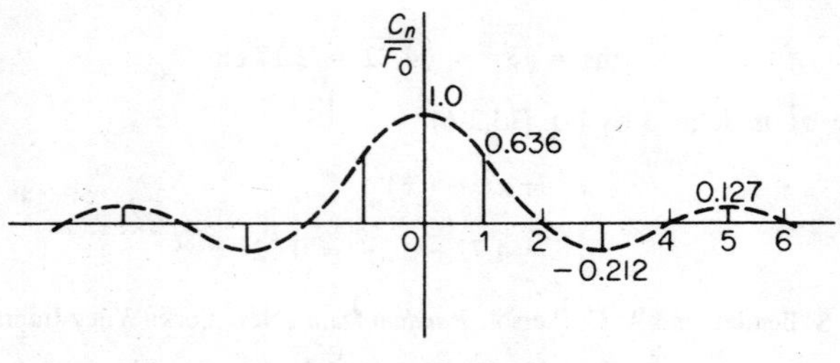

Figure 13.6-10. Fourier coefficients versus n.

n	$\frac{n\pi}{2}$	$\sin\frac{n\pi}{2}$	$\frac{1}{2}C_n$
0	0	0	$\frac{F_0}{2} = 1.0\frac{F_0}{2}$
1	$\frac{\pi}{2}$	1	$\left(\frac{2}{\pi}\right)\frac{F_0}{2} = 0.636\frac{F_0}{2}$
2	π	0	0
3	$3\frac{\pi}{2}$	-1	$\left(-\frac{2}{3\pi}\right)\frac{F_0}{2} = -0.212\frac{F_0}{2}$
4	2π	0	0
5	$5\frac{\pi}{2}$	1	$\left(\frac{2}{5\pi}\right)\frac{F_0}{2} = 0.127\frac{F_0}{2}$

The mean square value is determined from the equation

$$\overline{x^2} = \lim_{T\to\infty}\frac{1}{2T}\int_{-T}^{T}x^2(t)\,dt$$

$$= \lim_{T\to\infty}\frac{1}{2T}\int_{T}^{-T}\frac{1}{4}\left[\sum_n\left(C_n e^{in\omega_0 t} + C_n^* e^{-in\omega_0 t}\right)\right]^2 dt$$

$$= \sum_{n=1}^{\infty}\frac{C_n C_n^*}{2}$$

and since $\overline{x^2} = \int_{\infty}^{0} S_f(\omega)\,d\omega$, the spectral density function can be represented by a series of delta functions as

$$S_f(\omega) = \sum_{n=1}^{\infty}\frac{C_n C_n^*}{2}\delta(\omega - n\omega_0)$$

13.7 FOURIER TRANSFORMS

The discrete frequency spectrum of periodic functions becomes a continuous one when the period T is extended to infinity. Random vibrations are generally not periodic and the determination of its continuous frequency spectrum requires the use of the fourier integral, which can be regarded as a limiting case of the Fourier series as the period approaches infinity.

The Fourier transform has become the underlying operation for the modern time series analysis. In many of the modern instruments for spectral analysis, the calculation performed is that of determining the amplitude and phase of a given record.

The *Fourier integral* is defined by the equation

$$x(t) = \int_{-\infty}^{\infty} X(f)e^{i2\pi ft}\,df \qquad (13.7\text{-}1)$$

In contrast to the summation of the discrete spectrum of sinusoids in the

Fourier series, the Fourier integral may be regarded as a summation of the continuous spectrum of sinusoids. The quantity $X(f)$ in the above equation is called the *Fourier transform* of $x(t)$, which may be evaluated from the equation

$$X(f) = \int_{-\infty}^{\infty} x(t)e^{-i2\pi ft}\,dt \tag{13.7-2}$$

Like the Fourier coefficient C_n, $X(f)$ is a complex quantity which is a continuous function of f from $-\infty$ to $+\infty$. Equation (13.7-2) resolves the function $x(t)$ into harmonic components $X(f)$, whereas Eq. (13.7-1) synthesizes these harmonic components to the original time function $x(t)$. The two equations above are referred to as the *Fourier transform pair*.

Fourier transform (FT) of basic functions. To demonstrate the spectral character of the FT, we consider the FT of some basic functions.

Example 13.7-1

$$x(t) = Ae^{i2\pi f_n t} \tag{a}$$

From Eq. (13.7-1) we have

$$Ae^{i2\pi f_n t} = \int_{-\infty}^{\infty} X(f)\,e^{i2\pi ft}\,df$$

Recognizing the properties of a delta function, this equation is satisfied if

$$X(f) = A\delta(f - f_n) \tag{b}$$

By substituting into Eq. (13.7-2), we obtain

$$\delta(f - f_n) = \int_{-\infty}^{\infty} e^{-i2\pi(f-f_n)t}\,dt \tag{c}$$

The FT of $x(t)$ is displayed in Fig. 13.7-1, which demonstrates its spectral character.

Example 13.7-2

$$x(t) = a_n \cos 2\pi f_n t \tag{a}$$

Since

$$\cos 2\pi f_n t = \frac{1}{2}(e^{i2\pi f_n t} + e^{-i2\pi f_n t})$$

the result of Example 13.7-1 immediately gives

$$X(f) = \frac{a_n}{2}[\delta(f - f_n) + \delta(f + f_n)] \tag{b}$$

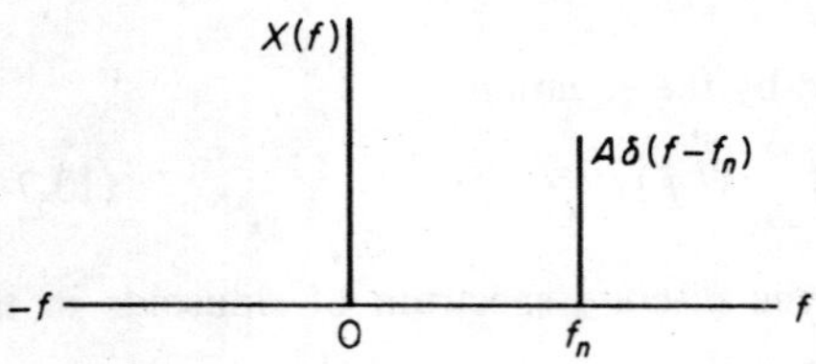

Figure 13.7-1. FT of $Ae^{i2\pi f_n t}$.

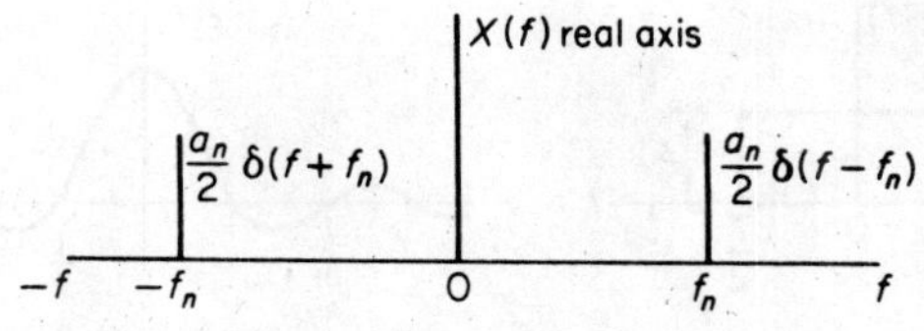

Figure 13.7-2. FT of $A_n \cos 2\pi f_n t$.

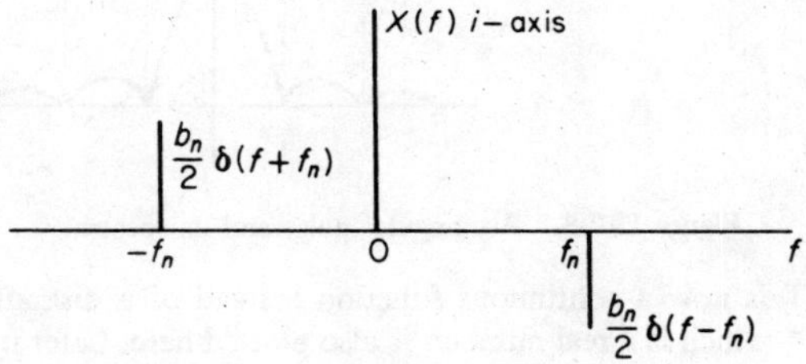

Figure 13.7-3. FT of $b_n \sin 2\pi f_n t$.

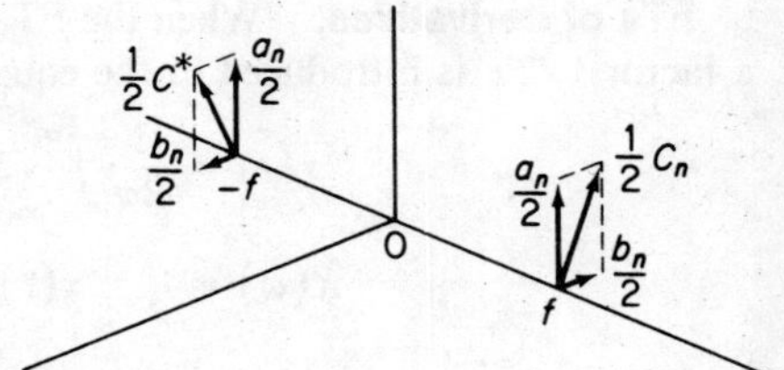

Figure 13.7-4. FT of $(a_n \cos 2\pi f_n t + b_n \sin 2\pi f_n t)$.

Figure 13.7-2 shows that $X(f)$ is a two-sided function of f.

In a similar manner, the FT of $b_n \sin 2\pi f_n t$ is

$$X(f) = -i\frac{b}{2}n[\delta(f - f_n) - \delta(f + f_n)]$$

which is shown on the imaginary plane of Fig. 13.7-3.

If we put the two FTs together in perpendicular planes, as shown in Fig. 13.7-4, we obtain the complex conjugate coefficients $C_n = a_n - ib_n$ and $C_n^* = a_n + ib_n$. Thus the product

$$\frac{C_n C_n^*}{4} = \frac{1}{4}\left(a_n^2 + b_n^2\right) = c_n c_n^*$$

is the square of the magnitude of the Fourier series, which is generally plotted at $\pm f$.

Example 13.7-3

We next determine the FT of a rectangular pulse, which is an example of an aperiodic function. (See Fig. 13.7-5.) Its FT is

$$X(f) = \int_{-\infty}^{\infty} x(t)e^{-i2\pi ft}\,dt = \int_{-T/2}^{T/2} Ae^{-i2\pi ft}\,dt = AT\left(\frac{\sin \pi fT}{\pi fT}\right)$$

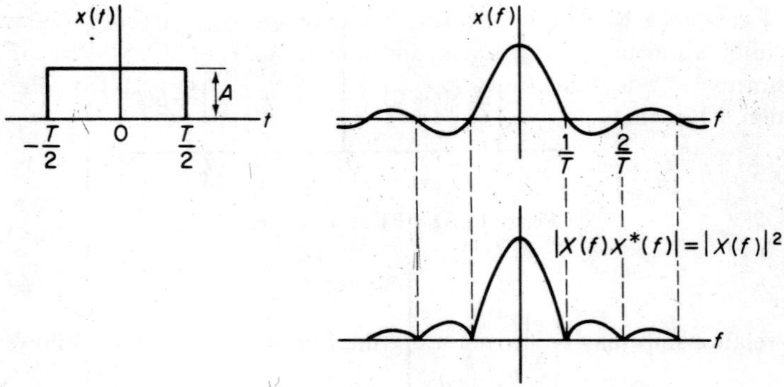

Figure 13.7-5. Rectangular pulse and its spectra.

Note that the FT is now a continuous function instead of a discontinuous function. The product XX^*, which is a real number, is also plotted here. Later it will be shown to be equal to the spectral density function.

FTs of derivatives. When the FT is expressed in terms of ω instead of f, a factor $1/2\pi$ is introduced in the equation for $x(t)$.

$$x(t) = \frac{1}{2\pi}\int_{-\infty}^{\infty} X(\omega)e^{i\omega t}\,d\omega \tag{13.7-3}$$

$$X(\omega) = \int_{-\infty}^{\infty} x(t)e^{-i\omega t}\,dt \tag{13.7-4}$$

This form is sometimes preferred in developing mathematical relationships. For example, if we differentiate Eq. (13.7-3) with respect to t, we obtain the FT pair

$$\dot{x}(t) = \frac{1}{2\pi}\int_{-\infty}^{\infty} [i\omega X(\omega)]\,e^{i\omega t}\,d\omega$$

$$i\omega X(\omega) = \int_{-\infty}^{\infty} \dot{x}(t)e^{-i\omega t}\,dt$$

Thus the FT of a derivative is simply the FT of the function multiplied by $i\omega$.

$$\text{FT}[\dot{x}(t)] = i\omega\text{FT}[x(t)] \tag{13.7-5}$$

Differentiating again, we obtain

$$\text{FT}[\ddot{x}(t)] = -\omega^2\text{FT}[x(t)] \tag{13.7-6}$$

These equations enable one to conveniently take the FT of differential equations. For example, if we take the FT of the differential equation

$$m\ddot{y} + c\dot{y} + ky = x(t)$$

we obtain

$$(-m\omega^2 + i\omega c + k)Y(\omega) = X(\omega)$$

where $X(\omega)$ and $Y(\omega)$ are the FT of $x(t)$ and $y(t)$, respectively.

Parseval's theorem. Parseval's theorem is a useful tool for converting time integration into frequency integration. If $X_1(f)$ and $X_2(f)$ are Fourier transforms of real time functions $x_1(t)$ and $x_2(t)$, respectively, Parseval's theorem states that

$$\begin{aligned}\int_{-\infty}^{\infty} x_1(t)x_2(t)\,dt &= \int_{-\infty}^{\infty} X_1(f)X_2^*(f)\,df \\ &= \int_{-\infty}^{\infty} X_1^*(f)X_2(f)\,df\end{aligned} \tag{13.7-7}$$

This relationship may be proved using the Fourier transform as follows:

$$x_1(t)x_2(t) = x_2(t)\int_{-\infty}^{\infty} X_1(f)e^{i2\pi ft}\,df$$

$$\begin{aligned}\int_{-\infty}^{\infty} x_1(t)x_2(t)\,dt &= \int_{-\infty}^{\infty} x_2(t)\int_{-\infty}^{\infty} X_1(f)e^{i2\pi ft}\,df\,dt \\ &= \int_{-\infty}^{\infty} X_1(f)\left[\int_{-\infty}^{\infty} x_2(t)e^{i2\pi ft}\,dt\right]df \\ &= \int_{-\infty}^{\infty} X_1(f)X_2^*(f)\,df\end{aligned}$$

All the previous formulas for the mean square value, autocorrelation, and cross correlation can now be expressed in terms of the Fourier transform by Parseval's theorem.

Example 13.7-4

Express the mean square value in terms of the Fourier transform. Letting $x_1(t) = x_2(t) = x(t)$, and averaging over T, which is allowed to go to ∞, we obtain

$$\overline{x^2} = \lim_{T\to\infty}\frac{1}{T}\int_{-T/2}^{T/2} x^2(t)\,dt = \int_{-\infty}^{\infty}\lim_{T\to\infty}\frac{1}{T}X(f)X^*(f)\,df$$

Comparing this with Eq. (13.6-6), we obtain the relationship

$$S(f_\pm) = \lim_{T\to\infty}\frac{1}{T}X(f)X^*(f) \tag{13.7-8}$$

where $S(f_\pm)$ is the spectral density function over positive and negative frequencies.

Example 13.7-5

Express the autocorrelation in terms of the Fourier transform. We begin with the Fourier transform of $x(t+\tau)$

$$x(t+\tau) = \int_{-\infty}^{\infty} X(f)e^{i2\pi f(t+\tau)}\,df$$

Substituting this into the expression for the autocorrelation, we obtain

$$R(\tau) = \lim_{T\to\infty} \frac{1}{T}\int_{-\infty}^{\infty} x(t)x(t+\tau)\,dt$$

$$= \lim_{T\to\infty} \frac{1}{T}\int_{-\infty}^{\infty} x(t)\int_{-\infty}^{\infty} X(f)\,e^{i2\pi ft}e^{i2\pi f\tau}\,df\,dt$$

$$= \int_{-\infty}^{\infty} \lim_{T\to\infty} \frac{1}{T}\left[\int_{-\infty}^{\infty} x(t)\,e^{i2\pi ft}\,dt\right] X(f)\,e^{i2\pi f\tau}\,df$$

$$= \int_{-\infty}^{\infty}\left[\lim_{T\to\infty} \frac{1}{T}X^*(f)\,X(f)\right]e^{i2\pi f\tau}\,df$$

Substituting from Eq. (13.7-8) the preceding equation becomes

$$R(\tau) = \int_{-\infty}^{\infty} S(f)\,e^{i2\pi f\tau}\,df \tag{13.7-9}$$

The inverse of the above equation is also available from the Fourier transform

$$S(f) = \int_{-\infty}^{\infty} R(\tau)\,e^{-i2\pi f\tau}\,d\tau \tag{13.7-10}$$

Since $R(\tau)$ is symmetric about $\tau = 0$, the last equation can also be written as

$$S(f) = 2\int_{0}^{\infty} R(\tau)\cos 2\pi f\tau\,d\tau \tag{13.7-11}$$

These are the *Wiener-Khintchine* equations, and they state that the spectral density function is the FT of the autocorrelation function.

As a parallel to the Wiener-Khintchine equations, we can define the cross correlation between two quantities $x(t)$ and $y(t)$ as

$$R_{xy}(\tau) = \langle x(t)\,y(t+\tau)\rangle = \lim_{T\to\infty}\frac{1}{T}\int_{-T/2}^{T/2} x(t)\,y(t+\tau)\,dt$$

$$= \int_{-\infty}^{\infty} \lim_{T\to\infty}\frac{1}{T}X^*(f)\,Y(f)\,e^{i2\pi f\tau}\,df \tag{13.7-12}$$

$$R_{xy}(\tau) = \int_{-\infty}^{\infty} S_{xy}(f)\,e^{i2\pi f\tau}\,df$$

where the cross-spectral density is defined as

$$S_{xy}(f) = \lim_{T\to\infty}\frac{1}{T}X^*(f)\,Y(f) \qquad -\infty \le f \le \infty$$

$$= \lim_{T\to\infty}\frac{1}{T}X(f)\,Y^*(f)$$

$$= S^*_{xy}(f) = S_{xy}(-f) \tag{13.7-13}$$

Its inverse from the Fourier transform is

$$S_{xy}(f) = \int_{-\infty}^{\infty} R_{xy}(\tau)\,e^{-i2\pi f\tau} \tag{13.7-14}$$

which is the parallel to Eq. (13.7-10). Unlike the autocorrelation, the cross-correlation and the cross-spectral density functions are, in general, not even functions; hence, the limits $-\infty$ to $+\infty$ are retained.

Example 13.7-6

Using the relationship

$$S(f) = 2\int_0^{\infty} R(\tau)\cos 2\pi f\tau\, d\tau$$

and the results of Example 13.5-1,

$$R(\tau) = A^2(T-\tau)$$

find $S(f)$ for the rectangular pulse.

Solution: Since $R(\tau) = 0$ for τ outside $\pm T$, we have .

$$\begin{aligned}
S(f) &= 2\int_0^T A^2(T-\tau)\cos 2\pi f\tau\, d\tau \\
&= 2A^2T\int_0^T \cos 2\pi f\tau\, d\tau - 2A^2\int_0^T \tau\cos 2\pi f\tau\, d\tau \\
&= 2A^2T\frac{\sin 2\pi f\tau}{2\pi f}\bigg|_0^T - 2A^2\left[\frac{\cos 2\pi f\tau}{(2\pi f)^2} + \frac{\tau}{2\pi f}\sin 2\pi f\tau\right]\bigg|_0^T \\
&= \frac{2A^2}{(2\pi f)^2}(1-\cos 2\pi fT) = A^2T^2\left(\frac{\sin \pi fT}{\pi fT}\right)^2
\end{aligned}$$

Thus the power spectral density of a rectangular pulse using Eq. (13.7-11), is

$$S(f) = A^2T^2\left(\frac{\sin \pi fT}{\pi fT}\right)^2$$

Note from Example 13.7-3 that this is also equal to $X(f)X^*(f) = |X(f)|^2$.

Example 13.7-7

Show that the frequency response function $H(\omega)$ is the Fourier transform of the impulse response function $h(t)$.

Solution: From the convolution integral, Eq. (4.2-1), the response equation in terms of the impulse response function is

$$x(t) = \int_{-\infty}^{t} f(\xi)h(t-\xi)\, d\xi$$

where the lower limit has been extended to $-\infty$ to account for all past excitations. By letting $\tau = (t-\xi)$, the last integral becomes

$$x(t) = \int_0^{\infty} f(t-\tau)h(\tau)\, d\tau$$

For a harmonic excitation $f(t) = e^{i\omega t}$, the preceding equation becomes

$$\begin{aligned}
x(t) &= \int_0^{\infty} e^{i\omega(t-\tau)}h(\tau)\, d\tau \\
&= e^{i\omega t}\int_0^{\infty} h(\tau)e^{-i\omega\tau}\, d\tau
\end{aligned}$$

Since the steady-state output for the input $y(t) = e^{i\omega t}$ is $x = H(\omega)e^{i\omega t}$, the frequency response function is

$$H(\omega) = \int_0^{\infty} h(\tau) e^{-i\omega\tau}\, d\tau = \int_{-\infty}^{\infty} h(\tau) e^{-i\omega\tau}\, d\tau$$

which is the FT of the impulse response function $h(t)$. The lower limit in the above integral has been changed from 0 to $-\infty$ since $h(t) = 0$ for negative t.

13.8 FTs AND RESPONSE

In engineering design, we often need to know the relationship between different points in the system. For example, how much of the roughness of a typical road is transmitted through the suspension system to the body of an automobile? (Here the term transfer function† is often used for the frequency response function.) Furthermore, it is often not possible to introduce a harmonic excitation to the input point of the system. It may be necessary to accept measurements $x(t)$ and $y(t)$ at two different points in the system for which the frequency response function is desired. The frequency response function for these points can be obtained by taking the FT of the input and output. The quantity $H(\omega)$ is then available from

$$H(\omega) = \frac{Y(\omega)}{X(\omega)} = \frac{\text{FT of output}}{\text{FT of input}} \tag{13.8-1}$$

where $X(\omega)$ and $Y(\omega)$ are the FT of $x(t)$ and $y(t)$.

If we multiply and divide the above equation by the complex conjugate $X^*(\omega)$, the result is

$$H(\omega) = \frac{Y(\omega)X^*(\omega)}{X(\omega)X^*(\omega)} \tag{13.8-2}$$

The denominator $X(\omega)X^*(\omega)$ is now a real quantity. The numerator is the cross spectrum $Y(\omega)X^*(\omega)$ between the input and the output and is a complex quantity. The phase of $H(\omega)$ is then found from the real and imaginary parts of the cross spectrum, which is simply

$$|Y(\omega)|\underline{/\phi_y} \cdot |X^*(\omega)|\underline{/\phi_x} = |Y(\omega)X^*(\omega)|\underline{/(\phi_y - \phi_x)} \tag{13.8-3}$$

Another useful relationship can be found by multiplying $H(\omega)$ by its conjugate $H^*(\omega)$. The result is

$$H(\omega)H^*(\omega) = \frac{Y(\omega)Y^*(\omega)}{X(\omega)X^*(\omega)}$$

or

$$Y(\omega)Y^*(\omega) = |H(\omega)|^2 X(\omega)X^*(\omega) \tag{13.8-4}$$

†Strictly speaking, the term transfer function is the ratio of the Laplace transform of the output to the Laplace transform of the input. In the frequency domain, however, the real part of $s = \alpha + i\omega$ is zero, and the LT becomes the FT.

Thus the output power spectrum is equal to the square of the system transfer function multiplied by the input power spectrum. Obviously, each side of the above equation is real and the phase does not enter in.

We wish now to examine the mean square value of the response. From Eq. (13.7-8) the mean square value of the input $x(t)$ is

$$\overline{x^2} = \int_{-\infty}^{\infty} S_x(f_\pm)\, df = \int_{-\infty}^{\infty} \lim_{T\to\infty} \frac{1}{T} X(f)X^*(f)\, df$$

The mean square value of the output $y(t)$ is

$$\overline{y^2} = \int_{-\infty}^{\infty} S_y(f_\pm)\, df = \int_{-\infty}^{\infty} \lim_{T\to\infty} \frac{1}{T} Y(f)Y^*(f)\, df$$

Substituting $YY^* = |H(f)|^2 XX^*$, we obtain

$$\begin{aligned}\overline{y^2} &= \int_{-\infty}^{\infty} |H(f)|^2 \left[\lim_{T\to\infty} \frac{1}{T} X(f)X^*(f)\right] df \\ &= \int_{-\infty}^{\infty} |H(f)|^2 S_x(f_\pm)\, df\end{aligned} \tag{13.8-5}$$

which is the mean square value of the response in terms of the system response function and the spectral density of the input.

In these expressions $S(f_\pm)$ are the two-sided spectral density functions over both the positive and negative frequencies. Also, $S(f_\pm)$ are even functions. In actual practice, it is desirable to work with spectral densities over only the positive frequencies. Equation (13.8-5) can then be written as

$$\overline{y^2} = \int_0^{\infty} |H(f)|^2 S_x(f_+)\, df \tag{13.8-6}$$

and since the two expressions must result in the same value for the mean square value, the relationship between the two must be

$$S(f_+) = S(f) = 2S(f_\pm) \tag{13.8-7}$$

Some authors also use the expression

$$\overline{y^2} = \int_0^{\infty} |H(\omega)|^2 S_x(\omega)\, d\omega \tag{13.8-8}$$

Again, the equations must result in the same mean square value so that

$$2\pi S(\omega) = S(f) \tag{13.8-9}$$

For a single-DOF system we have

$$H(f) = \frac{\frac{1}{k}}{\left[1 - (f/f_n)^2\right] + i\left[2\zeta(f/f_n)\right]} \tag{13.8-10}$$

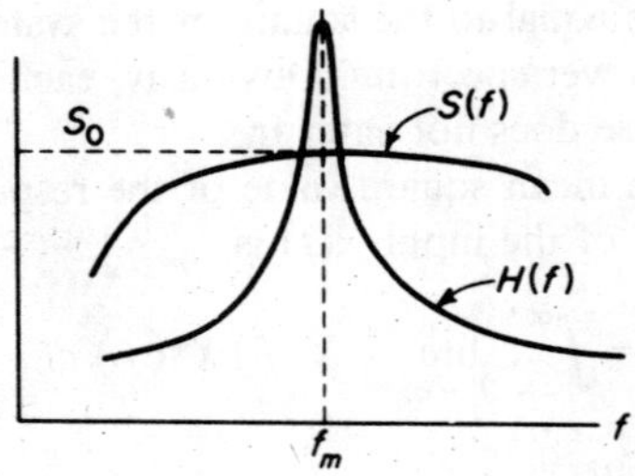

Figure 13.8-1. $S(f)$ and $H(f)$ leading to $\overline{y^2}$ of Equation 13.7-9.

If the system is lightly damped, the response function $H(f)$ is peaked steeply at resonance, and the system acts like a narrow-band filter. If the spectral density of the excitation is broad, as in Fig. 13.8-1, the mean square response for the single-DOF system can be approximated by the equation

$$\overline{y^2} \cong \frac{f_n}{k^2} S_x(f_n) \frac{\pi}{4\zeta} \tag{13.8-11}$$

where

$$\frac{\pi}{4\zeta} = \int_0^\infty \frac{d\left(\dfrac{f}{f_n}\right)}{\left[1 - \left(\dfrac{f}{f_n}\right)^2\right]^2 + \left(2\zeta \dfrac{f}{f_n}\right)^2}$$

and $S_x(f_n)$ is the spectral density of the excitation at frequency f_n.

Example 13.8-1

The response of any structure to a single point random excitation can be computed by a simple numerical procedure, provided the spectral density of the excitation and the frequency response curve of the structure are known. For example, consider the structure of Fig. 13.8-2(a), whose base is subjected to a random acceleration input with the power spectral density function shown in Fig. 13.8-2(b). It is desired to compute the

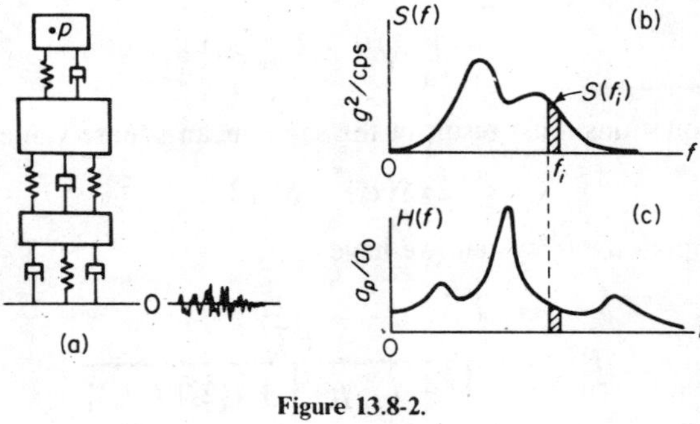

Figure 13.8-2.

response of the point p and establish the probability of exceeding any specified acceleration.

The frequency response function $H(f)$ for the point p may be obtained experimentally by applying to the base a variable frequency sinusoidal shaker with a constant acceleration input a_0, and measuring the acceleration response at p. Dividing the measured acceleration by a_0, $H(f)$ may appear as in Fig. 13.8-2(c).

The mean square response $\overline{a_p^2}$ at p is calculated numerically from the equation

$$\overline{a_p^2} = \sum_i S(f_i)|H(f_i)|^2 \Delta f_i$$

The following numerical table illustrates the computational procedure.

NUMERICAL EXAMPLE

f cps	Δf cps	$S(f_i)$ g^2/cps	$\|H(f_i)\|$ Nondimensional	$\|H(f_i)\|^2\Delta f$ cps	$S(f_i)\|H(f_i)\|^2\Delta f$ g^2 units
0	10	0.	1.0	10.	0.
10	10	0.	1.0	10.	0.
20	10	0.2	1.1	12.1	2.4
30	10	0.6	1.4	19.6	11.8
40	10	1.2	2.0	40.	48.0
50	10	1.8	1.3	16.9	30.5
60	10	1.8	1.3	16.9	30.5
70	10	1.1	2.0	40.	44.0
80	10	0.9	3.7	137.	123.
90	10	1.1	5.4	291.	320.
100	10	1.2	2.2	48.4	57.7
110	10	1.1	1.3	16.9	18.6
120	10	0.8	0.8	6.4	5.1
130	10	0.6	0.6	3.6	2.2
140	10	0.3	0.5	2.5	0.8
150	10	0.2	0.6	3.6	0.7
160	10	0.2	0.7	4.9	0.1
170	10	0.1	1.3	16.9	1.7
180	10	0.1	1.1	12.1	1.2
190	10	0.5	0.7	4.9	2.3
200	10	0.	0.5	2.5	0.
210	10	0.	0.4	1.6	0.

$$\overline{a^2} = 700.6g^2$$

$$\sigma = \sqrt{700.6g^2} = 26.6g$$

The probability of exceeding specified accelerations are

$$p[|a| > 26.6g] = 31.7\%$$

$$p\left[a_{\text{peak}} > 26.6g\right] = 60.7\%$$

$$p[|a| > 79.8g] = 0.3\%$$

$$p\left[a_{\text{peak}} > 79.8g\right] = 1.2\%$$

REFERENCES

[1] BENDAT, J. S. *Principles and Applications of Random Noise Theory*. New York: John Wiley & Sons, Inc., 1958.

[2] BENDAT, J. S., and PIERSOL, A. G. *Measurement and Analysis of Random Data*. New York: John Wiley & Sons, Inc., 1966.

[3] BLACKMAN, R. B., and TUKEY, J. W. *The Measurement of Power Spectra*. New York: Dover Publications, Inc., 1958.

[4] CLARKSON, B. L. "The Effect of Jet Noise on Aircraft Structures." *Aeronautical Quarterly*, Vol. 10, Part 2, May 1959.

[5] CRAMER, H. *The Elements of Probability Theory*. New York: John Wiley & Sons, Inc., 1955.

[6] CRANDALL, S. H. *Random Vibration*. Cambridge, Mass.: The Technology Press of M.I.T., 1948.

[7] CRANDALL, S. H. *Random Vibration*, Vol. 2. Cambridge, Mass.: The Technology Press of M.I.T., 1963.

[8] RICE, S. O. *Mathematical Analysis of Random Noise*. New York: Dover Publications, Inc., 1954.

[9] ROBSON, J. D. *Random Vibration*. Edinburgh University Press, 1964.

[10] THOMSON, W. T., and BARTON, M. V. "The Response of Mechanical Systems to Random Excitation." *J. Appl. Mech*., June 1957, pp. 248–51.

PROBLEMS

13-1 Give examples of random data and indicate classifications for each example.

13-2 Discuss the differences between nonstationary, stationary, and ergodic data.

13-3 Discuss what we mean by the expected value. What is the expected number of heads when 8 coins are thrown 100 times; 1000 times? What is the probability for tails?

13-4 Throw a coin 50 times, recording 1 for head and 0 for tail. Determine the probability of obtaining heads by dividing the cumulative heads by the number of throws and plot this number as a function of the number of throws. The curve should approach 0.5.

13-5 For the series of triangular waves shown in Fig. P13-5, determine the mean and the mean square values.

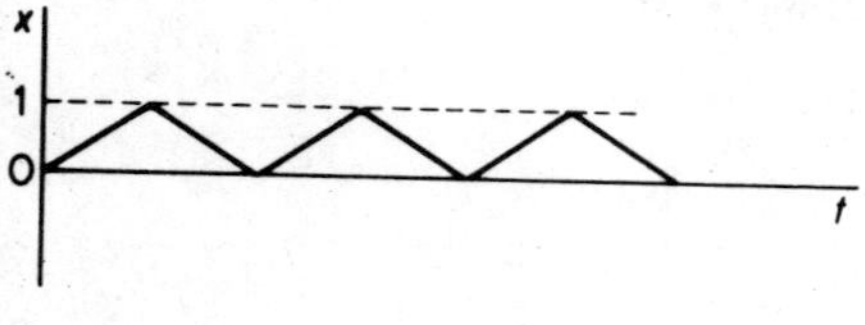

Figure P13-5.

13-6 A sine wave with a steady component has the equation

$$x = A_0 + A_1 \sin \omega t$$

Determine the expected values $E(x)$ and $E(x^2)$.

13-7 Determine the mean and mean square values for the rectified sine wave.

13-8 Discuss why the probability distribution of the peak values of a random function should follow the Rayleigh distribution or one similar in shape to it.

13-9 Show that for the Gaussian probability distribution $p(x)$ the central moments are given by

$$E(x^n) = \int_{-\infty}^{\infty} x^n p(x)\, dx = \begin{cases} 0 & \text{for } n \text{ odd} \\ 1 \cdot 3 \cdot 5 \cdots (n-1)\sigma^n & \text{for } n \text{ even} \end{cases}$$

13-10 Derive the equations for the cumulative probability and the probability density functions of the sine wave. Plot these results.

13-11 What would the cumulative probability and the probability density curves look like for the rectangular wave shown in Fig. P13-11.

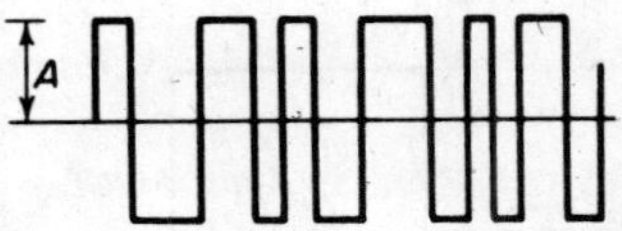

Figure P13-11.

13-12 Determine the autocorrelation of a cosine wave $x(t) = A \cos t$, and plot it against τ.

13-13 Determine the autocorrelation of the rectangular wave shown in Fig. P13-26.

13-14 Determine the autocorrelation of the rectangular pulse and plot it against τ.

13-15 Determine the autocorrelation of the binary sequence shown in Fig. P13-15. *Suggestion*: Trace the above wave on transparent graph paper and shift it through τ.

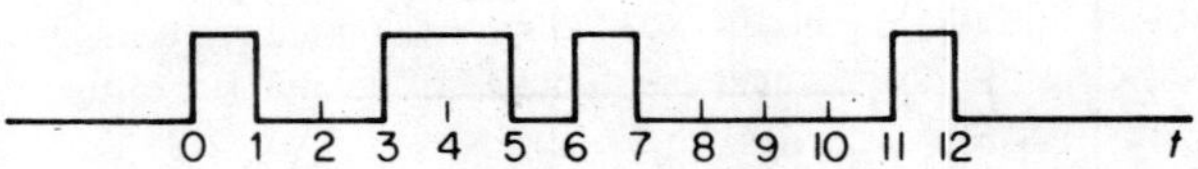

Figure P13-15.

13-16 Determine the autocorrelation of the triangular wave shown in Fig. P13-16.

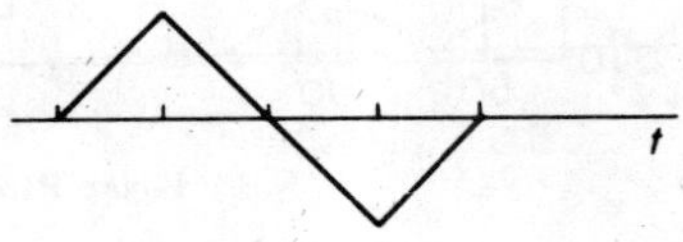

Figure P13-16.

13-17 Figure P13-17 shows the acceleration spectral density plot of a random vibration. Approximate the area by a rectangle and determine the rms value in m/s^2.

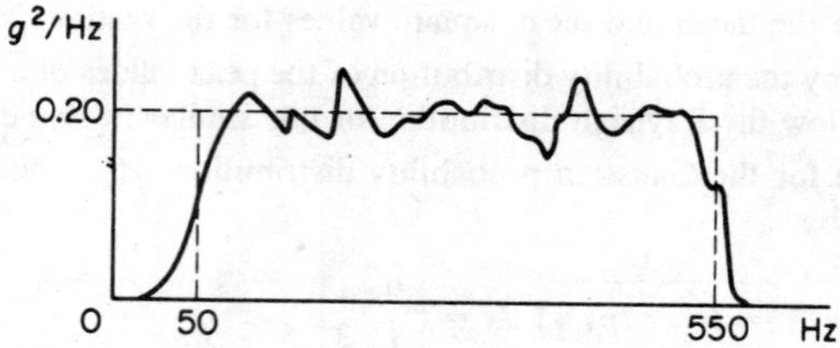

Figure P13-17.

13-18 Determine the rms value of the spectral density plot shown in Fig. P13-18.

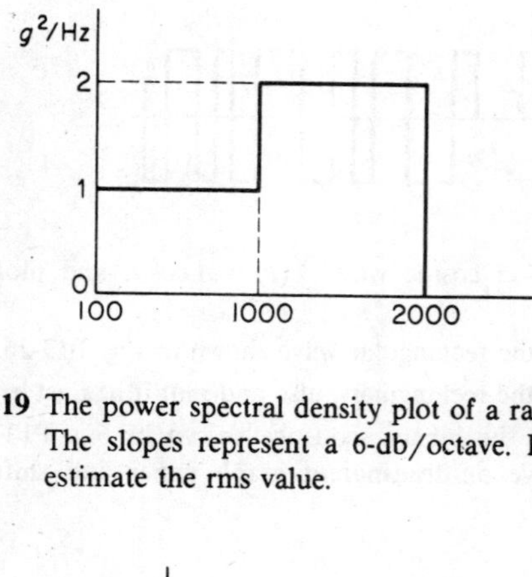

Figure P13-18.

13-19 The power spectral density plot of a random vibration is shown in Fig. P13-19. The slopes represent a 6-db/octave. Replot the result on a linear scale and estimate the rms value.

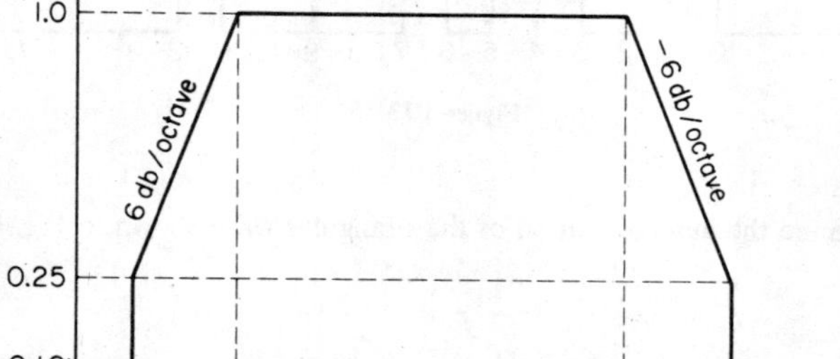

Figure P13-19.

13-20 Determine the spectral density function for the waves in Fig. P13-20.

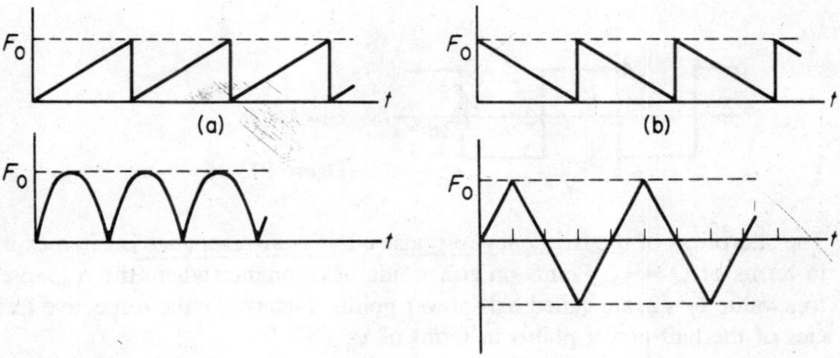

Figure P13-20.

13-21 A random signal is found to have a constant spectral density of $S(f) = 0.002$ in.^{2}/cps between 20 cps and 2000 cps. Outside this range, the spectral density is zero. Determine the standard deviation and the rms value if the mean value is 1.732 in. Plot this result.

13-22 Derive the equation for the coefficients C_n of the periodic function

$$f(t) = Re \sum_{n=0}^{\infty} C_n e^{in\omega_0 t}$$

13-23 Show that for Prob. 13-22, $C_{-n} = C_n^*$, and that $f(t)$ can be written as

$$f(t) = \sum_{n=-\infty}^{\infty} C_n e^{in\omega_0 t}$$

13-24 Determine the Fourier series for the saw tooth wave shown in Fig. P13-24 and plot its spectral density.

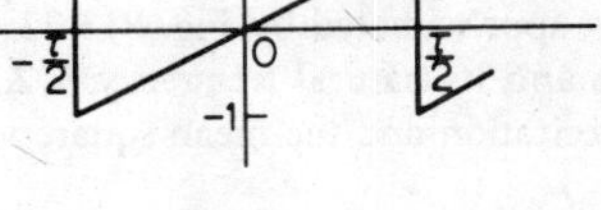

Figure P13-24.

13-25 Determine the complex form of the Fourier series for the wave shown in Fig. P13-25, and plot its spectral density.

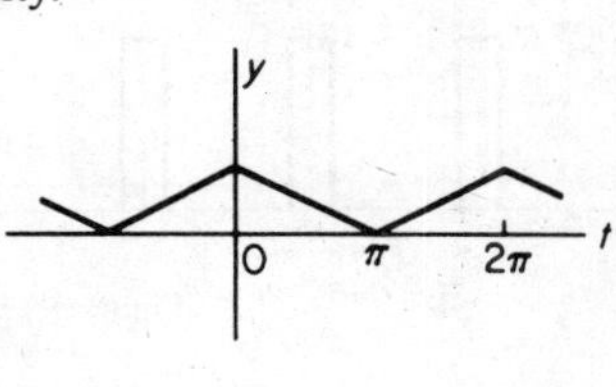

Figure P13-25.

13-26 Determine the complex form of the Fourier series for the rectangular wave shown in Fig. P13-26, and plot its spectral density.

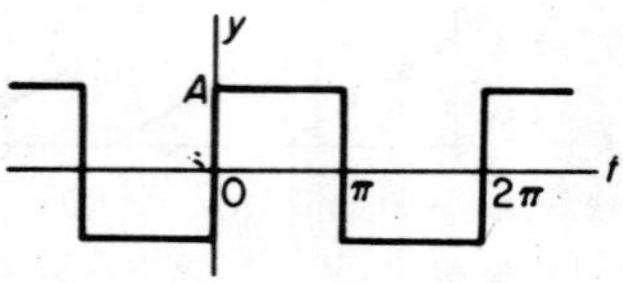

Figure P13-26.

13-27 The sharpness of the frequency response curve near resonance is often expressed in terms of $Q = \frac{1}{2}\zeta$. Points on either side of resonance where the response falls to a value $1/\sqrt{2}$ are called half-power points. Determine the respective frequencies of the half-power points in terms of ω_n and Q.

13-28 Show that

$$\int_0^\infty \frac{d\eta}{[1-\eta^2]^2 + [2\zeta\eta]^2} = \frac{\pi}{4\zeta} \qquad \text{for } \zeta \ll 1$$

13-29 The differential equation of a system with structural damping is given as

$$m\ddot{x} + k(1 + i\gamma)x = F(t)$$

Determine the frequency response function.

13-30 A single-DOF system with natural frequency ω_n and damping factor $\zeta = 0.10$ is excited by the force

$$F(t) = F\cos(0.5\omega_n t - \theta_1) + F\cos(\omega_n t - \theta_2) + F\cos(2\omega_n t - \theta_3)$$

Show that the mean square response is

$$\overline{y^2} = (1.74 + 25.0 + 0.110)\frac{1}{2}\left(\frac{F}{k}\right)^2 = 13.43\left(\frac{F}{k}\right)^2$$

13-31 In Example 13.7-3 what is the probability of the instantaneous acceleration exceeding a value $53.2g$? Of the peak value exceeding this value?

13-32 A large hydraulic press stamping out metal parts is operating under a series of forces approximated by Fig. P13-32. The mass of the press on its foundation is 40 kg and its natural frequency is 2.20 Hz. Determine the Fourier spectrum of the excitation and the mean square value of the response.

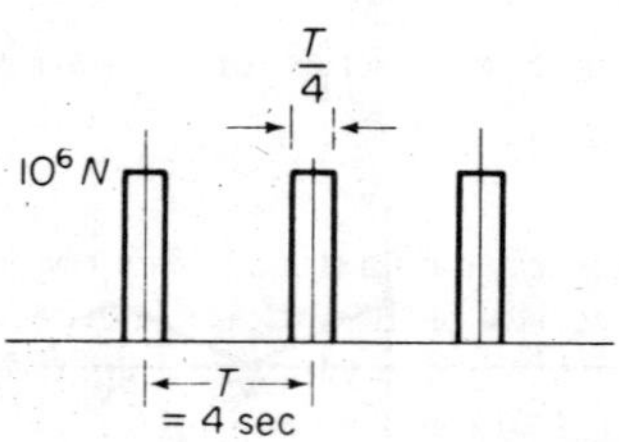

Figure P13-32.

13-33 For a single-DOF system, the substitution of Eq. (13.8-10) into Eq. (13.8-6) results in

$$\overline{y^2} = \int_0^\infty S_x(f_+)\frac{1}{k^2}\frac{df}{\left[1-\left(\dfrac{f}{f_n}\right)^2\right]^2+\left(2\zeta\dfrac{f}{f_n}\right)^2}$$

where $S_x(f_+)$ is the spectral density of the excitation force. When the damping ζ is small and the variation of $S_x(f_+)$ is gradual, the last equation becomes

$$\overline{y^2} \cong S_x(f_n)\frac{f_n}{k^2}\int_0^\infty \frac{d\left(\dfrac{f}{f_n}\right)}{\left[1-\left(\dfrac{f}{f_n}\right)^2\right]^2+\left(2\zeta\dfrac{f}{f_n}\right)^2} = S_x(f_n)\frac{f_n}{k^2}\frac{\pi}{4\zeta}$$

which is Eq. (13.8-11). Derive a similar equation for the mean square value of the relative motion z of a single-DOF system excited by the base motion, in terms of the spectral density $S_y(f_+)$ of the base acceleration. (See Sec. 3.5.) If the spectral density of the base acceleration is constant over a given frequency range, what must be the expression for $\overline{z^2}$?

13-34 Referring to Sec. 3.5 we can write the equation for the absolute acceleration of the mass undergoing base excitation as

$$\ddot{x} = \frac{k + i\omega c}{k - m\omega^2 + i\omega c}\cdot \ddot{y}$$

Determine the equation for the mean square acceleration $\overline{\ddot{x}^2}$. Establish a numerical integration technique for the computer evaluation of $\overline{\ddot{x}^2}$.

13-35 A radar dish with a mass of 60 kg is subject to wind loads with the spectral density shown in Fig. P13-35. The dish-support system has a natural frequency of 4 Hz. Determine the mean square response and the probability of the dish exceeding a vibration amplitude of 0.132 m. Assume $\zeta = 0.05$.

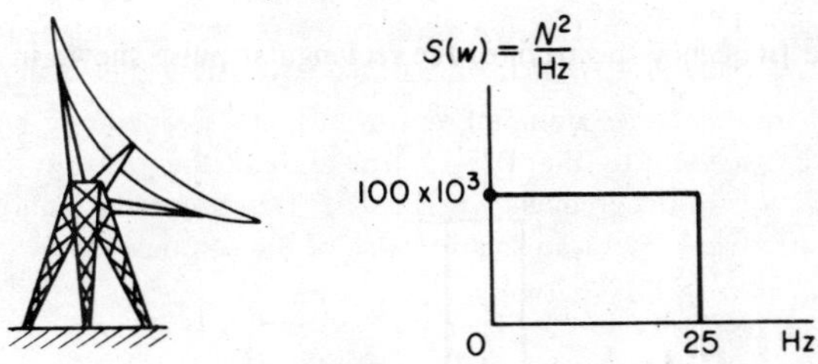

Figure P13-35.

13-36 A jet engine with a mass of 272 kg is tested on a stand, which results in a natural frequency of 26 Hz. The spectral density of the jet force under test is shown in Fig. P13-36. Determine the probability of the vibration amplitude in the axial direction of the jet thrust exceeding 0.012 m. Assume $\zeta = 0.10$.

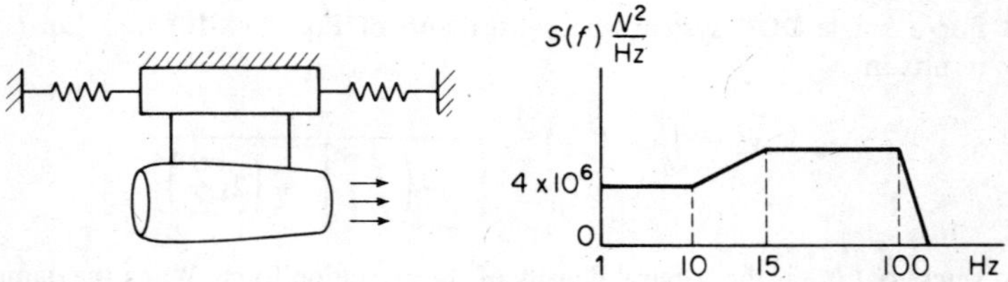

Figure P13-36.

13-37 An SDF system with viscous damping $\zeta = 0.03$ is excited by white noise excitation $F(t)$ having a constant power spectral density of $5 \times 10^6\ N^2/Hz$. The system has a natural frequency of $\omega_n = 30$ rad/s and a mass of 1500 kg. Determine σ. Assuming Rayleigh distribution for peaks, determine the probability that the maximum peak response will exceed 0.037 m.

13-38 Starting with the relationship

$$x(t) = \int_0^{\infty} f(t - \xi) h(\xi)\, d\xi$$

and using the FT technique, show that

$$X(i\omega) = F(i\omega) H(i\omega)$$

and

$$\overline{x^2} = \int_0^{\infty} S_F(\omega) |H(i\omega)|^2\, d\omega$$

where

$$S_F(\omega) = \lim_{T \to \infty} \frac{1}{2\pi T} F(i\omega) F^*(i\omega)$$

13-39 Starting with the relationship

$$H(i\omega) = |H(i\omega)| e^{i\phi(\omega)}$$

show that

$$\frac{H(i\omega)}{H^*(i\omega)} = e^{i2\phi(\omega)}$$

13-40 Find the frequency spectrum of the rectangular pulse shown in Fig P13-40.

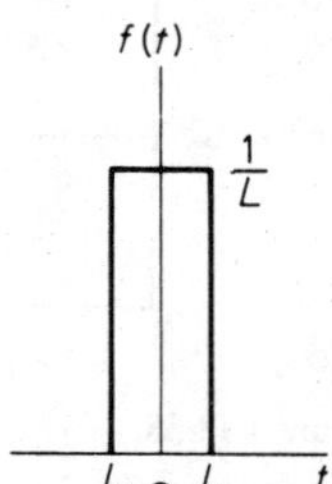

Figure P13-40.

13-41 Show that the unit step function has no Fourier transform. *Hint*: Examine

$$\int_{-\infty}^{\infty} |f(t)|\, dt$$

13-42 Starting with the equations

$$S_{FX}(\omega) = \lim_{T\to\infty} \frac{1}{2\pi T} F^*(i\omega)\, X(i\omega) = \lim_{T\to\infty} \frac{1}{2\pi T} F^*(FH) = S_F H$$

and

$$S_{XF}(\omega) = \lim_{T\to\infty} \frac{1}{2\pi T} X^* F = \lim_{T\to\infty} \frac{1}{2\pi T}(F^* H^*)\, F = S_F H^*$$

show that

$$\frac{S_{FX}(\omega)}{S_{XF}(\omega)} = e^{i2\phi(\omega)}$$

and

$$\frac{S_F(\omega)}{S_{XF}(\omega)} = \frac{S_{FX}(\omega)}{S_F(\omega)} = H(i\omega)$$

13-43 The differential equation for the longitudinal motion of a uniform slender rod is

$$\frac{\partial^2 u}{\partial t^2} = c^2 \frac{\partial^2 u}{\partial x^2}$$

Show that for an arbitrary axial force at the end $x = 0$, with the other end $x = l$ free, the Laplace transform of the response is

$$\bar{u}(x, s) = \frac{-c\bar{F}(s)\, e^{-s(l/c)}}{sAE(1 - e^{-2s(l/c)})} \{ e^{(s/c)(x-l)} + e^{-(s/c)(x-l)} \}$$

13-44 If the force in Prob. 13-43 is harmonic and equal to $F(t) = F_0 e^{i\omega t}$, show that

$$u(x, t) = \frac{cF_0 e^{i\omega t} \cos[(\omega l/c)(x/l - 1)]}{\omega AE \sin(\omega l/c)}$$

and

$$\sigma(x, t) = \frac{-\sin[(\omega l/c)(x/l - 1)]}{\sin(\omega l/c)} \frac{F_0}{A} e^{i\omega t}$$

where σ is the stress.

13-45 With $S(\omega)$ as the spectral density of the excitation stress at $x = 0$, show that the mean square stress in Prob. 13-43 is

$$\overline{\sigma^2} \cong \frac{2\pi}{\gamma} \sum_n \frac{c}{n\pi l} S(\omega_n) \sin^2 n\pi \frac{x}{l}$$

where structural damping is assumed. The normal modes of the problem are

$$\varphi_n(x) = \sqrt{2} \cos n\pi\left(\frac{x}{l} - 1\right), \qquad \omega_n = n\pi\left(\frac{c}{l}\right), \qquad c = \sqrt{\frac{AE}{m}}$$

13-46 Determine the FT of $x(t - t_0)$ and show that it is equal to $e^{-i2\pi f t_0} X(f)$ where $X(f) = \text{FT}[x(t)]$.

13-47 Prove that the FT of a convolution is the product of the separate FT.

$$\text{FT}[x(t) * y(t)] = X(f)Y(f)$$

13-48 Using the derivative theorem, show that the FT of the derivative of a rectangular pulse is a sine wave.

14

NONLINEAR VIBRATIONS

Linear system analysis serves to explain much of the behavior of oscillatory systems. However, there are a number of oscillatory phenomena that cannot be predicted or explained by the linear theory.

In the linear systems which we have studied, cause and effect are related linearly; i.e., if we double the load, the response is doubled. In a nonlinear system this relationship between cause and effect is no longer proportional. For example, the center of an oil can may move proportionally to the force for small loads, but at a certain critical load it will snap over to a large displacement. The same phenomenon is also encountered in the buckling of columns, electrical oscillations of circuits containing inductance with an iron core, and vibration of mechanical systems with nonlinear restoring forces.

The differential equation describing a nonlinear oscillatory system may have the general form

$$\ddot{x} + f(\dot{x}, x, t) = 0$$

Such equations are distinguished from linear equations in that the principle of superposition does not hold for their solution.

Analytical procedures for the treatment of nonlinear differential equations are difficult and require extensive mathematical study. Exact solutions that are known are relatively few, and a large part of the progress in the knowledge of nonlinear systems comes from approximate and graphical solutions and from studies made on computing machines. Much can be learned about a nonlinear system, however, by using the state space approach and studying the motion presented in the phase plane.

14.1 PHASE PLANE

In an *autonomous* system, the time t does not appear explicitly in the differential equation of motion. Thus only the differential of time, dt, will appear in such an equation.

We will first study an automonous system with the differential equation

$$\ddot{x} + f(x, \dot{x}) = 0 \tag{14.1-1}$$

where $f(x, \dot{x})$ may be a nonlinear function of x and $\dot{x}$. In the method of *state space*, we express the last equation in terms of two first order equations as follows:

$$\begin{aligned} \dot{x} &= y \\ \dot{y} &= -f(x, y) \end{aligned} \tag{14.1-2}$$

If x and y are Cartesian coordinates, the xy-plane is called the *phase plane*. The *state* of the system is defined by the coordinate x and $y = \dot{x}$, which represents a point on the phase plane. As the state of the system changes, the point on the phase plane moves, thereby generating a curve which is called the *trajectory*.

Another useful concept is the *state speed V* defined by the equation

$$V = \sqrt{\dot{x}^2 + \dot{y}^2} \tag{14.1-3}$$

When the state speed is zero, an *equilibrium state* is reached in that both the velocity of $\dot{x}$ and the acceleration $\dot{y} = \ddot{x}$ are zero.

Dividing the second of Eq. (14.1-2) by the first, we obtain the relation

$$\frac{dy}{dx} = \frac{-f(x, y)}{y} = \phi(x, y) \tag{14.1-4}$$

Thus for every point x, y in the phase plane for which $\phi(xy)$ is not indeterminate, there is a unique slope of the trajectory.

If $y = 0$ (i.e., points along the x-axis) and $f(x, y) \neq 0$, the slope of the trajectory is infinite. Thus all trajectories corresponding to such points must cross the x-axis at right angles.

If $y = 0$ and $f(x, y) = 0$, the slope is indeterminate. We define such points as *singular points*. Singular points correspond to a state of equilibrium in that both the velocity $y = \dot{x}$ and the force $\ddot{x} = \dot{y} = -f(x, y)$ are zero. Further discussion is required to establish whether the equilibrium represented by the singular point is stable or unstable.

Example 14.1-1

Determine the phase plane of a single-DOF oscillator

$$\ddot{x} + \omega^2 x = 0$$

Solution: With $y = \dot{x}$, the above equation is written in terms of two first-order equations

$$\begin{aligned} \dot{y} &= -\omega^2 x \\ \dot{x} &= y \end{aligned}$$

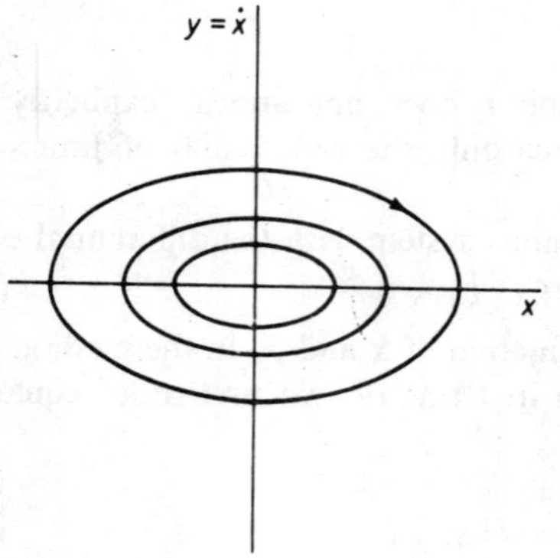

Figure 14.1-1.

Dividing we obtain

$$\frac{dy}{dx} = -\frac{\omega^2 x}{y}$$

Separating variables and integrating

$$y^2 + \omega^2 x^2 = C$$

which is a series of ellipses, the size of which is determined by C. The above equation is also that of conservation of energy

$$\tfrac{1}{2} m\dot{x}^2 + \tfrac{1}{2} kx^2 = C'$$

Since the singular point is at $x = y = 0$, the phase plane plot appears as in Fig. 14.1-1. If y/ω is plotted in place of y, the ellipses of Fig. 14.1-1 reduce to circles.

14.2 CONSERVATIVE SYSTEMS

In a conservative system the total energy remains constant. Summing the kinetic and potential energies per unit mass, we have

$$\tfrac{1}{2}\dot{x}^2 + U(x) = E = \text{constant} \tag{14.2-1}$$

Solving for $y = \dot{x}$, the ordinate of the phase plane is given by the equation

$$y = \dot{x} = \pm\sqrt{2[E - U(x)]} \tag{14.2-2}$$

It is evident from this equation that the trajectories of a conservative system must be symmetric about the x-axis.

The differential equation of motion for a conservative system can be shown to have the form

$$\ddot{x} = f(x) \tag{14.2-3}$$

Since $\ddot{x} = \dot{x}(d\dot{x}/dx)$, the last equation can be written as

$$\dot{x}\, d\dot{x} - f(x)\, dx = 0 \tag{14.2-4}$$

Integrating, we have

$$\frac{\dot{x}^2}{2} - \int_0^x f(x)\,dx = E \tag{14.2-5}$$

and by comparison with Eq. (14.2-1) we find

$$\begin{aligned} U(x) &= -\int_0^x f(x)\,dx \\ f(x) &= -\frac{dU}{dx} \end{aligned} \tag{14.2-6}$$

Thus for a conservative system the force is equal to the negative gradient of the potential energy.

With $y = \dot{x}$, Eq. (14.2-4) in the state space becomes

$$\frac{dy}{dx} = \frac{f(x)}{y} \tag{14.2-7}$$

We note from this equation that singular points correspond to $f(x) = 0$ and $y = \dot{x} = 0$, and hence are equilibrium points. Equation (14.2-6) then indicates that at the equilibrium points the slope of the potential energy curve $U(x)$ must be zero. It can be shown that the minima of $U(x)$ are stable equilibrium positions, whereas the saddle points corresponding to the maxima of $U(x)$ are positions of unstable equilibrium.

Stability of equilibrium. Examining Eq. (14.2-2) the value of E is determined by the initial conditions of $x(0)$ and $y(0) = \dot{x}(0)$. If the initial conditions are large, E will also be large. For every position x, there is a potential $U(x)$; for motion to take place, E must be greater than $U(x)$. Otherwise, Eq. (14.2-2) shows that the velocity $y = \dot{x}$ is imaginary.

Figure 14.2-1 shows a general plot of $U(x)$ and the trajectory y vs. x for various values of E computed from Eq. (14.2-2).

For $E = 7$, $U(x)$ lies below $E = 7$ only between $x = 0$ to 1.2, $x = 3.8$ to 5.9, and $x = 7$ to 8.7. The trajectories corresponding to $E = 7$ are closed curves and the period associated with them can be found from Eq. (14.2-2) by integration

$$\tau = 2\int_{x_1}^{x_2} \frac{dx}{\sqrt{2[E - U(x)]}}$$

where x_1 and x_2 are extreme points of the trajectory on the x-axis.

For smaller initial conditions, these closed trajectories become smaller. For $E = 6$ the trajectory about the equilibrium point $x = 7.5$ contracts to a point, whereas the trajectory about the equilibrium point $x = 5$ is a closed curve between $x = 4.2$ to 5.7.

For $E = 8$ one of the maxima of $U(x)$ at $x = 6.5$ is tangent to $E = 8$ and the trajectory at this point has four branches. The point $x = 6.5$ is a

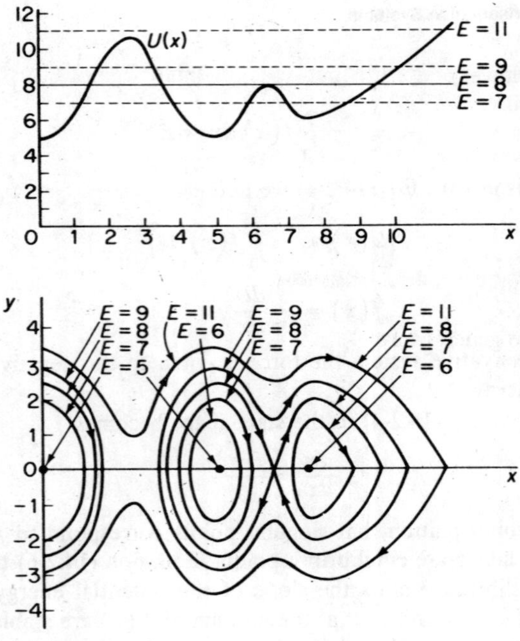

Figure 14.2-1.

COMPUTATION OF PHASE PLANE FOR $U(x)$ GIVEN IN FIG. 1.2-1

$$y = \pm\sqrt{2[E - U(x)]}$$

x	$U(x)$	$\pm y$ at ($E = 7$)	$\pm y$ at ($E = 8$)	$\pm y$ at ($E = 9$)	$\pm y$ at ($E = 11$)
0	5.0	2.0	2.45	2.83	3.46
1.0	6.3	1.18	1.84	2.32	
1.5	8.0	imag	0	1.41	2.45
2.0	9.6	imag	imag	imag	
3.0	10.0	imag	imag	imag	1.41
3.5	8.0	imag	0	1.41	2.45
4.0	6.5	1.0	1.73	2.24	
5.0	5.0	2.0	2.45	2.83	3.46
5.5	5.7	1.61	2.24	2.57	
6.0	7.2	imag	1.26	1.90	
6.5	8.0	imag	0	1.41	2.45
7.0	7.0	0	1.41	2.0	
7.5	6.0	1.41	2.0	2.45	3.16
8.0	6.3	1.18	1.84	2.32	
9.0	7.4	imag	1.09	1.79	
9.5	8.0	imag	0	1.41	
10.0	8.8	imag	imag	0.63	
11.5					0

saddle point for $E = 8$ and the motion is unstable. The saddle point trajectories are called *separatrices*.

For $E > 8$ the trajectories may or may not be closed. $E = 9$ shows a closed trajectory between $x = 3.3$ to 10.2. Note that at $x = 6.5$, $dU/dx = -f(x) = 0$ and $y = \dot{x} \neq 0$ for $E = 9$, and hence equilibrium does not exist.

14.3 STABILITY OF EQUILIBRIUM

Expressed in the general form

$$\frac{dy}{dx} = \frac{P(x, y)}{Q(x, y)} \tag{14.3-1}$$

the singular points (x_s, y_s) of the equation are identified by

$$P(x_s, y_s) = Q(x_s, y_s) = 0 \tag{14.3-2}$$

Equation (14.3-1), of course, is equivalent to the two equations

$$\begin{aligned} \frac{dx}{dt} &= Q(x, y) \\ \frac{dy}{dt} &= P(x, y) \end{aligned} \tag{14.3-3}$$

from which the time dt has been eliminated. A study of these equations in the neighborhood of the singular point provides us with answers as to the stability of equilibrium.

Recognizing that the slope dy/dx of the trajectories does not vary with translation of the coordinate axes, we translate the u, v axis to one of the singular points to be studied, as shown in Fig. 14.3-1. We then have

$$\begin{aligned} x &= x_s + u \\ y &= y_s + v \\ \frac{dy}{dx} &= \frac{dv}{du} \end{aligned} \tag{14.3-4}$$

If $P(x, y)$ and $Q(x, y)$ are now expanded in terms of the Taylor series about

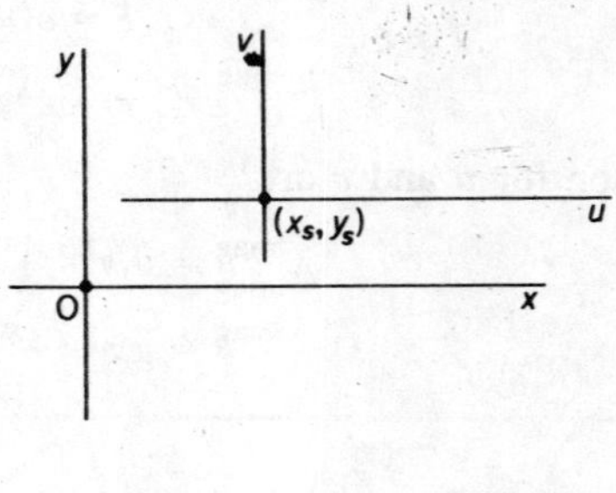

Figure 14.3-1.

the singular point (x_s, y_s), we obtain for $Q(x, y)$

$$Q(x, y) = Q(x_s, y_s) + \left(\frac{\partial Q}{\partial u}\right)_s u + \left(\frac{\partial Q}{\partial v}\right)_s v + \left(\frac{\partial^2 Q}{\partial u^2}\right)_s u^2 + \cdots \quad (14.3\text{-}5)$$

and a similar equation for $P(x, y)$. Since $Q(x_s, y_s)$ is zero and $(\partial Q/\partial u)_s$ and $(\partial Q/\partial v)_s$ are constants, Eq. (14.3-1) in the region of the singularity becomes

$$\frac{dv}{du} = \frac{cu + ev}{au + bv} \quad (14.3\text{-}6)$$

where the higher-order derivatives of P and Q have been omitted. Thus a study of the singularity at (x_s, y_s) is possible by studying Eq. (14.3-6) for small u and v.

Returning to Eq. (14.3-3) and taking note of Eqs. (14.3-4) and (14.3-5), Eq. (14.3-6) is seen to be equivalent to

$$\begin{aligned} \frac{du}{dt} &= au + bv \\ \frac{dv}{dt} &= cu + ev \end{aligned} \quad (14.3\text{-}7)$$

which may be rewritten in the matrix form

$$\begin{Bmatrix} \dot{u} \\ \dot{v} \end{Bmatrix} = \begin{bmatrix} a & b \\ c & e \end{bmatrix} \begin{Bmatrix} u \\ v \end{Bmatrix} \quad (14.3\text{-}8)$$

It was shown in Sec. 6.7 that if the eigenvalues and eigenvectors of a matrix equation such as Eq. (14.3-8) are known, a transformation

$$\begin{Bmatrix} u \\ v \end{Bmatrix} = [P] \begin{Bmatrix} \xi \\ \eta \end{Bmatrix} = \left[\begin{Bmatrix} u_1 \\ v_1 \end{Bmatrix} \begin{Bmatrix} u_2 \\ v_2 \end{Bmatrix} \right] \begin{Bmatrix} \xi \\ \eta \end{Bmatrix} \quad (14.3\text{-}9)$$

where $[P]$ is a modal matrix of the eigenvector columns, will decouple the equation to the form

$$\begin{Bmatrix} \dot{\xi} \\ \dot{\eta} \end{Bmatrix} = [\Lambda] \begin{Bmatrix} \xi \\ \eta \end{Bmatrix} = \begin{bmatrix} \lambda_1 & 0 \\ 0 & \lambda_2 \end{bmatrix} \begin{Bmatrix} \xi \\ \eta \end{Bmatrix} \quad (14.3\text{-}10)$$

Since Eq. (14.3-10) has the solution

$$\begin{aligned} \xi &= e^{\lambda_1 t} \\ \eta &= e^{\lambda_2 t} \end{aligned} \quad (14.3\text{-}11)$$

the solution for u and v are

$$\begin{aligned} u &= u_1 e^{\lambda_1 t} + u_2 e^{\lambda_2 t} \\ v &= v_1 e^{\lambda_1 t} + v_2 e^{\lambda_2 t} \end{aligned} \quad (14.3\text{-}12)$$

It is evident, then, that the stability of the singular point depends on the eigenvalues λ_1 and λ_2 determined from the characteristic equation

$$\begin{vmatrix} (a-\lambda) & b \\ c & (e-\lambda) \end{vmatrix} = 0$$

or

$$\lambda_{1,2} = \left(\frac{a+e}{2}\right) \pm \sqrt{\left(\frac{a+e}{2}\right)^2 - (ae - bc)} \qquad (14.3\text{-}13)$$

Thus

If $(ae - bc) > \left(\dfrac{a+e}{2}\right)^2$, the motion is oscillatory;
If $(ae - bc) < \left(\dfrac{a+e}{2}\right)^2$, the motion is aperiodic;
If $(a + e) > 0$, the system is unstable;
If $(a + e) < 0$, the system is stable.

The type of trajectories in the neighborhood of the singular point can be determined by first examining Eq. (14.3-10) in the form

$$\frac{d\xi}{d\eta} = \frac{\lambda_1}{\lambda_2}\frac{\xi}{\eta} \qquad (14.3\text{-}14)$$

which has the solution

$$\xi = (\eta)^{\lambda_1/\lambda_2}$$

and using the transformation of Eq. (14.3-9) to plot v versus u.

14.4 METHOD OF ISOCLINES

Consider the autonomous system with the equation

$$\frac{dy}{dx} = -\frac{f(x, y)}{y} = \phi(x, y) \qquad (14.4\text{-}1)$$

that was discussed in Sec. 14.1, Eq. (14.1-4). In the method of isoclines we fix the slope dy/dx by giving it a definite number α, and solve for the curve

$$\phi(x, y) = \alpha \qquad (14.4\text{-}2)$$

With a family of such curves drawn, it is possible to sketch in a trajectory starting at any point x, y as shown in Fig. 14.4-1.

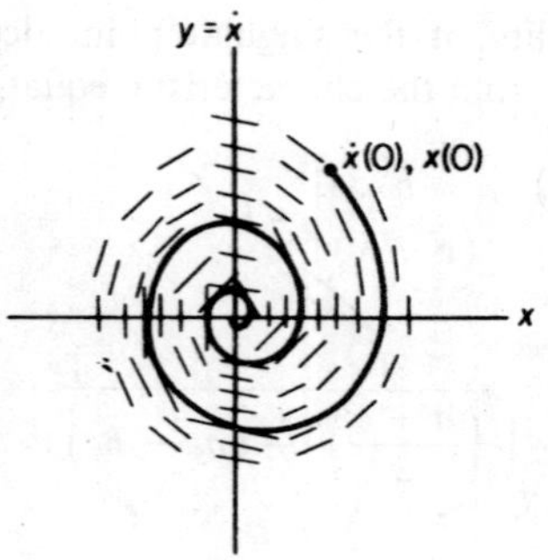

Figure 14.4-1.

Example 14.4-1

Determine the isoclines for the simple pendulum.

Solution: The equation for the simple pendulum is

$$\ddot{\theta} + \frac{g}{l}\sin\theta = 0 \tag{a}$$

Letting $x = \theta$ and $y = \dot{\theta} = \dot{x}$ we obtain

$$\frac{dy}{dx} = -\frac{g}{l}\frac{\sin x}{y} \tag{b}$$

Thus for $dy/dx = \alpha$, a constant, the equation for the isocline, Eq. (14.4-2), becomes

$$y = -\left(\frac{g}{l\alpha}\right)\sin x \tag{c}$$

It is evident from Eq. (b) that the singular points lie along the x-axis at $x = 0$, $\pm\pi$, $\pm 2\pi$, and so on. Figure 14.4-2 shows isoclines in the first quadrant that correspond to negative values of α. Starting at an arbitrary point $x(0)$, $y(0)$, the trajectory can be sketched in by proceeding tangentially to the slope segments.

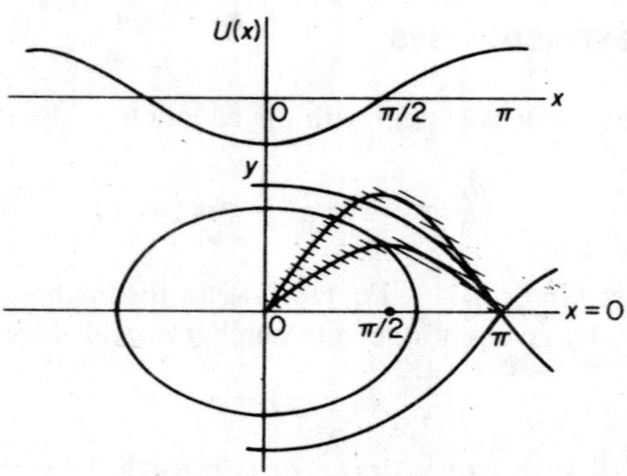

Figure 14.4-2. Isocline curves for the simple pendulum.

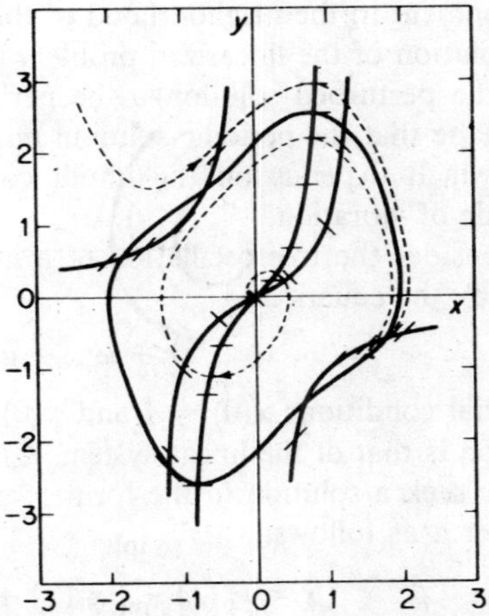

Figure 14.4-3. Isocline curves for van der Pol's equation with $\mu = 1.0$.

$$\frac{y^2}{2} - \frac{g}{l}\cos x = E$$

where E is a constant of integration corresponding to the total energy (see Eq. (14.2-1). We also have $U(x) = -(g/l)\cos x$ and the discussions of Sec. 14.2 apply. For the motion to exist, E must be greater than $-g/l$. $E = g/l$ corresponds to the separatrix and for $E > g/l$ the trajectory does not close. This means that the initial conditions are large enough to cause the pendulum to continue past $\theta = 2\pi$.

Example 14.4-2

One of the interesting nonlinear equations that has been studied extensively is the *van der Pol equation*

$$\ddot{x} - \mu\dot{x}(1 - x^2) + x = 0$$

The equation somewhat resembles that of free vibration of a spring-mass system with viscous damping; however, the damping term of this equation is nonlinear in that it depends on both the velocity and the displacement. For small oscillations ($x < 1$) the damping is negative, and the amplitude will increase with time. For $x > 1$ the damping is positive, and the amplitude will diminish with time. If the system is initiated with $x(0)$ and $\dot{x}(0)$, the amplitude will increase or decrease, depending on whether x is small or large, and it will finally reach a stable state known as the *limit cycle*, graphically displayed by the phase plane plot of Fig. 14.4-3.

14.5 PERTURBATION METHOD

The *perturbation method* is applicable to problems where a small parameter μ is associated with the nonlinear term of the differential equation. The solution is formed in terms of a series of the perturbation parameter μ, the result being a development in the neighborhood of the solution of the linearized problem. If the solution of the linearized problem is periodic, and if μ is small, we can expect the perturbed solution to be periodic also. We can reason from the phase plane that the periodic solution must represent a closed trajectory. The period which depends on the initial conditions is then a function of the amplitude of vibration.

Consider the free oscillation of a mass on a nonlinear spring, which is defined by the equation

$$\ddot{x} + \omega_n^2 x + \mu x^3 = 0 \tag{14.5-1}$$

with initial conditions $x(0) = A$ and $\dot{x}(0) = 0$. When $\mu = 0$, the frequency of oscillation is that of the linear system, $\omega_n = \sqrt{k/m}$.

We seek a solution in the form of an infinite series of the perturbation parameter μ as follows:

$$x = x_0(t) + \mu x_1(t) + \mu^2 x_2(t) + \cdots \tag{14.5-2}$$

Furthermore, we know that the frequency of the nonlinear oscillation will depend on the amplitude of oscillation as well as on μ. We express this fact also in terms of a series in μ.

$$\omega^2 = \omega_n^2 + \mu\alpha_1 + \mu^2\alpha_2 + \cdots \tag{14.5-3}$$

where the α_i are as yet undefined functions of the amplitude, and ω is the frequency of the nonlinear oscillations.

We will consider only the first two terms of Eqs. (14.5-2) and (14.5-3), which will adequately illustrate the procedure. Substituting these into Eq. (14.5-1), we obtain

$$\ddot{x}_0 + \mu\ddot{x}_1 + \left(\omega^2 - \mu\alpha_1\right)\left(x_0 + \mu x_1\right) + \mu\left(x_0^3 + 3\mu x_0^2 x_1 + \cdots\right) = 0 \tag{14.5-4}$$

Since the perturbation parameter μ could have been chosen arbitrarily, the coefficients of the various powers of μ must be equated to zero. This leads to a system of equations that can be solved successively:

$$\begin{aligned} \ddot{x}_0 + \omega^2 x_0 &= 0 \\ \ddot{x}_1 + \omega^2 x_1 &= \alpha_1 x_0 - x_0^3 \end{aligned} \tag{14.5-5}$$

The solution to the first equation, subject to the initial conditions $x(0) = A$, $\dot{x}(0) = 0$, is

$$x_0 = A\cos\omega t \tag{14.5-6}$$

which is called the *generating solution*. Substituting this into the right side of the second equation in Eq. (14.5-5), we obtain

$$\begin{aligned} \ddot{x}_1 + \omega^2 x_1 &= \alpha_1 A\cos\omega t - A^3\cos^3\omega t \\ &= \left(\alpha_1 - \frac{3}{4}A^2\right)A\cos\omega t - \frac{A^3}{4}\cos 3\omega t \end{aligned} \tag{14.5-7}$$

where $\cos^3 \omega t = \frac{3}{4}\cos \omega t + \frac{1}{4}\cos 3\omega t$ has been used. We note here that the forcing term $\cos \omega t$ would lead to a secular term $t \cos \omega t$ in the solution for x_1 (i.e., we have a condition of resonance). Such terms violate the initial stipulation that the motion is to be periodic; hence, we impose the condition

$$\left(\alpha_1 - \tfrac{3}{4}A^2\right) = 0$$

Thus α_1, which we stated earlier to be some function of the amplitude A, is evaluated to be equal to

$$\alpha_1 = \tfrac{3}{4}A^2 \tag{14.5-8}$$

With the forcing term $\cos \omega t$ eliminated from the right side of the equation, the general solution for x_1 is

$$\begin{aligned} x_1 &= C_1 \sin \omega t + C_2 \cos \omega t + \frac{A^3}{32\omega^2}\cos 3\omega t \\ \omega^2 &= \omega_n^2 + \frac{3}{4}\mu A^2 \end{aligned} \tag{14.5-9}$$

Imposing the initial conditions $x_1(0) = \dot{x}_1(0) = 0$, the constants C_1 and C_2 are evaluated as

$$C_1 = 0 \qquad C_2 = -\frac{A^3}{32\omega^2}$$

Thus

$$x_1 = \frac{A^3}{32\omega^2}(\cos 3\omega t - \cos \omega t) \tag{14.5-10}$$

and the solution at this point from Eq. (14.5-2) becomes

$$\begin{aligned} x &= A \cos \omega t + \mu \frac{A^3}{32\omega^2}(\cos 3\omega t - \cos \omega t) \\ \omega &= \omega_n \sqrt{1 + \frac{3}{4}\frac{\mu A^2}{\omega_0^2}} \end{aligned} \tag{14.5-11}$$

The solution is thus found to be periodic, and the fundamental frequency ω is found to increase with the amplitude, as expected for a hardening spring.

Mathieu equation. Consider the nonlinear equation

$$\ddot{x} + \omega_n^2 x + \mu x^3 = F \cos \omega t \tag{a}$$

and assume a perturbation solution

$$x = x_1(t) + \xi(t) \tag{b}$$†

†See Ref. [4], pp. 259–73.

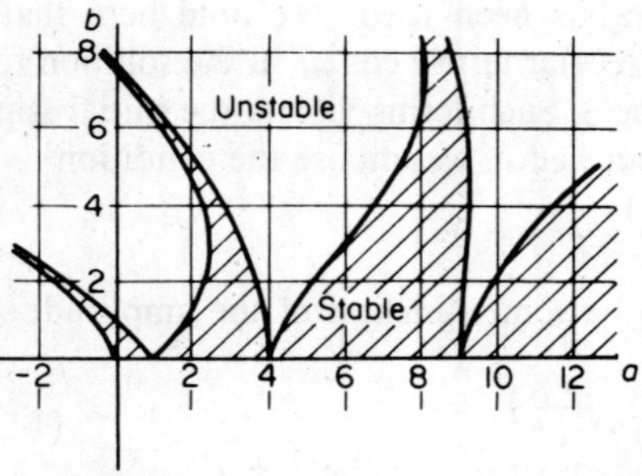

Figure 14.5-1. Stable region of Mathieu equation indicated by shaded area, which is symmetric about the horizontal axis.

Substituting Eq. (b) into (a), we obtain the following two equations:

$$\ddot{x}_1 + \omega_n^2 x_1 + \mu x_1^3 = F\cos\omega t \tag{c}$$

$$\ddot{\xi} + \left(\omega_n^2 + \mu 3x_1^2\right)\xi = 0 \tag{d}$$

If μ is assumed to be small, we can let

$$x_1 \cong A\sin\omega t \tag{e}$$

and substitute it into Eq. (d), which becomes

$$\ddot{\xi} + \left[\left(\omega_n^2 + \frac{3\mu}{2}A^2\right) - \frac{3\mu}{2}A^2\cos 2\omega t\right]\xi = 0 \tag{f}$$

This equation is of the form

$$\frac{d^2y}{dz^2} + (a - 2b\cos 2z)y = 0 \tag{g}$$

which is known as the *Mathieu equation*. The stable and unstable regions of the Mathieu equation depend on the parameters a and b, and are shown in Fig. 14.5-1.

14.6 METHOD OF ITERATION

Duffing† made an exhaustive study of the equation

$$m\ddot{x} + c\dot{x} + kx \pm \mu x^3 = F\cos\omega t$$

which represents a mass on a cubic spring, excited harmonically. The $\pm$ sign signifies a hardening or softening spring. The equation is nonautonomous in that the time t appears explicitly in the forcing term.

In this section we wish to examine a simpler equation where damping is zero, written in the form

$$\ddot{x} + \omega_n^2 x \pm \mu x^3 = F\cos\omega t \tag{14.6-1}$$

†See Ref. [6].

We seek only the steady state harmonic solution by the *method of iteration*, which is essentially a process of *successive approximation*. An assumed solution is substituted into the differential equation, which is integrated to obtain a solution of improved accuracy. The procedure may be repeated any number of times to achieve the desired accuracy.

For the first assumed solution, let

$$x_0 = A\cos\omega t \tag{14.6-2}$$

and substitute into the differential equation

$$\begin{aligned}\ddot{x} &= -\omega_n^2 A\cos\omega t \mp \mu A^3\left(\tfrac{3}{4}\cos\omega t + \tfrac{1}{4}\cos 3\omega t\right) + F\cos\omega t\\ &= \left(-\omega_n^2 A \mp \tfrac{3}{4}\mu A^3 + F\right)\cos\omega t \mp \tfrac{1}{4}\mu A^3\cos 3\omega t\end{aligned}$$

In integrating this equation it is necessary to set the constants of integration to zero if the solution is to be harmonic with period $\tau = 2\pi/\omega$. Thus, we obtain for the improved solution

$$x_1 = \frac{1}{\omega^2}\left(\omega_n^2 A \pm \frac{3}{4}\mu A^3 - F\right)\cos\omega t \mp \cdots \tag{14.6-3}$$

where the higher harmonic term is ignored.

The procedure may be repeated but we will not go any further. Duffing reasoned at this point that if the first and second approximations are reasonable solutions to the problem, then the coefficients of $\cos\omega t$ in the two equations (14.6-2) and (14.6-3) must not differ greatly. Thus, by equating these coefficients we obtain

$$A = \frac{1}{\omega^2}\left(\omega_n^2 A \pm \frac{3}{4}\mu A^3 - F\right) \tag{14.6-4}$$

which may be solved for ω^2

$$\omega^2 = \omega_n^2 \pm \frac{3}{4}\mu A^2 - \frac{F}{A} \tag{14.6-5}$$

It is evident from this equation that if the nonlinear parameter is zero, we obtain the exact result for the linear system

$$A = \frac{F}{\omega_n^2 - \omega^2}$$

For $\mu \neq 0$, the frequency ω is a function of μ, F, and A. It is evident that when $F = 0$, we obtain the frequency equation for free vibration

$$\frac{\omega^2}{\omega_n^2} = 1 \pm \frac{3}{4}\mu\frac{A^2}{\omega_n^2}$$

discussed in the previous section. Here we see that the frequency increases with the amplitude for the hardening spring (+), and decreased for the softening spring (−).

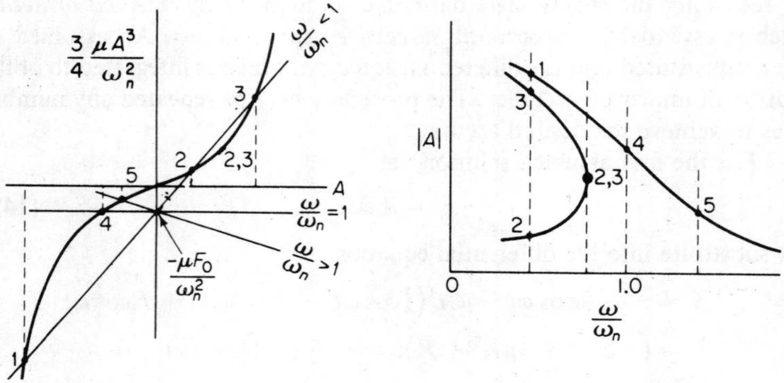

Figure 14.6-1. Solution of Equation 14.6-6.

For $\mu \neq 0$ and $F \neq 0$, it is convenient to hold both μ and F constant and plot $|A|$ against ω/ω_n. In the construction of these curves, it is helpful to rearrange Eq. (14.6-5) to

$$\frac{3}{4}\mu\frac{A^3}{\omega_n^2} = \left(1 - \frac{\omega^2}{\omega_n^2}\right)A - \frac{F}{\omega_n^2} \tag{14.6-6}$$

each side of which can be plotted against A, as shown in Fig. 14.6-1. The left side of this equation is a cubic, whereas the right side is a straight line of slope $(1 - \omega^2/\omega_n^2)$ and intercept $-F/\omega_n^2$. For $\omega/\omega_n < 1$, the two curves intersect at three points 1, 2, 3, which are also shown in the amplitude-frequency plot. As ω/ω_n increases towards unity, points 2 and 3 approach each other; after which only one value of the amplitude will satisfy Eq. (14.6-6). When $\omega/\omega_n = 1$, or when $\omega/\omega_n > 1$, these points are 4 or 5.

The jump phenomenon. In problems of this type, it is found that the amplitude A undergoes a sudden discontinuous jump near resonance. The *jump phenomenon* can be described as follows. For the softening spring, with increasing frequency of excitation, the amplitude gradually increases until point "a" in Fig. 14.6-2 is reached. It then suddenly jumps to a larger value indicated by the point b, and diminishes along the curve to its right. In decreasing the frequency from some point c, the amplitude continues to increase beyond b to point d, and suddenly drops to a smaller value at e. The shaded region in the amplitude-frequency plot is unstable; the extent of unstableness depends on a number of factors such as the amount of damping present, the rate of change of the exciting frequency, and so on. If a hardening spring had been chosen instead of the softening spring, the same type of analysis would be applicable and the result would be a curve of the type shown in Fig. 14.6-3.

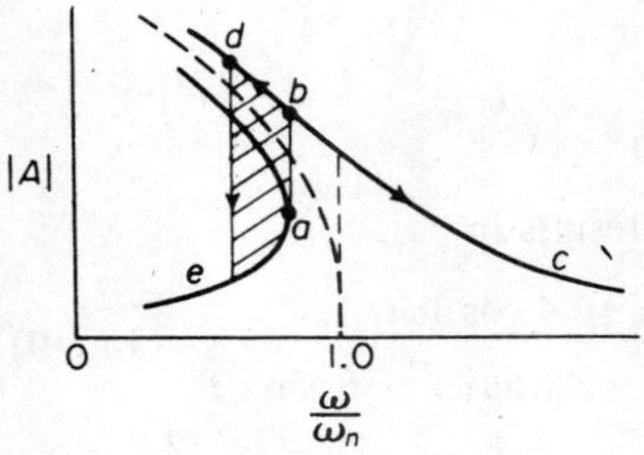

Figure 14.6-2. The jump phenomenon for the softening spring.

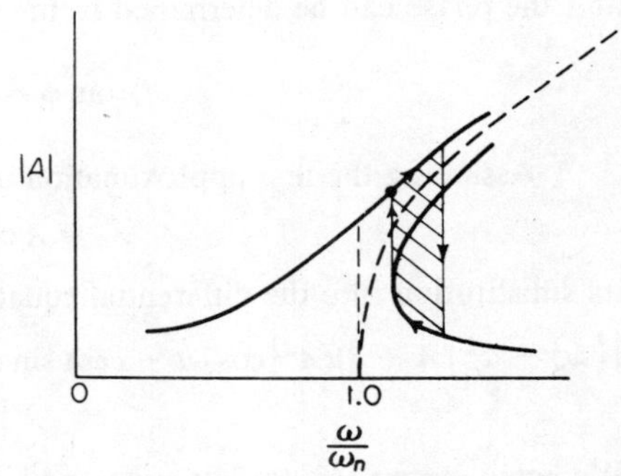

Figure 14.6-3. The jump phenomenon for the hardening spring.

Effect of damping. In the undamped case the amplitude-frequency curves approach the backbone curve (shown dotted) asymptotically. This is also the case for the linear system where the backbone curve is the vertical line at $\omega/\omega_n = 1.0$.

With a small amount of damping present, the behavior of the system cannot differ appreciably from that of the undamped system. The upper end of the curve, instead of approaching the backbone curve asymptotically, will cross in a continuous curve as shown in Fig. 14.6-4. The jump phenomenon is also present here but damping generally tends to reduce the size of the unstable region.

The method of successive approximation is also applicable to the damped vibration case. The major difference in its treatment lies in the phase angle between the force and the displacement, which is no longer 0° or 180° as in the undamped problem. It is found that by introducing the phase in the force term rather than the displacement, the algebraic work is somewhat simplified. The differential equation can then be written as

$$\begin{aligned} \ddot{x} + c\dot{x} + \omega_n^2 x + \mu x^3 &= F\cos(\omega t + \phi) \\ &= A_0 \cos \omega t - B_0 \sin \omega t \end{aligned} \tag{14.6-7}$$

where the magnitude of the force is

$$F = \sqrt{A_0^2 + B_0^2} \tag{14.6-8}$$

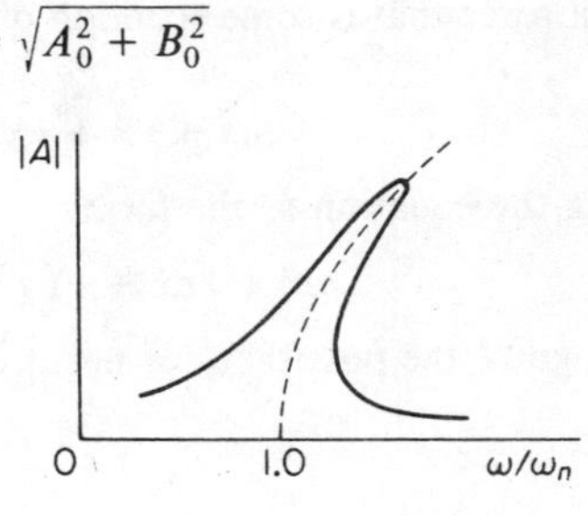

Figure 14.6-4.

and the phase can be determined from

$$\tan\phi = \frac{B_0}{A_0}$$

Assuming the first approximation to be

$$x_0 = A\cos\omega t$$

its substitution into the differential equation results in

$$\left[\left(\omega_n^2 - \omega^2\right)A + \tfrac{3}{4}\mu A^3\right]\cos\omega t - c\omega A\sin\omega t + \tfrac{1}{4}\mu A^3\cos 3\omega t = A_0\cos\omega t - B_0\sin\omega t \tag{14.6-9}$$

We again ignore the $\cos 3\omega t$ term and equate coefficients of $\cos\omega t$ and $\sin\omega t$ to obtain

$$\begin{aligned}\left(\omega_n^2 - \omega^2\right)A + \tfrac{3}{4}\mu A^3 &= A_0\\ c\omega A &= B_0\end{aligned} \tag{14.6-10}$$

Squaring and adding these results, the relationship between the frequency, amplitude, and force becomes

$$F^2 = \left[\left(\omega_n^2 - \omega^2\right)A + \tfrac{3}{4}A^3\right]^2 + \left[c\omega A\right]^2 \tag{14.6-11}$$

By fixing μ, c, and F, the frequency ratio ω/ω_n can be computed for assigned values of A.

14.7 SELF-EXCITED OSCILLATIONS

Oscillations that depend on the motion itself are called self-excited. The shimmy of automobile wheels, the flutter of airplane wings, and the oscillations of the van der Pol equation are some examples.

Self-excited oscillations may occur in a linear or a nonlinear system. The motion is induced by an excitation that is some function of the velocity or of the velocity and the displacement. If the motion of the system tends to increase the energy of the system, the amplitude will increase, and the system may become unstable.

As an example, consider a viscously damped single-DOF linear system excited by a force that is some function of the velocity. Its equation of motion is

$$m\ddot{x} + c\dot{x} + kx = F(\dot{x}) \tag{14.7-1}$$

Rearranging the equation to the form

$$m\ddot{x} + \left(c\dot{x} - F(\dot{x})\right) + kx = 0 \tag{14.7-2}$$

we can recognize the possibility of negative damping if $F(\dot{x})$ becomes greater than $c\dot{x}$.

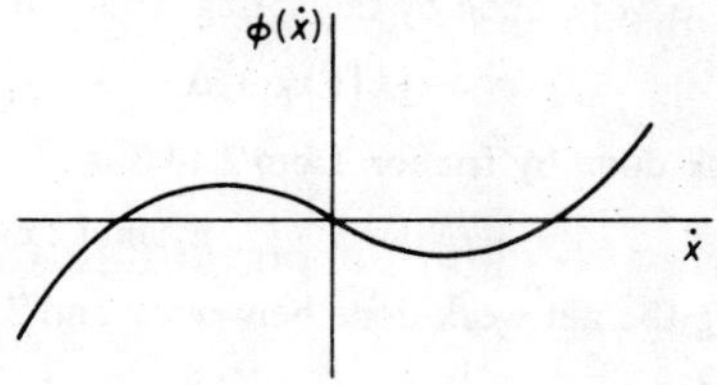

Figure 14.7-1. System with apparent damping $\phi(\dot{x}) = c\dot{x} - F(\dot{x})$.

Suppose that $\phi(\dot{x}) = c\dot{x} - F(\dot{x})$ in the preceding equations varies as in Fig. 14.7-1. For small velocities the apparent damping $\phi(\dot{x})$ is negative, and the amplitude of oscillation will increase. For large velocities the opposite is true, and hence the oscillations will tend to a limit cycle.

Example 14.7-1

The coefficient of kinetic friction μ_k is generally less than the coefficient of static friction μ_s, this difference increasing somewhat with the velocity. Thus if the belt of Fig. 14.7-2 is started, the mass will move with the belt until the spring force is balanced by the static friction.

$$kx_0 = \mu_s mg \tag{a}$$

At this point the mass will start to move back to the left, and the forces will again be balanced on the basis of kinetic friction when

$$k(x_0 - x) = \mu_{kl} mg$$

From these two equations, the amplitude of oscillation is found to be

$$x = x_0 - \mu_{kl}\frac{mg}{k} = \frac{(\mu_s - \mu_{kl})g}{\omega_n^2} \tag{b}$$

While the mass is moving to the left, the relative velocity between it and the belt is greater than when it is moving to the right, thus μ_{kl} is less than μ_{kr} where the subscripts l and r refer to left and right. It is evident then that the work done by the friction force while moving to the right is greater than that while moving to the left, so that more energy is put into the spring-mass system than taken out. This then represents one type of self-excited oscillation and the amplitude will continue to increase.

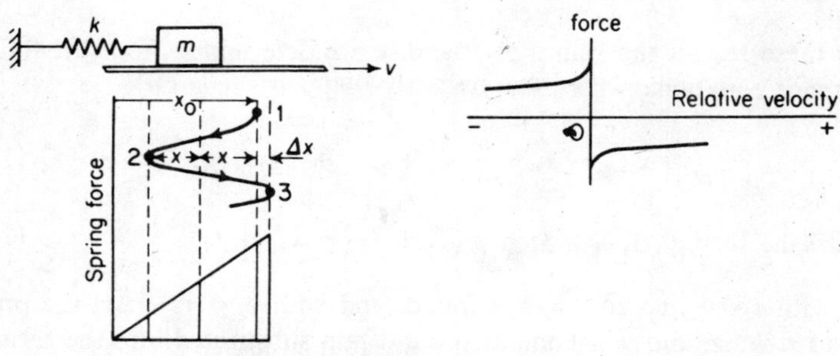

Figure 14.7-2. Coulomb friction between belt and mass.

The work done by the spring from 2 to 3 is

$$-\tfrac{1}{2}k[(x_0 + \Delta x) + (x_0 - 2x)](2x + \Delta x)$$

The work done by friction from 2 to 3 is

$$\mu_{kr} mg(2x + \Delta x)$$

Equating the net work done between 2 and 3 to the change in kinetic energy, which is zero,

$$-\tfrac{1}{2}k(2x_0 - 2x + \Delta x) + \mu_{kr} mg = 0 \tag{c}$$

Substituting (a) and (b) into (c), the increase in amplitude per cycle of oscillation is found to be

$$\Delta x = \frac{2g(\mu_{kr} - \mu_{kl})}{\omega_n^2} \tag{d}$$

14.8 RUNGE-KUTTA METHOD

The Runge-Kutta method discussed in Chapter 4 can be used to solve nonlinear differential equations. We consider the nonlinear equation

$$\frac{d^2x}{d\tau^2} + 0.4\frac{dx}{d\tau} + x + 0.5x^3 = 0.5\cos(0.5\tau) \tag{14.8-1}$$

and rewrite it in the first order form by letting $y = dx/d\tau$ as follows:

$$\frac{dy}{d\tau} = 0.5\cos(0.5\tau) - x - 0.5x^3 - 0.4y = F(\tau, x, y)$$

The computational equations to be used are programmed for the digital computer in the following order:

τ	x	y	F
$t_1 = \tau_1$	$k_1 = x_1$	$g_1 = y_1$	$f_1 = F(t_1, k_1, g_1)$
$t_2 = \tau_1 + h/2$	$k_2 = x_1 + g_1 h/2$	$g_2 = y_1 + f_1 h/2$	$f_2 = F(t_2, k_2, g_2)$
$t_3 = \tau_1 + h/2$	$k_3 = x_1 + g_1 h/2$	$g_3 = y_1 + f_2 h/2$	$f_3 = F(t_3, k_3, g_3)$
$t_4 = \tau + h$	$k_4 = x_1 + g_3 h$	$g_4 = y_1 + f_3 h$	$f_4 = F(t_4, k_4, g_4)$

From these results the values of x and y are determined from the following recurrence equations, where $h = \Delta t$.

$$x_{i+1} = x_i + \frac{h}{6}(g_1 + 2g_2 + 2g_3 + g_4) \tag{14.8-2}$$

$$y_{i+1} = y_i + \frac{h}{6}(f_1 + 2f_2 + 2f_3 + f_4) \tag{14.8-3}$$

Thus with $i = 1$, x_2 and y_2 are found, and with $\tau_2 = \tau_1 + \Delta\tau$, the previous table of t, k, g, and f is computed and again substituted into the recurrence equations to find x_3 and y_3.

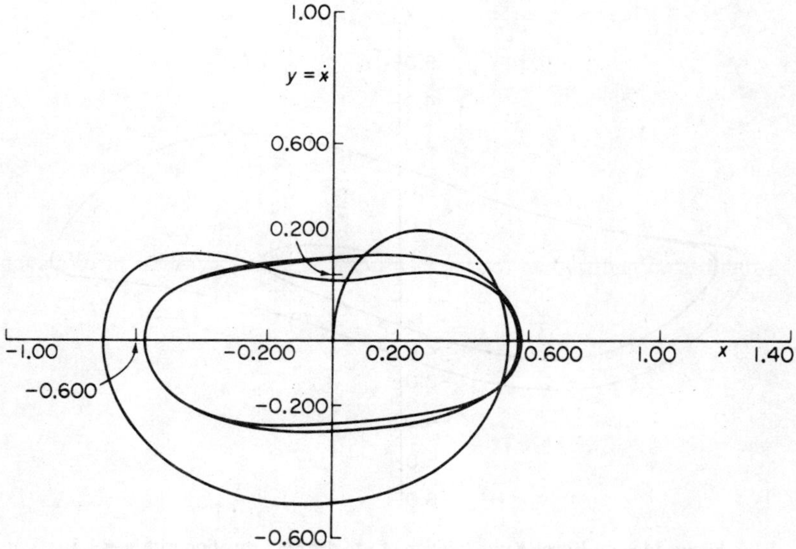

Figure 14.8-1. Runge-Kutta solution for nonlinear differential equation 14.8-1.

The error in the Runge-Kutta method is of order $h^5 = (\Delta\tau)^5$. Also, the method avoids the necessity of calculating derivatives and hence excellent accuracy is obtained.

Equation (14.8-1) was solved on the digital computer with the Runge-Kutta program and with $h = \Delta\tau = 0.1333$. The results were plotted out by the machine for the phase plane plot y vs. x in Fig. 14.8-1. It is evident that the limit cycle was reached in less than two cycles.

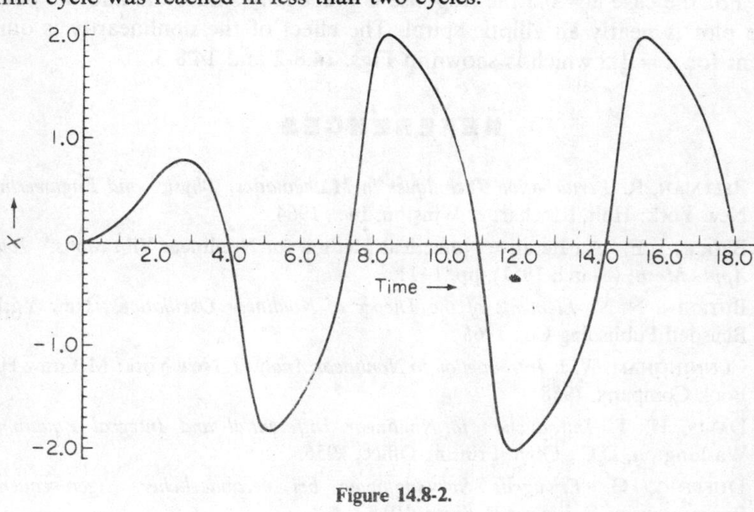

Figure 14.8-2.

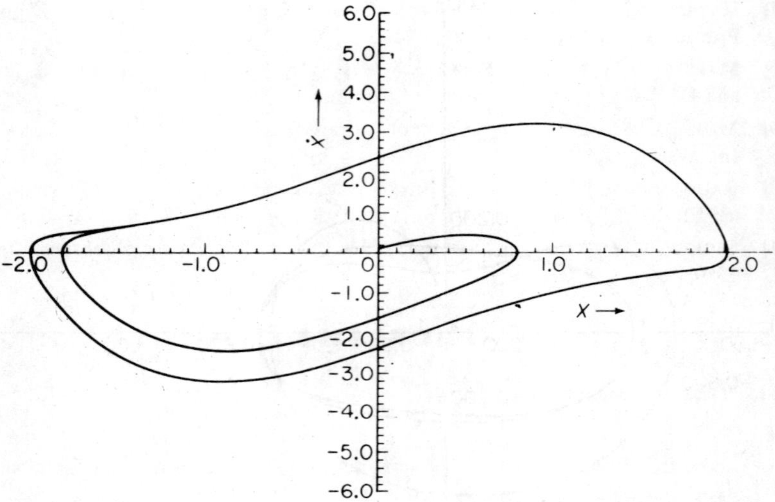

Figure 14.8-3. Runge-Kutta solution of van der Pol's equation with $\mu = 1.5$.

Using the digital computer, the van der Pol equation

$$\ddot{x} - \mu\dot{x}(1 - x^2) + x = 0$$

was solved by the Runge-Kutta method for $\mu = 0.2$, 0.7, 1.5, 3 and 4 with a small initial displacement. Both the phase plane and the time plots were automatically plotted.

For the case $\mu = 0.2$ the response is practically sinusoidal and the phase plane plot is nearly an elliptic spiral. The effect of the nonlinearity is quite evident for $\mu = 1.5$ which is shown in Figs. 14.8-2 and 14.8-3.

REFERENCES

[1] BELLMAN, R. *Perturbation Techniques in Mathematics, Physics and Engineering.* New York: Holt, Rinehart & Winston, Inc., 1964.

[2] BROCK, J. E. "An Iterative Numerical Method for Nonlinear Vibrations," *Jour. Appl. Mech.* (March 1951), pp. 1–11.

[3] BUTENIN, N. V. *Elements of the Theory of Nonlinear Oscillations.* New York: Blaisdell Publishing Co., 1965.

[4] CUNNINGHAM, W. J. *Introduction to Nonlinear Analysis.* New York: McGraw-Hill Book Company, 1958.

[5] DAVIS, H. T. *Introduction to Nonlinear Differential and Integral Equations.* Washington, D.C.: Govt. Printing Office, 1956.

[6] DUFFING, G. *Erwugene Schwingungen bei veranderlicher Eigenfrequenz.* Braunschweig: F. Vieweg u. Sohn, 1918.

[7] HAYASHI, C. *Forced Oscillations in Nonlinear Systems*. Osaka, Japan: Nippon Printing & Publishing Co., Ltd., 1953.

[8] MALKIN, I. G. *Some Problems in the Theory of Nonlinear Oscillations*, Books I and II. Washington, D.C.: Dept. of Commerce, 1959.

[9] MINORSKY, N. *Nonlinear Oscillations*. Princeton, N.J.: D. Van Nostrand Co., Inc., 1962.

[10] RAUSCHER, M. "Steady Oscillations of Systems with Nonlinear and Unsymmetrical Elasticity," *Jour. Appl. Mech.* (Dec. 1938), pp. A169–77.

[11] STOKER, J. J. *Nonlinear Vibrations*. New York: Interscience Publishers, Inc., 1950.

PROBLEMS

14-1 Using the nonlinear equation

$$\ddot{x} + x^3 = 0$$

show that if x_1 and x_2 are solutions satisfying the differential equation, their superposition $(x_1 + x_2)$ is not a solution.

14-2 A mass is attached to the midpoint of a string of length $2l$, as shown in Fig. P14-2. Determine the differential equation of motion for large deflection. Assume string tension to be T.

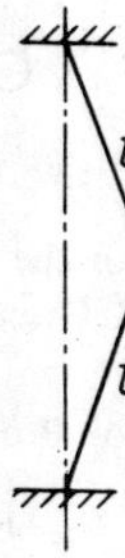

Figure P14-2.

14-3 A buoy is composed of two cones of diameter $2r$ and height h, as shown in Fig. P14-3. A weight attached to the bottom allows it to float in the equilibrium position x_0. Establish the differential equation of motion for vertical oscillation.

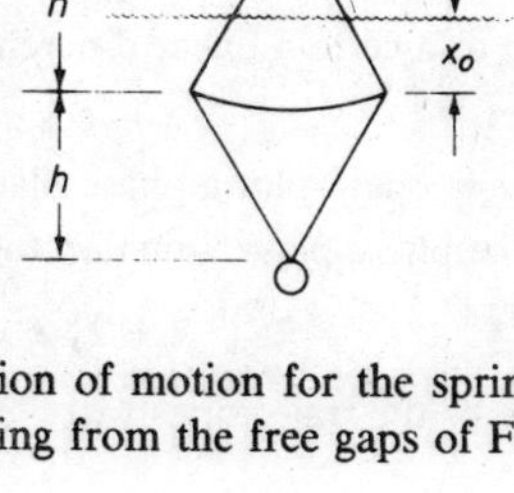

Figure P14-3.

14-4 Determine the differential equation of motion for the spring-mass system with the discontinuous stiffness resulting from the free gaps of Fig. P14-4.

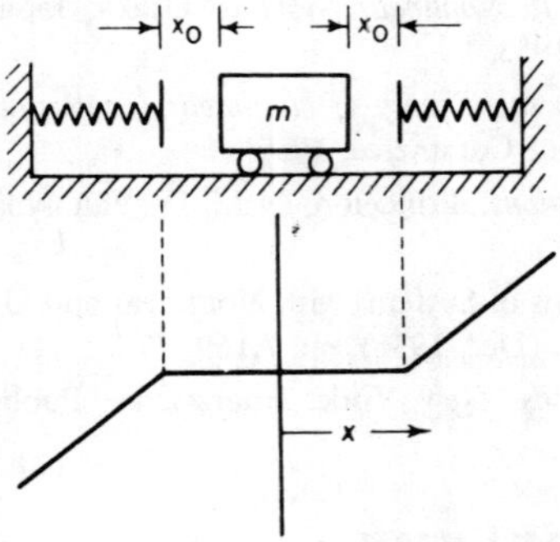

Figure P14-4.

14-5 The cord of a simple pendulum is wrapped around a fixed cylinder of radius R such that its length is l when in the vertical position as shown in Fig. P14-5. Determine the differential equation of motion.

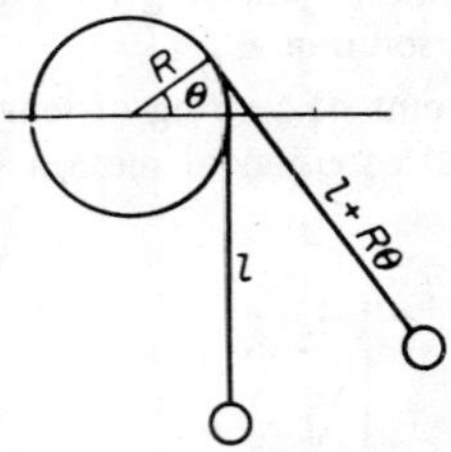

Figure P14-5.

14-6 Plot the phase plane trajectory for the undamped spring-mass system including the potential energy curve $U(x)$. Discuss the initial conditions associated with the plot.

14-7 From the plot of $U(x)$ vs. x of Prob. 14-6, determine the period from the equation

$$\tau = 4\int_0^{x_{\max}} \frac{dx}{\sqrt{2[E - U(x)]}}$$

(Remember that E in the text was for a unit mass.)

14-8 For the undamped spring-mass system with initial conditions $x(0) = A$ and $\dot{x}(0) = 0$ determine the equation for the state speed V and state under what condition the system is in equilibrium.

14-9 The solution to a certain linear differential equation is given as

$$x = \cos \pi t + \sin 2\pi t$$

Determine $y = \dot{x}$ and plot a phase plane diagram.

14-10 Determine the phase plane equation for the damped spring-mass system

$$\ddot{x} + 2\zeta\omega_n\dot{x} + \omega_n^2 x = 0$$

and plot one of the trajectories witl $= y/\omega_n$ and x as coordinates.

14-11 If the potential energy of a simple pendulum is given with the positive sign

$$U(\theta) = +\frac{g}{l}\cos\theta$$

determine which of the singular points are stable or unstable and explain their physical implications. Compare the phase plane with Fig. 14.4-2.

14-12 Given the potential $U(x) = 8 - 2\cos \pi x/4$, plot the phase plane trajectories for $E = 6, 7, 8, 10, 12$, and discuss the curves.

14-13 Determine the eigenvalues and eigenvectors of the equations

$$\dot{x} = 5x - y$$
$$\dot{y} = 2x + 2y$$

14-14 Determine the modal transformation of the equations of Prob. 14-13, which will decouple them to the form

$$\dot{\xi} = \lambda_1 \xi$$
$$\dot{\eta} = \lambda_2 \eta$$

14-15 Plot the ξ, η phase plane trajectories of Prob. 14-14 for $\lambda_1/\lambda_2 = 0.5$ and 2.0.

14-16 For $\lambda_1/\lambda_2 = 2.0$ in Prob. 14-15, plot the trajectory y versus x.

14-17 If λ_1 and λ_2 of Prob. 14-14 are complex conjugates $-\alpha \pm i\beta$, show that the equation in the u, v plane becomes

$$\frac{dv}{du} = \frac{\beta u + \alpha v}{\alpha u - \beta v}$$

14-18 Using the transformation $u = \rho\cos\theta$ and $v = \rho\sin\theta$ show that the phase plane equation for Prob. 14-17 becomes

$$\frac{d\rho}{\rho} = \frac{\alpha}{\beta}\,d\theta$$

with the trajectories identified as logarithmic spirals

$$\rho = e^{(\alpha/\beta)\theta}$$

14-19 Near a singular point in the xy plane, the trajectories appear as shown in Fig. P14-19. Determine the form of the phase plane equation and the corresponding trajectories in the $\xi\eta$-plane.

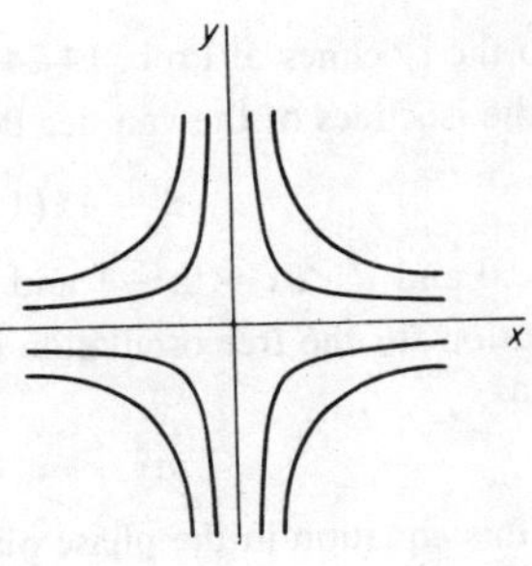

Figure P14-19.

14-20 The phase plane trajectories in the vicinity of a singularity of an over-damped system ($\zeta > 1$) are shown in Fig. P14-20. Identify the phase plane equation and plot the corresponding trajectories in the $\xi\eta$- plane.

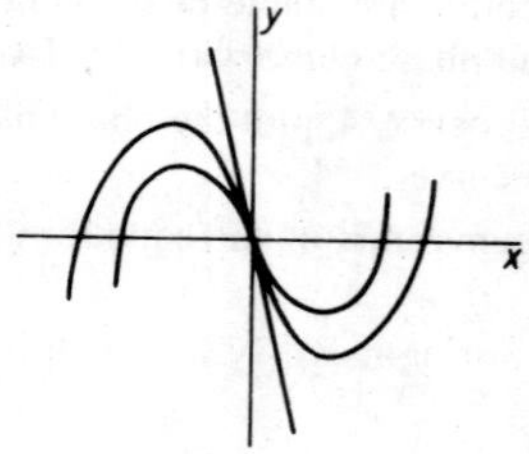

Figure P14-20.

14-21 Show that the solution of the equation

$$\frac{dy}{dx} = \frac{-x - y}{x + 3y}$$

is $x^2 + 2xy + 3y^2 = C$, which is a family of ellipses with axes rotated from the x, y coordinates. Determine the rotation of the semimajor axis and plot one of the ellipses.

14-22 Show that the isoclines of the linear differential equation of second order are straight lines.

14-23 Draw the isoclines for the equation

$$\frac{dy}{dx} = xy(y - 2)$$

14-24 Consider the nonlinear equation

$$\ddot{x} + \omega_n^2 x + \mu x^3 = 0$$

Replacing $\ddot{x}$ by $y(dy/dx)$, where $y = \dot{x}$, gives the integral

$$y^2 + \omega_n^2 x^2 + \tfrac{1}{2}\mu x^4 = 2E$$

With $y = 0$ when $x = A$, show that the period is available from

$$\tau = 4\int_0^A \frac{dx}{\sqrt{2[E - U(x)]}}$$

14-25 What do the isoclines of Prob. 14-24 look like?

14-26 Plot of the isoclines of the van der Pol's equation

$$\ddot{x} - \mu\dot{x}(1 - x^2) + x = 0$$

for $\mu = 2.0$ and $dy/dx = 0, -1$ and $+1$.

14-27 The equation for the free oscillation of a damped system with hardening spring is given as

$$m\ddot{x} + c\dot{x} + kx + \mu x^3 = 0$$

Express this equation in the phase plane form.

14-28 The following numerical values are given for the equation in Prob. 14-27

$$\omega_n^2 = \frac{k}{m} = 25, \qquad \frac{c}{m} = 2\zeta\omega_n = 2.0, \qquad \frac{\mu}{m} = 5$$

Plot the phase trajectory for the initial conditions $x(0) = 4.0$, $\dot{x}(0) = 0$.

14-29 Plot the phase plane trajectory for the simple pendulum with the initial conditions $\theta(0) = 60°$ anc $\dot{\theta}(0) = 0$.

14-30 Determine the period of the pendulum of Prob. 14-29 and compare with that of the linear system.

14-31 The equation of motion for a spring-mass system with constant Coulomb damping can be written as

$$\ddot{x} + \omega_n^2 x + C\,\mathrm{sgn}(\dot{x}) = 0$$

where $sgn(\dot{x})$ signifies either a positive or negative sign equal to that of the sign of $\dot{x}$. Express this equation in a form suitable for the phase plane.

14-32 A system with Coulomb damping has the following numerical values: $k = 3.60$ lb/in., $m = 0.10$ lb s^2 in.$^{-1}$, $\mu = 0.20$. Using the phase plane, plot the trajectory for $x(0) = 20''$, $\dot{x}(0) = 0$.

14-33 Consider the motion of a simple pendulum with viscous damping and determine the singular points. With the aid of Fig. 14.4-2, and the knowledge that the trajectories must spiral into the origin, sketch some approximate trajectories.

14-34 Apply the perturbation method to the simple pendulum with $\sin\theta$ replaced by $\theta - \frac{1}{6}\theta^3$. Use only the first two terms of the series for x and ω.

14-35 From the perturbation method, what is the equation for the period of the simple pendulum as a function of the amplitude?

14-36 For a given system the numerical values of Eq. (14.7-7) are given as

$$\ddot{x} + 0.15\dot{x} + 10x + x^3 = 5\cos(\omega t + \phi)$$

Plot A versus ω from Eq. (14.7-11) by first assuming a value of A and solving for ω^2.

14-37 Determine the phase angle ϕ versus ω for Prob. 14-36.

14-38 The supporting end of a simple pendulum is given a motion, as shown in Fig. P14-38. Show that the equation of motion is

$$\ddot{\theta} + \left(\frac{g}{l} - \frac{\omega^2 y_0}{l}\cos 2\omega t\right)\sin\theta = 0$$

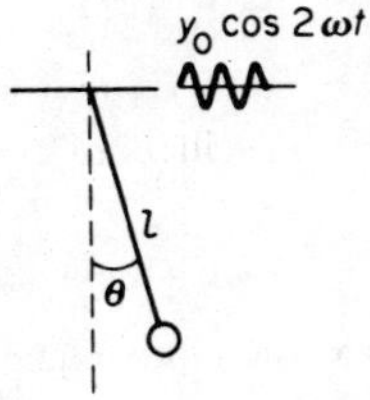

Figure P14-38.

14-39 For a given value of g/l, determine the frequencies of the excitation for which the simple pendulum of Prob. 14-38 with a stiff arm l will be stable in the inverted position.

14-40 Determine the perturbation solution for the system shown in Fig. P14-40 leading to a Mathieu equation. Use initial conditions $\dot{x}(0) = 0, x(0) = A$.

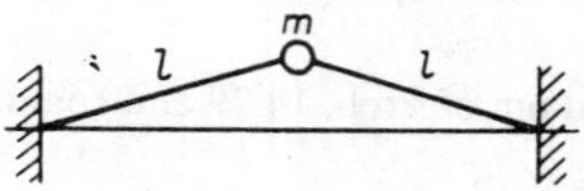

Figure P14-40.

14-41 Using the Runge-Kutta routine and $g/l = 1.0$, calculate the angle θ for the simple pendulum of Prob. 14-29.

14-42 With damping added to Prob. 14-41, the equation of motion is given as

$$\ddot{\theta} + 0.30\dot{\theta} + \sin\theta = 0$$

Solve by the Runge-Kutta method for the initial conditions $\theta(0) = 60°, \dot{\theta}(0) = 0$.

14-43 Obtain a numerical solution for the system of Prob. 14-40 by using (a) the central difference method and (b) the Runge-Kutta method.

A

SPECIFICATIONS OF VIBRATION BOUNDS

Specifications for vibrations are often based on harmonic motion.

$$x = x_0 \sin \omega t$$

The velocity and acceleration are then available from differentiation and the following relationships for the peak values can be written.

$$\dot{x}_0 = 2\pi f x_0$$
$$\ddot{x}_0 = -4\pi^2 f^2 x_0 = -2\pi f \dot{x}_0$$

These equations can be represented on the log-log paper by rewriting them in the form

$$\ln \dot{x}_0 = \ln x_0 + \ln 2\pi f$$
$$\ln \dot{x}_0 = -\ln \ddot{x}_0 - \ln 2\pi f$$

By letting x_0 = constant, the plot of $\ln \dot{x}_0$ against $\ln 2\pi f$ is a straight line of slope equal to +1. By letting $\ddot{x}_0$ = constant, the plot of $\ln \dot{x}_0$ versus $\ln 2\pi f$ is again a straight line of slope −1. These lines are shown graphically in Fig. A-1. The graph is often used to specify bounds for the vibration. Shown in heavy lines are bounds for a maximum acceleration of 10 g, minimum and maximum frequencies of 5 and 500 cps., and an upper limit for the displacement of 0.30 in.

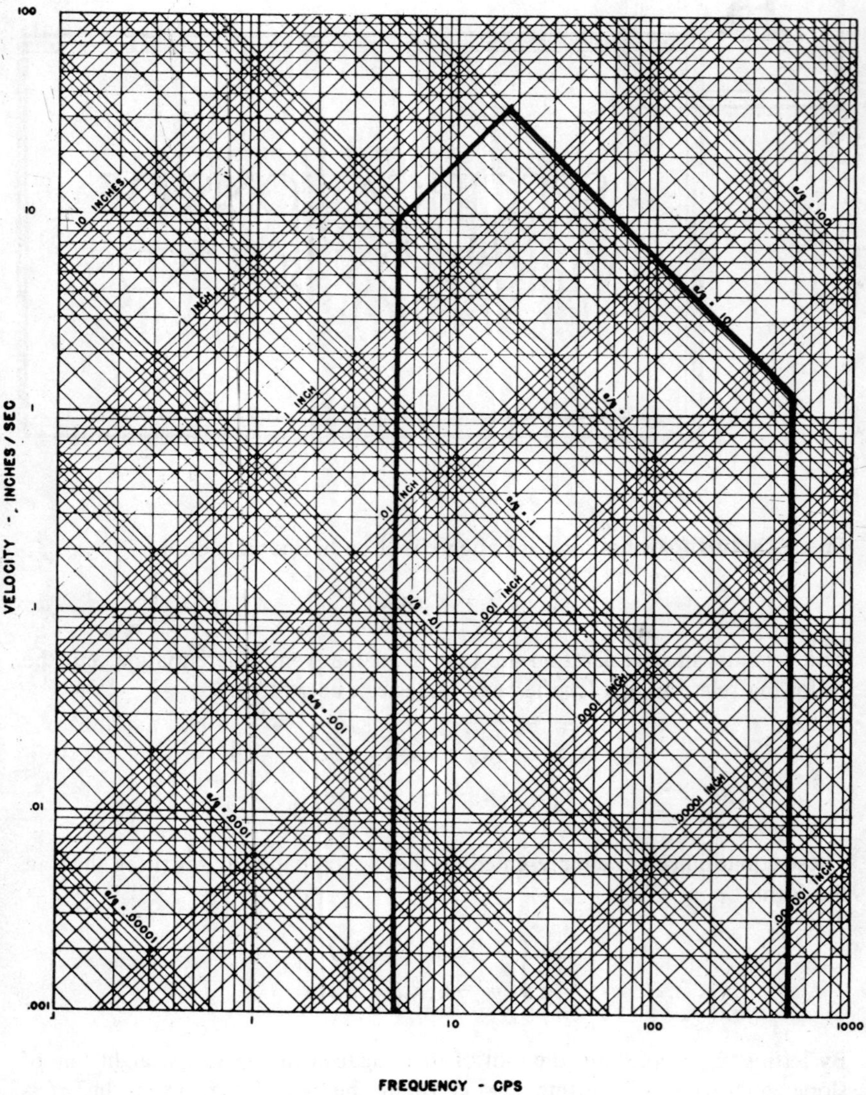
100
10
1
.1
.01
.001
VELOCITY - INCHES / SEC
10 INCHES
1 INCH
.1 INCH
.01 INCH
.001 INCH
.0001 INCH
.00001 INCH
.000001 INCH
a/g = 100
a/g = 10
a/g = 1
a/g = .1
a/g = .01
a/g = .001
a/g = .0001
a/g = .00001
.1
1
10
100
1000
FREQUENCY - CPS

Figure A-1.

B

INTRODUCTION TO LAPLACE TRANSFORMATION

Definition

If $f(t)$ is a known function of t for values of $t > 0$, its Laplace transform (LT) $\bar{f}(s)$ is defined by the equation

$$\bar{f}(s) = \int_0^\infty e^{-st} f(t)\, dt = \mathfrak{L} f(t) \tag{1}$$

where s is a complex variable. The integral exists for the real part of $s > 0$ provided $f(t)$ is an absolutely integrable function of t in the time interval 0 to ∞.

Example 1

Let $f(t)$ be a constant c for $t > 0$. Its LT is

$$\mathfrak{L} c = \int_0^\infty c e^{-st}\, dt = -\left.\frac{c e^{-st}}{s}\right|_0^\infty = \frac{c}{s}$$

which exists for $R(s) > 0$.

Example 2

Let $f(t) = t$. Its LT is found by integration by parts, letting

$$u = t \qquad du = dt$$

$$dv = e^{-st}\, dt \qquad v = -\frac{e^{-st}}{s}$$

The result is

$$\mathfrak{L} t = -\left.\frac{t e^{-st}}{s}\right|_0^\infty + \frac{1}{s}\int_0^\infty e^{-st}\, dt = \frac{1}{s^2} \qquad R(s) > 0$$

LT of derivatives. If $\mathfrak{L}f(t) = \bar{f}(s)$ exists, where $f(t)$ is continuous, then $f(t)$ tends to $f(0)$ as $t \to 0$ and the LT of its derivative $f'(t) = df(t)/dt$ is equal to

$$\mathfrak{L}f'(t) = s\bar{f}(s) - f(0) \tag{2}$$

The above relation is found by integration by parts

$$\int_0^\infty e^{-st}f'(t)\,dt = e^{-st}f(t)\Big|_0^\infty + s\int_0^\infty e^{-st}f(t)\,dt$$

Similarly the LT of the second derivative can be shown to be

$$\mathfrak{L}f''(t) = s^2\bar{f}(s) - sf(0) - f'(0) \tag{3}$$

Shifting Theorem

Consider the LT of the function $e^{at}x(t)$.

$$\mathfrak{L}e^{at}x(t) = \int_0^\infty e^{-st}\left[e^{at}x(t)\right]dt = \int_0^\infty e^{-(s-a)t}x(t)\,dt$$

We conclude from this expression that

$$\mathfrak{L}e^{at}x(t) = \bar{x}(s-a) \tag{4}$$

where $\mathfrak{L}x(t) = \bar{x}(s)$. Thus, the multiplication of $x(t)$ by e^{at} shifts the transform by a, where a may be any number, real or complex.

Transformation of Ordinary Differential Equations

Consider the differential equation

$$m\ddot{x} + c\dot{x} + kx = F(t) \tag{5}$$

Its LT is

$$m\left[s^2\bar{x}(s) - sx(0) - \dot{x}(0)\right] + c\left[s\bar{x}(s) - x(0)\right] + k\bar{x}(s) = \bar{F}(s)$$

which can be rearranged to

$$\bar{x}(s) = \frac{\bar{F}(s)}{ms^2 + cs + k} + \frac{(ms + c)x(0) + m\dot{x}(0)}{ms^2 + cs + k} \tag{6}$$

The last equation is called the subsidiary equation of the differential equation. The response $x(t)$ is found from the inverse transformation, the first term representing the forced response and the second term the response due to the initial conditions.

For the more general case, the subsidiary equation can be written in the form

$$\bar{x}(s) = \frac{A(s)}{B(s)} \tag{7}$$

where $A(s)$ and $B(s)$ are polynomials. $B(s)$ is in general of higher order than $A(s)$.

Transforms Having Simple Poles

Considering the subsidiary equation

$$\bar{x}(s) = \frac{A(s)}{B(s)}$$

we examine the case where $B(s)$ is factorable in terms of n roots a_k which are distinct (simple poles).

$$B(s) = (s - a_1)(s - a_2)\cdots(s - a_n)$$

The subsidiary equation can then be expanded in the following partial fractions:

$$\bar{x}(s) = \frac{A(s)}{B(s)} = \frac{C_1}{s - a_1} + \frac{C_2}{s - a_2} + \cdots + \frac{C_n}{s - a_n} \tag{8}$$

To determine the constants C_k, we multiply both sides of the preceding equation by $(s - a_k)$ and let $s = a_k$. Every term on the right will then be zero except C_k and we arrive at the result

$$C_k = \lim_{s \to a_k} (s - a_k)\frac{A(s)}{B(s)} \tag{9}$$

Since $\mathfrak{L}^{-1}C_k/(s - a_k) = C_k e^{a_k t}$, the inverse transform of $\bar{x}(s)$ becomes

$$x(t) = \sum_{k=1}^{n} \lim_{s \to a_k} (s - a_k)\frac{A(s)}{B(s)} e^{a_k t} \tag{10}$$

Another expression for the last equation becomes apparent by noting that

$$B(s) = (s - a_k)B_1(s)$$
$$B'(s) = (s - a_k)B_1'(s) + B_1(s)$$
$$\lim_{s \to a_k} B'(s) = B_1(a_k)$$

Since $(s - a_k)A(s)/B(s) = A(s)/B_1(s)$, it is evident that

$$x(t) = \sum_{k=1}^{n} \frac{A(a_k)}{B'(a_k)} e^{a_k t} \tag{11}$$

Transforms Having Poles of Higher Order

If in the subsidary equation

$$\bar{x}(s) = \frac{A(s)}{B(s)}$$

a factor in $B(s)$ is repeated m times, we say that $\bar{x}(s)$ has an mth order pole. Assuming that there is an mth order pole at a_1, $B(s)$ will have the form

$$B(s) = (s - a_1)^m (s - a_2)(s - a_3)\cdots$$

The partial fraction expansion of $\bar{x}(s)$ then becomes

$$\bar{x}(s) = \frac{C_{11}}{(s - a_1)^m} + \frac{C_{12}}{(s - a_1)^{m-1}} + \cdots + \frac{C_{1m}}{(s - a_1)} + \frac{C_2}{(s - a_2)} + \frac{C_3}{(s - a_3)} + \cdots \tag{12}$$

The coefficient C_{11} is determined by multiplying both sides of the equation by $(s - a_1)^m$ and letting $s = a_1$

$$(s - a_1)^m \bar{x}(s) = C_{11} + (s - a_1)C_{12} + \cdots + (s - a_1)^{m-1} C_{1m} + \frac{(s - a_1)^m}{(s - a_2)} C_2 + \cdots \tag{13}$$

$$\therefore C_{11} = \left[(s - a_1)^m \bar{x}(s)\right]_{s=a_1}$$

The coefficient C_{12} is determined by differentiating the equation for $(s - a_1)^m \bar{x}(s)$ with respect to s and then letting $s = a_1$

$$C_{12} = \left[\frac{d}{ds}(s - a_1)^m \bar{x}(s)\right]_{s=a_1} \tag{14}$$

It is evident then that

$$C_{1n} = \frac{1}{(n-1)!}\left[\frac{d^{n-1}}{ds^{n-1}}(s - a_1)^m \bar{x}(s)\right]_{s=a_1} \tag{15}$$

The remaining coefficients $C_2, C_3, \ldots,$ are evaluated as in the previous section for simple poles.

Since by the shifting theorem

$$\mathfrak{L}^{-1}\frac{1}{(s - a_1)^n} = \frac{t^{n-1}}{(n-1)!}{}^{a_1 t}$$

the inverse transform of $\bar{x}(s)$ becomes

$$x(t) = \left[C_{11}\frac{t^{m-1}}{(m-1)!} + C_{12}\frac{t^{m-2}}{(m-2)!} + \cdots\right]e^{a_1 t} + C_2 e^{a_2 t} + C_3 e^{a_3 t} + \cdots \tag{16}$$

Most ordinary differential equations can be solved by the elementary theory of LT. The tables here give the LT of simple functions. The table is also

SHORT TABLE OF LAPLACE TRANSFORMS

	$f(s)$	$f(t)$
(1)	1	$\delta(t)$ = unit impulse at $t = 0$
(2)	$\frac{1}{s}$	$\mathcal{U}(t)$ = unit step function at $t = 0$
(3)	$\frac{1}{s^n} (n = 1, 2, \cdots)$	$\frac{t^{n-1}}{(n-1)!}$
(4)	$\frac{1}{s+a}$	e^{-at}
(5)	$\frac{1}{(s+a)^2}$	te^{-at}
(6)	$\frac{1}{(s+a)^n} (n = 1, 2, \cdots)$	$\frac{1}{(n-1)!} t^{n-1} e^{-at}$
(7)	$\frac{1}{s(s+a)}$	$\frac{1}{a}(1 - e^{-at})$
(8)	$\frac{1}{s^2(s+a)}$	$\frac{1}{a^2}(e^{-at} + at - 1)$
(9)	$\frac{s}{s^2 + a^2}$	$\cos at$
(10)	$\frac{s}{s^2 - a^2}$	$\cosh at$
(11)	$\frac{1}{s^2 + a^2}$	$\frac{1}{a} \sin at$
(12)	$\frac{1}{s^2 - a^2}$	$\frac{1}{a} \sinh at$
(13)	$\frac{1}{s(s^2 + a^2)}$	$\frac{1}{a^2}(1 - \cos at)$
(14)	$\frac{1}{s^2(s^2 + a^2)}$	$\frac{1}{a^3}(at - \sin at)$
(15)	$\frac{1}{(s^2 + a^2)^2}$	$\frac{1}{2a^3}(\sin at - at \cos at)$
(16)	$\frac{s}{(s^2 + a^2)^2}$	$\frac{t}{2a} \sin at$
(17)	$\frac{s^2 - a^2}{(s^2 + a^2)^2}$	$t \cos at$
(18)	$\frac{1}{s^2 + 2\zeta\omega_0 s + \omega_0^2}$	$\frac{1}{\omega_0\sqrt{1-\zeta^2}} e^{-\zeta\omega_0 t} \sin \omega_0\sqrt{1-\zeta^2}\, t$

used to establish the inverse LT, since if

$$\mathfrak{L}f(t) = \bar{f}(s)$$

then

$$f(t) = \mathfrak{L}^{-1}\bar{f}(s).$$

REFERENCE

[1] THOMSON, W. T. *Laplace Transformation*, 2d ed., Englewood Cliffs, N.J.: Prentice-Hall, Inc., 1960.

C

DETERMINANTS AND MATRICES

C.1 Determinant

A determinant of the second order and its numerical evaluation are defined by the following notation and operation

$$D = \begin{vmatrix} a & b \\ c & d \end{vmatrix} = ad - bc$$

An nth-order determinant has n rows and n columns, and in order to identify the position of its elements, the following notation is adopted:

$$\begin{vmatrix} a_{11} & a_{12} & a_{13} & \cdots & a_{1n} \\ a_{21} & a_{22} & a_{23} & \cdots & a_{2n} \\ \vdots & \vdots & \vdots & & \vdots \\ a_{n1} & a_{n2} & a_{n3} & \cdots & a_{nn} \end{vmatrix}$$

Minors

A minor M_{ij} of the element a_{ij} is a determinant formed by deleting the ith row and the jth column from the original determinant.

Cofactor

The cofactor C_{ij} of the element a_{ij} is defined by the equation

$$C_{ij} = (-1)^{i+j} M_{ij}$$

Example

Given the third-order determinant

$$\begin{vmatrix} 2 & 1 & 5 \\ 4 & 2 & 1 \\ 2 & 0 & 3 \end{vmatrix}$$

The minor of the term $a_{21} = 4$ is

$$M_{21} \text{ of } \begin{vmatrix} 2 & 1 & 5 \\ 4 & 2 & 1 \\ 2 & 0 & 3 \end{vmatrix} = \begin{vmatrix} 1 & 5 \\ 0 & 3 \end{vmatrix} = 3$$

and its cofactor is

$$C_{21} = (-1)^{2+1}3 = -3$$

Expansion of a Determinant

The order of a determinant can be reduced by one by expanding any row or column in terms of its cofactors.

Example

The determinant of the previous example is expanded in terms of the second column as

$$D = \begin{vmatrix} 2 & 1 & 5 \\ 4 & 2 & 1 \\ 2 & 0 & 3 \end{vmatrix} = 1(-1)^{1+2}\begin{vmatrix} 4 & 1 \\ 2 & 3 \end{vmatrix} + 2(-1)^{2+2}\begin{vmatrix} 2 & 5 \\ 2 & 3 \end{vmatrix}$$

$$+0(-1)^{3+2}\begin{vmatrix} 2 & 5 \\ 4 & 1 \end{vmatrix}$$

$$= -10 - 8 = -18$$

Properties of Determinants

The following properties of determinants are stated without proof:

1. Interchange of any two columns or rows changes the sign of the determinant.
2. If two rows or two columns are identical, the determinant is zero.
3. Any row or column may be multiplied by a constant and added to another row or column without changing the value of the determinant.

C.2 MATRICES

Matrix. A rectangular array of terms arranged in m rows and n columns is called a matrix. For example

$$A = \begin{bmatrix} a_{11} & a_{12} & a_{13} & a_{14} \\ a_{21} & a_{22} & a_{23} & a_{24} \\ a_{31} & a_{32} & a_{33} & a_{34} \end{bmatrix}$$

is a 3×4 matrix.

Square matrix. A square matrix is one in which the number of rows is equal to the number of columns. It is referred to as an $n \times n$ matrix or a matrix of order n.

Symmetric matrix. A square matrix is said to be symmetric if the elements on the upper right half can be obtained by flipping the matrix about the diagonal.

$$A = \begin{bmatrix} 2 & 1 & 3 \\ 1 & 5 & 0 \\ 3 & 0 & 1 \end{bmatrix} = \text{symmetric matrix}$$

Trace. The sum of the diagonal elements of a square matrix is called the trace. For the matrix above

$$\text{Trace } A = 2 + 5 + 1 = 8$$

Singular matrix. If the determinant of a matrix is zero, the matrix is said to be singular.

Row matrix. A row matrix has $m = 1$.

$$B = [b_1 \quad b_2 \quad b_3]$$

Column matrix. A column matrix has $n = 1$.

$$C = \begin{Bmatrix} C_1 \\ C_2 \\ C_3 \end{Bmatrix}$$

Zero matrix. The zero matrix is defined as one in which all elements are zero.

$$0 = \begin{bmatrix} 0 & 0 & 0 \\ 0 & 0 & 0 \end{bmatrix}$$

Unit matrix. The unit matrix

$$I = \begin{bmatrix} 1 & 0 & 0 \\ 0 & 1 & 0 \\ 0 & 0 & 1 \end{bmatrix}$$

is a square matrix in which the diagonal elements from the top left to the bottom right are unity with all other elements equal to zero.

Diagonal matrix. A square matrix having elements a_{ii} along the diagonal with all other elements equal to zero is a diagonal matrix

$$[a_{ii}] = \begin{bmatrix} a_{11} & 0 & 0 \\ 0 & a_{22} & 0 \\ 0 & 0 & a_{33} \end{bmatrix}$$

Transpose. The transpose A' of a matrix A is one in which the rows and columns are interchanged. For example

$$A = \begin{bmatrix} a_{11} & a_{12} & a_{13} \\ a_{21} & a_{22} & a_{23} \end{bmatrix} \qquad A' = \begin{bmatrix} a_{11} & a_{21} \\ a_{12} & a_{22} \\ a_{13} & a_{23} \end{bmatrix}$$

The transpose of a column matrix is a row matrix

$$X = \begin{Bmatrix} x_1 \\ x_2 \\ x_3 \end{Bmatrix} \qquad X' = [x_1 x_2 x_3]$$

Minor. A minor M_{ij} of a matrix A is formed by deleting the ith row and the jth column from the determinant of the original matrix.

$$\text{Let } A = \begin{bmatrix} a_{11} & a_{12} & a_{13} \\ a_{21} & a_{22} & a_{23} \\ a_{31} & a_{32} & a_{33} \end{bmatrix}$$

$$M_{12} \begin{vmatrix} a_{11} & a_{12} & a_{13} \\ a_{21} & a_{22} & a_{23} \\ a_{31} & a_{32} & a_{33} \end{vmatrix} = \begin{vmatrix} a_{21} & a_{23} \\ a_{31} & a_{33} \end{vmatrix}$$

Cofactor. The cofactor C_{ij} is equal to the signed minor $(-1)^{i+j} M_{ij}$. From the previous example

$$C_{12} = (-1)^{1+2} M_{12} = -M_{12}$$

Adjoint matrix. An adjoint matrix of a square matrix A is a transpose of the matrix of cofactors of A.

Let cofactor matrix of A be

$$[C_{ij}] = \begin{bmatrix} C_{11} & C_{12} & C_{13} \\ C_{21} & C_{22} & C_{23} \\ C_{31} & C_{32} & C_{33} \end{bmatrix}$$

$$\text{adj } A = [C_{ij}]' = [C_{ji}] = \begin{bmatrix} C_{11} & C_{21} & C_{31} \\ C_{12} & C_{22} & C_{32} \\ C_{13} & C_{23} & C_{33} \end{bmatrix}$$

Inverse matrix. The inverse A^{-1} of a matrix A satisfies the relationship

$$A^{-1}A = AA^{-1} = I$$

Orthogonal matrix. An orthogonal matrix A satisfies the relationship

$$A'A = AA' = I$$

From the definition of an inverse matrix it is evident that for an orthogonal matrix $A' = A^{-1}$.

C.3 RULES OF MATRIX OPERATIONS

Addition. Two matrices having the same number of rows and columns may be added by summing the corresponding elements.

Example

$$\begin{bmatrix} 1 & 3 & 2 \\ 4 & 1 & 1 \end{bmatrix} + \begin{bmatrix} 2 & 0 & 4 \\ 1 & -2 & -3 \end{bmatrix} = \begin{bmatrix} 3 & 3 & 6 \\ 5 & -1 & -2 \end{bmatrix}$$

Multiplication. The product of two matrices A and B is another matrix C.

$$AB = C$$

The element C_{ij} of C is determined by multiplying the elements of the ith row in A by the elements of the jth column in B according to the rule

$$c_{ij} = \sum_k a_{ik} b_{kj}$$

Example

$$\text{Let } A = \begin{bmatrix} 1 & 1 & 1 \\ 1 & 2 & 2 \\ 1 & 2 & 3 \end{bmatrix} \qquad B = \begin{bmatrix} 2 & 0 \\ 0 & 1 \\ 3 & -1 \end{bmatrix}$$

$$AB = \begin{bmatrix} 1 & 1 & 1 \\ 1 & 2 & 2 \\ 1 & 2 & 3 \end{bmatrix} \begin{bmatrix} 2 & 0 \\ 0 & 1 \\ 3 & -1 \end{bmatrix} = \begin{bmatrix} 5 & 0 \\ 8 & 0 \\ 11 & -1 \end{bmatrix} = C$$

i.e.,

$$c_{21} = 1 \times 2 + 2 \times 0 + 2 \times 3 = 8$$

It is evident that the number of columns in A must equal the number of rows in B, or that the matrices must be conformable. We also note that $AB \neq BA$.

The postmultiplication of a matrix by a column matrix results in a column matrix.

Example

$$\begin{bmatrix} 1 & 1 & 1 \\ 1 & 5 & 2 \\ 2 & 1 & 3 \end{bmatrix} \begin{Bmatrix} 1 \\ 3 \\ 2 \end{Bmatrix} = \begin{Bmatrix} 6 \\ 20 \\ 11 \end{Bmatrix}$$

Premultiplication of a matrix by a row matrix (or transpose of a column matrix) results in a row matrix.

Example

$$\begin{bmatrix} 1 & 3 & 2 \end{bmatrix} \begin{bmatrix} 1 & 1 & 1 \\ 1 & 5 & 2 \\ 2 & 1 & 3 \end{bmatrix} = \begin{bmatrix} 8 & 18 & 13 \end{bmatrix}$$

The transpose of a product $AB = C$ is $C' = B'A'$

Example

$$\text{Let } A\begin{bmatrix} 1 & 1 \\ 2 & 3 \end{bmatrix} \qquad B = \begin{bmatrix} 2 & 1 \\ 1 & 1 \end{bmatrix}$$

$$C = AB = \begin{bmatrix} 3 & 2 \\ 7 & 5 \end{bmatrix} \qquad C' = B'A' = \begin{bmatrix} 2 & 1 \\ 1 & 1 \end{bmatrix}\begin{bmatrix} 1 & 2 \\ 1 & 3 \end{bmatrix} = \begin{bmatrix} 3 & 7 \\ 2 & 5 \end{bmatrix}$$

Inversion of a Matrix. Consider a set of equations

$$\begin{aligned} a_{11}x_1 + a_{12}x_2 + a_{13}x_3 &= y_1 \\ a_{21}x_1 + a_{22}x_2 + a_{23}x_3 &= y_2 \\ a_{31}x_1 + a_{32}x_2 + a_{33}x_3 &= y_3 \end{aligned} \tag{1}$$

which can be expressed in the matrix form

$$AX = Y \tag{2}$$

Premultiplying by the inverse A^{-1}, we obtain the solution

$$X = A^{-1}Y \tag{3}$$

We can identify the term A^{-1} by Cramer's rule as follows. The solution for x_1 is

$$\begin{aligned} x_1 &= \frac{1}{|A|}\begin{vmatrix} y_1 & a_{12} & a_{13} \\ y_2 & a_{22} & a_{23} \\ y_3 & a_{32} & a_{33} \end{vmatrix} \\ &= \frac{1}{|A|}\left\{ y_1\begin{vmatrix} a_{22} & a_{23} \\ a_{32} & a_{33} \end{vmatrix} - y_2\begin{vmatrix} a_{12} & a_{13} \\ a_{32} & a_{33} \end{vmatrix} + y_3\begin{vmatrix} a_{12} & a_{13} \\ a_{22} & a_{23} \end{vmatrix}\right\} \\ &= \frac{1}{|A|}\{ y_1C_{11} + y_2C_{21} + y_3C_{31} \} \end{aligned}$$

where A is the determinant of the coefficient matrix A, and C_{11}, C_{21}, and C_{31} are the cofactors of A corresponding to elements 11, 21, and 31. We can also write similar expressions for x_2 and x_3 by replacing the second and third columns by the y column, respectively. Thus the complete solution can be written in matrix form as

$$\begin{Bmatrix} x_1 \\ x_2 \\ x_3 \end{Bmatrix} = \frac{1}{|A|}\begin{bmatrix} C_{11} & C_{21} & C_{31} \\ C_{12} & C_{22} & C_{32} \\ C_{13} & C_{23} & C_{33} \end{bmatrix}\begin{Bmatrix} y_1 \\ y_2 \\ y_3 \end{Bmatrix} \tag{4}$$

or

$$\{x\} = \frac{1}{|A|}[C_{ji}]\{y\} = \frac{1}{|A|}[\text{adj } A]\{y\}$$

Thus, by comparison with Eq. (3) we arrive at the result

$$A^{-1} = \frac{1}{|A|}\text{adj } A \tag{5}$$

Example

Find the inverse of the matrix

$$A = \begin{bmatrix} 1 & 1 & 1 \\ 1 & 2 & 2 \\ 1 & 0 & 3 \end{bmatrix}$$

(a) The determinant of A is $|A| = 3$.
(b) The minors of A are

$$M_{11} = \begin{vmatrix} 2 & 2 \\ 0 & 3 \end{vmatrix} = 6, \qquad M_{12} = \begin{vmatrix} 1 & 2 \\ 1 & 3 \end{vmatrix} = 1, \cdots$$

(c) Supply the signs $(-1)^{i+j}$ to the minors to form the cofactors

$$\left[C_{ij}\right] = \begin{vmatrix} 6 & -1 & -2 \\ -3 & 2 & 1 \\ 0 & -1 & 1 \end{vmatrix}$$

(d) The adjoint matrix is the transpose of the cofactor matrix, or $[C_{ij}]' = [C_{ji}]$. Thus the inverse A^{-1} is found to be

$$A^{-1} = \frac{1}{|A|} \operatorname{adj} A = \frac{1}{3} \begin{bmatrix} 6 & -3 & 0 \\ -1 & 2 & -1 \\ -2 & 1 & 1 \end{bmatrix}$$

(e) The result can be checked as follows:

$$A^{-1}A = \frac{1}{3} \begin{bmatrix} 6 & -3 & 0 \\ -1 & 2 & -1 \\ -2 & 1 & 1 \end{bmatrix} \begin{bmatrix} 1 & 1 & 1 \\ 1 & 2 & 2 \\ 1 & 0 & 3 \end{bmatrix}$$
$$= \frac{1}{3} \begin{bmatrix} 3 & 0 & 0 \\ 0 & 3 & 0 \\ 0 & 0 & 3 \end{bmatrix} = \begin{bmatrix} 1 & 0 & 0 \\ 0 & 1 & 0 \\ 0 & 0 & 1 \end{bmatrix}$$

It should be noted that for an inverse to exist, the determinant $|A|$ must not be zero.

Equation (5) for the inverse offers another means of evaluating a determinant. Premultiply Eq. (5) by A:

$$AA^{-1} = \frac{A}{|A|} \operatorname{adj} A = I$$

Thus we obtain the expression

$$|A|I = A \operatorname{adj} A \tag{6}$$

Transpose of a Product

The following operations are given without proof:

$$\begin{aligned} (AB)' &= B'A' \\ (A + B)' &= A' + B' \end{aligned} \tag{7}$$

Orthogonal transformation. A matrix P is orthogonal if

$$P^{-1} = P'$$

The determinant of an orthogonal matrix is equal to ± 1. If A = symmetric matrix, then

$$P^{-1}AP = D = P'AP \qquad \text{a diagonal matrix} \tag{8}$$

If A is a symmetric matrix, then

$$\begin{aligned} P'A &= AP \\ \{x\}'A &= A\{x\} \end{aligned} \tag{9}$$

Partitioned Matrices

A matrix may be partitioned into submatrices by horizontal and vertical lines, as shown by the following example:

$$\left[\begin{array}{cc|c} 2 & 4 & -1 \\ 0 & -3 & 4 \\ 1 & 2 & 2 \\ \hline 3 & -1 & -5 \end{array}\right] = \left[\begin{array}{c|c} [A] & [B] \\ \hline [C] & [D] \end{array}\right]$$

where the submatrices are

$$A = \begin{bmatrix} 2 & 4 \\ 0 & -3 \\ 1 & 2 \end{bmatrix} \qquad B = \begin{Bmatrix} -1 \\ 4 \\ 2 \end{Bmatrix}$$

$$C = [3 \quad -1] \qquad D = [-5]$$

Partitioned matrices obey the normal rules of matrix algebra and can be added, subtracted, and multiplied as though the submatrices were ordinary matrix elements. Thus

$$\left[\begin{array}{c|c} A & B \\ \hline C & D \end{array}\right] \left\{\begin{array}{c} x \\ \hline y \end{array}\right\} = \begin{bmatrix} A\{x\} + B\{y\} \\ C\{x\} + D\{y\} \end{bmatrix}$$

$$\left[\begin{array}{c|c} A & B \\ \hline C & D \end{array}\right] \left[\begin{array}{c|c} E & F \\ \hline G & H \end{array}\right] = \left[\begin{array}{c|c} AE + BG & AF + BH \\ \hline CE + DG & CF + DH \end{array}\right]$$

C.4 DETERMINATION OF EIGENVECTORS

The eigenvector X_i corresponding to the eigenvalue λ_i can be found from the cofactors of any row of the characteristic equation.

Let $[A - \lambda_i I]X_i = 0$ be written out for a third-order system as

$$\begin{bmatrix} (a_{11} - \lambda_i) & a_{12} & a_{13} \\ a_{21} & (a_{22} - \lambda_i) & a_{23} \\ a_{31} & a_{32} & (a_{33} - \lambda_i) \end{bmatrix} \begin{Bmatrix} x_1 \\ x_2 \\ x_3 \end{Bmatrix}_i = 0 \tag{1}$$

Its characteristic equation $|A - \lambda_i I| = 0$ written out in determinant form is

$$\begin{vmatrix} (a_{11} - \lambda_i) & a_{12} & a_{13} \\ a_{21} & (a_{22} - \lambda_i) & a_{23} \\ a_{31} & a_{32} & (a_{33} - \lambda_i) \end{vmatrix} = 0 \tag{2}$$

The determinant expanded in terms of the cofactors of the first row is

$$(a_{11} - \lambda_i)C_{11} + a_{12}C_{12} + a_{13}C_{13} = 0 \tag{3}$$

Next replace the first row of the determinant by the second row, leaving the other two rows unchanged. The value of the determinant is still zero because of two identical rows

$$\begin{vmatrix} a_{21} & (a_{22} - \lambda_i) & a_{23} \\ a_{21} & (a_{22} - \lambda_i) & a_{23} \\ a_{31} & a_{32} & (a_{33} - \lambda_i) \end{vmatrix} = 0 \tag{4}$$

Again expand in terms of the cofactors of the first row, which are identical to the cofactors of the previous determinant.

$$a_{21}C_{11} + (a_{22} - \lambda_i)C_{12} + a_{23}C_{13} = 0 \tag{5}$$

Finally replace the first row by the third row and expand in terms of the first row of the new determinant.

$$\begin{vmatrix} a_{31} & a_{32} & (a_{33} - \lambda_i) \\ a_{21} & (a_{22} - \lambda_i) & a_{23} \\ a_{31} & a_{32} & (a_{33} - \lambda_i) \end{vmatrix} = 0 \tag{6}$$

$$a_{31}C_{11} + a_{32}C_{12} + (a_{33} - \lambda_i)C_{13} = 0 \tag{7}$$

Equations (3), (5), and (7) can now be assembled in a single matrix equation:

$$\begin{bmatrix} (a_{11} - \lambda_i) & a_{12} & a_{13} \\ a_{21} & (a_{22} - \lambda_i) & a_{23} \\ a_{31} & a_{32} & (a_{33} - \lambda_i) \end{bmatrix} \begin{Bmatrix} C_{11} \\ C_{12} \\ C_{13} \end{Bmatrix} = 0 \tag{8}$$

Comparison of Eqs. (1) and (8) indicates that the eigenvector X_i may be determined from the cofactors of the characteristic equation with $\lambda = \lambda_i$. Since the eigenvectors are relative to a normalized coordinate, the column of cofactors may differ by a multiplying factor.

$$\begin{Bmatrix} x_1 \\ x_2 \\ x_3 \end{Bmatrix}_i = \alpha \begin{Bmatrix} C_{11} \\ C_{12} \\ C_{13} \end{Bmatrix}$$

Instead of the first row, any other row may have been used for the determination of the cofactors.

C.5 CHOLESKY'S METHOD OF SOLUTION†

The matrix equation

$$[A]\{X\} = \{C\} \tag{1}$$

may be solved for $\{X\}$ by premultiplying the equation by the inverse of $[A]$

$$\{X\} = [A]^{-1}\{C\}$$

Cholesky's method avoids the necessity of inverting the matrix $[A]$, the elements of $\{X\}$ being available by successive algebraic steps.

Cholesky's method depends on converting the original equation, Eq. (1), to the form

$$[T]\{X\} = \{k\} \tag{2}$$

where (for a 3×3 matrix)

$$[T] = \begin{bmatrix} 1 & t_{12} & t_{13} \\ 0 & 1 & t_{23} \\ 0 & 0 & 1 \end{bmatrix} \tag{3}$$

is an upper triangular matrix with unit diagonal elements. For example, consider a 3×3 matrix

$$\begin{bmatrix} 1 & t_{12} & t_{13} \\ 0 & 1 & t_{23} \\ 0 & 0 & 1 \end{bmatrix} \begin{Bmatrix} x_1 \\ x_2 \\ x_3 \end{Bmatrix} = \begin{Bmatrix} k_1 \\ k_2 \\ k_3 \end{Bmatrix}$$

The elements of $\{X\}$ from the above equation are simply found by a backward substitution as follows:

$$x_3 = k_3 \qquad \therefore\ x_3 = k_3$$

$$x_2 + t_{23}x_3 = k_2 \qquad \therefore\ x_2 = k_2 - t_{23}x_3$$

$$x_1 + t_{12}x_2 + t_{13}x_3 = k_3 \qquad \therefore\ x_1 = k_3 - t_{12}x_3 - t_{13}x_3$$

Thus if $[T]$ and $\{k\}$ are known, the solution for $\{X\}$ in Eq. (1) is available.

To determine $[T]$ and $\{k\}$, multiply Eq. (2) by a lower triangular matrix

$$[L] = \begin{bmatrix} l_{11} & 0 & 0 \\ l_{21} & l_{22} & 0 \\ l_{31} & l_{32} & l_{33} \end{bmatrix} \tag{4}$$

as follows:

$$[L][T]\{X\} = [L]\{k\} \tag{5}$$

For this equation to equal the original equation, the following relationships

†Salvadori and Baron, *Numerical Methods in Engineering*. Englewood Cliffs, N.J.: Prentice-Hall, Inc., 1952, pp. 23–28.

must exist:

$$[A] = [L][T] \tag{6}$$

$$\{C\} = [L]\{k\} \tag{7}$$

Writing out the above equations in terms of their elements, we have

$$\begin{bmatrix} a_{11} & a_{12} & a_{13} \\ a_{21} & a_{22} & a_{23} \\ a_{31} & a_{32} & a_{33} \end{bmatrix} = \begin{bmatrix} l_{11} & l_{11}t_{12} & l_{11}t_{13} \\ l_{21} & (l_{21}t_{12} + l_{22}) & (l_{21}t_{13} + l_{22}t_{23}) \\ l_{31} & (l_{31}t_{12} + l_{32}) & (l_{31}t_{13} + l_{32}t_{23} + l_{33}) \end{bmatrix}$$

$$\begin{Bmatrix} c_1 \\ c_2 \\ c_3 \end{Bmatrix} = \begin{Bmatrix} l_{11}k_1 \\ l_{21}k_1 + l_{22}k_2 \\ l_{31}k_1 + l_{32}k_2 + l_{33}k_3 \end{Bmatrix}$$

By equating the elements in these equations we have

$$a_{11} = l_{11}$$

$$a_{21} = l_{21}$$

$$a_{31} = l_{31}$$

$$a_{12} = l_{11}t_{12} \qquad \therefore t_{12} = \frac{a_{12}}{l_{11}}$$

$$a_{13} = l_{11}t_{13} \qquad \therefore t_{13} = \frac{a_{13}}{l_{11}}$$

$$a_{22} = (l_{21}t_{12} + l_{22}) \qquad \therefore l_{22} = a_{22} - l_{21}t_{12}$$

$$a_{23} = (l_{21}t_{13} + l_{22}t_{23}) \qquad \therefore t_{23} = \frac{1}{l_{22}}(a_{23} - l_{21}t_{13})$$

$$a_{32} = (l_{31}t_{12} + l_{32}) \qquad \therefore l_{32} = a_{32} - l_{31}t_{12}$$

$$a_{33} = (l_{31}t_{13} + l_{32}t_{23} + l_{33}) \qquad \therefore l_{33} = a_{33} - l_{31}t_{13} - l_{32}t_{23}$$

$$c_1 = l_{11}k_1 \qquad \therefore k_1 = \frac{c_1}{l_{11}}$$

$$c_2 = l_{21}k_1 + l_{22}k_2 \qquad \therefore k_2 = \frac{1}{l_{22}}(c_2 - l_{21}k_1)$$

$$c_3 = l_{31}k_1 + l_{32}k_2 + l_{33}k_3 \qquad \therefore k_3 = \frac{1}{l_{33}}(c_3 - l_{31}k_1 - l_{32}k_2)$$

Thus the elements of the matrices $[L]$, $[T]$ and $\{k\}$ are now available in terms of the known elements of $[A]$ and $[C]$, and Eq. (1) may be solved without inverting the matrix $[A]$.

D

NORMAL MODES OF UNIFORM BEAMS

We assume the free vibrations of a uniform beam to be governed by Euler's differential equation.

$$EI\frac{\partial^4 y}{\partial x^4} + m\frac{\partial^2 y}{\partial t^2} = 0 \tag{1}$$

To determine the normal modes of vibration, the solution in the form

$$y(x, t) = \phi_n(x)e^{i\omega_n t} \tag{2}$$

is substituted into Eq. (1) to obtain the equation

$$\frac{d^4\phi_n(x)}{dx^4} - \beta_n^4\phi_n(x) = 0 \tag{3}$$

where

$\phi_n(x)$ = characteristic function describing the deflection of the nth mode

m = mass density per unit length

$\beta_n^4 = m\omega_n^2/EI$

$\omega_n = (\beta_n l)^2\sqrt{EI/ml^4}$ = natural frequency of the nth mode

The characteristic functions $\phi_n(x)$ and the normal mode frequencies ω_n depend on the boundary conditions and have been tabulated by Young and Felgar. An abbreviated summary taken from this work† is presented here.

†Young, D. and R. P. Felgar, Jr., *Tables of Characteristic Functions Representing Normal Modes of Vibration of a Beam*. The University of Texas Publication No. 4913, July 1, 1949.

1 CLAMPED-CLAMPED BEAM

n	$\beta_n l$	$(\beta_n l)^2$	ω_n/ω_1
1	4.7300	22.3733	1.0000
2	7.8532	61.6728	2.7565
3	10.9956	120.9034	5.4039

2 FREE-FREE BEAM

The natural frequencies of the free-free beam are equal to those of the clamped-clamped beam. The characteristic functions of the free-free beam are related to those of the clamped-clamped beam as follows:

free-free		clamped-clamped
ϕ_n	=	ϕ_n''
ϕ_n'	=	ϕ_n'''
ϕ_n''	=	ϕ_n
ϕ_n'''	=	ϕ_n'

3 CLAMPED-FREE BEAM

n	$\beta_n l$	$(\beta_n l)^2$	ω_n/ω_1
1	1.8751	3.5160	1.0000
2	4.6941	22.0345	6.2669
3	7.8548	61.6972	17.5475

4 CLAMPED-PINNED BEAM

n	$\beta_n L$	$(\beta_n l)^2$	ω_n/ω_1
1	3.9266	15.4182	1.0000
2	7.0686	49.9645	3.2406
3	10.2101	104.2477	6.7613

5 FREE-PINNED BEAM

The natural frequencies of the free-pinned beam are equal to those of the clamped-pinned beam. The characteristic functions of the free-pinned beam are related to those of the clamped-pinned beam as follows:

free-pinned		clamped-pinned
ϕ_n	=	ϕ_n''
ϕ_n'	=	ϕ_n'''
ϕ_n''	=	ϕ_n
ϕ_n'''	=	ϕ_n'

TABLE 1.
CHARACTERISTIC FUNCTIONS AND DERIVATIVES
CLAMPED-CLAMPED BEAM
FIRST MODE

$\frac{x}{l}$	ϕ_1	$\phi_1' = \frac{1}{\beta_1}\frac{d\phi_1}{dx}$	$\phi_1'' = \frac{1}{\beta_1^2}\frac{d^2\phi_1}{dx^2}$	$\phi_1''' = \frac{1}{\beta_1^3}\frac{d^3\phi_1}{dx^3}$
0.00	0.00000	0.00000	2.00000	−1.96500
0.04	0.03358	0.34324	1.62832	−1.96285
0.08	0.12545	0.61624	1.25802	−1.94862
0.12	0.26237	0.81956	0.89234	−1.91254
0.16	0.43126	0.95451	0.53615	−1.84732
0.20	0.61939	1.02342	0.19545	−1.74814
0.24	0.81459	1.02986	−0.12305	−1.61250
0.28	1.00546	0.97870	−0.41240	−1.44017
0.32	1.18168	0.87608	−0.66581	−1.23296
0.36	1.33419	0.72992	−0.87699	−0.99452
0.40	1.45545	0.54723	−1.04050	−0.73007
0.44	1.53962	0.33897	−1.15202	−0.44611
0.48	1.58271	0.11478	−1.20854	−0.15007
0.52	1.58271	−0.11478	−1.20854	0.15007
0.56	1.53962	−0.33897	−1.15202	0.44611
0.60	1.45545	−0.54723	−1.04050	0.73007
0.64	1.33419	−0.72992	−0.87699	0.99452
0.68	1.18168	−0.87608	−0.66581	1.23296
0.72	1.00546	−0.97870	−0.41240	1.44017
0.76	0.81459	−1.02986	−0.12305	1.61250
0.80	0.61939	−1.02342	0.19545	1.74814
0.84	0.43126	−0.95451	0.53615	1.84732
0.88	0.26237	−0.81956	0.89234	1.91254
0.92	0.12545	−0.61624	1.25802	1.94862
0.96	0.03358	−0.34324	1.62832	1.96285
1.00	0.00000	0.00000	2.00000	1.96500

TABLE 1.
CHARACTERISTIC FUNCTIONS AND DERIVATIVES
CLAMPED-CLAMPED BEAM
SECOND MODE

$\frac{x}{l}$	ϕ_2	$\phi_2' = \frac{1}{\beta_2}\frac{d\phi_2}{dx}$	$\phi_2'' = \frac{1}{\beta_2^2}\frac{d^2\phi_2}{dx^2}$	$\phi_2''' = \frac{1}{\beta_2^3}\frac{d^3\phi_2}{dx^3}$
0.00	0.00000	0.00000	2.00000	−2.00155
0.04	0.08834	0.52955	1.37202	−1.99205
0.08	0.31214	0.86296	0.75386	−1.93186
0.12	0.61058	1.00644	0.16713	−1.78813
0.16	0.92602	0.97427	−0.35923	−1.54652
0.20	1.20674	0.79030	−0.79450	−1.21002
0.24	1.41005	0.48755	−1.11133	−0.79651
0.28	1.50485	0.10660	−1.28991	−0.33555
0.32	1.47357	−0.30736	−1.32106	0.13566
0.36	1.31314	−0.70819	−1.20786	0.57665
0.40	1.03457	−1.05271	−0.96605	0.94823
0.44	0.66150	−1.30448	−0.62296	1.21670
0.48	0.22751	−1.43728	−0.21508	1.35744
0.52	−0.22751	−1.43728	0.21508	1.35744
0.56	−0.66150	−1.30448	0.62296	1.21670
0.60	−1.03457	−1.05271	0.96605	0.94823
0.64	−1.31314	−0.70819	1.20786	0.57665
0.68	−1.47357	−0.30736	1.32106	0.13566
0.72	−1.50485	0.10660	1.28991	−0.33555
0.76	−1.41005	0.48755	1.11133	−0.79651
0.80	−1.20674	0.70930	0.79450	−1.21002
0.84	−0.92602	0.97427	0.35923	−1.54652
0.88	−0.61058	1.00644	−0.16713	−1.78813
0.92	−0.31214	0.86296	−0.75386	−1.93186
0.96	−0.08834	0.52955	−1.37202	−1.99205
1.00	0.00000	0.00000	−2.00000	−2.00155

TABLE 1.
CHARACTERISTIC FUNCTIONS AND DERIVATIVES
CLAMPED-CLAMPED BEAM
THIRD MODE

$\frac{x}{l}$	ϕ_3	$\phi_3' = \frac{1}{\beta_3}\frac{d\phi_3}{dx}$	$\phi_3'' = \frac{1}{\beta_3^2}\frac{d^2\phi_3}{dx^2}$	$\phi_3''' = \frac{1}{\beta_3^3}\frac{d^3\phi_3}{dx^3}$
0.00	0.00000	0.00000	2.00000	−1.99993
0.04	0.16510	0.68646	1.12323	−1.97469
0.08	0.54804	0.99303	0.28189	−1.82280
0.12	0.98720	0.95006	−0.45252	−1.48447
0.16	1.34190	0.62285	−0.99738	−0.96698
0.20	1.50782	0.11050	−1.28572	−0.33199
0.24	1.42971	−0.46573	−1.28637	0.32333
0.28	1.10719	−0.98087	−1.01443	0.88956
0.32	0.59186	−1.32694	−0.53145	1.26880
0.36	−0.02445	−1.43171	0.06438	1.39529
0.40	−0.62837	−1.27099	0.65569	1.24912
0.44	−1.10739	−0.87257	1.12747	0.86096
0.48	−1.37174	−0.31031	1.38852	0.30669
0.52	−1.37174	0.31031	1.38852	−0.30669
0.56	−1.10739	0.87257	1.12747	−0.86096
0.60	−0.62837	1.27099	0.65569	−1.24912
0.64	−0.02445	1.43171	0.06438	−1.39529
0.68	0.59186	1.32694	−0.53145	−1.26880
0.72	1.10719	0.98087	−1.01443	−0.88956
0.76	1.42971	0.46573	−1.28637	−0.32333
0.80	1.50782	−0.11050	−1.28572	0.33199
0.84	1.34190	−0.62285	−0.99738	0.96698
0.88	0.98720	−0.95006	−0.45252	1.48447
0.92	0.54804	−0.99303	0.28189	1.82280
0.96	0.16510	−0.68646	1.12323	1.97469
1.00	0.00000	0.00000	2.00000	1.99993

TABLE 2.
CHARACTERISTIC FUNCTIONS AND DERIVATIVES
CLAMPED-FREE BEAM
FIRST MODE

$\frac{x}{l}$	ϕ_1	$\phi_1' = \frac{1}{\beta_1}\frac{d\phi_1}{dx}$	$\phi_1'' = \frac{1}{\beta_1^2}\frac{d^2\phi_1}{dx^2}$	$\phi_1''' = \frac{1}{\beta_1^3}\frac{d^3\phi_1}{dx^3}$
0.00	0.00000	0.00000	2.00000	−1.46819
0.04	0.00552	0.14588	1.88988	−1.46805
0.08	0.02168	0.28350	1.77980	−1.46710
0.12	0.04784	0.41286	1.66985	−1.46455
0.16	0.08340	0.53400	1.56016	−1.45968
0.20	0.12774	0.64692	1.45096	−1.45182
0.24	0.18024	0.75167	1.34247	−1.44032
0.28	0.24030	0.84832	1.23500	−1.42459
0.32	0.30730	0.93696	1.23889	−1.40410
0.36	0.38065	1.01771	1.02451	−1.37834
0.40	0.45977	1.09070	0.92227	−1.34685
0.44	0.54408	1.15612	0.82262	−1.30924
0.48	0.63301	1.21418	0.72603	−1.26512
0.52	0.72603	1.26512	0.63301	−1.21418
0.56	0.82262	1.30924	0.54408	−1.15612
0.60	0.92227	1.34685	0.45977	−1.09070
0.64	1.02451	1.37834	0.38065	−1.01771
0.68	1.12889	1.40410	0.30730	−0.93696
0.72	1.23500	1.42459	0.24030	−0.84832
0.76	1.34247	1.44032	0.18024	−0.75167
0.80	1.45096	1.45182	0.12774	−0.64692
0.84	1.56016	1.45968	0.08340	−0.53400
0.88	1.66985	1.46455	0.04784	−0.41286
0.92	1.77980	1.46710	0.02168	−0.28350
0.96	1.88988	1.46805	0.00552	−0.14588
1.00	2.00000	1.46819	0.00000	0.00000

TABLE 2.
CHARACTERISTIC FUNCTIONS AND DERIVATIVES
CLAMPED-FREE BEAM
SECOND MODE

$\frac{x}{l}$	ϕ_2	$\phi_2' = \frac{1}{\beta_2}\frac{d\phi_2}{dx}$	$\phi_2'' = \frac{1}{\beta_2^2}\frac{d^2\phi_2}{dx^2}$	$\phi_2''' = \frac{1}{\beta_2^3}\frac{d^3\phi_2}{dx^3}$
0.00	0.00000	0.00000	2.00000	−2.03693
0.04	0.03301	0.33962	1.61764	−2.03483
0.08	0.12305	0.60754	1.23660	−2.02097
0.12	0.25670	0.80728	0.86004	−1.98590
0.16	0.42070	0.93108	0.49261	−1.92267
0.20	0.60211	0.99020	0.14007	−1.82682
0.24	0.78852	0.98502	−0.19123	−1.69625
0.28	0.96827	0.92013	−0.49475	−1.53113
0.32	1.13068	0.80136	−0.76419	−1.33373
0.36	1.26626	0.63565	−0.99384	−1.10821
0.40	1.36694	0.43094	−1.17895	−0.86040
0.44	1.42619	0.19593	−1.31600	−0.59748
0.48	1.43920	−0.06012	−1.40289	−0.32772
0.52	1.40289	−0.32772	−1.43920	−0.06012
0.56	1.31600	−0.59748	−1.42619	0.19593
0.60	1.17895	−0.86040	−1.36694	0.43094
0.64	0.99384	−1.10821	−1.26626	0.63565
0.68	0.76419	−1.33373	−1.13068	0.80136
0.72	0.49475	−1.53113	−0.96827	0.92013
0.76	0.19123	−1.69625	−0.78852	0.98502
0.80	−0.14007	−1.82682	−0.60211	0.99020
0.84	−0.49261	−1.92267	−0.42070	0.93108
0.88	−0.86004	−1.98590	−0.25670	0.80428
0.92	−1.23660	−2.02097	−0.12305	0.60754
0.96	−1.61764	−2.03483	−0.03301	0.33962
1.00	−2.00000	−2.03693	0.00000	0.00000

TABLE 2.
CHARACTERISTIC FUNCTIONS AND DERIVATIVES
CLAMPED-FREE BEAM
THIRD MODE

$\frac{x}{l}$	ϕ_3	$\phi_3' = \frac{1}{\beta_3}\frac{d\phi_3}{dx}$	$\phi_3'' = \frac{1}{\beta_3^2}\frac{d^2\phi_3}{dx^2}$	$\phi_3''' = \frac{1}{\beta_3^3}\frac{d^3\phi_3}{dx^3}$
0.00	0.00000	0.00000	2.00000	−1.99845
0.04	0.08839	0.52979	1.37287	−1.98892
0.08	0.31238	0.86367	0.75558	−1.92871
0.12	0.61120	1.00785	0.16974	−1.78480
0.16	0.92728	0.97665	−0.35563	−1.54286
0.20	1.20901	0.79394	−0.78975	−1.20575
0.24	1.41376	0.49285	−1.10515	−0.79124
0.28	1.51056	0.11405	−1.28189	−0.32872
0.32	1.48203	−0.29711	−1.31055	0.14479
0.36	1.32534	−0.69422	−1.19398	0.58908
0.40	1.05185	−1.03374	−0.94753	0.96533
0.44	0.68568	−1.27881	−0.59802	1.24030
0.48	0.26103	−1.40247	−0.18130	1.39004
0.52	−0.18130	−1.39004	0.26103	1.40247
0.56	−0.59802	−1.24030	0.68568	1.27881
0.60	−0.94753	−0.96533	1.05185	1.03374
0.64	−1.19398	−0.58908	1.32534	0.69422
0.68	−1.31055	−0.14479	1.48203	0.29711
0.72	−1.28189	0.32872	1.51056	−0.11405
0.76	−1.10515	0.79124	1.41376	−0.49285
0.80	−0.78975	1.20575	1.20901	−0.79394
0.84	−0.35563	1.54236	0.92728	−0.97665
0.88	0.16974	1.78480	0.61120	−1.00785
0.92	0.75558	1.92871	0.31238	−0.86367
0.96	1.37287	1.98892	0.08829	−0.52979
1.00	2.00000	1.99845	0.00000	0.00000

TABLE 3.
CHARACTERISTIC FUNCTIONS AND DERIVATIVES
CLAMPED-PINNED BEAM
FIRST MODE

$\frac{x}{l}$	ϕ_1	$\phi_1' = \frac{1}{\beta_1}\frac{d\phi_1}{dx}$	$\phi_1'' = \frac{1}{\beta_1^2}\frac{d^2\phi_1}{dx^2}$	$\phi_1''' = \frac{1}{\beta_1^3}\frac{d^3\phi_1}{dx^3}$
0.00	0.00000	0.00000	2.00000	−2.00155
0.04	0.02338	0.28944	1.68568	−2.00031
0.08	0.08834	0.52955	1.37202	−1.99203
0.12	0.18715	0.72055	1.06060	−1.97079
0.16	0.31214	0.86296	0.75386	−1.93187
0.20	0.45574	0.95776	0.45486	−1.87177
0.24	0.61058	1.00643	0.16712	−1.78812
0.28	0.76958	1.01105	−0.10554	−1.67975
0.32	0.92601	0.97427	−0.35923	−1.54652
0.36	1.07363	0.89940	−0.59009	−1.38932
0.40	1.20675	0.79029	−0.79450	−1.21002
0.44	1.32032	0.65138	−0.96918	−1.01128
0.48	1.41006	0.48755	−1.11133	−0.79652
0.52	1.47245	0.30410	−1.21875	−0.56977
0.56	1.50485	0.10661	−1.28992	−0.33555
0.60	1.50550	−0.09916	−1.32402	−0.09872
0.64	1.47357	−0.30736	−1.32106	0.13566
0.68	1.40913	−0.51224	−1.28180	0.36247
0.72	1.31313	−0.70820	−1.20786	0.57666
0.76	1.18741	−0.88996	−1.10157	0.77340
0.80	1.03457	−1.05270	−0.96606	0.94823
0.84	0.85795	−1.19210	−0.80507	1.09714
0.88	0.66151	−1.30448	−0.62295	1.21670
0.92	0.44974	−1.38693	−0.42455	1.30414
0.96	0.22752	−1.43727	−0.21507	1.35743
1.00	0.00000	−1.45420	0.00000	1.37533

TABLE 3.
CHARACTERISTIC FUNCTIONS AND DERIVATIVES
CLAMPED-PINNED BEAM
SECOND MODE

$\frac{x}{l}$	ϕ_2	$\phi_2' = \frac{1}{\beta_2}\frac{d\phi_2}{dx}$	$\phi_2'' = \frac{1}{\beta_2^2}\frac{d^2\phi_2}{dx^2}$	$\phi_2''' = \frac{1}{\beta_2^3}\frac{d^3\phi_2}{dx^3}$
0.00	0.00000	0.00000	2.00000	−2.00000
0.04	0.07241	0.48557	1.43502	−1.99300
0.08	0.25958	0.81207	0.87658	−1.94824
0.12	0.51697	0.98325	0.33937	−1.83960
0.16	0.80176	1.00789	−0.15633	−1.65333
0.20	1.07449	0.90088	−0.58802	−1.38736
0.24	1.30078	0.68345	−0.93412	−1.05012
0.28	1.45308	0.38242	−1.17673	−0.65879
0.32	1.51208	0.02894	−1.30380	−0.23724
0.36	1.46765	−0.34350	−1.31068	0.18649
0.40	1.31923	−0.70122	−1.20092	0.58286
0.44	1.07550	−1.01270	−0.98634	0.92349
0.48	0.75348	−1.25090	−0.68631	1.18364
0.52	0.37700	−1.39515	−0.32640	1.34442
0.56	−0.02536	−1.43265	0.06348	1.39438
0.60	−0.42268	−1.35944	0.45136	1.33056
0.64	−0.78413	−1.18058	0.80569	1.15876
0.68	−1.08158	−0.90972	1.09776	0.89319
0.72	−1.29186	−0.56793	1.30395	0.55537
0.76	−1.39858	−0.18205	1.40755	0.17245
0.80	−1.39351	0.21752	1.40010	−0.22494
0.84	−1.27726	0.59923	1.28198	−0.60506
0.88	−1.05919	0.93288	1.06244	−0.93759
0.92	−0.75676	1.19208	0.75879	−1.19604
0.96	−0.39406	1.35629	0.39504	−1.35983
1.00	0.00000	1.41251	0.00000	−1.41592

TABLE 3.
CHARACTERISTIC FUNCTIONS AND DERIVATIVES
CLAMPED-PINNED BEAM
THIRD MODE

$\frac{x}{l}$	ϕ_3	$\phi_3' = \frac{1}{\beta_e}\frac{d\phi_3}{dx}$	$\phi_3'' = \frac{1}{\beta_3^2}\frac{d^2\phi_3}{dx^2}$	$\phi_3''' = \frac{1}{\beta_3^3}\frac{d^3\phi_3}{dx^3}$
0.00	0.00000	0.00000	2.00000	−2.00000
0.04	0.14410	0.65020	1.18532	−1.97961
0.08	0.48626	0.97168	0.39742	−1.85535
0.12	0.89584	0.98593	−0.30845	−1.57331
0.16	1.25604	0.74002	−0.86560	−1.13046
0.20	1.47476	0.30725	−1.21523	−0.56678
0.24	1.49419	−0.21934	−1.32168	0.04683
0.28	1.29662	−0.73864	−1.18195	0.62397
0.32	0.90489	−1.15556	−0.82867	1.07934
0.36	0.37703	−1.39512	−0.32637	1.34445
0.40	−0.20439	−1.41364	0.23807	1.37996
0.44	−0.74658	−1.20525	0.76897	1.18287
0.48	−1.16223	−0.80234	1.17711	0.78746
0.52	−1.38422	−0.26994	1.39411	0.26005
0.56	−1.37687	0.30522	1.38344	−0.31179
0.60	−1.14194	0.82907	1.14631	−0.83344
0.64	−0.71844	1.21582	0.72134	−1.21873
0.68	−0.17628	1.40210	0.17821	−1.40403
0.72	0.39519	1.35742	−0.39391	−1.35870
0.76	0.90188	2.08924	−0.90103	−1.09010
0.80	1.26035	0.64175	−1.25980	−0.64233
0.84	1.41160	0.08860	−1.41124	−0.08900
0.88	1.33072	−0.47918	−1.33049	0.47891
0.92	1.03098	−0.96820	−1.03085	0.96800
0.96	0.56168	−1.29798	−0.56162	1.29782
1.00	0.00000	−1.41429	0.00000	1.41414

ANSWERS TO SELECTED PROBLEMS

CHAPTER 1

1-1 $\dot{x}_{max} = 8.38$ cm/s; $\ddot{x}_{max} = 350.9$ cm/s^2

1-3 $x_{max} = 7.27$ cm, $\tau = 0.10$ s, $\ddot{x}_{max} = 278.1$ m/s^2

1-5 $z = 5e^{0.6435i}$

1-8 $R = 8.697$, $\theta = 13.29°$

1-9 $x(t) = \dfrac{4}{\pi}\left(\sin \omega_1 t + \dfrac{1}{3}\sin 3\omega_1 t + \dfrac{1}{5}\sin 5\omega_1 t + \cdots\right)$

1-11 $x(t) = \dfrac{1}{2} + \dfrac{4}{\pi^2}\left(\cos \omega_1 t + \dfrac{1}{3^2}\cos 3\omega_1 t + \dfrac{1}{5^2}\cos 5\omega_1 t + \cdots\right)$

1-13 $\sqrt{\overline{x^2}} = A/2$

1-14 $\overline{x^2} = 1/3$

1-16 $a_0 = 1/3$, $b_n = \dfrac{1}{n\pi}\left(1 - \cos\dfrac{2\pi n}{3}\right)$, $a_n = \dfrac{1}{n\pi}\sin\dfrac{2\pi n}{3}$

1-18 RMS = $0.3162A$

1-20 Error = ± 0.148 mm

1-22 $x_{peak}/x_{1000} = 39.8$

CHAPTER 2

2-1 5.62 Hz

2-3 0.159 s

2-5 $x(t) = \dfrac{m_2 g}{k}(1 - \cos \omega t) + \dfrac{m_2\sqrt{2gh}}{\sqrt{k(m_1 + m_2)}}\sin \omega t$

2-7 $J_0 = 9.30$ lb in s^2

2-9 $\kappa = 0.4507$ m

2-11 $\omega = \sqrt{\dfrac{k}{m + J_0/r^2}}$

2-13 $\tau = 1.97$ s

2-15 $\tau = 2\pi\sqrt{\dfrac{J}{Wh}}$

2-17 $\tau = 2\pi \dfrac{L}{a}\sqrt{\dfrac{h}{3g}}$

2-19 $f = \dfrac{1}{2\pi}\sqrt{\dfrac{gab}{h\kappa^2}}$

2-21 $\tau = 2\pi\sqrt{\dfrac{l}{2g}}$

2-23 $m_{\text{eff}} = \dfrac{3}{8} ml$ for each column, ml = mass of column

2-25 $m_{\text{eff}} = \dfrac{33}{140} ml$

2-27 $K_{\text{eff}} = \dfrac{K_1 K_2}{K_1 + K_2} + K_2$

2-29 $J_{\text{eff}} = J_1 + J_2\left(\dfrac{r_1}{r_2}\right)^2$

2-31 $M = 0.0289$ kg

2-33 $\zeta = 1.45$

2-35 $\omega_n = 27.78, \quad \delta = 0.0202, \quad \zeta = 0.003215, \quad c = 0.405$

2-39 $\omega_d = \sqrt{\dfrac{k}{m}\left(\dfrac{b}{a}\right)^2 - \left(\dfrac{c}{2m}\right)^2}, \quad c_{\text{cr}} = \dfrac{2b}{a}\sqrt{km}$

2-41 $\omega_d = \dfrac{a}{l}\sqrt{\dfrac{3k}{m}}\sqrt{1 - \dfrac{3}{4km}\left(\dfrac{cl}{a}\right)^2}, \quad c_{\text{cr}} = \dfrac{2}{3}\dfrac{a}{l}\sqrt{3km}$

2-43 $\dot{x}_{\max} = 92.66$ ft/s, $\quad t = 0.214$ s

2-45 $\zeta_1 = 0.59 \quad x_{\text{overshoot}} = 0.379$

2-48 Flexibility $= \dfrac{4}{243}\dfrac{l^3}{EI}$

2-53 $(0.854ml + .5625M)\ddot{x} + 0.5625kx + 2/3c\dot{x} = 0$

2-54 $(Ml^2 + \dfrac{2}{3}ml^2)\ddot{\theta} + (kl^2 + 2K)\theta = 0$

CHAPTER 3

3-1 $c = 61.3$ Ns/m

3-3 $X = 0.797$ cm, $\quad \phi = 51.43°$

3-5 $\zeta = 0.1847$

3-9 $f_m = 15$ Hz, $\quad \zeta = .023505, \quad X = .231$ cm $\quad \phi = 175.4°$

3-12 $f = 1028$ rpm

3-14 $F = 1273$ N, $F = 241.1$ N for $d = 1.905$ cm

3-16 $V = \dfrac{L}{2\pi}\sqrt{\dfrac{k}{m}}$

3-21 $k = 18.8$ lb/in each spring

3-24 $X = 0.01105$ cm, $F_T = 42.0$ N

3-25 $\omega^2 X = 3.166\ \text{cm/s}^2$

3-34 $c_{eq} = 4D/\pi\omega x$

3-38 (a) 15.9 cps (b) 7.45 cps

3-41 (a) 624.5 m.v. (b) 3.123 m.v.

3-44 $E = 25.7$ m.v./g

CHAPTER 4

4-5 $x = \dfrac{F_0}{k}[\cos\omega_n(t - t_0) - \cos\omega_n t]$, $t > t_0$

4-10 $z = \dfrac{100}{\omega_n^2}(1 - \cos\omega_n t) - \dfrac{20}{\omega_n}\sin\omega_n t$

$$z_{\max} = \frac{100}{\omega_n^2}\left[1 - \frac{5}{\sqrt{25 + \omega_n^2}}\right] - \frac{20}{\omega_n}\frac{\omega_n}{\sqrt{25 + \omega_n^2}}$$

4-13 $\tan\omega_n t = \dfrac{\sqrt{mgs/k}}{s - mg/4k}$

4-14 $x_{\max} = 12.08''$, $t = 0.392$ s

4-20 $x_{\max} = 2.34''$

4-29 $x(t) = \dfrac{F_0}{c\omega_n}\left\{\dfrac{e^{-\zeta\omega_n t}}{\sqrt{1-\zeta^2}}\sin\left(\sqrt{1-\zeta^2}\,\omega_n t + \sin^{-1}\sqrt{1-\zeta^2}\right) - \cos\omega_n t\right\}$

4-42 $y(t) = y(0)\cos\omega t + \dfrac{\dot{y}(0)}{\omega}\sin\omega t + \dfrac{V}{\omega}(\omega t - \sin\omega t) - \dfrac{g}{\omega^2}(1 - \cos\omega t)$

CHAPTER 5

5-2 $\omega_1^2 = k/m \quad (X_1/X_2)_1 = 1$
$\omega_2^2 = 3k/m \quad (X_1/X_2)_2 = -1$

5-4 $\omega_1^2 = 0.570k/m \quad (X_1/X_2)_1 = 3.43$
$\omega_2^2 = 4.096k/m \quad (X_1/X_2)_2 = -0.096$

5-8 $\omega_n = 15.72$ rad/s

5-10 $\ddot{\theta}_1 + 2g/l\theta_1 - g/l\theta_2 = 0$
$\ddot{\theta}_1 + \ddot{\theta}_2 + g/l\theta_2 = 0$

5-13 $\omega_1 = 0.796\sqrt{T/ml} \quad (Y_1/Y_2)_1 = 1.365$
$\omega_2 = 1.538\sqrt{T/ml} \quad (Y_1/Y_2)_2 = -0.366$

5-15 $\omega = \sqrt{\dfrac{g}{l} + \dfrac{k}{ml^2}(1 \pm 1)}$, beat period = 53.02 s

5-20 $\begin{bmatrix} m & 0 \\ 0 & J \end{bmatrix}\begin{Bmatrix} \ddot{x} \\ \ddot{\theta} \end{Bmatrix} + \begin{bmatrix} 2k & kl/4 \\ kl/4 & 5kl^2/16 \end{bmatrix}\begin{Bmatrix} x \\ \theta \end{Bmatrix} = \{0\}$ x down, θ clockwise

5-22 both static and dynamic coupling present

5-24 $f_1 = 0.963$ Hz node 10.9 ft forward of cg
$f_2 = 1.33$ Hz node 1.48 ft aft of cg

5-26 $\omega_1 = 31.6$ rad/s $(X_1/X_2)_1 = 0.50$
$\omega_2 = 63.4$ rad/s $(X_1/X_2)_2 = -1.00$

5-29 $x_1 = \frac{8}{9}\cos\omega_1 t + \frac{1}{9}\cos\omega_2 t;\quad x_2 = \frac{4}{9}\cos\omega_1 t - \frac{1}{9}\cos\omega_2 t$

5-30 shear ratio 1st/2nd story = 2.0

5.34 $(\omega/\omega_h)_2 = 2.73,\quad (Y_1/Y_0)_2 = -0.74$

5.36 $V_1 = 43.3$ ft/s, $V_2 = 60.3$ ft/s

5.41 $d_2 = 1/2''$

5-43 $w = 11.4$ lb, $k = 17.9$ lb/in

5-45 $\zeta_0 = 0.105,\quad \omega/\omega_n = 0.943$

5-52 $$\begin{bmatrix} J_1 & 0 \\ 0 & J_2 \end{bmatrix}\begin{Bmatrix} \ddot{\theta}_1 \\ \ddot{\theta}_2 \end{Bmatrix} + l^2\begin{bmatrix} \frac{9}{16}k_1 & -\frac{9}{16}k_1 \\ -\frac{9}{16}k_1 & \left(\frac{9}{16}k_1 + \frac{1}{4}k_2\right) \end{bmatrix}\begin{Bmatrix} \theta_1 \\ \theta_2 \end{Bmatrix} = \begin{Bmatrix} 0 \\ 0 \end{Bmatrix}$$

$$J_1 = m_1\frac{l^2}{3},\quad J_2 = \frac{7}{48}m_2 l^2$$

CHAPTER 6

6-1 $a_{11} = \dfrac{k_2 + k_3}{\Sigma k_i k_j},\quad a_{21} = a_{12} = \dfrac{k_2}{\Sigma k_i k_j},\quad a_{22} = \dfrac{k_1 + k_2}{\Sigma k_i k_j}$

6-3 $a_{11} = 0.0114 l^3/EI,\quad a_{21} = a_{12} = 0.0130 l^3/EI,\quad a_{22} = 0.0192 l^3/EI$

6-6 $$[K] = \begin{bmatrix} (K_1 + K_2) & -K_2 & 0 \\ -K_2 & (K_2 + K_3) & -K_3 \\ 0 & -K_3 & K_3 \end{bmatrix},$$

$$[a] = \begin{bmatrix} 1/K_1 & 1/K_1 & 1/K_1 \\ 1/K_1 & (1/K_1 + 1/K_2) & (1/K_1 + 1/K_2) \\ 1/K_1 & (1/K_1 + 1/K_2) & (1/K_1 + 1/K_2 + 1/K_3) \end{bmatrix}$$

6-7 $$[a] = \frac{l^3}{EI}\begin{bmatrix} 7/96 & 1/8 \\ 1/8 & 5/6 \end{bmatrix}$$

6-8 $$[a] = \frac{l^3}{12EI}\begin{bmatrix} 1 & 1 & 1 & 1 \\ 1 & 2 & 2 & 2 \\ 1 & 2 & 3 & 3 \\ 1 & 2 & 3 & 4 \end{bmatrix}$$

6-11 $$\begin{Bmatrix} F_1 \\ M_2 \\ M_3 \end{Bmatrix} = \begin{bmatrix} 24EI_1/l_1^3 & -6EI_1/l_1^2 & -6EI_1/l_1^2 \\ -6EI_1/l_1^2 & (4EI_1/l_1 + 4EI_2/l_2) & 2EI_2/l_2 \\ -6EI_1/l_1^2 & 2EI_2/l_2 & (4EI_1/l_1 + 4EI_2/l_2) \end{bmatrix}\begin{Bmatrix} u_1 \\ \theta_1 \\ \theta_2 \end{Bmatrix}$$

6-17 $P = \begin{bmatrix} 1.44/l & -8.40/l \\ 1.00 & 1.00 \end{bmatrix}$

6-20 $\tilde{P} = \begin{bmatrix} 0.207 & -1.208 \\ 0.293 & 1.707 \end{bmatrix}$

6-22 $$\begin{bmatrix} (m_1 + m_2) & 0 & 0 \\ 0 & J_1 & 0 \\ 0 & 0 & J_2 \end{bmatrix} \begin{Bmatrix} \ddot{x} \\ -\ddot{\theta}_1 \\ \ddot{\theta}_2 \end{Bmatrix} + \frac{EI}{l} \begin{bmatrix} 6/l^2 & 3/l & -3/l \\ 3/l & 7 & 2 \\ -3/l & 2 & 7 \end{bmatrix} \begin{Bmatrix} x \\ -\theta_1 \\ \theta_2 \end{Bmatrix} = \{0\}$$

$F \rightarrow \qquad M_1 \curvearrowleft \qquad M_2 \curvearrowleft$

6-24 $[K] = k\begin{bmatrix} 2 & -1 \\ -1 & 2 \end{bmatrix}$, $[C] = c\begin{bmatrix} 2 & -1 \\ -1 & 1 \end{bmatrix}$ $\therefore$ not proportional

6-31 $\ddot{q}_2 + 0.8902\zeta_2\sqrt{\dfrac{k}{m}}\,\dot{q}_2 + 0.1981\dfrac{k}{m}q_2 = 0.4068\ddot{u}_0(t)$

$\ddot{q}_3 + 1.4614\zeta_3\sqrt{\dfrac{k}{m}}\,\dot{q}_3 + 0.5339\dfrac{k}{m}q_3 = -0.3268\ddot{u}_0(t)$

6-32 $\left|\dfrac{kx(10)}{ma_0}\right| = 1.90 + \sqrt{.610^2 + .369^2} = 2.61$

6-37 $\alpha = \dfrac{2\omega_1\omega_2(\zeta_1\omega_2 - \zeta_2\omega_1)}{\omega_2^2 - \omega_1^2}$, $\beta = \dfrac{2(\zeta_2\omega_2 - \zeta_1\omega_1)}{\omega_2^2 - \omega_1^2}$,

$\zeta_i = \dfrac{\alpha + \beta\omega_i^2}{2\omega_i}$, $\zeta_3 = 0.1867$

6-39 $C_1 = 0.8985$, $C_2 = -0.1477$, $C_3 = 0.3886$

6-41 $\bar{a}_{11} = \dfrac{l^3}{3EI} + \dfrac{1}{k} + \dfrac{l^2}{K}$

$\bar{a}_{12} = \bar{a}_{21} = \dfrac{l^2}{2EI} + \dfrac{l}{K}$

$\bar{a}_{22} = \dfrac{l}{EI} + \dfrac{1}{K}$

CHAPTER 7

7-1 Constraints $u_2 = u_5 = 0$

$u_1 = u_3$

$u_4 = u_6$

System has 2 DOF.

7-2 Let $u_3 = q_1$, $u_6 = q_2$ $\begin{Bmatrix} u_3 \\ u_6 \end{Bmatrix} = \begin{bmatrix} 1 & 0 \\ 0 & 1 \end{bmatrix} \begin{Bmatrix} q_1 \\ q_2 \end{Bmatrix}$

7-3 $\tan\theta = \left(\dfrac{l_1}{l_2}\right)^2$

7-5 $\tan\theta = \left(\dfrac{m_2 - m_1}{m_1 + m_2}\right)\dfrac{l}{\sqrt{(2R)^2 + l^2}}$

7-7 $\tan\theta = \dfrac{1}{\mu}$

7-9 $\sin\theta = \dfrac{3}{4}\dfrac{h}{l} - \dfrac{mg}{2kl}$

7-11 $\ddot{\theta}_{\sim} + \frac{3}{2}g\left(\frac{l_2^2\cos\theta_0 + l_1^2\sin\theta_0}{l_1^3 + l_2^3}\right)\theta_{\sim} = 0$

where $\tan\theta_0 = (l_1/l_2)^2$

7-12 $(m_1 + m_2)R^2\ddot{\theta}_{\sim}$

$+ g\left[(m_1 + m_2)\sqrt{R^2 + (l/2)^2}\cos\theta_0 + (m_2 - m_1)\frac{l}{2}\sin\theta_0\right]\theta_{\sim} = 0$

7-15 $m_0(\ddot{r} - r\dot{\theta}^2) + k(r - r_0) = m_0 g\cos\theta$

$m_0 r(r\ddot{\theta} + 2\dot{r}\dot{\theta}) + m_{\text{rod}}\frac{l^2}{3}\ddot{\theta} + m_0 g(r - r_0)\sin\theta + m_{\text{rod}}g\frac{l}{2}\sin\theta = 0$

7-18 $[J_1 + (m_1 + m_2)4l^2]\ddot{q}_1 + [K + l^2(k_1 + 4k_2)]q_1 + 4l^2k_2q_2 = 0$

$J_2\ddot{q}_2 + 4l^2k_2(q_1 + q_2) = 0$

7-20 $[k] = \frac{EI}{l^3}\begin{bmatrix} 20.43 & -5.25\,l \\ -5.25\,l & 7.0\,l^2 \end{bmatrix}$

7-22 $R = \frac{P\left(l_1^3/3 + l_1^2 l_2/2\right) + M\left(l_1^2/2 + l_1 l_2\right)}{\left(l_1^3/3 + 2l_2^3/3 + l_1^2 l_2 + l_1 l_2^2\right)}$

$M_1 = R(l_1 + l_2) - Pl_1 - M$

CHAPTER 8

8-2 $f = \frac{n}{2l}\sqrt{\frac{T}{l}}, \quad n = 1, 2, 3, \cdots$

8-3 $\tan\frac{\omega l}{c} = -\left(\frac{T}{kl}\right)\frac{\left(\frac{\omega l}{c}\right)}{1 - \left(\frac{\omega}{\omega_n}\right)^2}, \quad \omega_n = \sqrt{\frac{k}{m}}$

8-5 4.792×10^3 m/s

8-15 $\omega_n = (2n - 1)\frac{\pi}{l}\sqrt{\frac{G}{\rho}}, \quad n = 1, 2, 3, \cdots$

8-16 $\tan\frac{\omega l}{c} = \frac{2\left(\frac{J_0}{J_s}\frac{\omega l}{c}\right)}{\left(\frac{J_0}{J_s}\frac{\omega l}{c}\right)^2 - 1}$

8-20 $E = 3.48 \times 10^6$ lb/in^2

8-23 $\omega = \beta^2\sqrt{EI/\rho}$ where β is determined from

$(1 + \cosh\beta l \cdot \cos\beta l) = \beta l\frac{W_0}{W_b}(\sinh\beta l \cdot \cos\beta l - \cosh\beta l \cdot \sin\beta l)$

CHAPTER 9

9-3 $\Gamma_i = \frac{p_0}{l}\int_0^l \phi_i(x)\,dx$

9-8 $y(x, t) = \frac{4p_0 l}{\pi M\omega_2^2}\sin\frac{2x\pi}{l}(1 - \cos\omega_2 t)$

9-10 Modes absent are $2nd$, $5th$, $8th$, etc.

9-11 $\Gamma = \sqrt{2}\cos(2n-1)\pi/6, \quad D_n = (1-\cos\omega_n t)$

$$u = \frac{2F_0 l}{AE}\left[\frac{\cos(\pi/6)\cos(\pi/2)(x/l)}{(\pi/2)^2}D_1 + \frac{\cos(5\pi/6)\cos(5\pi/2)(x/l)}{(5\pi/2)^2}D_2 + \cdots\right]$$

9-14
$$\Gamma_1 = \frac{1}{l}\int_0^l \phi_1(x)\,dx = 0.784$$
$$\Gamma_2 = \frac{1}{l}\int_0^l \phi_2(x)\,dx = 0.434$$
$$\Gamma_3 = \frac{1}{l}\int_0^l \phi_3(x)\,dx = 0.254$$

9-19
$$\left\{1 + \frac{K\varphi_2'^2(0)}{M\omega_1^2\left[1-(\omega/\omega_1)^2\right]}\right\}\left\{1 + \frac{K\varphi_2'^2(0)}{M\omega_2^2\left[1-(\omega/\omega_2)^2\right]}\right\} = \left\{\frac{K\varphi_1'(0)\varphi_2'(0)}{M\omega_1^2\left[1-(\omega/\omega_1)^2\right]}\right\}\left\{\frac{K\varphi_1'(0)\varphi_2'(0)}{M\omega_2^2\left[1-(\omega/\omega_2)^2\right]}\right\}$$

$\varphi_1 = \sqrt{2}\sin\frac{\pi x}{l}, \quad \varphi_1' = \frac{\pi}{l}\sqrt{2}\cos\frac{\pi x}{l}$, etc.

One mode approximation gives

$$\left(\frac{\omega}{\omega_1}\right)^2 = 1 + \frac{2K}{M\omega_1^2}\left(\frac{\pi}{l}\right)^2, \quad \omega_1 = \pi^2\sqrt{\frac{EI}{Ml^3}}$$

9-20 One mode approximation

$$\left(\frac{\omega}{\omega_1}\right)^2 = 1 + \frac{4K}{M\omega_1^2}\left(\frac{\pi}{l}\right)^2$$

9-21 Using one free-free mode and translation mode of M_0

$$\left(\frac{\omega}{\omega_1}\right)^2 = \frac{M_1}{M_1 + M_0\varphi_1^2(0) - \left[M_0^2\varphi_1^2(0)/(M_0 + 2ml)\right]}$$

where $M_1 = \int \varphi_1^2(x) m\,dx = 2ml, \quad \omega_1 = 22.4\sqrt{\frac{EI}{m(2l)^4}}$

CHAPTER 10

10-2
$$\omega_1 = 1.403\sqrt{\frac{EA_2}{M_2 l}} \qquad \begin{Bmatrix} u_1 \\ u_2 \end{Bmatrix}_1 = \begin{Bmatrix} .577 \\ 1.000 \end{Bmatrix}$$
$$\omega_2 = 3.648\sqrt{\frac{EA_2}{M_2 l}} \qquad \begin{Bmatrix} u_1 \\ u_2 \end{Bmatrix}_2 = \begin{Bmatrix} -.526 \\ 1.000 \end{Bmatrix}$$

10-3
$$\frac{M}{18}\begin{bmatrix} 4 & 1 & 0 \\ 1 & 4 & 1 \\ 0 & 1 & 2 \end{bmatrix}\begin{Bmatrix} \ddot{u}_1 \\ \ddot{u}_2 \\ \ddot{u}_3 \end{Bmatrix} + \frac{3EA}{l}\begin{bmatrix} 2 & -1 & 0 \\ -1 & 2 & -1 \\ 0 & -1 & 1 \end{bmatrix}\begin{Bmatrix} u_1 \\ u_2 \\ u_3 \end{Bmatrix} = \begin{Bmatrix} 0 \\ 0 \\ 0 \end{Bmatrix}$$

10-5 $\omega_1 = 1.611\sqrt{\frac{GA}{Ml}}, \quad \omega_2 = 5.629\sqrt{\frac{GA}{Ml}}$

10-8 $\omega_1 = 2.368\sqrt{\frac{E}{\rho l^2}}, \quad \omega_2 = 8.664\sqrt{\frac{E}{\rho l^2}}, \quad \rho =$ mass density

10-11 $$\begin{Bmatrix} F_{3x} = 0 \\ F_{3y} = 0 \\ F_{4x} = P \\ F_{4y} = 0 \end{Bmatrix} = \frac{EA}{l}\begin{bmatrix} \left(1 + \frac{.5}{\sqrt{2}}\right) & \frac{.5}{\sqrt{2}} & -1 & 0 \\ \frac{.5}{\sqrt{2}} & \left(1 + \frac{.5}{\sqrt{2}}\right) & 0 & 0 \\ -1 & 0 & 1 & 0 \\ 0 & 0 & 0 & -1 \end{bmatrix}\begin{Bmatrix} \bar{u}_3 \\ \bar{v}_3 \\ \bar{u}_4 \\ \bar{v}_4 \end{Bmatrix}$$

10-13 $\bar{v}_2 = -\dfrac{Pl^3}{192EI}, \quad \theta_2 = -\dfrac{Ml}{16EI}$

10-14 $\omega_1 = 22.74\sqrt{\dfrac{EI}{ml^4}}, \quad \omega_2 = 81.67\sqrt{\dfrac{EI}{ml^4}}$

10-17 $$\begin{Bmatrix} -P \\ 0 \\ M \end{Bmatrix} = \frac{EI}{8l^3}\begin{bmatrix} (2.5R + 6) & .5(R - 12) & 8.484l \\ .5(R - 12) & (.5R + 30) & 15.52l \\ 8.484l & 15.52l & 40.0l^2 \end{bmatrix}\begin{Bmatrix} \bar{u}_2 \\ \bar{v}_2 \\ \theta_2 \end{Bmatrix}$$

$$R = \left(\frac{4Al^2}{I}\right)$$

10-18 (e) $$\begin{Bmatrix} 0 \\ -M \\ 0 \\ 0 \end{Bmatrix} = \frac{EI}{l^3}\begin{bmatrix} 12 & -6l & 0 & 0 \\ -6l & 8l^2 & 6l & 2l^2 \\ 0 & 6l & \left(12 + \frac{kl^3}{EI}\right) & 6l \\ 0 & 2l^2 & 6l & 4l^2 \end{bmatrix}\begin{Bmatrix} \bar{v}_2 \\ \theta_2 \\ \bar{u}_3 \\ \theta_3 \end{Bmatrix}$$

10-20 (a) $$\begin{Bmatrix} \bar{F}_2 \\ \bar{M}_2 \\ \bar{F}_3 \\ \bar{M}_3 \end{Bmatrix} = -p_0 l\begin{Bmatrix} .2875 \\ -.2271l \\ .0375 \\ .0146l \end{Bmatrix}$$

10-20 (b) $$\bar{v}_2 = -\frac{.2031}{192}\frac{pl^4}{EI}$$

$$\theta_2 = -\frac{.0286}{16}\frac{pl^3}{EI}$$

CHAPTER 11

11-2 $\omega_1 = 4.63\sqrt{\dfrac{EI}{Ml^3}}$

11-3 $\omega_1 = 1.62\sqrt{\dfrac{EI}{Ml^3}}$

11-9 $\omega_1^2 = \dfrac{3EI/l^3 + 2k}{\frac{33}{140}ml + \frac{2}{3}m_0}$

11-11 $\omega_1 = 9.96\sqrt{\dfrac{EI}{(0.2188m_0 l)l^3}}$ where $0.2188\ m_0 l$ = total mass

11-12 $\omega_1^2 = \left(\dfrac{3EI}{l^3}\right)\dfrac{1}{27m_1 + 8m_2 + m_3}$

11-15 $f_{11} = 495.2$ cps

11-16 $\left(\frac{1}{6} - \frac{1}{\pi^2}\right)(ml\omega^2)^2 - \left(\frac{\pi^4}{6}\frac{EI}{l^3} + \frac{k}{2}\right)(ml\omega^2) + \frac{\pi^4}{2}\frac{EI}{l^3}k = 0$

11-18 $\left(\frac{ml}{2}\omega^2\right)^2 - \left[5lEA\left(\frac{\pi}{2l}\right)^2 + 2k_0\right]\left(\frac{ml}{2}\omega^2\right) + \left[\frac{lEA}{2}\left(\frac{\pi}{2l}\right)^2 + k_0\right]$

$\left[\frac{lEA}{2}\left(\frac{3\pi}{2l}\right)^2 + k_0\right] - k_0^2 = 0$

11-23 $\omega_1 = 0.584\sqrt{\frac{k}{m}}, \quad \omega_2 = 1.200\sqrt{\frac{k}{m}}, \quad \omega_3 = 1.642\sqrt{\frac{k}{m}}$

CHAPTER 12

12-2 $\omega_1 = 0.629\sqrt{\frac{K}{J}}, \qquad \begin{Bmatrix}\theta_1\\ \theta_2\\ \theta_3\end{Bmatrix}_1 = \begin{Bmatrix}1.000\\ 0.604\\ 0.287\end{Bmatrix}$

12-4 $\omega_1 = 0.445\sqrt{\frac{k}{m}}, \qquad \begin{Bmatrix}x_1\\ x_2\\ x_3\end{Bmatrix}_1 = \begin{Bmatrix}1.000\\ 0.802\\ 0.445\end{Bmatrix}$

$\omega_2 = 1.247\sqrt{\frac{k}{m}}, \qquad \begin{Bmatrix}x_1\\ x_2\\ x_3\end{Bmatrix}_2 = \begin{Bmatrix}-1.000\\ 0.555\\ 1.247\end{Bmatrix}$

$\omega_3 = 1.802\sqrt{\frac{k}{m}}, \qquad \begin{Bmatrix}x_1\\ x_2\\ x_3\end{Bmatrix}_3 = \begin{Bmatrix}1.000\\ -2.247\\ 1.802\end{Bmatrix}$

12-9 $\omega_1 = \sqrt{\frac{6EI}{Ml^3}\left(1 + \frac{n}{2}\right)}, \qquad y_1/y_2 = -n/2$

12-10 $\omega_1 = 1.651\sqrt{\frac{EI}{ml^3}}, \qquad \begin{Bmatrix}y_1\\ y_2\end{Bmatrix}_1 = \begin{Bmatrix}0.320\\ 1.000\end{Bmatrix}$

12-12 $u_{43} - \frac{u_{41}u_{23}}{u_{21}} = 0$

12-17 $\omega = 22.7$

12-18 $\omega_1 = 22.5 \quad \omega_2 = 52.3$

12-21 $\omega_1 = 101.2 \quad \omega_2 = 1836$

12-22 $\omega_1 = 0.5375\sqrt{\frac{K}{J}} \quad \omega_3 = 1.805\sqrt{\frac{K}{J}}$

$\begin{Bmatrix}\theta_1\\ \theta_2\\ \theta_3\\ \theta_4\end{Bmatrix}_1 = \begin{Bmatrix}1.000\\ 0.714\\ 0.239\\ -0.326\end{Bmatrix} \qquad \begin{Bmatrix}\theta_1\\ \theta_2\\ \theta_3\\ \theta_4\end{Bmatrix}_3 = \begin{Bmatrix}1.000\\ -2.270\\ 1.870\\ -0.101\end{Bmatrix}$

12-26 $\omega_k = 2\sqrt{\frac{K}{J}}\sin\frac{(2k-1)\pi}{2(2N+1)}$

12-28 $\omega_n = 2\sqrt{\frac{k}{m}}\sin\frac{n\pi}{2(N+1)}$

12-31 $-2\cos\beta\left(N+\frac{1}{2}\right)\quad \sin\beta/2=\frac{K_N}{k}\sin\beta N$

CHAPTER 13

13-5 $\bar{x}=0.50,\quad \overline{x^2}=0.333$

13-6 $\bar{x}=A_0,\quad \overline{x^2}=A_0^2+\frac{1}{2}A_1^2$

13-14 A triangle of twice the base, symmetric about $t=0$

13-15 $R(\tau)=5$ at $\tau=0$ and linearly decrease to $R(1)=1$.

13-18 RMS $=53.85g=528.3\ m/s^2$

13-21 RMS $=1.99$ in, $\sigma=0.9798$

13-24 $f(t)=\frac{2}{\pi}\left[\sin\omega_1 t-\frac{1}{2}\sin 2\omega_1 t+\frac{1}{3}\sin 3\omega_1 t-\cdots\right]$

$S(\omega)=\sum_n \frac{1}{2}C_n^2=\frac{2}{\pi^2}\sum_n\frac{1}{n^2}$

13-26 $x(t)=\frac{2A}{\pi}\sum_{n=-\infty}^{\infty}\frac{1}{in}e^{in\omega_1 t}\qquad n-\text{odd}$

$=\frac{4A}{\pi}\left[\sin\omega_1 t+\frac{1}{3}\sin 3\omega_1 t+\frac{1}{5}\sin 5\omega_1 t+\cdots\right]$

$S(\omega_n)=\frac{C_nC_n^*}{2}=\frac{4A^2}{n^2\pi^2}$

13-27 $f_{1,2}=f_n\left(1\mp\frac{1}{2Q}\right)$

13-31 $53.2g=2\sigma,\quad P[x>2\sigma]=4.6\%$
$P[X>2\sigma]=13.5\%$

13-32 $F(t)=10^6\left[\frac{1}{4}+\sum_{n=1}^{\infty}\frac{2}{n\pi}\sin\frac{n\pi}{4}\cdot\cos n\frac{2\pi}{T}t\right]$

$S_F(\omega_n)=\frac{1}{2}\times10^{12}\left[\left(\frac{1}{16}\right)_{n=0}+\sum_{n=1}^{\infty}\left(\frac{2}{n\pi}\right)^2\sin^2\frac{n\pi}{4}\right]$

$\overline{y^2}=\sum\frac{1}{k^2}\frac{S_F(\omega_n)}{\left[1-\left(\frac{n\omega_1}{2\pi}T\right)^2\right]^2},\quad k=\left(\frac{2\pi}{T}\right)^2 m$

13-35 $\overline{y^2}=\sigma^2\cong\frac{S_0}{k^2}\frac{f_n\pi}{4\zeta}=.00438,\quad \sigma=.0662m$

$P[|y|>0.132]=2\sigma=4.6\%$

13-36 $\sigma=.0039m\quad P[|y|>0.012]=0.3\%$

CHAPTER 14

14-2 $m\ddot{x}+\frac{2T_0}{l_0}x\left[1+\frac{1}{2}\left(\frac{EA}{T_0}-1\right)\left(\frac{x}{l}\right)^2\right]=0$

14-3 $m\ddot{x} + \dfrac{\pi r_0^2 \rho}{3(h - x_0)^2}[(h - x)^3 - (h - x_0)^3]$

r_0 = radius of circle at water line

$\rho = wt/$vol of water

14-8 $V = \sqrt{y^2 + \omega_n^4 x^2}$ $\quad x = y = 0$ for equilibrium

14-11 Shift origin of phase plane to π in Fig. 11.5-2

14-13 $\lambda_{1,2} = 3, 4 \quad \begin{Bmatrix} x \\ y \end{Bmatrix}_1 = \begin{Bmatrix} .50 \\ 1.00 \end{Bmatrix}_1 \quad \begin{Bmatrix} x \\ y \end{Bmatrix}_2 = \begin{Bmatrix} 1 \\ 1 \end{Bmatrix}$

14-14 $\begin{Bmatrix} x \\ y \end{Bmatrix} = \begin{bmatrix} 0.5 & 1 \\ 1 & 1 \end{bmatrix} \begin{Bmatrix} \xi \\ \eta \end{Bmatrix}$

14-25 $y = \dfrac{-x(\omega_n^2 + \mu x^2)}{c}, \quad \dfrac{dy}{dx} = c$

14-27 $\dfrac{dy}{dx} = \dfrac{-(x + \delta)}{y}$ where $\delta = \left(\dfrac{c}{m}y + \dfrac{\mu}{\omega_n^2 m}x^3\right)$

$$\omega_n^2 = k/m$$

$$\tau = \omega_n t, \quad y = \frac{dx}{d\tau}$$

14-30 $\tau = 4\sqrt{\dfrac{l}{g}} \displaystyle\int_0^{60^\circ} \dfrac{d\theta}{\sqrt{2(\cos\theta - \cos\theta_0)}} = 4\sqrt{\dfrac{l}{g}} \int_0^{\pi/2} \dfrac{d\theta}{\sqrt{1 - k^2 \sin^2\phi}}$

where $k = \sin\dfrac{\theta_0}{2}, \quad \sin\dfrac{\theta}{2} = k \sin\phi$

14-34 $\omega \cong \sqrt{\dfrac{g}{l}}\left(1 + \dfrac{1}{16}\theta_0^2\right)$

14-40 $m\ddot{x}_1 + \left(\dfrac{2T_0}{l_0}\right)x_1 + \left(\dfrac{2K}{l_0^2}\right)x_1^3 = 0, \quad T = T_0 + K\dfrac{x^2}{l_0}$

second approximation

$$m\ddot{x}_2 + \left(\frac{2T_0}{l_0}\right)x_2 + 3\alpha x_1^2 x_2 = 0, \quad \alpha = \frac{2K}{l_0^2}$$

$$m\ddot{x}_2 + \left[\left(\frac{2T_0}{l_0} + \frac{3}{2}\alpha A^2\right) + \frac{3}{2}\alpha A^2 \cos 2\omega_n t\right]x_2 = 0$$

INDEX

A

B

E

F

G

H

I

J

K

L

M

N

O

P

Q

R

S

T

U

V

W

Y